SCHAUM'S OUTLINE OF

THEORY AND PROBLEMS

OF

COLLEGE MATHEMATICS
Second Edition

Algebra

Discrete Mathematics

Trigonometry

Geometry

Introduction to Calculus

FRANK AYRES, Jr., Ph.D.

*Formerly Professor and Head
Department of Mathematics
Dickinson College*

PHILIP A. SCHMIDT, Ph.D.

*Dean of the School of Education
S.U.N.Y., The College at New Paltz*

SCHAUM'S OUTLINE SERIES
McGRAW-HILL
New York   San Francisco   Washington, D.C.   Auckland   Bogotá   Caracus   Lisbon
London   Madrid   Mexico City   Milan   Montreal   New Dehli
San Juan   Singapore   Sydney   Tokyo   Toronto

FRANK AYRES, Jr., Ph.D., was formerly Professor and Head of the Department of Mathematics at Dickinson College, Carlisle, Pennsylvania. He is the author or coauthor of eight Schaum's Outlines, including *Calculus, Trigonometry, Differential Equations*, and *Modern Abstract Algebra*.

PHILIP A. SCHMIDT, Ph.D., has a B.S. from Brooklyn College (with a major in mathematics), an M.A. in mathematics, and a Ph.D. in mathematics education from Syracuse University. He is currently Dean of the School of Education at SUNY College at New Paltz. He is the author of *3000 Solved Problems in Precalculus* and *2500 Solved Problems in College Algebra and Trigonometry* as well as numerous journal articles. He has also completed a revision of the late Barnett Rich's *Geometry*.

Schaum's Outline of Theory and Problems of
COLLEGE MATHEMATICS

ISBN 0-07-002664-5

Sponsoring Editor: John Aliano
Production Supervisor: Leroy Young
Editing Supervisors: Meg Tobin, Maureen Walker

Library of Congress Cataloging-in-Publication Data

Ayres, Frank, date
    Schaum's outline of theory and problems of college mathematics. —
2nd ed. / Frank Ayres, Jr., Philip A. Schmidt.
      p.   cm. — (Schaum's outline series)
    Rev. ed. of: First year college mathematics / Frank Ayres, Jr.
c 1958.
    Includes index.
    ISBN 0-07-002664-5
    1. Mathematics. 2. Mathematics—Problems, exercises, etc.
I. Schmidt, Philip A. II. Ayres, Frank, date First year college
mathematics. III. Title. IV. Title: Theory and problems of college
mathematics. V. Title: College mathematics.
QA37.2.A948 1992
510—dc20
                                 91-19897
                                    CIP

McGraw-Hill
A Division of The *McGraw-Hill* Companies

Preface

As Dr. Ayres indicates in his preface to the First Edition, this book will help students become proficient in the mathematics commonly presented in the first year (or two) of college. In updating this text, I have left that point-of-view intact; however, I have deleted material that is no longer presented (for example, logarithmic solutions of the right triangle). Additionally, I have added material which has been placed in the curriculum since the publication of the first edition and I have "modernized" many of the problems and exercises. The notation has been changed when necessary and discrete mathematics has been reemphasized.

My thanks must be expressed to Professor Ayres: He has provided me (and so many students) with the very finest of review materials. My thanks, as always, go to John Aliano, Executive Editor of the Schaum Division at McGraw-Hill and to Maureen Walker for her handling of the manuscript and proofs. Cathy Decker-Coffey typed all revisions with her usual meticulous care. Finally, my family has provided me with "quiet time" at home and without that contribution, this revision would have been impossible.

PHILIP A. SCHMIDT
New Paltz, NY
January 1992

Preface to the First Edition

This book is designed primarily to assist students in acquiring a more thorough knowledge and proficiency in basic college mathematics. It includes a thorough coverage of algebra, plane trigonometry, and plane analytic geometry together with selected topics in solid analytic geometry and a brief introduction to the calculus, in that order. In addition to the use of the book by students taking a formal course in first year college mathematics, it should also be of considerable value to those who wish to review the fundamental principles and applications in anticipation of further work in mathematics.

Each chapter begins with a clear statement of the pertinent definitions, principles, and theorems, together with illustrative and descriptive material. This is followed by carefully graded sets of solved and supplementary problems. The solved problems have been selected and solutions arranged so that a study of each will be rewarding. They serve to illustrate and amplify the theory, provide the repetition of basic principles so vital to effective teaching, and bring into sharp focus those fine points without which the student continually feels on unsafe ground. Derivations of formulas and proofs of theorems are included among the solved problems. The supplementary problems offer a complete review of the material of each chapter.

Although in many texts some degree of unification of the material has been achieved, it seemed best to make no attempt in that direction here. However, the reader will find that the material has been so divided into chapters and the problems in these chapters so arranged as to make the book a useful supplement to all current standard texts.

Considerably more material has been included here than can be covered in most first courses. This has been done to make the book more flexible, to provide a more useful book of reference, and to stimulate further interest in the topics.

The author gratefully acknowledges his indebtedness to Mr. Henry Hayden for painstaking work in the preparation of all drawings and for typographical arrangement.

FRANK AYRES, JR.
Carlisle, Pa.
June 1958

Contents

PART I. ALGEBRA

PART II. TOPICS IN DISCRETE MATHEMATICS

PART III. TRIGONOMETRY

PART IV. ANALYTIC GEOMETRY

PART V. INTRODUCTION TO CALCULUS

Part I

ALGEBRA

Chapter 1

The Number System of Algebra

ELEMENTARY MATHEMATICS is concerned mainly with certain elements called *numbers* and with certain operations defined on them.

The unending set of symbols 1, 2, 3, 4, 5, 6, 7, 8, 9, 10, 11, 12, ... used in counting are called *natural numbers*.

In adding two of these numbers, say 5 and 7, we begin with 5 (or with 7) and count to the *right* seven (or five) numbers to get 12. The sum of two natural numbers is a natural number, that is, the sum of two members of the above set is a member of the set.

In subtracting 5 from 7, we begin with 7 and count to the *left* five numbers to 2. It is clear, however, that 7 cannot be subtracted from 5 since there are only four numbers to the left of 5.

INTEGERS.   In order that subtraction be always possible, it is necessary to increase our set of numbers. We prefix each natural number with a + sign (in practice, it is more convenient not to write the sign) to form the *positive integers*, we prefix each natural number with a − sign (the sign must always be written) to form the *negative integers*, and we create a new symbol 0, read *zero*. On the set of *integers*

$$\ldots, -8, -7, -6, -5, -4, -3, -2, -1, 0, +1, +2, +3, +4, +5, +6, +7, +8, \ldots$$

the operations of addition and subtraction are possible without exception.

To add two integers as +7 and −5, we begin with +7 and count to the left (indicated by the sign of −5) five numbers to +2 or we begin with −5 and count to the right (indicated by the sign of +7) seven numbers to +2. How would you add −7 and −5?

To subtract +7 from −5 we begin with −5 and count to the left (opposite to the direction indicated by +7) seven numbers to −12. To subtract −5 from +7 we begin with +7 and count to the right (opposite to the direction indicated by −5) five numbers to +12. How would you subtract +7 from +5? −7 from −5? −5 from −7?

If one is to operate easily with integers it is necessary to avoid the process of counting. To do this we memorize an addition table and establish certain rules of procedure. We note that each of the numbers +7 and −7 is seven steps from 0 and indicate this fact by saying that the *numerical value* of each of the numbers +7 and −7 is 7. We may state:

Rule 1.   To add two numbers having like signs, add their numerical values and prefix their common sign.

Rule 2.   To add two numbers having unlike signs, subtract the smaller numerical value from the larger, and prefix the sign of the number having the larger numerical value.

Rule 3.   To subtract a number, change its sign and add. Since $3 \cdot 2 = 2 + 2 + 2 = 3 + 3 = 6$, we assume

$$(+3)(+2) = +6 \qquad (-3)(+2) = (+3)(-2) = -6 \qquad \text{and} \qquad (-3)(-2) = +6$$

Rule 4.   To multiply or divide two numbers (never divide by 0!), multiply or divide the numerical values, prefixing a + sign if the two numbers have like signs and a − sign if the two numbers have unlike signs. (See Problem 1.1.)

If m and n are integers then $m + n$, $m - n$, and $m \cdot n$ are integers but $m \div n$ may not be an integer. (Common fractions will be treated in the next section.) Moreover, there exists a unique integer x such that $m + x = n$. If $x = 0$, then $m = n$; if x is positive ($x > 0$), then m is less than n ($m < n$); if x is negative ($x < 0$), then m is greater than n ($m > n$).

The integers may be made to correspond one-to-one with equally spaced points on a straight line as in Fig. 1-1. Then $m > n$ indicates that the point on the scale corresponding to m lies to the right of the point corresponding to n. There will be no possibility of confusion if we write the point m rather than the point which corresponds to m and we shall do so hereafter. Then $m < n$ indicates that the point m lies to the left of n. (See Problems 1.2–1.4.)

Fig. 1-1

Every positive integer m is divisible by ± 1 and $\pm m$. A positive integer $m > 1$ is called a prime if its only factors or divisors are ± 1 and $\pm m$; otherwise, m is called *composite*. For example, $2, 7, 19$, are primes while $6 = 2 \cdot 3$, $18 = 2 \cdot 3 \cdot 3$, and $30 = 2 \cdot 3 \cdot 5$ are composites. In these examples, the composite numbers have been expressed as products of prime factors, that is, factors which are prime numbers. Clearly, if $m = r \cdot s \cdot t$ is such a factorization of m, then $-m = (-1)r \cdot s \cdot t$ is a factorization of $-m$. (See Problems 1.5–1.6.)

THE RATIONAL NUMBERS. The set of rational numbers consists of all numbers of the form m/n, where m and $n \neq 0$ are integers. Thus, the rational numbers include the integers and common fractions.

Every rational number has an infinitude of representations; for example, the integer 1 may be represented by $\frac{1}{1}, \frac{2}{2}, \frac{3}{3}, \frac{4}{4}, \ldots$ and the fraction $\frac{2}{3}$ may be represented by $\frac{4}{6}, \frac{6}{9}, \frac{8}{12}, \ldots$ A fraction is said to be expressed in lowest terms by the representation m/n when m and n have no common prime factor. The most useful rule concerning rational numbers is, therefore,

Rule 5. The value of a rational number is unchanged if both the numerator and denominator are multiplied or divided by the same nonzero number.

Caution. We use Rule 5 with division to reduce a fraction to lowest terms. For example, we write

$\dfrac{15}{21} = \dfrac{\cancel{3} \cdot 5}{\cancel{3} \cdot 7} = \dfrac{5}{7}$ and speak of canceling the 3's. Now canceling is not an operation on numbers. We cancel

or strike out the 3's as a safety measure, that is, to be sure that they will not be used in computing the final result. The operation is division and Rule 5 states that we may divide the numerator by 3 provided we also divide the denominator by 3. This point is belabored here because of the all too common error

$\dfrac{12\cancel{a} - 5}{7\cancel{a}}$. The fact is that $\dfrac{12a - 5}{7a}$ cannot be further simplified for if we divide $7a$ by a we must also

divide $12a$ and 5 by a. This would lead to the more cumbersome $\dfrac{12 - 5/a}{7}$. (See Problems 1.7–1.8.)

The rational numbers may be associated in a one-to-one manner with points on a straight line as in Fig. 1-2. Here the point associated with the rational number m is m units from that point (called the origin) associated with 0, the distance between the points 0 and 1 being the unit of measure.

Fig. 1-2

If two rational numbers have representations r/n and s/n, where n is a positive integer, then $r/n > s/n$ if $r > s$, $r/n = s/n$ if $r = s$, and $r/n < s/n$ if $r < s$. Thus, in comparing two rational numbers it is necessary to express them with the same denominator. Of the many denominators (positive integers) there is always a least one, called the *least common denominator*. For the fractions $\frac{3}{5}$ and $\frac{2}{3}$, the least common denominator is 15. We conclude that $\frac{3}{5} < \frac{2}{3}$ since $\frac{3}{5} = \frac{9}{15} < \frac{10}{15} = \frac{2}{3}$. (See Problems 1.9–1.10.)

Rule 6. The sum (difference) of two rational numbers expressed with the same denominator is a rational number whose denominator is the common denominator and whose numerator is the sum (difference) of the numerators.

Rule 7. The product of two or more rational numbers is a rational number whose numerator is the product of the numerators and whose denominator is the product of the denominators of the several factors.

Rule 8. The quotient of two rational numbers can be evaluated by the use of Rule 5 with the least common denominator of the two numbers as the multiplier.

(See Problems 1.11–1.13.)

If a and b are rational numbers, $a + b$, $a - b$, and $a \cdot b$ are rational numbers. Moreover, if a and b are $\neq 0$, there exists a rational number x, unique except for its representation, such that

$$ax = b \qquad (1.1)$$

When a or b or both are zero, we have the following situations:

$b = 0$ and $a \neq 0$: (1.1) becomes $a \cdot x = 0$ and $x = 0$, that is, $0/a = 0$ when $a \neq 0$.
$a = 0$ and $b \neq 0$: (1.1) becomes $0 \cdot x = b$; then $b/0$, when $b \neq 0$, is without meaning since $0 \cdot x = 0$.
$a = 0$ and $b = 0$: (1.1) becomes $0 \cdot x = 0$; then $0/0$ is indeterminate since every number x satisfies the equation.
In brief: $0/a = 0$ when $a \neq 0$, but division by 0 is never permitted.

DECIMALS. In writing numbers we use a positional system, that is, the value given any particular digit depends upon its position in the sequence. For example, in 423 the positional value of the digit 4 is 4(100) while in 234 the positional value of the digit 4 is 4(1). Since the positional value of a digit involves the number 10, this system of notation is called the *decimal system*. In this system the number 4238.75 means

$$4(1000) + 2(100) + 3(10) + 8(1) + 7(\tfrac{1}{10}) + 5(\tfrac{1}{100})$$

It is interesting to note that from this example certain definitions to be made in a later study of exponents may be anticipated. Since $1000 = 10^3$, $100 = 10^2$, $10 = 10^1$ it would seem natural to define $1 = 10^0$, $\frac{1}{10} = 10^{-1}$, $\frac{1}{100} = 10^{-2}$.

By the process of division, any rational number can be expressed as a decimal; for example, $\frac{70}{33} = 2.121212\ldots$. This is termed a *repeating decimal* since the digits 12, called the cycle, are repeated without end. It will be seen later that every repeating decimal represents a rational number.

In operating with decimals, it is necessary to "round off" a decimal representation to a prescribed number of decimal places. For example, $\frac{1}{3} = 0.3333\ldots$ is written as 0.33 to two decimal places and $\frac{2}{3} = 0.6666\ldots$ is written as 0.667 to three decimal places. In rounding off, use will be made of the Computer's Rule:

(*a*) Increase the last digit retained by 1 if the digits rejected exceed the sequence $50000\ldots$; for example, $2.384629\ldots$ becomes 2.385 to three decimal places.

(*b*) Leave the last digit retained unchanged if the digits rejected are less than $5000\ldots$; for example, $2.384629\ldots$ becomes 2.38 to two decimal places.

(*c*) Make the last digit retained even if the digit rejected is exactly 5; for example, to three decimal places 11.3865 becomes 11.386 and 9.3815 becomes 9.382.

(See Problem 1.14.)

PERCENTAGE.   The symbol %, read percent, means per hundred; thus 5% is equivalent to $\frac{5}{100}$ or 0.05.

Any number, when expressed in decimal notation, can be written as a percent by multiplying by 100 and adding the symbol %. For example, $0.0125 = 100(0.0125)\% = 1.25\% = 1\frac{1}{4}\%$, $2.3 = 230\%$, and $\frac{7}{20} = 0.35 = 35\%$.

Conversely, any percentage may be expressed in decimal form by dropping the symbol % and dividing by 100. For example, $42.5\% = 42.5/100 = 0.425$, $3.25\% = 0.0325$, and $2000\% = 20$.

When reckoning percentages, express the percent as a decimal and, when possible, as a simple fraction. For example, $4\frac{1}{4}\%$ of $48 = 0.0425 \times 48 = 2.04$ and $12\frac{1}{2}\%$ of $5.28 = \frac{1}{8}$ of $5.28 = 0.66$. (See Problems 1.15–1.18.)

THE IRRATIONAL NUMBERS.   The existence of numbers other than the rational numbers may be inferred from either of the following considerations:

(*a*)   We may conceive of a nonrepeating decimal constructed in endless time by setting down a succession of digits chosen at random.

(*b*)   The length of the diagonal of a square of side 1 is not a rational number, that is, there exists no rational number a such that $a^2 = 2$. Numbers such as $\sqrt{2}$, $\sqrt[3]{2}$, $\sqrt[5]{-3}$, and π (but not $\sqrt{-3}$ or $\sqrt[4]{-5}$) are called *irrational numbers*. The first three of these are called *radicals*. The radical $\sqrt[n]{a}$ is said to be of order n; n is called the *index* and a is called the *radicand*.

(See Problems 1.19–1.21.)

THE REAL NUMBERS.   The set of *real numbers* consists of the rational and irrational numbers. The real numbers may be ordered by comparing their decimal representations. For example, $\sqrt{2} = 1.4142\ldots$; then $\frac{7}{5} = 1.4 < \sqrt{2}$, $\frac{3}{2} = 1.5 > \sqrt{2}$, etc.

We assume that the totality of real numbers may be placed in one-to-one correspondence with the totality of points on a straight line. See Fig. 1-3.

Fig. 1-3

The number associated with a point on the line, called the *coordinate* of the point, gives its distance and direction from that point (called the origin) associated with the number 0. If a point A has coordinate a, we shall speak of it as the point $A\,(a)$.

The directed distance from point A (a) to point B (b) on the real number scale is given by $AB = b - a$. The midpoint of the segment AB has coordinate $\frac{1}{2}(a + b)$. (See Problems 1.22–1.25.)

THE COMPLEX NUMBERS.   In the set of real numbers there is no number whose square is -1. If there is to be such a number, say $\sqrt{-1}$ then by definition $(\sqrt{-1})^2 = -1$. Note carefully that $(\sqrt{-1})^2 = \sqrt{-1}\,\sqrt{-1} = \sqrt{(-1)(-1)} = \sqrt{1} = 1$ is incorrect. In order to avoid this error, the symbol i with the following properties is used:

$$\text{If } a > 0, \qquad \sqrt{-a} = i\sqrt{a}; \qquad i^2 = -1$$

Then
$$(\sqrt{-2})^2 = \sqrt{-2}\,\sqrt{-2} = (i\sqrt{2})(i\sqrt{2}) = i^2 \cdot 2 = -2$$

and
$$\sqrt{-2}\,\sqrt{-3} = (i\sqrt{2})(i\sqrt{3}) = i^2\sqrt{6} = -\sqrt{6}$$

Numbers of the form $a + bi$, where a and b are real numbers, are called *complex numbers*. In the complex number $a + bi$, a is called the *real part* and bi is called the *imaginary part*. Numbers of the form ci, where c is real, are called *imaginary numbers*.

The complex number $a + bi$ is a real number when $b = 0$ and a pure imaginary number when $a = 0$. When a complex number is not a real number it is called *imaginary*.

Complex numbers will be considered in more detail in a later chapter. Only the following operations will be considered here:

To add (subtract) two complex numbers, add (subtract) the real parts and add (subtract) the pure imaginary parts.

To multiply two complex numbers, form the product treating i as an ordinary number and then replace i^2 by -1.

(See Problems 1.26–1.27.)

Solved Problems

1.1 Give the results when the following operations are performed on each of the numbers $-9, -6, -3, 0, 3, 6, 9, 12, 15$: (a) add -4, (b) subtract 6, (c) subtract -2, (d) multiply by -5, (e) divide by 3, (f) divide by -1, (g) divide by -3.

(a) $-13, -10, -7, -4, -1, 2, 5, 8, 11$ (e) $-3, -2, -1, 0, 1, 2, 3, 4, 5$

(b) $-15, -12, -9, -6, -3, 0, 3, 6, 9$ (f) $9, 6, 3, 0, -3, -6, -9, -12, -15$

(c) $-7, -4, -1, 2, 5, 8, 11, 14, 17$ (g) $3, 2, 1, 0, -1, -2, -3, -4, -5$

(d) $45, 30, 15, 0, -15, -30, -45, -60, -75$

1.2 Arrange the integers in each set so that they may be separated by $<$ and again so that they may be separated by $>$.

(a) $3, 15, 12, 20, 0$ *Ans.* $0 < 3 < 12 < 15 < 20;\ 20 > 15 > 12 > 3 > 0$

(b) $3, -3, 5, 0, -2$ *Ans.* $-3 < -2 < 0 < 3 < 5;\ 5 > 3 > 0 > -2 > -3$

(c) $-7, -5, -10, -8$ *Ans.* $-10 < -8 < -7 < -5;\ -5 > -7 > -8 > -10$

1.3 Let x be an integer. By means of Fig. 1-1, interpret each of the following:

(a) $x < 10$ *Ans.* x is to the left of 10.

(b) $x > -2$ *Ans.* x is to the right of -2.

(c) $x \geq 5$ *Ans.* x is 5 or is to the right of 5.

(d) $2 < x < 6$ *Ans.* x is to the right of 2 but to the left of 6.

(e) $10 > x > -3$ *Ans.* x is to the left of 10 but to the right of -3.

1.4 List all integral values of x when

(a) $2 < x < 6$ *Ans.* $3, 4, 5$ (d) $2 \leq x < 5$ *Ans.* $2, 3, 4$

(b) $2 > x > -3$ *Ans.* $-2, -1, 0, 1$ (e) $-4 < x \leq -1$ *Ans.* $-3, -2, -1$

(c) $-5 < x < 0$ *Ans.* $-4, -3, -2, -1$ (f) $2 \geq x \geq -3$ *Ans.* $-3, -2, -1, 0, 1, 2$

1.5 List the first 15 primes.

$2, 3, 5, 7, 11, 13, 17, 19, 23, 29, 31, 37, 41, 43, 47.$

1.6   Express each of the following integers as a product of primes: (*a*) 6930, (*b*) 23,595.

(*a*)   A systematic procedure is to test the primes 2, 3, 5, ... in order. When a factor is found, we then repeat the procedure, using in order all primes not already rejected, on the quotient. Thus, 6930 = 2.3465. Since 3465 is not divisible by 2, we try 3 and obtain $6930 = 2 \cdot 3 \cdot 1155$. Using 3 again, we find $6930 = 2 \cdot 3 \cdot 3 \cdot 385$. Since 385 is not divisible by 3, we try 5 and obtain $6930 = 2 \cdot 3 \cdot 3 \cdot 5 \cdot 77 = 2 \cdot 3 \cdot 3 \cdot 5 \cdot 7 \cdot 11$.

(*b*) $23,595 = 3 \cdot 5 \cdot 11 \cdot 11 \cdot 13$.

1.7   Express $\frac{5}{6}$ as a fraction having denominator (*a*) 12, (*b*) 36, (*c*) 84, (*d*) 126.

(*a*) $\dfrac{5}{6} = \dfrac{5 \cdot 2}{6 \cdot 2} = \dfrac{10}{12}$ (*b*) $\dfrac{5}{6} = \dfrac{5 \cdot 6}{6 \cdot 6} = \dfrac{30}{36}$ (*c*) $\dfrac{5}{6} = \dfrac{5 \cdot 14}{6 \cdot 14} = \dfrac{70}{84}$ (*d*) $\dfrac{5}{6} = \dfrac{5 \cdot 21}{6 \cdot 21} = \dfrac{105}{126}$

1.8   Reduce to lowest terms: (*a*) $\frac{6}{24}$, (*b*) $\frac{30}{42}$, (*c*) $\frac{27}{45}$, (*d*) $\frac{60}{96}$.

(*a*) $\dfrac{6}{24} = \dfrac{2 \cdot 3}{2 \cdot 2 \cdot 2 \cdot 3} = \dfrac{1}{2 \cdot 2} = \dfrac{1}{4}$ (*c*) $\dfrac{27}{45} = \dfrac{3 \cdot 3 \cdot 3}{3 \cdot 3 \cdot 5} = \dfrac{3}{5}$

(*b*) $\dfrac{30}{42} = \dfrac{2 \cdot 3 \cdot 5}{2 \cdot 3 \cdot 7} = \dfrac{5}{7}$ (*d*) $\dfrac{60}{96} = \dfrac{2 \cdot 2 \cdot 3 \cdot 5}{2 \cdot 2 \cdot 2 \cdot 2 \cdot 2 \cdot 3} = \dfrac{5}{2 \cdot 2 \cdot 2} = \dfrac{5}{8}$

1.9   In each of the following find the lowest common denominator (LCD) of the several fractions: (*a*) $\frac{3}{4}, \frac{5}{6}$, (*b*) $\frac{1}{6}, \frac{2}{9}, \frac{5}{24}$, (*c*) $\frac{1}{12}, \frac{7}{60}, \frac{2}{25}$, (*d*) $\frac{7}{72}, \frac{4}{75}, \frac{9}{80}$.

To find the LCD: Express each of the several denominators as the product of prime factors, write each distinct factor the greatest number of times it occurs in any denominator, and form the product.

(*a*)   Here $4 = 2 \cdot 2$ and $6 = 2 \cdot 3$; LCD $= (2 \cdot 2)(3) = 12$.

(*b*)   Here $6 = 2 \cdot 3$, $9 = 3 \cdot 3$, $24 = 2 \cdot 2 \cdot 2 \cdot 3$; LCD $= (2 \cdot 2 \cdot 2)(3 \cdot 3) = 72$.

(*c*)   Here $12 = 2 \cdot 2 \cdot 3$, $60 = 2 \cdot 2 \cdot 3 \cdot 5$, $25 = 5 \cdot 5$, LCD $= (2 \cdot 2)(3)(5 \cdot 5) = 300$.

(*d*)   Here $72 = 2 \cdot 2 \cdot 2 \cdot 3 \cdot 3$, $75 = 3 \cdot 5 \cdot 5$, $80 = 2 \cdot 2 \cdot 2 \cdot 2 \cdot 5$; LCD $= 2 \cdot 2 \cdot 2 \cdot 2 \cdot 3 \cdot 3 \cdot 5 \cdot 5 = 3600$.

1.10   Arrange each set of rational numbers so that they may be separated by <: (*a*) $1, -\frac{1}{2}, \frac{1}{3}, \frac{3}{4}, -\frac{5}{3}$; (*b*) $-\frac{2}{3}, -\frac{7}{8}, -\frac{5}{6}, -\frac{11}{6}, -\frac{5}{12}$.

(*a*)   Since $1 = \frac{12}{12}$, $-\frac{1}{2} = -\frac{6}{12}$, $\frac{1}{3} = \frac{4}{12}$, $\frac{3}{4} = \frac{9}{12}$, $-\frac{5}{3} = -\frac{20}{12}$, then $-\frac{5}{3} < -\frac{1}{2} < \frac{1}{3} < \frac{3}{4} < 1$.

(*b*)   Since $-\frac{2}{3} = -\frac{16}{24}$, $-\frac{7}{8} = -\frac{21}{24}$, $-\frac{5}{6} = -\frac{20}{24}$, $-\frac{11}{6} = -\frac{44}{24}$, $-\frac{5}{12} = -\frac{10}{24}$, then $-\frac{11}{6} < -\frac{7}{8} < -\frac{5}{6} < -\frac{2}{3} < -\frac{5}{12}$.

1.11   Perform the indicated operations:

(*a*) $\dfrac{1}{2} + \dfrac{2}{5} = \dfrac{5}{10} + \dfrac{4}{10} = \dfrac{5+4}{10} = \dfrac{9}{10}$

(*b*) $\dfrac{2}{3} - \dfrac{1}{2} = \dfrac{4-3}{6} = \dfrac{1}{6}$

(*c*) $\dfrac{11}{12} + \dfrac{4}{3} - \dfrac{1}{6} = \dfrac{11 + 16 - 2}{12} = \dfrac{25}{12}$

(*d*) $\dfrac{15}{64} - \dfrac{17}{32} + \dfrac{1}{8} = \dfrac{15 - 34 + 8}{64} = -\dfrac{11}{64}$

(*e*) $\dfrac{2}{3} \times \dfrac{4}{5} = \dfrac{2 \cdot 4}{3 \cdot 5} = \dfrac{8}{15}$

(*f*) $\dfrac{5}{8} \times \dfrac{16}{15} = \dfrac{5 \cdot 16}{8 \cdot 15} = \dfrac{2}{3}$

(*g*) $\dfrac{15}{7}\left(-\dfrac{21}{10}\right)\left(\dfrac{4}{9}\right) = -\dfrac{15 \cdot 21 \cdot 4}{7 \cdot 10 \cdot 9} = -\dfrac{3 \cdot 5 \cdot 3 \cdot 7 \cdot 2 \cdot 2}{7 \cdot 2 \cdot 5 \cdot 3 \cdot 3} = -2$

(*h*) $\dfrac{5}{7} \div \dfrac{11}{14} = 14\dfrac{5}{7} \div 14\dfrac{11}{14} = 10 \div 11 = \dfrac{10}{11}$

(*i*) $\dfrac{16}{5} \div \dfrac{7}{10} = 10\dfrac{16}{5} \div 10\dfrac{7}{10} = \dfrac{32}{7}$

(*j*) $8 \div \left(-\dfrac{2}{3}\right) = 3 \cdot 8 \div 3\left(-\dfrac{2}{3}\right) = -\dfrac{24}{2} = -12$

(*k*) $\dfrac{4 - \frac{3}{4}}{2 + \frac{1}{2}} = \dfrac{4 \cdot 4 - 4(\frac{3}{4})}{4 \cdot 2 + 4(\frac{1}{2})} = \dfrac{16 - 3}{8 + 2} = \dfrac{13}{10}$

(*l*) $\dfrac{\frac{1}{2} - \frac{7}{4}}{1 - \frac{3}{8}} = \dfrac{8(\frac{1}{2}) - 8(\frac{7}{4})}{8 \cdot 1 - 8(\frac{3}{8})} = \dfrac{4 - 14}{8 - 3} = \dfrac{-10}{5} = -2$

1.12 The product of five factors each equal to 2, that is, $2 \cdot 2 \cdot 2 \cdot 2 \cdot 2$, is denoted by 2^5 and read the fifth power of 2. We call 2 the *base* and 5 the *exponent*. Show that the solutions of Solved Problem 8(a), (c), (d) may be written as follows using exponents:

(a) $\dfrac{6}{24} = \dfrac{2 \cdot 3}{2^3 \cdot 3} = \dfrac{1}{2^2} = \dfrac{1}{4}$ (c) $\dfrac{27}{45} = \dfrac{3^3}{3^2 \cdot 5} = \dfrac{3}{5}$ (d) $\dfrac{60}{96} = \dfrac{2^2 \cdot 3 \cdot 5}{2^5 \cdot 3} = \dfrac{5}{2^3} = \dfrac{5}{8}$

1.13 Verify

(a) $a^4 \cdot a^2 = (a \cdot a \cdot a \cdot a)(a \cdot a) = a^{4+2} = a^6$,   (b) $\dfrac{a^5}{a^3} = \dfrac{a \cdot a \cdot a \cdot a \cdot a}{a \cdot a \cdot a} = a^{5-3} = a^2$,

(c) $\dfrac{a^3}{a^5} = \dfrac{a \cdot a \cdot a}{a \cdot a \cdot a \cdot a \cdot a} = \dfrac{1}{a^{5-3}} = \dfrac{1}{a^2}$,

(d) $(a \cdot b)^4 = (a \cdot b)(a \cdot b)(a \cdot b)(a \cdot b) = (a \cdot a \cdot a \cdot a)(b \cdot b \cdot b \cdot b) = a^4 b^4$.

The general rules are: If m and n are positive integers, then $a^m a^n = a^{m+n}$; $a^m/a^n = a^{m-n}$ if $m > n$; $a^m/a^n = 1$ if $m = n$; $a^m/a^n = 1/a^{n-m}$ if $m < n$; $(a \cdot b)^m = a^m b^m$.

1.14 Express $\frac{2}{7}$ as a decimal to (a) five, (b) four, (c) three, (d) two decimal places.

By division, $\frac{2}{7} = 0.285714\ldots$. Then we have for a) 0.28571, b) 0.2857, c) 0.286, d) 0.29.

1.15 Compute:

(a) 6% of 400 $= 0.06 \times 400 = 24$ (c) 135% of 500 $= 1.35 \times 500 = 675$

(b) $4\frac{1}{2}$% of 1200 $= 0.045 \times 1200 = 54$ (d) 2% of 6% of 8000 $= 0.02 \times 0.06 \times 8000 = 9.6$

1.16 What percent (a) of 75 is 15? (b) of 112 is 14? (c) of 72 is 3.96? (d) of 0.44 is 1.034?

(a) $\frac{15}{75} = \frac{1}{5} = 20\%$ (c) $3.96/72 = 0.055 = 5\frac{1}{2}\%$

(b) $\frac{14}{112} = \frac{1}{8} = 12\frac{1}{2}\%$ (d) $1.034/0.44 = 2.35 = 235\%$

1.17 Find the number, given (a) 5% of it is 32, (b) 8% of it is 8.4, (c) 210% of it is 54.6, (d) 0.5% of it is 2.3.

(a) 1% of the number is $\frac{32}{5} = 6.4$; 100% of the number is $100 \times 6.4 = 640$ or $32/0.05 = 640$.

(b) $8.4/0.08 = 105$ (c) $54.6/2.1 = 26$ (d) $2.3/0.005 = 460$

1.18 Express the *percentage* strength of each of the following solutions (by a 10% silver nitrate solution is meant 10 grams of silver nitrate in 100 grams of solution): (a) 200 grams of solution containing a 0.5-gram tablet of bichloride of mercury; (b) 50 grams of solution containing 0.8 gram of salt.

(a) $0.5/200 = 0.0025 = 0.25\% = \frac{1}{4}\%$ (b) $0.8/50 = 0.016 = 1.6\%$

1.19 Simplify each of the following radicals:

(a) $\sqrt{700} = \sqrt{100 \cdot 7} = 10\sqrt{7}$ (e) $\dfrac{2}{\sqrt{3}} = \dfrac{2\sqrt{3}}{\sqrt{3}\sqrt{3}} = \dfrac{2}{3}\sqrt{3}$

(b) $\sqrt[3]{24} = \sqrt[3]{8 \cdot 3} = 2\sqrt[3]{3}$ (f) $\sqrt{\dfrac{2}{3}} = \dfrac{\sqrt{2}}{\sqrt{3}} = \dfrac{\sqrt{2}\sqrt{3}}{\sqrt{3}\sqrt{3}} = \dfrac{\sqrt{2 \cdot 3}}{3} = \dfrac{\sqrt{6}}{3}$

(c) $\sqrt{72} = \sqrt{36 \cdot 2} = 6\sqrt{2}$ (g) $\sqrt{\dfrac{5}{8}} = \sqrt{\dfrac{5 \cdot 2}{8 \cdot 2}} = \sqrt{\dfrac{10}{16}} = \dfrac{\sqrt{10}}{4}$

(d) $\sqrt[5]{-64} = \sqrt[5]{-32 \cdot 2} = -2\sqrt[5]{2}$

1.20   Perform the indicated operations.

(a) $4\sqrt{2} + 3\sqrt{2} - 2\sqrt{2} = (4 + 3 - 2)\sqrt{2} = 5\sqrt{2}$

(b) $6\sqrt{3} - \sqrt{27} = 6\sqrt{3} - 3\sqrt{3} = 3\sqrt{3}$

(c) $2\sqrt[3]{5} - \sqrt[3]{135} + 4\sqrt[6]{25} = 2\sqrt[3]{5} - 3\sqrt[3]{5} + 4\sqrt[3]{5} = 3\sqrt[3]{5}$

(d) $2 \cdot \dfrac{1}{\sqrt{7}} + 3\sqrt{28} - \sqrt{63} = \frac{2}{7}\sqrt{7} + 6\sqrt{7} - 3\sqrt{7} = \frac{23}{7}\sqrt{7}$

(e) $\sqrt[4]{64} - 5\sqrt[6]{\frac{1}{8}} = 2\sqrt{2} - \frac{5}{2}\sqrt{2} = -\frac{1}{2}\sqrt{2}$

(f) $\sqrt{3} \cdot \sqrt{15} = \sqrt{45} = 3\sqrt{5}$

(g) $\sqrt[3]{18} \cdot \sqrt[3]{4} = \sqrt[3]{72} = 2\sqrt[3]{9}$

(h) $(2\sqrt{5} + 3)(3\sqrt{5} - 4) = 30 - 8\sqrt{5} + 9\sqrt{5} - 12 = 18 + \sqrt{5}$

(i) $(2\sqrt{3} - 3\sqrt{2})(5\sqrt{3} + \sqrt{2}) = 30 + 2\sqrt{6} - 15\sqrt{6} - 6 = 24 - 13\sqrt{6}$

(j) $i^5 = i^4 \cdot i = 1 \cdot i = i$

1.21   Simplify each of the following:

(a) $\dfrac{3 + 5\sqrt{2}}{2\sqrt{2}} = \dfrac{(3 + 5\sqrt{2})\sqrt{2}}{(2\sqrt{2})\sqrt{2}} = \dfrac{3\sqrt{2} + 10}{4}$

(b) $\dfrac{4}{\sqrt{2} + \sqrt{3}} = \dfrac{4}{\sqrt{2} + \sqrt{3}} \cdot \dfrac{\sqrt{2} - \sqrt{3}}{\sqrt{2} - \sqrt{3}} = \dfrac{4\sqrt{2} - 4\sqrt{3}}{2 - 3} = 4\sqrt{3} - 4\sqrt{2}$

(c) $\dfrac{2\sqrt{3} + 3\sqrt{2}}{3\sqrt{5} - 5\sqrt{3}} = \dfrac{2\sqrt{3} + 3\sqrt{2}}{3\sqrt{5} - 5\sqrt{3}} \cdot \dfrac{3\sqrt{5} + 5\sqrt{3}}{3\sqrt{5} + 5\sqrt{3}} = \dfrac{6\sqrt{15} + 30 + 9\sqrt{10} + 15\sqrt{6}}{45 - 75}$

$= \dfrac{30 + 15\sqrt{6} + 9\sqrt{10} + 6\sqrt{15}}{-30} = -\dfrac{10 + 5\sqrt{6} + 3\sqrt{10} + 2\sqrt{15}}{10}$

(d) $\dfrac{1 + 4\sqrt[3]{2}}{\sqrt[3]{2}} = \dfrac{(1 + 4\sqrt[3]{2})\sqrt[3]{4}}{\sqrt[3]{2} \cdot \sqrt[3]{4}} = \dfrac{\sqrt[3]{4} + 4\sqrt[3]{8}}{\sqrt[3]{8}} = \dfrac{\sqrt[3]{4} + 8}{2}$

1.22   The numerical value of a real number N ($|N|$) is defined as follows:

$$|N| = N \quad \text{if} \quad N > 0; \qquad |N| = 0 \quad \text{if} \quad N = 0; \qquad |N| = -N \quad \text{if} \quad N < 0.$$

Arrange each set of numbers so that they may be separated by $<$. (a) $|-12|, |\frac{5}{2}|, |-9|, |1|, |-2|, |6|$; (b) $|3 + 4|$, $|9 - 6|, |2 - 8|, |-3 - 6|$.

(a) $|1| < |-2| < |\frac{5}{2}| < |6| < |-9| < |-12|$ since $1 < 2 < \frac{5}{2} < 6 < 9 < 12$.

(b) $|9 - 6| < |2 - 8| < |3 + 4| < |-3 - 6|$ since $3 < 6 < 7 < 9$.

1.23   Find the directed distance AB, given (a) $A(2)$, $B(6)$; (b) $A(3)$, $B(-7)$; (c) $A(-2)$, $B(-8)$; (d) $A(-10)$, $B(2)$; (e) $A(-9)$, $B(-2)$; (f) $A(1)$, $B(x)$; (g) $A(x_1)$, $B(3)$; (h) $A(x_1)$, $B(x_2)$.

(a) $AB = 6 - 2 = 4$ (c) $AB = -8 - (-2) = -6$ (e) $AB = -2 - (-9) = 7$ (g) $AB = 3 - x_1$

(b) $AB = -7 - 3 = -10$ (d) $AB = 2 - (-10) = 12$ (f) $AB = x - 1$ (h) $AB = x_2 - x_1$

1.24   (a) On a number scale locate the points $A(-5)$, $B(1)$, $C(7)$ and show that $AB + BC + CA = 0$. (b) Relabel the above points, reading from left to right, $B(-5)$, $C(1)$, $A(7)$ and show that $AB + BC + CA = 0$.

(a) $AB + BC + CA = [1 - (-5)] + (7 - 1) + (-5 - 7) = 6 + 6 - 12 = 0$

(b) $AB + BC + CA = (-5 - 7) + [1 - (-5)] + (7 - 1) = 0$

1.25   Find the coordinate of the midpoint of the segments AB in Problem 1.23.

(a) $\frac{1}{2}(2 + 6) = 4$ (c) $\frac{1}{2}[-2 + (-8)] = -5$ (e) $\frac{1}{2}(-9 - 2) = -\frac{11}{2}$ (g) $\frac{1}{2}(x_1 + 3)$

(b) $\frac{1}{2}[3 + (-7)] = -2$ (d) $\frac{1}{2}(-10 + 2) = -4$ (f) $\frac{1}{2}(1 + x)$ (h) $\frac{1}{2}(x_1 + x_2)$

1.26   Rewrite each of the following, using i:

(a) $\sqrt{-5} = i\sqrt{5}$ (b) $\sqrt{-4} = 2i$ (c) $\sqrt{-a^2} = ai$ (d) $\sqrt{-32} = 4i\sqrt{2}$

(e) $3 - \sqrt{-9} = 3 - 3i$ (f) $\dfrac{8 + \sqrt{-16}}{2} = \dfrac{8 + 4i}{2} = 4 + 2i$

(g) $\dfrac{6 - \sqrt{-128}}{12} = \dfrac{6 - 8i\sqrt{2}}{12} = \dfrac{1}{2} - \dfrac{2\sqrt{2}}{3}\, i$

1.27   Perform the indicated operations.

(a) $(2 - 5i) + (4 + 3i) = (2 + 4) + (-5 + 3)i = 6 - 2i$

(b) $(3 + 2i) - (-6 - 3i) = [3 - (-6)] + [2 - (-3)]i = 9 + 5i$

(c) $(-5 + 2\sqrt{-4}) + (1 - \sqrt{-9}) = (-5 + 4i) + (1 - 3i) = -4 + i$

(d) $(2 + \sqrt{-27}) - (4 - \sqrt{-3}) = (2 + 3i\sqrt{3}) - (4 - i\sqrt{3}) = -2 + 4i\sqrt{3}$

(e) $(-2 - \sqrt{-8}) - (5 + \sqrt{-27}) = (-2 - 2i\sqrt{2}) - (5 + 3i\sqrt{3}) = -7 - (2\sqrt{2} + 3\sqrt{3})i$

(f) $(2 + 3i) + (2 - 3i) = 4$

(g) $(2 + 3i) - (2 - 3i) = 6i$

(h) $(2 - 5i)(4 + 3i) = 8 + 6i - 20i - 15i^2 = 8 - 14i + 15 = 23 - 14i$

(i) $(2 + 3i)(2 - 3i) = 4 - 6i + 6i - 9i^2 = 13$

Supplementary Problems

1.28   Arrange each of the following so that they may be separated by $<$.

(a) $\frac{2}{3}, -\frac{3}{4}, \frac{5}{6}, -1, \frac{4}{5}, -\frac{4}{3}, -\frac{1}{4}$ (b) $\frac{3}{2}, 2, \frac{7}{5}, \frac{4}{3}, 3$ (c) $\frac{3}{2}, \sqrt{3}, -\frac{1}{2}, -\sqrt{5}, 0$

1.29   Determine the greater of each pair.

(a) $|4 + (-2)|$ and $|-4| + |-2|$ (b) $|4 + (-2)|$ and $|4| + |-2|$ (c) $|4 - (-2)|$ and $|4| - |-2|$

1.30   Convert each of the following fractions into equivalent fractions having the indicated denominator:

(a) $\frac{3}{5}, 15$ (b) $-\frac{3}{5}, 20$ (c) $\frac{7}{3}, 42$ (d) $\frac{5}{7}, 35$ (e) $\frac{12}{13}, 156$

1.31   Perform the indicated operations.

(a) $(-2)(3)(-5)$

(b) $3(-2)(4) + (-5)(2)(0)$

(c) $-8 - (-6) + 2$

(d) $\frac{3}{4} + \frac{2}{3}$

(e) $\frac{3}{4} - \frac{2}{3}$

(f) $\frac{5}{6} - \frac{1}{2} - \frac{2}{3}$

(g) $\frac{3}{4} - \frac{7}{12} - \frac{1}{3}$

(h) $(\frac{1}{2})(\frac{8}{9})(\frac{6}{5})$

(i) $\frac{3}{8} \times 5\frac{1}{3}$

(j) $2\frac{1}{4} \times 2\frac{2}{3} \times 1\frac{1}{5} \times 2\frac{1}{7}$

(k) $\frac{25}{32} \div \frac{35}{64}$

(l) $3\frac{1}{3} \div \frac{7}{10}$

(m) $(1\frac{1}{2} \times 2\frac{1}{4}) \div 1\frac{1}{8}$

(n) $\dfrac{3 - \frac{2}{3}}{5 + \frac{5}{6}}$

(o) $\dfrac{\frac{2}{3} + \frac{3}{4}}{\frac{5}{6} - \frac{7}{8}}$

(p) $\dfrac{1\frac{1}{2} - 2\frac{2}{3}}{3\frac{1}{5} - 1\frac{1}{4}}$

1.32   Perform the indicated operations.

(a) $5\sqrt{3} + 2\sqrt{3} - 8\sqrt{3}$

(b) $5\sqrt{2} + \sqrt{32} - 3\sqrt{8}$

(c) $\sqrt[3]{12} \cdot \sqrt[3]{36}$

(d) $(1 + \sqrt{2})(3 - \sqrt{2})$

(e) $(2\sqrt{3} + 3\sqrt{2})(2\sqrt{3} - 3\sqrt{2})$

(f) $(4\sqrt{3} - 3\sqrt{5})(2\sqrt{3} + \sqrt{5})$

(g) $\dfrac{4 - 2\sqrt{3}}{5\sqrt{3}}$

(h) $\dfrac{2\sqrt{5} - 3\sqrt{2}}{3\sqrt{5} + 4\sqrt{2}}$

(i) $\dfrac{3\sqrt{2} - 4\sqrt{3}}{4\sqrt{2} - 3\sqrt{3}}$

1.33   Perform the indicated operations.

(a) $i\sqrt{12} + i\sqrt{75} - i\sqrt{108}$ (f) $\sqrt{-12} \cdot \sqrt{-27}$ (k) $(2\sqrt{3} + i\sqrt{2})(3\sqrt{3} - 5i\sqrt{2})$

(b) $i\sqrt{50} - i\sqrt{32} + i\sqrt{8}$ (g) $3i(2 + i)$ (l) $(3 - 2i)(1 + 5i)(-2 - i)$

(c) $4\sqrt{-27} + 2\sqrt{-12} - 5\sqrt{-48}$ (h) $(-3 + 5i) + (4 - 2i)$ (m) i^{73}

(d) $3\sqrt{-20} + 5\sqrt{-80} - \sqrt{-45}$ (i) $(3 + 5i) - (-4 - 2i)$

(e) $\sqrt{-9} \cdot \sqrt{-16}$ (j) $(3 + 5i)(2 - 7i)$

ANSWERS TO SUPPLEMENTARY PROBLEMS

1.28. (a) $-\frac{4}{3}, -1, -\frac{3}{4}, -\frac{1}{4}, \frac{2}{3}, \frac{4}{5}, \frac{5}{6}$ (b) $\frac{4}{3}, \frac{7}{5}, \frac{3}{2}, 2, 3$ (c) $-\sqrt{5}, -\frac{1}{2}, 0, \frac{3}{2}, \sqrt{3}$

1.29. (a) second   (b) second   (c) first

1.30. (a) $\frac{9}{15}$ (b) $-\frac{12}{20}$ (c) $\frac{98}{42}$ (d) $\frac{25}{35}$ (e) $\frac{144}{156}$

1.31. (a) 30   (c) 0   (e) $\frac{1}{12}$ (g) $-\frac{1}{6}$ (i) 2   (k) $\frac{10}{7}$ (m) 3   (o) -34

(b) -24 (d) $\frac{17}{12}$ (f) $-\frac{1}{3}$ (h) $\frac{8}{15}$ (j) 18   (l) $\frac{100}{21}$ (n) $\frac{2}{5}$ (p) $-\frac{70}{117}$

1.32. (a) $-\sqrt{3}$ (c) $6\sqrt[3]{2}$ (e) -6 (g) $\dfrac{4\sqrt{3} - 6}{15}$ (h) $\dfrac{54 - 17\sqrt{10}}{13}$ (i) $-\dfrac{12 + 7\sqrt{6}}{5}$

(b) $3\sqrt{2}$ (d) $1 + 2\sqrt{2}$ (f) $9 - 2\sqrt{15}$

1.33. (a) $i\sqrt{3}$ (d) $-17i\sqrt{5}$ (g) $-3 + 6i$ (j) $41 - 11i$ (m) i

(b) $3i\sqrt{2}$ (e) -12 (h) $1 + 3i$ (k) $28 - 7i\sqrt{6}$

(c) $-4i\sqrt{3}$ (f) -18 (i) $7 + 7i$ (l) $-13 - 39i$

Chapter 2

Elements of Algebra

IN ARITHMETIC the numbers used are always known numbers; a typical problem is to convert 5 hours and 35 minutes to minutes. This is done by multiplying 5 by 60 and adding 35; thus, $5 \cdot 60 + 35 = 335$ minutes.

In algebra some of the numbers used may be known but others are either unknown or not specified, that is, they are represented by letters. For example, convert h hours and m minutes into minutes. This is done in precisely the same manner as in the paragraph above, by multiplying h by 60 and adding m, thus, $h \cdot 60 + m = 60h + m$. We call $60h + m$ an *algebraic expression*. (See Problem 2.1.)

Since algebraic expressions are numbers, they may be added, subtracted, etc., following the same laws as govern these operations on known numbers. For example, the sum of $5 \cdot 60 + 35$ and $2 \cdot 60 + 35$ is $(5 + 2) \cdot 60 + 2 \cdot 35$; similarly, the sum of $h \cdot 60 + m$ and $k \cdot 60 + m$ is $(h + k) \cdot 60 + 2m$. (See Problems 2.2–2.6.)

POSITIVE INTEGRAL EXPONENTS. If a is any number and n is any positive integer, the product of the n factors $a \cdot a \cdot a \cdots a$ is denoted by a^n. To distinguish between the letters, a is called the *base* and n is called the *exponent*.

If a and b are any bases and m and n are any positive integers, we have the following laws of exponents:

(1) $a^m \cdot a^n = a^{m+n}$

(2) $(a^m)^n = a^{mn}$

(3) $\dfrac{a^m}{a^n} = a^{m-n}, \quad a \neq 0, \quad m > n; \qquad \dfrac{a^m}{a^n} = \dfrac{1}{a^{n-m}}, \quad a \neq 0, \quad m < n$

(4) $(a \cdot b)^n = a^n b^n$

(5) $\left(\dfrac{a}{b}\right)^n = \dfrac{a^n}{b^n}, \quad b \neq 0$

(See Problem 2.7.)

LET n BE A POSITIVE INTEGER and a and b be two numbers such that $b^n = a$; then b is called an nth *root of a*. Every number $a \neq 0$ has exactly n distinct nth roots.

If a is imaginary, all of its nth roots are imaginary; this case will be excluded here and treated later.

If a is real and n is odd, then exactly one of the nth roots of a is real. For example, 2 is the real cube root of 8, $(2^3 = 8)$, and -3 is the real fifth root of -243, $[(-3)^5 = -243]$.

If a is real and n is even, then there are exactly two real nth roots of a when $a > 0$, but no real nth roots of a when $a < 0$. For example, $+3$ and -3 are the square roots of 9; $+2$ and -2 are the real sixth roots of 64.

THE PRINCIPAL nth ROOT OF a is the positive real nth root of a when a is positive and the real nth root of a, if any, when a is negative. The principal nth root of a is denoted by $\sqrt[n]{a}$, called a *radical*. The integer n is called the *index* of the radical and a is called the *radicand*. For example,

$$\sqrt{9} = 3 \qquad \sqrt[6]{64} = 2 \qquad \sqrt[5]{-243} = -3$$

(See Problem 2.8.)

ZERO, FRACTIONAL, AND NEGATIVE EXPONENTS. When s is a positive integer, r is any integer, and p is any rational number, the following extend the definition of a^n in such a way that the laws (1)-(5) are satisfied when n is any rational number.

DEFINITIONS	EXAMPLES
(6) $a^0 = 1, \ a \neq 0$	$2^0 = 1, \ (\frac{1}{100})^0 = 1, \ (-8)^0 = 1$
(7) $a^{r/s} = \sqrt[s]{a^r} = (\sqrt[s]{a})^r$	$3^{1/2} = \sqrt{3}, \ (64)^{5/6} = (\sqrt[6]{64})^5 = 2^5 = 32, \ 3^{-2/1} = 3^{-2} = \frac{1}{9}$
(8) $a^{-p} = 1/a^p, \ a \neq 0$	$2^{-1} = \frac{1}{2}, \ 3^{-1/2} = 1/\sqrt{3}$

[NOTE: Without attempting to define them, we shall assume the existence of numbers such as $a^{\sqrt{2}}, a^{\pi}, \dots,$ in which the exponent is irrational. We shall also assume that these numbers have been defined in such a way that the laws (1)-(5) are satisfied.] (See Problems 2.9-2.10.)

Solved Problems

2.1   For each of the following statements, write the equivalent algebraic expression: (a) the sum of x and 2, (b) the sum of a and $-b$, (c) the sum of $5a$ and $3b$, (d) the product of $2a$ and $3a$, (e) the product of $2a$ and $5b$, (f) the number which is 4 more than 3 times x, (g) the number which is 5 less than twice y, (h) the time required to travel 250 miles at x miles per hour, (i) the cost (in cents) of x eggs at 65¢ per dozen.

(a) $x + 2$ (d) $(2a)(3a) = 6a^2$ (g) $2y - 5$

(b) $a + (-b) = a - b$ (e) $(2a)(5b) = 10ab$ (h) $250/x$

(c) $5a + 3b$ (f) $3x + 4$ (i) $65(x/12)$

2.2   Let x be the present age of a father. (a) Express the present age of his son who 2 years ago was one-third his father's age. (b) Express the age of his daughter who 5 years from today will be one-fourth her father's age.

(a)   Two years ago the father's age was $x - 2$ and the son's age was $(x - 2)/3$. Today the son's age is $2 + (x - 2)/3$.

(b)   Five years from today the father's age will be $x + 5$ and his daughter's age will be $\frac{1}{4}(x + 5)$. Today the daughter's age is $\frac{1}{4}(x + 5) - 5$.

2.3   A pair of parentheses may be inserted or removed at will in an algebraic expression if the first parenthesis of the pair is preceded by a + sign. If, however, this sign is −, the signs of all terms within the parentheses must be changed.

(a) $5a + 3a - 6a = (5 + 3 - 6)a = 2a$ (b) $\frac{1}{2}a + \frac{1}{4}b - \frac{1}{4}a + \frac{3}{4}b = \frac{1}{4}a + b$

(c) $(13a^2 - b^2) + (-4a^2 + 3b^2) - (6a^2 - 5b^2) = 13a^2 - b^2 - 4a^2 + 3b^2 - 6a^2 + 5b^2 = 3a^2 + 7b^2$

(d) $(2ab - 3bc) - [5 - (4ab - 2bc)] = 2ab - 3bc - [5 - 4ab + 2bc]$
$$= 2ab - 3bc - 5 + 4ab - 2bc = 6ab - 5bc - 5$$

(e) $(2x + 5y - 4)3x = (2x)(3x) + (5y)(3x) - 4(3x) = 6x^2 + 15xy - 12x$

(f) $5a - 2$ (g) $2x - 3y$ (h) $3a^2 + 2a - 1$

 $\underline{3a + 4}$ $\underline{5x + 6y}$ $\underline{2a - 3}$

 $15a^2 - \ 6a$ $10x^2 - 15xy$ $6a^3 + 4a^2 - 2a$

 $(+)\underline{ 20a - 8}$ $(+)\underline{ 12xy - 18y^2}$ $(+)\underline{ -9a^2 - 6a + 3}$

 $15a^2 + 14a - 8$ $10x^2 - 3xy - 18y^2$ $6a^3 - 5a^2 - 8a + 3$

(i)

$$
\begin{array}{r}
x^2 + 4x - 2 \\
x - 3 \overline{)\ x^3 +\ \ x^2 - 14x + 6} \\
(-)\underline{x^3 - 3x^2} \\
4x^2 - 14x \\
(-)\underline{4x^2 - 12x} \\
-2x + 6 \\
(-)\underline{-2x + 6}
\end{array}
$$

(j)

$$
\begin{array}{r}
x^2 - 2x - 1 \\
x^2 + 3x - 2 \overline{)\ x^4 +\ \ x^3 - 9x^2 + x + 5} \\
(-)\underline{x^4 + 3x^3 - 2x^2} \\
-2x^3 - 7x^2 + x \\
(-)\underline{-2x^3 - 6x^2 + 4x} \\
-x^2 - 3x + 5 \\
(-)\underline{-x^2 - 3x + 2} \\
3
\end{array}
$$

$$\frac{x^3 + x^2 - 14x + 6}{x - 3} = x^2 + 4x - 2 \qquad \frac{x^4 + x^3 - 9x^2 + x + 5}{x^2 + 3x - 2} = x^2 - 2x - 1 + \frac{3}{x^2 + 3x - 2}$$

2.4 The problems below involve the following types of factoring:

$$ab + ac - ad = a(b + c - d) \qquad\qquad a^2 \pm 2ab + b^2 = (a \pm b)^2$$

$$a^3 + b^3 = (a + b)(a^2 - ab + b^2) \qquad a^2 - b^2 = (a - b)(a + b)$$

$$acx^2 + (ad + bc)x + bd = (ax + b)(cx + d) \qquad a^3 - b^3 = (a - b)(a^2 + ab + b^2)$$

(a) $5x - 10y = 5(x - 2y)$

(b) $\frac{1}{2}gt^2 - \frac{1}{2}g^2t = \frac{1}{2}gt(t - g)$

(c) $x^2 + 4x + 4 = (x + 2)^2$

(d) $x^2 + 5x + 4 = (x + 1)(x + 4)$

(e) $x^2 - 3x - 4 = (x - 4)(x + 1)$

(f) $4x^2 - 12x + 9 = (2x - 3)^2$

(g) $12x^2 + 7x - 10 = (4x + 5)(3x - 2)$

(h) $x^3 - 8 = (x - 2)(x^2 + 2x + 4)$

(i) $2x^4 - 12x^3 + 10x^2 = 2x^2(x^2 - 6x + 5) = 2x^2(x - 1)(x - 5)$

2.5 Simplify.

(a) $\dfrac{8}{12x + 20} = \dfrac{4 \cdot 2}{4 \cdot 3x + 4 \cdot 5} = \dfrac{2}{3x + 5}$

(b) $\dfrac{9x^2}{12xy - 15xz} = \dfrac{3x \cdot 3x}{3x \cdot 4y - 3x \cdot 5z} = \dfrac{3x}{4y - 5z}$

(c) $\dfrac{5x - 10}{7x - 14} = \dfrac{5(x - 2)}{7(x - 2)} = \dfrac{5}{7}$

(d) $\dfrac{4x - 12}{15 - 5x} = \dfrac{4(x - 3)}{5(3 - x)} = \dfrac{4(x - 3)}{-5(x - 3)} = -\dfrac{4}{5}$

(e) $\dfrac{x^2 - x - 6}{x^2 + 7x + 10} = \dfrac{(x + 2)(x - 3)}{(x + 2)(x + 5)} = \dfrac{x - 3}{x + 5}$

(f) $\dfrac{6x^2 + 5x - 6}{2x^2 - 3x - 9} = \dfrac{(2x + 3)(3x - 2)}{(2x + 3)(x - 3)} = \dfrac{3x - 2}{x - 3}$

(g) $\dfrac{3a^2 - 11a + 6}{a^2 - a - 6} \cdot \dfrac{4 - 4a - 3a^2}{36a^2 - 16} = \dfrac{(3a - 2)(a - 3)(2 - 3a)(2 + a)}{(a - 3)(a + 2)4(3a + 2)(3a - 2)} = -\dfrac{3a - 2}{4(3a + 2)}$

2.6 Combine as indicated.

(a) $\dfrac{2a + b}{10} + \dfrac{a - 6b}{15} = \dfrac{3(2a + b) + 2(a - 6b)}{30} = \dfrac{8a - 9b}{30}$

(b) $\dfrac{2}{x} - \dfrac{3}{2x} + \dfrac{5}{4} = \dfrac{2 \cdot 4 - 3 \cdot 2 + 5 \cdot x}{4x} = \dfrac{2 + 5x}{4x}$

(c) $\dfrac{2}{3a - 1} - \dfrac{3}{2a + 1} = \dfrac{2(2a + 1) - 3(3a - 1)}{(3a - 1)(2a + 1)} = \dfrac{5 - 5a}{(3a - 1)(2a + 1)}$

(d) $\dfrac{3}{x + y} - \dfrac{5}{x^2 - y^2} = \dfrac{3}{x + y} - \dfrac{5}{(x + y)(x - y)} = \dfrac{3(x - y) - 5}{(x + y)(x - y)} = \dfrac{3x - 3y - 5}{(x + y)(x - y)}$

(e) $\dfrac{a - 2}{6a^2 - 5a - 6} + \dfrac{2a + 1}{9a^2 - 4} = \dfrac{a - 2}{(2a - 3)(3a + 2)} + \dfrac{2a + 1}{(3a + 2)(3a - 2)}$

$$= \dfrac{(a - 2)(3a - 2) + (2a + 1)(2a - 3)}{(2a - 3)(3a + 2)(3a - 2)} = \dfrac{7a^2 - 12a + 1}{(2a - 3)(3a + 2)(3a - 2)}$$

2.7   Perform the indicated operations.

(a) $3^4 = 3 \cdot 3 \cdot 3 \cdot 3 = 81$ (f) $2^6 \cdot 2^4 = 2^{6+4} = 2^{10} = 1024$ (k) $a^{10}/a^4 = a^{10-4} = a^6$

(b) $-3^4 = -81$ (g) $(\tfrac{1}{2})^3(\tfrac{1}{2})^2 = (\tfrac{1}{2})^5$ (l) $a^4/a^{10} = 1/a^{10-4} = 1/a^6$

(c) $(-3)^4 = 81$ (h) $a^{n+3}a^{m+2} = a^{m+n+5}$ (m) $(-2)^8/(-2)^5 = (-2)^3 = -8$

(d) $-(-3)^4 = -81$ (i) $(a^2)^5 = a^{2 \cdot 5} = a^{10}$ (n) $a^{2n}b^{5m}/a^{3n}b^{2m} = b^{3m}/a^n$

(e) $-(-3)^3 = 27$ (j) $(a^{2n})^3 = a^{6n}$ (o) $36^{x+3}/6^{x-1} = 6^{2x+6}/6^{x-1} = 6^{x+7}$

2.8   Evaluate.

(a) $81^{1/2} = \sqrt{81} = 9$ (d) $(-27)^{1/3} = \sqrt[3]{-27} = -3$

(b) $81^{3/4} = (\sqrt[4]{81})^3 = 3^3 = 27$ (e) $(-32)^{4/5} = (\sqrt[5]{-32})^4 = (-2)^4 = 16$

(c) $(\tfrac{16}{49})^{3/2} = (\sqrt{\tfrac{16}{49}})^3 = (\tfrac{4}{7})^3 = \tfrac{64}{343}$ (f) $-400^{1/2} = -\sqrt{400} = -20$

2.9   Evaluate.

(a) $4^0 = 1$ (c) $(4a)^0 = 1$ (e) $4^{-1} = \tfrac{1}{4}$ (g) $125^{-1/3} = 1/125^{1/3} = \tfrac{1}{5}$

(b) $4a^0 = 4 \cdot 1 = 4$ (d) $4(3+a)^0 = 4 \cdot 1 = 4$ (f) $5^{-2} = (\tfrac{1}{5})^2 = \tfrac{1}{25}$ (h) $(-125)^{-1/3} = -\tfrac{1}{5}$

(i) $-(\tfrac{1}{64})^{5/6} = -(\sqrt[6]{\tfrac{1}{64}})^5 = -(\tfrac{1}{2})^5 = -\tfrac{1}{32}$

(j) $-(-\tfrac{1}{32})^{4/5} = -(\sqrt[5]{-\tfrac{1}{32}})^4 = -(-\tfrac{1}{2})^4 = -\tfrac{1}{16}$

2.10   Perform each of the following operations and express the result without negative or zero exponents:

(a) $\left(\dfrac{81a^4}{b^8}\right)^{-1/4} = \dfrac{3^{-1}a^{-1}}{b^{-2}} = \dfrac{b^2}{3a}$ (b) $(a^{1/2} + a^{-1/2})^2 = a + 2a^0 + a^{-1} = a + 2 + \dfrac{1}{a}$

(c) $(a - 3b^{-2})(2a^{-1} - b^2) = 2a^0 - ab^2 - 6a^{-1}b^{-2} + 3b^0 = 5 - ab^2 - 6/ab^2$

(d) $\dfrac{a^{-2} + b^{-2}}{a^{-1} - b^{-1}} = \dfrac{(a^{-2} + b^{-2})(a^2b^2)}{(a^{-1} - b^{-1})(a^2b^2)} = \dfrac{b^2 + a^2}{ab^2 - a^2b}$

(e) $\left(\dfrac{a^2}{b}\right)^7\left(-\dfrac{b^2}{a^3}\right)^6 = \dfrac{a^{14} \cdot b^{12}}{b^7 \cdot a^{18}} = \dfrac{b^5}{a^4}$

(f) $\left(\dfrac{a^{1/2}b^{2/3}}{c^{3/4}}\right)^6\left(\dfrac{c^{1/2}}{a^{1/4}b^{1/3}}\right)^9 = \dfrac{a^3b^4}{c^{9/2}} \cdot \dfrac{c^{9/2}}{a^{9/4}b^3} = a^{3/4}b$

Supplementary Problems

2.11   Combine.

(a) $2x + (3x - 4y)$ (c) $[(s + 2t) - (s + 3t)] - [(2s + 3t) - (-4s + 5t)]$

(b) $5a + 4b - (-2a + 3b)$ (d) $8x^2y - \{3x^2y + [2xy^2 + 4x^2y - (3xy^2 - 4x^2y)]\}$

2.12   Perform the indicated operations.

(a) $4x(x - y + 2)$ (c) $(5x^2 - 4y^2)(-x^2 + 3y^2)$ (e) $(2x^3 + 5x^2 - 33x + 20) \div (2x - 5)$

(b) $(5x + 2)(3x - 4)$ (d) $(x^2 - 3x + 5)(2x - 7)$ (f) $(2x^3 + 5x^2 - 22x + 10) \div (2x - 3)$

2.13   Factor.

(a) $8x + 12y$ (e) $16a^2 - 8ab + b^2$ (i) $(x - y)^2 + 6(x - y) + 5$

(b) $4ax + 6ay - 24az$ (f) $25x^2 + 30xy + 9y^2$ (j) $4x^2 - 8x - 5$

(c) $a^2 - 4b^2$ (g) $x^2 - 4x - 12$ (k) $40a^2 + ab - 6b^2$

(d) $50ab^4 - 98a^3b^2$ (h) $a^2 + 23ab - 50b^2$ (l) $x^4 + 24x^2y^2 - 25y^4$

2.14   Simplify.

(a) $\dfrac{a^2 - b^2}{2ax + 2bx}$ (d) $\dfrac{16a^2 - 25}{2a - 10} \times \dfrac{a^2 - 10a + 25}{4a + 5}$

(b) $\dfrac{x^2 + 4x + 3}{1 - x^2}$ (e) $\dfrac{x^2 + xy - 6y^2}{2x^3 + 6x^2y} \times \dfrac{8x^2y}{x^2 - 5xy + 6y^2}$

(c) $\dfrac{1 - x - 12x^2}{1 + x - 6x^2}$

2.15   Perform the indicated operations.

(a) $\dfrac{5x}{18} + \dfrac{4x}{18}$ (c) $\dfrac{3a}{4b} - \dfrac{4b}{3a}$ (e) $x + 5 - \dfrac{x^2}{x - 5}$ (g) $\dfrac{2x + 3}{18x^2 - 27x} - \dfrac{2x - 3}{18x^2 + 27x}$

(b) $\dfrac{3a}{x} + \dfrac{5a}{2x}$ (d) $\dfrac{2a - 3b}{a^2 - b^2} + \dfrac{1}{a - b}$ (f) $\dfrac{a + 2}{2a - 6} - \dfrac{a - 2}{2a + 6}$

2.16   Simplify.

(a) $\dfrac{a}{2 - \dfrac{3}{a}}$ (b) $\dfrac{4 - \dfrac{x}{3}}{\dfrac{x}{6}}$ (c) $\dfrac{x - \dfrac{4}{x}}{1 - \dfrac{2}{x}}$ (d) $\dfrac{\dfrac{1}{x} + \dfrac{1}{y}}{\dfrac{x + y}{y} + \dfrac{x + y}{x}}$

ANSWERS TO SUPPLEMENTARY PROBLEMS

2.11.   (a) $5x - 4y$ (b) $7a + b$ (c) $t - 6s$ (d) $xy(y - 3x)$

2.12.   (a) $4x^2 - 4xy + 8x$ (c) $-5x^4 + 19x^2y^2 - 12y^4$ (e) $x^2 + 5x - 4$

(b) $15x^2 - 14x - 8$ (d) $2x^3 - 13x^2 + 31x - 35$ (f) $x^2 + 4x - 5 - 5/(2x - 3)$

2.13.   (a) $4(2x + 3y)$ (e) $(4a - b)^2$ (i) $(x - y + 1)(x - y + 5)$

(b) $2a(2x + 3y - 12z)$ (f) $(5x + 3y)^2$ (j) $(2x + 1)(2x - 5)$

(c) $(a - 2b)(a + 2b)$ (g) $(x - 6)(x + 2)$ (k) $(5a + 2b)(8a - 3b)$

(d) $2ab^2(5b - 7a)(5b + 7a)$ (h) $(a + 25b)(a - 2b)$ (l) $(x - y)(x + y)(x^2 + 25y^2)$

2.14.   (a) $\dfrac{a - b}{2x}$ (b) $-\dfrac{x + 3}{x - 1}$ (c) $\dfrac{4x - 1}{2x - 1}$ (d) $\frac{1}{2}(4a^2 - 25a + 25)$ (e) $\dfrac{4y}{x - 3y}$

2.15.   (a) $\frac{1}{2}x$ (b) $\dfrac{11a}{2x}$ (c) $\dfrac{9a^2 - 16b^2}{12ab}$ (d) $\dfrac{3a - 2b}{a^2 - b^2}$ (e) $-\dfrac{25}{x - 5}$

(f) $\dfrac{5a}{a^2 - 9}$ (g) $\dfrac{8}{12x^2 - 27}$

2.16.   (a) $\dfrac{a^2}{2a - 3}$ (b) $\dfrac{24 - 2x}{x}$ (c) $x + 2$ (d) $\dfrac{1}{x + y}$

Chapter 3

Functions

A VARIABLE IS A SYMBOL selected to represent any one of a given set of numbers, here assumed to be real numbers. Should the set consist of just one number, the symbol representing it is called a *constant*.

The *range* of a variable consists of the totality of numbers of the set which it represents. For example, if x is a day in September, the range of x is the set of positive integers $1, 2, 3, \ldots, 30$; if x (ft) is the length of rope cut from a piece 50 ft long, the range of x is the set of numbers greater than 0 and less than 50.

Examples of ranges of a real variable, together with special notations and graphical representations, are given in Problem 3.1.

FUNCTION. A correspondence (x, y) between two sets of numbers which pairs to an arbitrary number y of the first set exactly one number x of the second set is called a *function*. In this case, it is customary to speak of y as a *function of x*. The variable x is called the *independent variable* and y is called the *dependent variable*.

A function may be defined

(a) By a table of correspondents or table of values, as in Table 3.1.

Table 3.1

x	1	2	3	4	5	6	7	8	9	10
y	3	4	5	6	7	8	9	10	11	12

(b) By an equation or formula, as $y = x + 2$.

For each value assigned to x, the above relation yields a correspondent. Note that the table above is a table of values for this function.

A FUNCTION IS CALLED *single-valued* if, to each value of y in its range, there corresponds just one value of x; otherwise, the function is called *multivalued*. For example, $y = x + 3$ defines y as a single-valued function of x while $y = x^2$ defines y as a multivalued (here, two-valued) function of x.

At times it will be more convenient to label a given function of x as $f(x)$, to be read "the f function of x" or simply "f of x." (Note carefully that this is not to be confused with "f times x.") If there are two functions one may be labeled $f(x)$ and the other $g(x)$. Also, if $y = f(x) = x^2 - 5x + 4$ the statement "the value of the function is -2 when $x = 3$" can be replaced by "$f(3) = -2$." (See Problem 3.2.)

Let $y = f(x)$. The set of values of the independent variable x is called the *domain of definition* of the function while the range of the dependent variable is called the *range of the function*. For example, $y = x^2$ defines a function whose domain of definition consists of all (real) numbers and whose range is all nonnegative numbers, that is, zero and the positive numbers; $f(x) = 3/(x - 2)$ defines a function whose domain of definition consists of all numbers except 2 (why?) and whose range is all numbers except 0. (See Problems 3.3–3.8.)

A VARIABLE w (dependent) is said to be a function of the (independent) variables $x, y, z, \ldots$ if when a value of each of the variables $x, y, z, \ldots$ is known, there corresponds exactly one value of w. For

18

example, the volume V of a rectangular parallelipiped of dimensions x, y, z is given by $V = xyz$. Here V is a function of three independent variables. (See Problems 3.9–3.10.)

Solved Problems

3.1   Represent graphically each of the following ranges.

(a) $x > -2$

(b) $x < 5$

(c) $x \le -1$

(d) $-3 < x < 4$

(e) $-2 < x < 2$ or $|x| < 2$

(f) $|x| > 3$

(g) $-3 \le x \le 5$

(h) $x \le -3, x \ge 4$

3.2   Given $f(x) = x^2 - 5x + 4$, find

(a) $f(0) = 0^2 - 5 \cdot 0 + 4 = 4$

(b) $f(2) = 2^2 - 5 \cdot 2 + 4 = -2$

(c) $f(-3) = (-3)^2 - 5(-3) + 4 = 28$

(d) $f(a) = a^2 - 5a + 4$

(e) $f(-x) = x^2 + 5x + 4$

(f) $f(b + 1) = (b + 1)^2 - 5(b + 1) + 4 = b^2 - 3b$

(g) $f(3x) = (3x)^2 - 5(3x) + 4 = 9x^2 - 15x + 4$

(h) $f(x + a) - f(a) = [(x + a)^2 - 5(x + a) + 4] - (a^2 - 5a + 4) = x^2 + 2ax - 5x$

(i) $\dfrac{f(x + a) - f(x)}{a} = \dfrac{[(x + a)^2 - 5(x + a) + 4] - (x^2 - 5x + 4)}{a} = \dfrac{2ax - 5a + a^2}{a} = 2x - 5 + a$

3.3   In each of the following, state the domain of definition.

(a) $y = 5x$

(b) $y = -5x$

(c) $y = \dfrac{1}{x + 5}$

(d) $y = \dfrac{x - 2}{(x - 3)(x + 4)}$

(e) $y = \dfrac{1}{x}$

(f) $y = \sqrt{25 - x^2}$

(g) $y = \sqrt{x^2 - 9}$

(h) $y = \dfrac{1}{16 - x^2}$

(i) $y = \dfrac{1}{16 + x^2}$

Ans. (a), (b), all real numbers; (c) $x \ne -5$; (d) $x \ne 3, -4$; (e) $x \ne 0$; (f) $-5 \le x \le 5$ or $|x| \le 5$; (g) $x \le -3, x \ge 3$ or $|x| \ge 3$; (h) $x \ne \pm 4$; (i) all real numbers.

3.4   A piece of wire 30 in. long is bent to form a rectangle. If one of its dimensions is x in., express the area as a function of x.

Since the semiperimeter of the rectangle is $\frac{1}{2} \cdot 30 = 15$ in. and one dimension is x in., the other is $(15 - x)$ in. Thus, $A = x(15 - x)$.

3.5   An open box is to be formed from a rectangular sheet of tin 20×32 in. by cutting equal squares, x in. on a side, from the four corners and turning up the sides. Express the volume of the box as a function of x.

From Fig. 3-1, it is seen that the base of the box has dimensions $(20 - 2x)$ by $(32 - 2x)$ in. and the height is x in. Then

$$V = x(20 - 2x)(32 - 2x) = 4x(10 - x)(16 - x)$$

3.6   A closed box is to be formed from the sheet of tin of Problem 3.5 by cutting equal squares, x cm on a
side, from two corners of the short side and two equal rectangles of width x cm from the other two corners,
and folding along the dotted lines shown in Fig. 3-2. Express the volume of the box as a function of x.

One dimension of the base of the box is $(20 - 2x)$ cm; let y cm be the other. Then $2x + 2y = 32$ and
$y = 16 - x$. Thus,

$$V = x(20 - 2x)(16 - x) = 2x(10 - x)(16 - x)$$

Fig. 3-1   **Fig. 3-2**   **Fig. 3-3**

3.7   A farmer has 600 ft of woven wire fencing available to enclose a rectangular field and to divide it into
three parts by two fences parallel to one end. If x ft of stone wall is used as one side of the field, express
the area enclosed as a function of x when the dividing fences are parallel to the stone wall. Refer to
Fig. 3-3.

The dimensions of the field are x and y ft where $3x + 2y = 600$. Then $y = \frac{1}{2}(600 - 3x)$ and the required
area is

$$A = xy = x \cdot \frac{1}{2}(600 - 3x) = \frac{3}{2}x(200 - x)$$

3.8   A right circular cylinder is said to be inscribed in a sphere if the circumferences of the bases of the cylinder
are in the surface of the sphere. If the sphere has radius R, express the volume of the inscribed right
circular cylinder as a function of the radius r of its base.

Let the altitude of the cylinder be denoted by $2h$. From Fig. 3-4, $h = \sqrt{R^2 - r^2}$ and the required volume
is

$$V = \pi r^2 \cdot 2h = 2\pi r^2 \sqrt{R^2 - r^2}$$

Fig. 3-4

3.9   Given $z = f(x, y) = 2x^2 + 3y^2 - 4$; find

(a) $f(0, 0) = 2(0)^2 + 3(0)^2 - 4 = -4$ (b) $f(2, -3) = 2(2)^2 + 3(-3)^2 - 4 = 31$

(c) $f(-x, -y) = 2(-x)^2 + 3(-y)^2 - 4 = 2x^2 + 3y^2 - 4 = f(x, y)$

3.10   Given $f(x, y) = \dfrac{x^2 + y^2}{x^2 - y^2}$, find

(a) $f(x, -y) = \dfrac{x^2 + (-y)^2}{x^2 - (-y)^2} = \dfrac{x^2 + y^2}{x^2 - y^2} = f(x, y)$

(b) $f\left(\dfrac{1}{x}, \dfrac{1}{y}\right) = \dfrac{(1/x)^2 + (1/y)^2}{(1/x)^2 - (1/y)^2} = \dfrac{1/x^2 + 1/y^2}{1/x^2 - 1/y^2} = \dfrac{y^2 + x^2}{y^2 - x^2} = -f(x, y)$

Supplementary Problems

3.11   Represent graphically each of the following ranges:

(a) $x > -3$ (c) $x \geq 0$ (e) $|x| < 2$ (g) $-4 \leq x \leq 4$

(b) $x < 5$ (d) $-3 < x < -1$ (f) $|x| \geq 0$ (h) $x < -3, x \geq 5$

3.12   In the three angles A, B, C of a triangle, angle B exceeds twice angle A by 15°. Express angle C in terms of angle A.

Ans. $C = 165° - 3A$

3.13   A grocer has two grades of coffee, selling at 90¢ and $1.05 per kilogram respectively. In making a mixture of 100 kg, he uses x kg of the $1.05 coffee. (a) How many kilograms of the 90¢ coffee does he use? (b) What is the value in cents of the mixture? (c) At what price per kilogram should he offer the mixture?

Ans.   (a) $100 - x$ (b) $90(100 - x) + 105x$ (c) $90 + 0.15x$

3.14   In a purse are nickels, dimes, and quarters. The number of dimes is twice the number of quarters and the number of nickels is three less than twice the number of dimes. If there are x quarters, find the sum (in cents) in the purse.

Ans. $65x - 15$

3.15   A and B start from the same place. A walks 4 mi/hr and B walks 5 mi/hr. (a) How far (in miles) will each walk in x hr? (b) How far apart will they be after x hr if they leave at the same time and move in opposite directions? (c) How far apart will they be after A has walked $x > 2$ hours if they move in the same direction but B leaves 2 hr after A? (d) In (c), for how many hours would B have to walk in order to overtake A?

Ans.   (a) $A, 4x; B, 5x$ (b) $9x$ (c) $|4x - 5(x - 2)|$ (d)   8

3.16   A motor boat, which moves at x mi/hr in still water, is on a river whose current is $y < x$ mi/hr. (a) What is the rate (mi/hr) of the boat when moving upstream? (b) What is the rate of the boat when moving downstream? (c) How far (miles) will the boat travel upstream in 8 hr? (d) How long (hours) will it take the boat moving downstream to cover 20 mi if the motor dies after the first 15 mi?

Ans.   (a) $x - y$ (b) $x + y$ (c) $8(x - y)$ (d) $\dfrac{15}{x + y} + \dfrac{5}{y}$

3.17 Given $f(x) = \dfrac{x-3}{x+2}$; find $f(0)$, $f(1)$, $f(-3)$, $f(3)$, $f(a)$, $f(3y)$, $f(x+a)$, $\dfrac{f(x+a)-f(x)}{a}$.

  Ans. $-\frac{3}{2}$, $-\frac{2}{3}$, 6, 0, $\dfrac{a-3}{a+2}$, $\dfrac{3y-3}{3y+2}$, $\dfrac{x+a-3}{x+a+2}$, $\dfrac{5}{(x+2)(x+a+2)}$

3.18 A ladder 25 ft long leans against a vertical wall with its foot on level ground 7 ft from the base of the wall. If the foot is pulled away from the wall at the rate 2 ft/s, express the distance (y ft) of the top of the ladder above the ground as a function of the time t seconds in moving.

  Ans. $y = 2\sqrt{144 - 7t - t^2}$

3.19 A boat is tied to a dock by means of a cable 60 m long. If the dock is 20 m above the water and if the cable is being drawn in at the rate 10 m/min, express the distance y meters of the boat from the dock after t min.

  Ans. $y = 10\sqrt{t^2 - 12t + 32}$

3.20 A train leaves a station at noon and travels east at the rate 30 mi/hr. At 2 P.M. of the same day a second train leaves the station and travels south at the rate 25 mi/hr. Express the distance d (miles) between the trains as a function of t (hours), the time the second train has been traveling.

  Ans. $d = 5\sqrt{61t^2 + 144t + 144}$

Chapter 4

Functions and Loci

A FUNCTION $y = f(x)$, by definition, yields a collection of number pairs $(x, f(x))$ or (x, y) in which x is any element in the domain of definition of the function and $f(x)$ or y is the corresponding value of the function.

EXAMPLE 1. Obtain ten number pairs for the function $y = 3x - 2$.

The domain of definition of the function is the set of real numbers. We may choose at random any ten real numbers as values of x. For one such choice, we obtain the chart in Table 4.1.

Table 4.1

x	-2	$-\frac{4}{3}$	$-\frac{1}{2}$	0	$\frac{1}{3}$	1	2	$\frac{5}{2}$	3	4
y	-8	-6	$-\frac{7}{2}$	-2	-1	1	4	$\frac{11}{2}$	7	10

(See Problem 4.1.)

THE RECTANGULAR CARTESIAN COORDINATE SYSTEM in a plane is a device by which there is established a one-to-one correspondence between the points of the plane and ordered pairs of real numbers (a, b).

Consider two real number scales intersecting at right angles in O, the origin of each (see Fig. 4-1), and having the positive direction on the horizontal scale (now called the x *axis*) directed to the right and the positive direction on the vertical scale (now called the y axis) directed upward.

Let P be any point distinct from O in the plane of the two axes and join P to O by a straight line. Let the projection of OP on the x axis be $OM = a$ and the projection of OP on the y axis be $ON = b$. Then the pair of numbers (a, b) in that order are called the plane rectangular cartesian coordinates (briefly, the rectangular coordinates) of P. In particular, the coordinates of O, the *origin* of the coordinate system, are $(0, 0)$.

The first coordinate, giving the directed distance of P from the y axis, is called the *abscissa* of P while the second coordinate, giving the directed distance of P from the x axis, is called the *ordinate* of P. Note carefully that the points $(3, 4)$ and $(4, 3)$ are distinct points.

The axes divide the plane into four sections, called *quadrants*. Figure 4-1 shows the customary numbering of the quadrants and the respective signs of the coordinates of a point in each quadrant. (See Problems 4.1–4.4.)

Fig. 4-1

THE GRAPH OF A FUNCTION $f(x)$ consists of the totality of points (x, y) whose coordinates satisfy the relation $y = f(x)$.

EXAMPLE 2.   Graph the function $3x - 2$.

After plotting the points whose coordinates (x, y) are given in Table 4.1, it appears that they lie on a straight line. See Fig. 4-2. Figure 4-2 is not the complete locus since $(1000, 2998)$ is one of its points and is not shown. Moreover, although we have joined the points by a straight line, we have not proved that every point on the line has as coordinates a number pair given by the function. These matters as well as such questions as: What values of x should be chosen? How many values of x are needed? will become clearer as we proceed with the study of functions. At present,

(1) Build a table of values.
(2) Plot the corresponding points.
(3) Pass a smooth curve through these points, moving from left to right.

Fig. 4-2

It is helpful to picture the curve in your mind before attempting to trace it on paper. If there is doubt about the curve between two plotted points, determine other points in the interval.

ANY VALUE OF x for which the corresponding value of a function $f(x)$ is zero is called a *zero* of the function. Such values of x are also called *roots* of the equation $f(x) = 0$. The real roots of an equation $f(x) = 0$ may be approximated by estimating from the graph of $f(x)$ the abscissas of its points of intersection with the x axis. (See Problems 4.9–4.11.)

Algebraic methods for finding the roots of equations will be treated in later chapters.

Solved Problems

4.1   (*a*)   Show that the points $A(1, 2)$, $B(0, -3)$, and $C(2, 7)$ are on the graph of $y = 5x - 3$.

(*b*)   Show that the points $D(0, 0)$ and $E(-1, -2)$ are not on the graph of $y = 5x - 3$.

   (*a*)   The point $A(1, 2)$ is on the graph since $2 = 5(1) - 3$, $B(0, -3)$ is on the graph since $-3 = 5(0) - 3$, and $C(2, 7)$ is on the graph since $7 = 5(2) - 3$.

   (*b*)   The point $D(0, 0)$ is not on the graph since $0 \neq 5(0) - 3$, and $E(-1, -2)$ is not on the graph since $-2 \neq 5(-1) - 3$.

4.2   Sketch the graph of the function $2x$. Refer to Table 4.2.

Table 4.2

x	0	1	2
$y = f(x)$	0	2	4

Fig. 4-3

This is a linear function and its graph is a straight line. For this locus only two points are necessary. Three points are used to provide a check. See Fig. 4-3. The equation of the line is $y = 2x$.

4.3.   Sketch the graph of the function $6 - 3x$. Refer to Table 4.3.

Table 4.3

x	0	2	3
$y = f(x)$	6	0	-3

See Fig. 4-4. The equation of the line is $y = 6 - 3x$.

4.4   Sketch the graph of the function x^2. Refer to Table 4.4.

Table 4.4

x	3	1	0	-2	-3
$y = f(x)$	9	1	0	4	9

See Fig. 4-5. The equation of this locus, called a *parabola*, is $y = x^2$. Note for $x \neq 0$, $x^2 > 0$. Thus, the curve is never below the x axis. Moreover as $|x|$ increases, x^2 increases, that is, as we move from the origin along the x axis in either direction, the curve moves farther and farther from the axis. Hence, in sketching parabolas sufficient points must be plotted so that its U shape can be seen.

Fig. 4-4   **Fig. 4-5**   **Fig. 4-6**

4.5   Sketch the graph of the function $x^2 + x - 12$. Refer to Table 4.5.

Table 4.5

x	4	3	1	0	-1	-4	-5
$y = f(x)$	8	0	-10	-12	-12	0	8

The equation of the parabola is $y = x^2 + x - 12$. Note that the points $(0, -12)$ and $(-1, -12)$ are *not* joined by a straight line segment. Check that the value of the function is $-12\frac{1}{4}$ when $x = -\frac{1}{2}$. See Fig. 4-6.

4.6   Sketch the graph of the function $-2x^2 + 4x + 1$. Refer to Table 4.6.

Table 4.6

x	3	2	1	0	-1
$y = f(x)$	-5	1	3	1	-5

See Fig. 4-7.

4.7   Sketch the graph of the function $(x + 1)(x - 1)(x - 2)$. Refer to Table 4.7.

Table 4.7

x	3	2	$\frac{3}{2}$	1	0	-1	-2
$y = f(x)$	8	0	$-\frac{5}{8}$	0	2	0	-12

This is a *cubic* curve of the equation $y = (x + 1)(x - 1)(x - 2)$. It crosses the x axis where $x = -1, 1$, and 2. See Fig. 4-8.

Fig. 4-7

Fig. 4-8

Fig. 4-9

4.8   Sketch the graph of the function $(x + 2)(x - 3)^2$. Refer to Table 4.8.

Table 4.8

x	5	4	$\frac{7}{2}$	3	2	1	0	-1	-2	-3
$y = f(x)$	28	6	$\frac{11}{8}$	0	4	12	18	16	0	-36

This cubic crosses the x axis where $x = -2$ and is tangent to the x axis where $x = 3$. Note that for $x > -2$, the value of the function is positive except for $x = 3$, where it is 0. Thus, to the right of $x = -2$, the curve is *never* below the x axis. See Fig. 4-9.

4.9   Sketch the graph of the function $x^2 + 2x - 5$ and by means of it determine the real roots of $x^2 + 2x - 5 = 0$. Refer to Table 4.9.

Table 4.9

x	2	1	0	-1	-2	-3	-4
$y = f(x)$	3	-2	-5	-6	-5	-2	3

The parabola cuts the x axis at a point whose abscissa is between 1 and 2 (the value of the function changes sign) and at a point whose abscissa is between -3 and -4.

Reading from the graph in Fig. 4-10, the roots are $x = 1.5$ and $x = -3.5$, approximately.

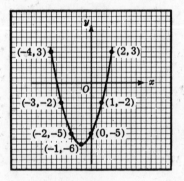

Fig. 4-10

Supplementary Problems

4.10   Sketch the graph of each of the following functions:

(a) $3x - 2$ (c) $x^2 - 1$ (e) $x^2 - 4x + 4$ (g) $(x - 2)(x + 1)^2$

(b) $2x + 3$ (d) $4 - x^2$ (f) $(x + 2)(x - 1)(x - 3)$

4.11   From the graph of each function $f(x)$ determine the real roots, if any, of $f(x) = 0$.

(a) $x^2 - 4x + 3$ (b) $2x^2 + 4x + 1$ (c) $x^2 - 2x + 4$

Ans. (a) 1, 3   (b) $-0.3, -1.7$ (c) none

4.12 If A is a point on the graph of $y = f(x)$, the function being restricted to the type considered in this chapter, and if all points of the graph sufficiently near A are higher than A (that is, lie above the horizontal drawn through A) then A is called a *relative minimum point* of the graph. (a) Verify that the origin is the relative minimum point of the graph of Problem 4.4. (b) Verify that the graph of Problem 4.5 has a relative minimum at a point whose abscissa is between $x = -1$ and $x = 0$ (at $x = -\frac{1}{2}$), the graph of Problem 4.7 has a relative minimum at a point whose abscissa is between $x = 1$ and $x = 2$ (approximately $x = 1.5$), and the graph of Problem 4.8 has $(3, 0)$ as relative minimum point.

4.13 If B is a point on the graph of $y = f(x)$ and if all points of the graph sufficiently near B are lower than B (that is, lie below the horizontal drawn through B) then B is called a *relative maximum point* of the graph. (a) Verify that $(1, 3)$ is the relative maximum point of the graph of Problem 4.6. (b) Verify that the graph of Problem 4.7 has a relative maximum at a point whose abscissa is between $x = -1$ and $x = 1$ (approximately $x = -0.2$), and that the graph of Problem 4.8 has a relative maximum between $x = -1$ and $x = 0$ (at $x = -\frac{1}{3}$).

4.14 Verify that the graphs of the functions of Problem 4.11 have relative minimums at $x = 2$, $x = -1$, and $x = 1$, respectively.

4.15 From the graph of the function of Problem 3.4 in Chapter 3 read that the area of the rectangle is a relative maximum when $x = \frac{15}{2}$.

4.16 From the graph of the function of Problem 3.7 in Chapter 3 read that the area enclosed is a relative maximum when $x = 100$.

Chapter 5

The Linear Equation

AN EQUATION is a statement, such as (*a*) $2x - 6 = 4 - 3x$, (*b*) $y^2 + 3y = 4$, (*c*) $2x + 3y = 4xy + 1$, that two expressions are equal. An equation is linear in an unknown if the highest degree of that unknown in the equation is one. An equation is quadratic in an unknown if the highest degree of that unknown is two. The first is a *linear* equation in one unknown, the second is a *quadratic* in one unknown, and the third is linear in each of the two unknowns but is of degree two in the two unknowns.

Any set of values of the unknowns for which the two members of an equation are equal is called a *solution* of the equation. Thus, $x = 2$ is a solution of (*a*) since $2(2) - 6 = 4 - 3(2)$; $y = 1$ and $y = -4$ are solutions of (*b*); and $x = 1$, $y = 1$ is a solution of (*c*). A solution of an equation in one unknown is also called a *root* of the equation.

TO SOLVE A LINEAR EQUATION in one unknown perform the same operations on both members of the equation in order to obtain the unknown alone in the left member.

EXAMPLE 1. Solve: $2x - 6 = 4 - 3x$.

Add 6: $\qquad 2x = 10 - 3x \qquad$ Check: $\quad 2(2) - 6 = 4 - 3(2)$

Add $3x$: $\qquad 5x = 10 \qquad\qquad\qquad\qquad -2 = -2$

Divide by 5: $\quad x = 2$

EXAMPLE 2. Solve: $\frac{1}{3}x - \frac{1}{2} = \frac{3}{4}x + \frac{5}{6}$.

Multiply by LCD = 12: $\quad 4x - 6 = 9x + 10 \qquad$ Check: $\quad \frac{1}{3}(-\frac{16}{5}) - \frac{1}{2} = \frac{3}{4}(-\frac{16}{5}) + \frac{5}{6}$

Add $6 - 9x$: $\qquad\qquad\qquad -5x = 16$

Divide by -5: $\qquad\qquad\qquad x = -\frac{16}{5} \qquad\qquad\qquad\qquad -\frac{47}{30} = -\frac{47}{30}$

(See Problems 5.1–5.3.)

An equation which contains fractions having the unknown in one or more denominators may sometimes reduce to a linear equation when cleared of fractions. When the resulting equation is solved, the solution *must* be checked since it may or may not be a root of the original equation. (See Problems 5.4–5.8.)

RATIO AND PROPORTION. The ratio of two quantities is their quotient. The ratio of 1 inch to 1 foot is 1/12 or 1:12, a pure number; the ratio of 30 miles to 45 minutes is $30/45 = 2/3$ mile per minute.

The expressed equality of two ratios, as $\dfrac{a}{b} = \dfrac{c}{d}$ is called a *proportion*. (See Problems 5.11–5.12.)

VARIATION. A variable y is said to vary *directly* as another variable x (or y is proportional to x) if y is equal to some constant c times x, that is, if $y = cx$.

A variable y is said to vary *inversely* as another variable x if y varies directly as the reciprocal of x, that is, if $y = c/x$.

A variable z is said to vary *jointly* as x and y if z varies directly as the product xy, that is, if $z = cxy$. (See Problems 5.13–5.14.)

Solved Problems

Solve and check the following equations. The check has been omitted in certain problems.

5.1 $x - 2(1 - 3x) = 6 + 3(4 - x)$.

$$x - 2 + 6x = 6 + 12 - 3x$$
$$7x - 2 = 18 - 3x$$
$$10x = 20$$
$$x = 2$$

5.2 $ay + b = cy + d$.

$$ay - cy = d - b$$
$$(a - c)y = d - b$$
$$y = \frac{d - b}{a - c}$$

5.3 $\dfrac{3x - 2}{5} = 4 - \dfrac{1}{2}x$.

Multiply by 10: $6x - 4 = 40 - 5x$ Check: $\dfrac{3(4) - 2}{5} = 4 - \dfrac{1}{2}(4)$

$$11x = 44$$ $2 = 2$
$$x = 4$$

5.4 $\dfrac{3x + 1}{3x - 1} = \dfrac{2x + 1}{2x - 3}$.   Here the LCD is $(3x - 1)(2x - 3)$.

Multiply by LCD: $(3x + 1)(2x - 3) = (2x + 1)(3x - 1)$

$$6x^2 - 7x - 3 = 6x^2 + x - 1$$
$$-8x = 2$$
$$x = -\tfrac{1}{4}$$

Check: $\dfrac{3(-\frac{1}{4}) + 1}{3(-\frac{1}{4}) - 1} = \dfrac{2(-\frac{1}{4}) + 1}{2(-\frac{1}{4}) - 3}$,

$\dfrac{-3 + 4}{-3 - 4} = \dfrac{-2 + 4}{-2 - 12}$, $-\dfrac{1}{7} = -\dfrac{1}{7}$

5.5 $\dfrac{1}{x - 3} - \dfrac{1}{x + 1} = \dfrac{3x - 2}{(x - 3)(x + 1)}$.   Here the LCD is $(x - 3)(x + 1)$.

$$(x + 1) - (x - 3) = 3x - 2$$
$$-3x = -6$$
$$x = 2$$

Check: $-1 - \dfrac{1}{3} = \dfrac{6 - 2}{-3}$, $-\dfrac{4}{3} = -\dfrac{4}{3}$

5.6 $\dfrac{1}{x-3}+\dfrac{1}{x-2}=\dfrac{3x-8}{(x-3)(x-2)}$.   The LCD is $(x-3)(x-2)$.

$$(x-2)+(x-3)=3x-8$$
$$2x-5=3x-8$$
$$x=3$$

Check: When $x=3$, $\dfrac{1}{x-3}$ is without meaning. The given equation has no root. The value $x=3$ is called *extraneous*.

5.7 $\dfrac{x^2-2}{x-1}=x+1-\dfrac{1}{x-1}$.

$$x^2-2=(x+1)(x-1)-1$$
$$=x^2-2$$

The given equation is satisfied by all values of x except $x=1$. It is called an identical equation or *identity*.

5.8 $\dfrac{1}{x-1}+\dfrac{1}{x-3}=\dfrac{2x-5}{(x-1)(x-3)}$.

$$(x-3)+(x-1)=2x-5$$
$$2x-4=2x-5$$

There is no solution. The given equation and the resulting equation are examples of *false equations*.

5.9   One number is 5 more than another and the sum of the two is 71. Find the numbers.

Let x be the smaller number and $x+5$ be the larger. Then $x+(x+5)=71$, $2x=66$, and $x=33$. The numbers are 33 and 38.

5.10   A father is now three times as old as his son. Twelve years ago he was six times as old as his son. Find the present age of each.

Let $x=$ the age of the son and $3x=$ the age of the father. Twelve years ago, the age of the son was $x-12$ and the age of the father was $3x-12$.

Then $3x-12=6(x-12)$, $3x=60$, and $x=20$. The present age of the son is 20 and that of the father is 60.

5.11   When two pulleys are connected by a belt, their angular velocities (revolutions per minute) are *inversely* proportional to their diameters, that is, $\omega_1:\omega_2=d_2:d_1$. Find the velocity of a pulley 15 cm in diameter when it is connected to a pulley 12 cm in diameter and rotating at 100 rev./cm.

Let ω_1 be the unknown velocity; then $d_1=15$, $\omega_2=100$, and $d_2=12$. The given formula becomes

$$\frac{\omega_1}{100}=\frac{12}{15}\qquad\text{and}\qquad\omega_1=\frac{12}{15}(100)=80\text{ rev/min}$$

5.12   Bleaching powder is obtained through the reaction of chlorine and slaked lime, 74.10 kg of lime and 70.91 kg of chlorine producing 127.00 kg of bleaching powder and 18.01 kg of water. How many kg of lime will be required to produce 1000 kg of bleaching powder?

Let $x=$ the number of kg of lime required. Then

$$\frac{x(\text{kg of lime})}{1000(\text{kg of powder})}=\frac{74.10(\text{kg of lime})}{127(\text{kg of powder})}\qquad 127x=74\,100\qquad\text{and}\qquad x=583.46\text{ kg}$$

5.13   The pressure of a gas in a container at constant temperature varies inversely as the volume. If $p = 30$ when $v = 45$, find p when $v = 25$.

   First Solution.   Here $p = c/v$, $30 = c/45$, $c = 30 \cdot 45$;   thus $p = 30 \cdot 45/v$.   When $v = 25$, $p = 30 \cdot 45/25 = 54$.

   Second Solution.   From $p_1 = c/v_1$ and $p_2 = c/v_2$, we obtain $\dfrac{p_1}{p_2} = \dfrac{v_2}{v_1}$. Taking $v_1 = 25$, $p_2 = 30$, $v_2 = 45$, we obtain as before $p_1 = 54$.

5.14   The safe load of a horizontal beam supported at both ends varies jointly as the breadth and the square of the depth, and inversely as the length of the beam. If a 2×4 cm beam 8 m long will support 500 kg safely, what is the safe load for a 4×8 cm beam of the same material 20 m long?

$$S_1 = c\frac{b_1 d_1^2}{l_1} \qquad S_2 = c\frac{b_2 d_2^2}{l_2}$$

Then

$$\frac{S_1}{S_2} = \frac{b_1 d_1^2 l_2}{b_2 d_2^2 l_1} \qquad \frac{S_1}{500} = \frac{4 \cdot 8^2 \cdot 8}{2 \cdot 4^2 \cdot 20} \qquad \text{and} \qquad S_1 = 1600 \text{ kg}$$

Supplementary Problems

5.15   Solve for x and check each of the following:

   (a) $2x - 7 = 29 - 4x$ (c) $\dfrac{x + 3}{x - 3} = 3$ (e) $\dfrac{2x + 1}{4} - \dfrac{1}{x - 1} = \dfrac{x}{2}$

   (b) $2(x - 1) - 3(x - 2) + 4(x - 3) = 0$ (d) $\dfrac{4}{x - 4} = \dfrac{2}{2x - 5}$ (f) $a(x + 3) + b(x - 2) = c(x - 1)$

   Ans.   (a) 6   (b) $\frac{8}{3}$ (c) 6   (d) 2   (e) 5   (f) $\dfrac{2b - 3a - c}{a + b - c}$

5.16   A piece of wire $11\frac{2}{3}$ m long is to be divided into two parts such that one part is $\frac{2}{3}$ that of the other. Find the length of the shorter piece.

   Ans. $4\frac{2}{3}$ m

5.17   A train leaves a station and travels at the rate of 40 mi/hr. Two hours later a second train leaves the station and travels at the rate of 60 mi/hr. Where will the second train overtake the first?

   Ans.   240 mi from the station

5.18   A tank is drained by two pipes. One pipe can empty the tank in 30 min, the other can empty it in 25 min. If the tank is $\frac{5}{6}$ filled and both pipes are open, in what time will the tank be emptied?

   Ans. $11\frac{4}{11}$ min

5.19   A man invests $\frac{1}{3}$ of his capital at 6% and the remainder at 8%. What is his capital if his total income is \$4400?

   Ans.   \$60,000

5.20 *A* can do a piece of work in 10 days. After he has worked 2 days, *B* comes to help him and together they finish it in 3 days. In how many days could *B* alone have done the work?

    Ans. 6 days

5.21 When two resistances R_1 and R_2 are placed in parallel, the resultant resistance R is given by $1/R = 1/R_1 + 1/R_2$. Find R when $R_1 = 80$ and $R_2 = 240$.

    Ans. 60

5.22 How soon after noon are the hands of a clock together again?

    Ans. 1 hr, $5\frac{5}{11}$ min

5.23 How much water will be produced in making the 1000 kg of bleaching powder in Problem 5.12?

    Ans. 141.81 kg

5.24 The reaction of 65.4 g of zinc and 72.9 g of hydrochloric acid produces 136.3 g of zinc chloride and 2 g of hydrogen. Find the weight of hydrochloric acid necessary for a complete reaction with 300 g of zinc and the weight of hydrogen produced.

    Ans. 334.4 g, 9.2 g

5.25 How much water must be used to prepare a 1:5000 solution of bichloride of mercury from a 0.5-g tablet?

    Ans. 2500 g

5.26 The volume of a cone varies jointly as the altitude and square of the radius. When the radius is 4 and the altitude is 6, the volume is 32π. What must the altitude be if the volume is 12π when the radius is 2?

    Ans. 9

5.27 Newton's law of gravitation states that the force F of attraction between two bodies varies jointly as their masses m_1 and m_2 and inversely as the square of the distance between them. Two bodies whose centers are 5000 mi apart attract each other with a force of 15 lb. What would be the force of attraction if their masses were tripled and the distance between their centers was doubled?

    Ans. $33\frac{3}{4}$ lb

5.28 If a body weighs 20 lb on the earth's surface, what would it weigh 2000 mi above the surface? (Assume the radius of the earth to be 4000 mi.)

    Ans. $8\frac{8}{9}$ lb

5.29 The horsepower of a steam engine varies jointly as the average pressure in the cylinder and the speed of rotation. When the average pressure is 400 lb/in^2 and the engine is making 750 rev/min, the horsepower is 100. What is the horsepower when the average pressure is 300 lb/in^2 and the engine is making 600 rev/min?

    Ans. 60

Chapter 6

Simultaneous Linear Equations

TWO LINEAR EQUATIONS IN TWO UNKNOWNS. Let the system of equations be

$$\begin{cases} a_1 x + b_1 y + c_1 = 0 \\ a_2 x + b_2 y + c_2 = 0 \end{cases}$$

Each equation has an unlimited number of solutions (x, y) corresponding to the unlimited number of points on the locus (straight line) which it represents. Our problem is to find all solutions common to the two equations or the coordinates of all points common to the two lines. There are three cases:

| Fig. 6-1 | Fig. 6-2 | Fig. 6-3 |

1. The system has one and only one solution, that is, the two lines have one and only one point in common. The equations are said to be *consistent* (have common solutions) *and independent.* See Fig. 6-1 indicating two distinct intersecting lines.
2. The system has an unlimited number of solutions, that is, the two equations are equivalent or the two lines are coincident. The equations are said to be *consistent and dependent.* See Fig. 6-2 indicating that the two equations represent the same line.
3. The system has no solution, that is, the two lines are parallel and distinct. The equations are said to be *inconsistent.* See Fig. 6-3 indicating that the two equations result in two parallel lines.

GRAPHICAL SOLUTION. We plot the graphs of the two equations on the same axes and scale off the coordinates of the point of intersection. The defect of this method is that, in general, only approximate solutions are obtained. (See Problem 6.1.)

ALGEBRAIC SOLUTION. A system of two consistent and independent equations in two unknowns may be solved algebraically by eliminating one of the unknowns.

EXAMPLE 1. Solve the system

$$\begin{cases} 3x - 6y = 10 & (6.1) \\ 9x + 15y = -14 & (6.2) \end{cases}$$

ELIMINATION BY SUBSTITUTION

Solve (6.1) for x:

$$x = \tfrac{10}{3} + 2y \tag{6.3}$$

Substitute in (6.2):

$$9(\tfrac{10}{3} + 2y) + 15y = -14$$

$$30 + 18y + 15y = -14 \qquad 33y = -44 \qquad y = -\tfrac{4}{3}$$

Substitute for y in (6.3):

$$x = \tfrac{10}{3} + 2(-\tfrac{4}{3}) = \tfrac{2}{3}$$

Check: Using (6.2),

$$9(\tfrac{2}{3}) + 15(-\tfrac{4}{3}) = -14$$

EXAMPLE 2. Solve the system

$$\begin{cases} 2x - 3y = 10 & (6.4) \\ 3x - 4y = 8 & (6.5) \end{cases}$$

ELIMINATION BY ADDITION

Multiply (6.4) by -3 and (6.5) by 2:

$$-6x + 9y = -30$$
$$\underline{6x - 8y = 16}$$
$$\text{Add:} \qquad y = -14$$

Substitute for x in (6.4):

$$2x + 42 = 10 \qquad \text{or} \qquad x = -16$$

Check: Using (6.5),

$$3(-16) - 4(-14) = 8$$

(See Problems 6.2–6.4.)

THREE LINEAR EQUATIONS IN THREE UNKNOWNS. A system of three consistent and independent equations in three unknowns may be solved algebraically by deriving from it a system of two equations in two unknowns.

EXAMPLE 3. Solve the system

$$\begin{cases} 2x + 3y - 4z = 1 & (6.6) \\ 3x - y - 2z = 4 & (6.7) \\ 4x - 7y - 6z = -7 & (6.8) \end{cases}$$

We shall eliminate y.

$$\begin{array}{ll} \text{Rewrite } (6.6): & 2x + 3y - 4z = 1 \\ 3 \times (6.7): & \underline{9x - 3y - 6z = 12} \\ \text{Add:} & 11x - 10z = 13 \end{array} \tag{6.9}$$

$$\begin{array}{ll} \text{Rewrite } (6.8): & 4x - 7y - 6z = -7 \\ -7 \times (6.8): & \underline{-21x + 7y + 14z = -28} \\ \text{Add:} & -17x + 8z = -35 \end{array} \tag{6.10}$$

Next, solve (*6.9*) and (*6.10*).

$$
\begin{array}{rl}
4 \times (6.9): & 44x - 40z = 52 \\
5 \times (6.10): & \underline{-85x + 40z = -175} \\
\text{Add:} & -41x = -123 \\
& x = 3
\end{array}
$$

From (*6.9*): $11(3) - 10z = 13 \qquad z = 2$
From (*6.6*): $2(3) + 3y - 4(2) = 1 \qquad y = 1$
Check: Using (*6.7*),

$$3(3) - 1 - 2(2) = 4$$

(See Problems 6.5-6.6.)

SOLUTIONS OF LINEAR SYSTEMS USING DETACHED COEFFICIENTS.   In Example 4 below, a variation of the method of addition and subtraction is used to solve a system of linear equations. On the left the equations themselves are used while on the right the same moves are made on the rectangular array (called a *matrix*) of the coefficients and constant terms. The numbering $(1), (2), (3), \ldots$ refers both to the equations and to the rows of the matrices.

EXAMPLE 4.   Solve the system

$$
\begin{cases}
2x - 3y = 2 \\
4x + 7y = -9
\end{cases}
$$

USING EQUATIONS   USING MATRICES

$$
\begin{array}{ll}
2x - 3y = 2 & (1) \\
4x + 7y = -9 & (2)
\end{array}
\qquad
\left(\begin{array}{cc|c} 2 & -3 & 2 \\ 4 & 7 & -9 \end{array} \right)
$$

Multiply (*1*) by $\frac{1}{2}$ and write as (*3*). Multiply (*1*) by -2 and add to (*2*) to obtain (*4*).

$$
\begin{array}{ll}
x - \frac{3}{2}y = 1 & (3) \\
13y = -13 & (4)
\end{array}
\qquad
\left(\begin{array}{cc|c} 1 & -\frac{3}{2} & 1 \\ 0 & 13 & -13 \end{array} \right)
$$

Multiply (*4*) by $\frac{3}{2}/13 = \frac{3}{26}$ and add to (*3*) to obtain (*5*). Multiply (*4*) by $\frac{1}{13}$ to obtain (*6*).

$$
\begin{array}{ll}
x = -\frac{1}{2} & (5) \\
y = -1 & (6)
\end{array}
\qquad
\left(\begin{array}{cc|c} 1 & 0 & -\frac{1}{2} \\ 0 & 1 & -1 \end{array} \right)
$$

The required solution is $x = -\frac{1}{2}$, $y = -1$.

EXAMPLE 5.   Solve, using matrices, the system

$$
\begin{cases}
2x - 3y + 2z = 14 \\
4x + 4y - 3z = 6 \\
3x + 2y - 3z = -2
\end{cases}
$$

The matrix of the system

$$
\left(\begin{array}{ccc|c} 2 & -3 & 2 & 14 \\ 4 & 4 & -3 & 6 \\ 3 & 2 & -3 & -2 \end{array} \right)
$$

is formed by writing in order the coefficients of x, y, z, and the constant terms.

There are, in essence, three moves:

   (*a*)   Multiply the elements of a row by a nonzero number. This move is used only to obtain an element 1 in a prescribed position.

(b) Multiply the elements of a row by a nonzero number and add to the corresponding elements of another row. This move is used to obtain an element 0 in a prescribed position.

(c) Exchange two rows when required.

The first attack must be planned to yield a matrix of the form

$$\begin{pmatrix} 1 & * & * & | & * \\ 0 & * & * & | & * \\ 0 & * & * & | & * \end{pmatrix}$$

in which only the elements of the first column are prescribed.

Multiply first row by $\frac{1}{2}$:

Multiply first row by -2 and add to second row:

Multiply first row by $-\frac{3}{2}$ and add to third row:

$$\begin{pmatrix} 1 & -\frac{3}{2} & 1 & | & 7 \\ 0 & 10 & -7 & | & -22 \\ 0 & \frac{13}{2} & -6 & | & -23 \end{pmatrix}$$

The second attack must be planned to yield a matrix of the form

$$\begin{pmatrix} 1 & 0 & * & | & * \\ 0 & 1 & * & | & * \\ 0 & 0 & * & | & * \end{pmatrix}$$

in which the elements of the first two columns are prescribed.

Multiply second row by $\frac{3}{20}$ and add to first row:

Multiply second row by $\frac{1}{10}$:

Multiply second row by $-\frac{13}{20}$ and add to third row:

$$\begin{pmatrix} 1 & 0 & -\frac{1}{20} & | & \frac{37}{10} \\ 0 & 1 & -\frac{7}{10} & | & -\frac{11}{5} \\ 0 & 0 & -\frac{29}{20} & | & -\frac{87}{10} \end{pmatrix}$$

The final attack must be planned to yield a matrix of the form

$$\begin{pmatrix} 1 & 0 & 0 & | & * \\ 0 & 1 & 0 & | & * \\ 0 & 0 & 1 & | & * \end{pmatrix}$$

in which the elements of the first three columns are prescribed.

Multiply third row by $-\frac{1}{29}$ and add to first row:

Multiply third row by $-\frac{14}{29}$ and add to second row:

Multiply third row by $-\frac{20}{29}$:

$$\begin{pmatrix} 1 & 0 & 0 & | & 4 \\ 0 & 1 & 0 & | & 2 \\ 0 & 0 & 1 & | & 6 \end{pmatrix}$$

The solution is $x = 4$, $y = 2$, $z = 6$.

Solved Problems

6.1 Solve graphically the systems:

(a) $\begin{cases} x + 2y = 5 \\ 3x - y = 1 \end{cases}$, (b) $\begin{cases} x + y = 1 \\ 2x + 3y = 0 \end{cases}$, (c) $\begin{cases} 3x - 6y = 10 \\ 9x + 15y = -14 \end{cases}$.

(a) $x = 1$, $y = 2$ (b) $x = 3$, $y = -2$ (c) $x = 0.7$, $y = -1.3$ (See Fig. 6-4.)

(a) (b) (c)

Fig. 6-4

6.2   Solve algebraically:

(a) $\begin{cases} x + 2y = 5 & (1) \\ 3x - y = 1 & (2) \end{cases}$ (b) $\begin{cases} 3x + 2y = 2 & (1) \\ 5x + 6y = 4 & (2) \end{cases}$ (c) $\begin{cases} 2x + 3y = 3 & (1) \\ 5x - 9y = -4 & (2) \end{cases}$

(a) Rewrite (1): $x + 2y = 5$
      Multiply (2) by 2: $6x - 2y = 2$
                  Add: $\overline{7x \qquad = 7}$
                          $x = 1$

Substitute for x in (1):

$$1 + 2y = 5, \qquad y = 2.$$

Check: Using (2),

$$3(1) - 2 = 1.$$

(b) Multiply (1) by -5: $-15x - 10y = -10$
      Multiply (2) by 3: $\underline{15x + 18y = 12}$
                  Add: $8y = 2$
                                  $y = \frac{1}{4}$

Substitute for y in (1):

$$3x + 2(\tfrac{1}{4}) = 2, \qquad x = \tfrac{1}{2}.$$

Check: Using (2),

$$5(\tfrac{1}{2}) + 6(\tfrac{1}{4}) = 4.$$

(c) Multiply (1) by 3: $6x + 9y = 9$
      Rewrite (2): $\underline{5x - 9y = -4}$
                  Add: $\overline{11x \qquad = 5}$
                          $x = \frac{5}{11}$

Substitute in (1):

$$3y = 3 - 2(\tfrac{5}{11}) = \tfrac{23}{11}, \qquad y = \tfrac{23}{33}.$$

Check: Using (2),

$$5(\tfrac{5}{11}) - 9(\tfrac{23}{33}) = -4.$$

6.3 If the numerator of a fraction is increased by 2, the fraction is $\frac{1}{4}$; if the denominator is decreased by 6, the fraction is $\frac{1}{6}$. Find the fraction.

Let $\frac{x}{y}$ be the original fraction. Then

$$\frac{x+2}{y} = \frac{1}{4} \quad\text{or}\quad 4x - y = -8 \quad (1)$$

$$\frac{x}{y-6} = \frac{1}{6} \quad\text{or}\quad 6x - y = -6 \quad (2)$$

Subtract (1) from (2):

$$2x = 2 \quad\text{and}\quad x = 1$$

Substitute $x = 1$ in (1):

$$4 - y = -8 \quad\text{and}\quad y = 12$$

The fraction is $\frac{1}{12}$.

6.4 A man can row downstream 6 mi in 1 hr and return in 2 hr. Find his rate in still water and the rate of the river.

Let x = rate in still water in mi/hr, y = rate of the river in mi/hr. Then $x + y$ = rate downstream and $x - y$ = rate upstream. Now

$$x + y = 6 \qquad (1)$$
$$x - y = \tfrac{6}{2} = 3 \qquad (2)$$

Add (1) and (2):

$$2x = 9 \quad\text{and}\quad x = 4\tfrac{1}{2}$$

Subtract (2) from (1):

$$2y = 3 \quad\text{and}\quad y = 1\tfrac{1}{2}$$

The rate in still water is $4\tfrac{1}{2}$ mi/hr and the rate of the river is $1\tfrac{1}{2}$ mi/hr.

6.5 Solve the system

$$\begin{cases} x - 5y + 3z = 9 & (1) \\ 2x - y + 4z = 6 & (2) \\ 3x - 2y + z = 2 & (3) \end{cases}$$

Eliminate z.

$$\begin{array}{ll} \text{Rewrite } (1): & x - 5y + 3z = 9 \\ \text{Multiply } (3) \text{ by } -3: & \underline{-9x + 6y - 3z = -6} \\ \text{Add:} & -8x + y = 3 \quad (4) \end{array}$$

$$\begin{array}{ll} \text{Rewrite } (2): & 2x - y + 4z = 6 \\ \text{Multiply } (3) \text{ by } -4: & \underline{-12x + 8y - 4z = -8} \\ \text{Add:} & -10x + 7y = -2 \quad (5) \end{array}$$

$$\begin{array}{ll} \text{Multiply } (4) \text{ by } -7: & 56x - 7y = -21 \\ \text{Rewrite } (5): & \underline{-10x + 7y = -2} \\ \text{Add:} & 46x = -23 \\ & x = -\tfrac{1}{2} \end{array}$$

Substitute $x = -\tfrac{1}{2}$ in (4):

$$-8\left(-\tfrac{1}{2}\right) + y = 3 \quad\text{and}\quad y = -1$$

Substitute $x = -\tfrac{1}{2}$, $y = -1$ in (1):

$$-\tfrac{1}{2} - 5(-1) + 3z = 9 \quad\text{and}\quad z = \tfrac{3}{2}$$

Check: Using (2), $2\left(-\tfrac{1}{2}\right) - (-1) + 4\left(\tfrac{3}{2}\right) = -1 + 1 + 6 = 6$.

6.6   A parabola $y = ax^2 + bx + c$ passes through the points $(1, 0)$, $(2, 2)$, and $(3, 10)$. Determine its equation.

Since $(1, 0)$ is on the parabola:

$$a + b + c = 0 \quad (1)$$

Since $(2, 2)$ is on the parabola:

$$4a + 2b + c = 2 \quad (2)$$

Since $(3, 10)$ is on the parabola:

$$9a + 3b + c = 10 \quad (3)$$

Subtract (1) from (2):

$$3a + b = 2 \quad (4)$$

Subtract (1) from (3):

$$8a + 2b = 10 \quad (5)$$

Multiply (4) by -2 and add to (5):

$$2a = 6 \quad \text{and} \quad a = 3$$

Substitute $a = 3$ in (4):

$$3(3) + b = 2 \quad \text{and} \quad b = -7$$

Substitute $a = 3$, $b = -7$ in (1):

$$3 - 7 + c = 0 \quad \text{and} \quad c = 4$$

The equation of the parabola is $y = 3x^2 - 7x + 4$.

6.7   Solve, using matrices, the system

$$\begin{cases} x - 5y + 3z = 9 \\ 2x - y + 4z = 6 \\ 3x - 2y + z = 2 \end{cases} \quad \text{(See Problem 6.5.)}$$

Begin with the matrix:

$$\begin{pmatrix} 1 & -5 & 3 & 9 \\ 2 & -1 & 4 & 6 \\ 3 & -2 & 1 & 2 \end{pmatrix}$$

Rewrite first row (since first element is 1):
Multiply first row by -2 and add to second row:
Multiply first row by -3 and add to third row:

$$\begin{pmatrix} 1 & -5 & 3 & 9 \\ 0 & 9 & -2 & -12 \\ 0 & 13 & -8 & -25 \end{pmatrix}$$

Multiply second row by $\frac{5}{9}$ and add to first row:
Multiply second row by $\frac{1}{9}$:
Multiply second row by $-\frac{13}{9}$ and add to third row:

$$\begin{pmatrix} 1 & 0 & \frac{17}{9} & \frac{7}{3} \\ 0 & 1 & -\frac{2}{9} & -\frac{4}{3} \\ 0 & 0 & -\frac{46}{9} & -\frac{23}{3} \end{pmatrix}$$

Multiply third row by $\frac{17}{46}$ and add to first row:
Multiply third row by $-\frac{1}{23}$ and add to second row:
Multiply third row by $-\frac{9}{46}$:

$$\begin{pmatrix} 1 & 0 & 0 & -\frac{1}{2} \\ 0 & 1 & 0 & -1 \\ 0 & 0 & 1 & \frac{3}{2} \end{pmatrix}$$

The solution is $x = -\frac{1}{2}$, $y = -1$, $z = \frac{3}{2}$.

6.8   Solve, using matrices, the system

$$\begin{cases} 2x + 2y + 3z = 2 \\ 3x - y - 6z = 4 \\ 8x + 4y + 3z = 8 \end{cases}$$

Begin with the matrix:

$$\begin{pmatrix} 2 & 2 & 3 & | & 2 \\ 3 & -1 & -6 & | & 4 \\ 8 & 4 & 3 & | & 8 \end{pmatrix}$$

Multiply first row by $\frac{1}{2}$:
Multiply first row by $-\frac{3}{2}$ and add to second row:
Multiply first row by -4 and add to third row:

$$\begin{pmatrix} 1 & 1 & \frac{3}{2} & | & 1 \\ 0 & -4 & -\frac{21}{2} & | & 1 \\ 0 & -4 & -9 & | & 0 \end{pmatrix}$$

Multiply second row by $\frac{1}{4}$ and add to first row:
Multiply second row by $-\frac{1}{4}$:
Subtract second row from third row:

$$\begin{pmatrix} 1 & 0 & -\frac{9}{8} & | & \frac{5}{4} \\ 0 & 1 & \frac{21}{8} & | & -\frac{1}{4} \\ 0 & 0 & \frac{3}{2} & | & -1 \end{pmatrix}$$

Multiply third row by $\frac{3}{4}$ and add to first row:
Multiply third row by $-\frac{7}{4}$ and add to second row:
Multiply third row by $\frac{2}{3}$:

$$\begin{pmatrix} 1 & 0 & 0 & | & \frac{1}{2} \\ 0 & 1 & 0 & | & \frac{3}{2} \\ 0 & 0 & 1 & | & -\frac{2}{3} \end{pmatrix}$$

The solution is $x = \frac{1}{2}$, $y = \frac{3}{2}$, $z = -\frac{2}{3}$.

Supplementary Problems

6.9   Solve graphically the systems.

(a) $\begin{cases} x + y = 5 \\ 2x - y = 1 \end{cases}$ (b) $\begin{cases} x - 3y = 1 \\ x - 2y = 0 \end{cases}$ (c) $\begin{cases} x + y = -1 \\ 3x - y = 3 \end{cases}$

Ans.   (a) $(2,3)$ (b) $(-2,-1)$ (c) $(\frac{1}{2}, -\frac{3}{2})$

6.10   Solve algebraically the systems.

(a) $\begin{cases} 3x + 2y = 2 \\ x - y = 9 \end{cases}$ (c) $\begin{cases} 3x - y = 1 \\ 2x + 5y = 41 \end{cases}$ (e) $\begin{cases} 1/x + 2/y = 2 \\ 2/x - 2/y = 1 \end{cases}$

(b) $\begin{cases} 3x - 5y = 5 \\ 7x + y = 75 \end{cases}$ (d) $\begin{cases} x + ay = b \\ 2x - by = a \end{cases}$ (f) $\begin{cases} 1/4x + 7/2y = \frac{5}{4} \\ 1/2x - 3/y = -\frac{5}{14} \end{cases}$

Hint: In (e) and (f) solve first for $1/x$ and $1/y$.

Ans.   (a) $x = 4$, $y = -5$ (c) $x = \frac{46}{17}$, $y = \frac{121}{17}$ (e) $x = 1$, $y = 2$

(b) $x = 10$, $y = 5$ (d) $x = \dfrac{b^2 + a^2}{2a + b}$, $y = \dfrac{2b - a}{2a + b}$ (f) $x = 1$, $y = \frac{7}{2}$

6.11 A and B are 30 km apart. If they leave at the same time and travel in the same direction A overtakes B in 60 hours. If they walk toward each other they meet in 5 hours. What are their rates?

Ans. A, $3\frac{1}{4}$ km/hr; B, $2\frac{3}{4}$ km/hr

6.12   Two trains, each 400 ft long, run on parallel tracks. When running in the same direction, they pass in 20 s; when running in the opposite direction, they pass in 5 s. Find the speed of each train.

     Ans.   100 ft/s, 60 ft/s

6.13   One alloy contains 3 times as much copper as silver, another contains 5 times as much silver as copper. How much of each alloy must be used to make 14 kg in which there is twice as much copper as silver?

     Ans.   12 kg of first, 2 kg of second

6.14   If a field is enlarged by making it 10 m longer and 5 m wider, its area is increased by 1050 square meters. If its length is decreased by 5 m and its width is decreased by 10 m its area is decreased by 1050 square meters. Find the original dimensions of the field.

     Ans.   80 m × 60 m

6.15   It takes a boat $1\frac{1}{2}$ hr to go 12 miles downstream and 6 hr to return. Find the rate of the current and the rate of the boat in still water.

     Ans.   3 mi/hr, 5 mi/hr

6.16   Solve each of the following systems:

$$(a) \begin{cases} x + y + z = 3 \\ 2x + y - z = -6 \\ 3x - y + z = 11 \end{cases} \quad (c) \begin{cases} 3x + y + 4z = 6 \\ 2x - 3y - 5z = 2 \\ 3x - 4y + 3z = 8 \end{cases} \quad (e) \begin{cases} 4x - 3y + 3z = 8 \\ 2x + 3y + 24z = 1 \\ 6x - y + 6z = -1 \end{cases}$$

$$(b) \begin{cases} 4x + 4y - 3z = 3 \\ 2x + 3y + 2z = -4 \\ 3x - y + 4z = 4 \end{cases} \quad (d) \begin{cases} 2x - 3y - 3z = 9 \\ x + 3y + 2z = 3 \\ 3x - 4y - z = 4 \end{cases} \quad (f) \begin{cases} 6x + 2y + 4z = 2 \\ 4x - y + 2z = -3 \\ 7x - 2y - 3z = 5 \end{cases}$$

     Ans.   (a) $x = 1, y = -3, z = 5$ (c) $x = \frac{3}{2}, y = -\frac{1}{2}, z = \frac{1}{2}$ (e) $x = -\frac{3}{2}, y = -4, z = \frac{2}{3}$

               (b) $x = 2, y = -2, z = -1$ (d) $x = 3, y = 2, z = -3$ (f) $x = \frac{2}{3}, y = \frac{7}{3}, z = -\frac{5}{3}$

6.17   Find the equation of the parabola $y = ax^2 + bx + c$ which passes through the points $(1, 6), (4, 0), (3, 4)$.

     Ans. $y = -x^2 + 3x + 4$

Chapter 7

Quadratic Functions and Equations

THE GRAPH OF THE QUADRATIC FUNCTION $y = ax^2 + bx + c$, $a \neq 0$, is a parabola. If $a > 0$, the parabola opens upward (Fig. 7-1); if $a < 0$, the parabola opens downward (Fig. 7-2). The lowest point of the parabola of Fig. 7-1 and the highest point of the parabola of Fig. 7-2 are called *vertices*. The abscissa of the vertex is given by $x = -b/2a$. (See Problem 7.1.)

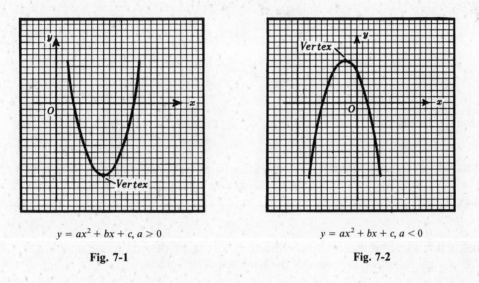

$y = ax^2 + bx + c, a > 0$	$y = ax^2 + bx + c, a < 0$
Fig. 7-1	**Fig. 7-2**

A QUADRATIC EQUATION in one unknown x is of the form

$$ax^2 + bx + c = 0 \qquad a \neq 0 \qquad\qquad (7.1)$$

Frequently a quadratic equation may be solved by *factoring*. (See Problem 7.2.)

Every quadratic equation *(7.1)* can be solved by the following process, known as *completing the square*:

    (*a*)   Subtract the constant term c from both members.

    (*b*)   Divide both members by a, the coefficient of x^2.

    (*c*)   Add to each member the square of one-half the coefficient of the term in x.

    (*d*)   Set the square root of the left member (a perfect square) equal to ± the square root of the right member and solve for x.

EXAMPLE 1. Solve $3x^2 - 8x - 4 = 0$ by completing the square.

    (*a*) $3x^2 - 8x - 4$,   (*b*) $x^2 - \dfrac{8}{3}x = \dfrac{4}{3}$,

    (*c*) $x^2 - \dfrac{8}{3}x + \dfrac{16}{9} = \dfrac{4}{3} + \dfrac{16}{9} = \dfrac{28}{9}$, $\left[\dfrac{1}{2}\left(-\dfrac{8}{3}\right)\right]^2 = \left(-\dfrac{4}{3}\right)^2 = \dfrac{16}{9}$,

    (*d*) $x - \dfrac{4}{3} = \pm\dfrac{2\sqrt{7}}{3}$.   Then $x = \dfrac{4}{3} \pm \dfrac{2\sqrt{7}}{3} = \dfrac{4 \pm 2\sqrt{7}}{3}$.

(See Problem 7.3.)

43

Every quadratic equation *(7.1)* can be solved by means of the quadratic formula

$$x = \frac{-b \pm \sqrt{b^2 - 4ac}}{2a}$$

(See Problems 7.4–7.5.)

It should be noted that it is possible that the roots may be complex numbers.

EQUATIONS IN QUADRATIC FORM.   An equation is in *quadratic form* if it is quadratic in some function of the unknown.

EXAMPLE 2.   Solve $x^4 + x^2 - 12 = 0$. This is a quadratic in x^2.
Factor:

$$x^4 + x^2 - 12 = (x^2 - 3)(x^2 + 4) = 0$$

Then

$$x^2 - 3 = 0 \qquad\qquad x^2 + 4 = 0$$
$$x = \pm\sqrt{3} \qquad\qquad x = \pm 2i$$

(See Problems 7.11–7.12.)

EQUATIONS INVOLVING RADICALS may sometimes reduce to quadratic equations after squaring to remove the radicals. All solutions of this quadratic equation *must* be tested since some may be extraneous. (See Problems 7.13–7.16.)

THE DISCRIMINANT of the quadratic equation *(7.1)* is, by definition, the quantity $b^2 - 4ac$. When a, b, c are rational numbers, the roots of the equation are

Real and unequal if and only if $b^2 - 4ac > 0$.
Real and equal if and only if $b^2 - 4ac = 0$.
Complex if and only if $b^2 - 4ac < 0$.

(See Problems 7.17–7.18.)

SUM AND PRODUCT OF THE ROOTS.   If x_1 and x_2 are the roots of the quadratic equation *(7.1)*, then $x_1 + x_2 = -b/a$ and $x_1 \cdot x_2 = c/a$.

A quadratic equation whose roots are x_1 and x_2 may be written in the form

$$x^2 - (x_1 + x_2)x + x_1 \cdot x_2 = 0$$

Solved Problems

7.1   Sketch the parabolas: (*a*) $y = x^2 - 2x - 8$, (*b*) $y = -2x^2 + 9x - 9$. Determine the coordinates of the vertex $V(x, y)$ of each.
   Vertex:

(*a*) $\qquad\qquad x = -\dfrac{b}{2a} = -\dfrac{-2}{2 \cdot 1} = 1 \qquad y = 1^2 - 2 \cdot 1 - 8 = -9$

hence $V(1, -9)$. See Fig. 7.3(*a*).

(b) $\qquad x = -\dfrac{9}{2(-2)} = \dfrac{9}{4} \qquad y = -2\left(\dfrac{9}{4}\right)^2 + 9\left(\dfrac{9}{4}\right) - 9 = \dfrac{9}{8}$

hence $V(\frac{9}{4}, \frac{9}{8})$. See Fig. 7-3($b$).

(a) $\qquad\qquad\qquad\qquad\qquad$ (b)

Fig. 7-3

7.2   Solve by factoring:

(a) $4x^2 - 5x = x(4x - 5) = 0$ $\qquad$ (e) $4x^2 + 20x + 25 = (2x + 5)(2x + 5) = 0$

(b) $4x^2 - 9 = (2x - 3)(2x + 3) = 0$ $\qquad$ (f) $6x^2 + 13x + 6 = (3x + 2)(2x + 3) = 0$

(c) $x^2 - 4x + 3 = (x - 1)(x - 3) = 0$ $\qquad$ (g) $3x^2 + 8ax - 3a^2 = (3x - a)(x + 3a) = 0$

(d) $x^2 - 6x + 9 = (x - 3)(x - 3) = 0$ $\qquad$ (h) $10ax^2 + (15 - 8a^2)x - 12a = (2ax + 3)(5x - 4a) = 0$

$\quad$ *Ans.*   (a) $0, \frac{5}{4}$ $\quad$ (c) $1, 3$ $\quad$ (e) $-\frac{5}{2}, -\frac{5}{2}$ $\quad$ (g) $a/3, -3a$

$\qquad\qquad\quad$ (b) $\frac{3}{2}, -\frac{3}{2}$ $\quad$ (d) $3, 3$ $\quad$ (f) $-\frac{2}{3}, -\frac{3}{2}$ $\quad$ (h) $-\frac{3}{2}a, 4a/5$

7.3   Solve by completing the square: (a) $x^2 - 2x - 1 = 0$, (b) $3x^2 + 8x + 7 = 0$.

(a) $x^2 - 2x = 1;\quad x^2 - 2x + 1 = 1 + 1 = 2;\quad x - 1 = \pm\sqrt{2};\quad x = 1 \pm \sqrt{2}$.

(b) $3x^2 + 8x = -7;\quad x^2 + \dfrac{8}{3}x = -\dfrac{7}{3};\quad x^2 + \dfrac{8}{3}x + \dfrac{16}{9} = -\dfrac{7}{3} + \dfrac{16}{9} = -\dfrac{5}{9};$

$\qquad x + \dfrac{4}{3} = \pm\sqrt{\dfrac{-5}{9}} = \pm\dfrac{i\sqrt{5}}{3};\quad x = \dfrac{-4 \pm i\sqrt{5}}{3}.$

7.4   Solve $ax^2 + bx + c = 0$, $a \neq 0$, by completing the square.

$\qquad$ Proceeding as in Problem 7.3, we have

$$x^2 + \frac{b}{a}x = -\frac{c}{a}, \qquad x^2 + \frac{b}{a}x + \frac{b^2}{4a^2} = \frac{b^2}{4a^2} - \frac{c}{a} = \frac{b^2 - 4ac}{4a^2},$$

$$x + \frac{b}{2a} = \pm\sqrt{\frac{b^2 - 4ac}{4a^2}} = \pm\frac{\sqrt{b^2 - 4ac}}{2a}, \quad \text{and} \quad x = \frac{-b \pm \sqrt{b^2 - 4ac}}{2a}$$

7.5   Solve the equations of Problem 7.3 using the quadratic formula.

(a) $\qquad x = \dfrac{-b \pm \sqrt{b^2 - 4ac}}{2a} = \dfrac{-(-2) \pm \sqrt{(-2)^2 - 4(1)(-1)}}{2 \cdot 1} = \dfrac{2 \pm \sqrt{4 + 4}}{2} = \dfrac{2 \pm 2\sqrt{2}}{2} = 1 \pm \sqrt{2}$

(b) $\qquad x = \dfrac{-(8) \pm \sqrt{8^2 - 4 \cdot 3 \cdot 7}}{2 \cdot 3} = \dfrac{-8 \pm \sqrt{64 - 84}}{6} = \dfrac{-8 \pm \sqrt{-20}}{6} = \dfrac{-8 \pm 2i\sqrt{5}}{6} = \dfrac{-4 \pm i\sqrt{5}}{3}$

7.6   An open box containing 24 cm^3 is to be made from a square piece of tin by cutting 2 cm squares from each corner and turning up the sides. Find the dimension of the piece of tin required.

Let x = the required dimension. The resulting box will have dimensions $(x - 4)$ by $(x - 4)$ by 2, and its volume will be $2(x - 4)(x - 4)$. See Fig. 7-4. Then

$$2(x - 4)^2 = 24, \qquad x - 4 = \pm 2\sqrt{3}, \qquad \text{and} \qquad x = 4 \pm 2\sqrt{3} = 7.464,\ 0.536$$

The required square of tin is 7.464 cm on a side.

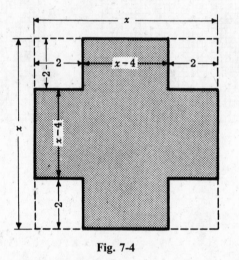

Fig. 7-4

7.7   Two pipes together can fill a reservoir in 6 hr 40 min. Find the time each alone will take to fill the reservoir if one of the pipes can fill it in 3 hr less time than the other.

Let x = time (hours) required by smaller pipe, $x - 3$ = time required by larger pipe. Then

$$\frac{1}{x} = \text{part filled in 1 hr by smaller pipe} \qquad \frac{1}{x - 3} = \text{part filled in 1 hr by larger pipe.}$$

Since the two pipes together fill $\dfrac{1}{\frac{20}{3}} = \dfrac{3}{20}$ of the reservoir in 1 hr,

$$\frac{1}{x} + \frac{1}{x - 3} = \frac{3}{20}, \qquad 20(x - 3) + 20x = 3x(x - 3),$$

$$3x^2 - 49x + 60 = (3x - 4)(x - 15) = 0 \qquad \text{and} \qquad x = \tfrac{4}{3},\ 15$$

The smaller pipe will fill the reservoir in 15 hr and the larger pipe in 12 hr.

7.8   Express each of the following in the form $a(x - h)^2 \pm b(y - k)^2 = c$.

(a) $x^2 + y^2 - 6x - 9y + 2 = 0$.

$$(x^2 - 6x) + (y^2 - 9y) = -2 \qquad (x^2 - 6x + 9) + (y^2 - 9y + \tfrac{81}{4}) = -2 + 9 + \tfrac{81}{4} = \tfrac{109}{4}$$

$$(x - 3)^2 + (y - \tfrac{9}{2})^2 = \tfrac{109}{4}$$

(b) $3x^2 + 4y^2 + 6x - 16y - 21 = 0$.

$$3(x^2 + 2x) + 4(y^2 - 4y) = 21 \qquad 3(x^2 + 2x + 1) + 4(y^2 - 4y + 4) = 21 + 3(1) + 4(4) = 40$$

$$3(x + 1)^2 + 4(y - 2)^2 = 40$$

7.9   Transform each of the following into the form $a\sqrt{(x-h)^2 + k}$ or $a\sqrt{k - (x-h)^2}$.

(a) $\sqrt{4x^2 - 8x + 9} = 2\sqrt{x^2 - 2x + \frac{9}{4}} = 2\sqrt{(x^2 - 2x + 1) + \frac{5}{4}} = 2\sqrt{(x-1)^2 + \frac{5}{4}}$

(b) $\sqrt{8x - x^2} = \sqrt{16 - (x^2 - 8x + 16)} = \sqrt{16 - (x-4)^2}$

(c) $\sqrt{3 - 4x - 2x^2} = \sqrt{2} \cdot \sqrt{\frac{3}{2} - 2x - x^2} = \sqrt{2} \cdot \sqrt{\frac{5}{2} - (x^2 + 2x + 1)} = \sqrt{2} \cdot \sqrt{\frac{5}{2} - (x+1)^2}$

7.10   If an object is thrown directly upward with initial speed v ft/s, its distance s ft above the ground after t s is given by

$$s = vt - \tfrac{1}{2}gt^2$$

Taking $g = 32.2$ ft/s^2 and initial speed 120 ft/s, find (a) when the object is 60 ft above the ground, (b) when it is highest in its path and how high.

   The equation of motion is $s = 120t - 16.1t^2$.

(a)   When $s = 60$:

$$60 = 120t - 16.1t^2 \qquad \text{or} \qquad 16.1t^2 - 120t + 60 = 0$$

$$t = \frac{120 \pm \sqrt{(120)^2 - 4(16.1)60}}{32.2} = \frac{120 \pm \sqrt{10536}}{32.2} = \frac{120 \pm 102.64}{32.2} = 6.91, 0.54$$

   After $t = 0.54$ s the object is 60 ft above the ground and rising. After $t = 6.91$ s, the object is 60 ft above the ground and falling.

(b)   The object is at its highest point when

$$t = \frac{-b}{2a} = \frac{-(-120)}{2(16.1)} = 3.73 \text{ s.} \quad \text{Its height is given by } 120t - 16.1t^2 = 120(3.73) - 16.1(3.73)^2 = 223.6 \text{ ft.}$$

7.11   Solve $9x^4 - 10x^2 + 1 = 0$.

   Factor: $(x^2 - 1)(9x^2 - 1) = 0$. Then $x^2 - 1 = 0$, $9x^2 - 1 = 0$; $x = \pm 1$, $x = \pm\frac{1}{3}$.

7.12   Solve $x^4 - 6x^3 + 12x^2 - 9x + 2 = 0$.

   Complete the square on the first two terms: $(x^4 - 6x^3 + 9x^2) + 3x^2 - 9x + 2 = 0$

   or $(x^2 - 3x)^2 + 3(x^2 - 3x) + 2 = 0$

   Factor: $[(x^2 - 3x) + 2][(x^2 - 3x) + 1] = 0$

   Then $x^2 - 3x + 2 = (x-2)(x-1) = 0$ and $x = 1, 2$

   $x^2 - 3x + 1 = 0$ and $x = \dfrac{3 \pm \sqrt{9 - 4}}{2} = \dfrac{3 \pm \sqrt{5}}{2}$

7.13   Solve $\sqrt{5x - 1} - \sqrt{x} = 1$.

   Transpose one of the radicals: $\sqrt{5x - 1} = \sqrt{x} + 1$

   Square: $5x - 1 = x + 2\sqrt{x} + 1$

   Collect terms: $4x - 2 = 2\sqrt{x}$ or $2x - 1 = \sqrt{x}$

   Square:

$$4x^2 - 4x + 1 = x, \qquad 4x^2 - 5x + 1 = (4x - 1)(x - 1) = 0, \qquad \text{and} \qquad x = \tfrac{1}{4}, 1$$

   For $x = \frac{1}{4}$: $\sqrt{5(\frac{1}{4}) - 1} - \sqrt{\frac{1}{4}} = 0 \neq 1$.

   For $x = 1$: $\sqrt{5(1) - 1} - \sqrt{1} = 1$.   The root is $x = 1$.

7.14   Solve $\sqrt{6x+7} - \sqrt{3x+3} = 1$.

Transpose one of the radicals: $\sqrt{6x+7} = 1 + \sqrt{3x+3}$

Square: $6x + 7 = 1 + 2\sqrt{3x+3} + 3x + 3$

Collect terms: $3x + 3 = 2\sqrt{3x+3}$

Square:

$$9x^2 + 18x + 9 = 4(3x+3) = 12x + 12, \qquad 9x^2 + 6x - 3 = 3(3x-1)(x+1) = 0, \qquad \text{and} \qquad x = \tfrac{1}{3}, -1$$

For $x = \tfrac{1}{3}$: $\sqrt{6(\tfrac{1}{3})+7} - \sqrt{3(\tfrac{1}{3})+3} = 3 - 2 = 1$.

For $x = -1$: $1 - 0 = 1$.   The roots are $x = \tfrac{1}{3}, -1$.

7.15   Solve $\dfrac{\sqrt{x+1} + \sqrt{x-1}}{\sqrt{x+1} - \sqrt{x-1}} = 3$.

Multiply the numerator and denominator of the fraction by $(\sqrt{x+1} + \sqrt{x-1})$:

$$\frac{(\sqrt{x+1} + \sqrt{x-1})(\sqrt{x+1} + \sqrt{x-1})}{(\sqrt{x+1} - \sqrt{x-1})(\sqrt{x+1} + \sqrt{x-1})} = \frac{(x+1) + 2\sqrt{x^2-1} + (x-1)}{(x+1) - (x-1)} = x + \sqrt{x^2-1} = 3$$

Then $x - 3 = -\sqrt{x^2-1}$, $x^2 - 6x + 9 = x^2 - 1$,   and $x = \tfrac{5}{3}$

Check: $\dfrac{\sqrt{\tfrac{8}{3}} + \sqrt{\tfrac{2}{3}}}{\sqrt{\tfrac{8}{3}} - \sqrt{\tfrac{2}{3}}} = \dfrac{2\sqrt{\tfrac{2}{3}} + \sqrt{\tfrac{2}{3}}}{2\sqrt{\tfrac{2}{3}} - \sqrt{\tfrac{2}{3}}} = \dfrac{3\sqrt{\tfrac{2}{3}}}{\sqrt{\tfrac{2}{3}}} = 3$

7.16   Solve $3x^2 - 5x + \sqrt{3x^2 - 5x + 4} = 16$.

Note that the unknown enters alike in both the expression free of radicals and under the radical. Add 4 to both sides:

$$3x^2 - 5x + 4 + \sqrt{3x^2 - 5x + 4} = 20$$

Let $y = \sqrt{3x^2 - 5x + 4}$. Then

$$y^2 + y - 20 = (y+5)(y-4) = 0 \qquad \text{and} \qquad y = 4, -5$$

Now $\sqrt{3x^2 - 5x + 4} = -5$ is impossible. From $\sqrt{3x^2 - 5x + 4} = 4$ we have

$$3x^2 - 5x + 4 = 16, \qquad 3x^2 - 5x - 12 = (3x+4)(x-3) = 0 \qquad \text{and} \qquad x = 3, -\tfrac{4}{3}$$

The reader will show that both $x = 3$ and $x = -\tfrac{4}{3}$ are solutions.

7.17   Without solving, determine the character of the roots of

(a) $x^2 - 8x + 9 = 0$. Here $b^2 - 4ac = 28$; the roots are irrational and unequal.

(b) $3x^2 - 8x + 9 = 0$. Here $b^2 - 4ac = -44$; the roots are imaginary and unequal.

(c) $6x^2 - 5x - 6 = 0$. Here $b^2 - 4ac = 169$; the roots are rational and unequal.

(d) $4x^2 - 4\sqrt{3}x + 3 = 0$. Here $b^2 - 4ac = 0$; the roots are real and equal.

(NOTE.   Although the discriminant is the square of a rational number, the roots $\tfrac{1}{2}\sqrt{3}, \tfrac{1}{2}\sqrt{3}$ are not rational. Why?)

7.18 Without sketching, state whether the graph of each of the following functions crosses the x axis, is tangent to it, or lies wholly above or below it.

(a) $3x^2 + 5x - 2$. $b^2 - 4ac = 25 + 24 > 0$; the graph crosses the x axis.

(b) $2x^2 + 5x + 4$. $b^2 - 4ac = 25 - 32 < 0$ and the graph is either wholly above or wholly below the x axis. Since $f(0) > 0$ (the value of the function for any other value of x would do equally well), the graph lies wholly above the x axis.

(c) $4x^2 - 20x + 25$. $b^2 - 4ac = 400 - 400 = 0$; the graph is tangent to the x axis.

(d) $2x - 9 - 4x^2$. $b^2 - 4ac = 4 - 144 < 0$ and $f(0) < 0$; the graph lies wholly below the x axis.

7.19 Find the sum and product of the roots of

(a) $x^2 + 5x - 8 = 0$.   *Ans.* Sum $= -\dfrac{b}{a} = -5$, product $= \dfrac{c}{a} = -8$.

(b) $8x^2 - x - 2 = 0$ or $x^2 - \frac{1}{8}x - \frac{1}{4} = 0$.   *Ans.* Sum $= \frac{1}{8}$, product $= -\frac{1}{4}$.

(c) $5 - 10x - 3x^2 = 0$ or $x^2 + \frac{10}{3}x - \frac{5}{3} = 0$.   *Ans.* Sum $= -\frac{10}{3}$, product $= -\frac{5}{3}$.

7.20 Form the quadratic equation whose roots x_1 and x_2 are:

(a) $3, \frac{2}{5}$. Here $x_1 + x_2 = \frac{17}{5}$ and $x_1 \cdot x_2 = \frac{6}{5}$. The equation is $x^2 - \frac{17}{5}x + \frac{6}{5} = 0$ or $5x^2 - 17x + 6 = 0$.

(b) $-2 + 3\sqrt{5}, -2 - 3\sqrt{5}$. Here $x_1 + x_2 = -4$ and $x_1 \cdot x_2 = 4 - 45 = -41$. The equation is $x^2 + 4x - 41 = 0$.

(c) $\dfrac{3 - i\sqrt{2}}{2}, \dfrac{3 + i\sqrt{2}}{2}$. The sum of the roots is 3 and the product is $\frac{11}{4}$. The equation is $x^2 - 3x + \frac{11}{4} = 0$ or $4x^2 - 12x + 11 = 0$.

7.21 Determine k so that the given equation will have the stated property, and write the resulting equation.

(a) $x^2 + 4kx + k + 2 = 0$ has one root 0. Since the product of the roots is to be 0, $k + 2 = 0$ and $k = -2$. The equation is $x^2 - 8x = 0$.

(b) $4x^2 - 8kx - 9 = 0$ has one root the negative of the other. Since the sum of the roots is to be 0, $2k = 0$ and $k = 0$. The equation is $4x^2 - 9 = 0$.

(c) $4x^2 - 8kx + 9 = 0$ has roots whose difference is 4. Denote the roots by r and $r + 4$. Then $r + (r + 4) = 2r + 4 = 2k$ and $r(r + 4) = \frac{9}{4}$. Solving for $r = k - 2$ in the first and substituting in the second, we have $(k - 2)(k + 2) = \frac{9}{4}$; then $4k^2 - 16 = 9$ and $k = \pm\frac{5}{2}$. The equations are $4x^2 + 20x + 9 = 0$ and $4x^2 - 20x + 9 = 0$.

(d) $2x^2 - 3kx + 5k = 0$ has one root twice the other. Let the roots be r and $2r$. Then $r + 2r = 3r = \frac{3}{2}k$, $r = \frac{1}{2}k$, and $r(2r) = 2r^2 = \frac{5}{2}k$. Thus $k = 0, 5$. The equations are $2x^2 = 0$ with roots $0, 0$ and $2x^2 - 15x + 25 = 0$ with roots $\frac{5}{2}, 5$.

Supplementary Problems

7.22 Locate the vertex of each of the parabolas of Problem 4.11. Compare the results with those of Problem 4.14.

7.23 Solve for x by factoring.

(a) $3x^2 + 4x = 0$ (c) $x^2 + 2x - 3 = 0$ (e) $10x^2 - 9x + 2 = 0$

(b) $16x^2 - 25 = 0$ (d) $2x^2 + 9x - 5 = 0$ (f) $2x^2 - (a + 4b)x + 2ab = 0$

    Ans. (a) $0, -\frac{4}{3}$ (b) $\pm\frac{5}{4}$ (c) $1, -3$ (d) $\frac{1}{2}, -5$ (e) $\frac{1}{2}, \frac{2}{5}$ (f) $\frac{1}{2}a, 2b$

7.24 Solve for x by completing the square.

(a) $2x^2 + x - 5 = 0$ (c) $3x^2 + 2x - 2 = 0$ (e) $15x^2 - (16m - 14)x + 4m^2 - 8m + 3 = 0$

(b) $2x^2 - 4x - 3 = 0$ (d) $5x^2 - 4x + 2 = 0$

Ans. (a) $\frac{1}{4}(-1 \pm \sqrt{41})$ (b) $\frac{1}{2}(2 \pm \sqrt{10})$ (c) $\frac{1}{3}(-1 \pm \sqrt{7})$ (d) $\frac{1}{5}(2 \pm i\sqrt{6})$

(e) $\frac{1}{3}(2m - 1), \frac{1}{5}(2m - 3)$

7.25 Solve the equations of Problem 7.24 using the quadratic formula.

7.26 Solve $6x^2 + 5xy - 6y^2 + x + 8y - 2 = 0$ for (a) y in terms of x, (b) x in terms of y.

Ans. (a) $\frac{1}{2}(3x + 2), \frac{1}{3}(1 - 2x)$ (b) $\frac{1}{2}(1 - 3y), \frac{2}{3}(y - 1)$

7.27 Solve.

(a) $x^4 - 29x^2 + 100 = 0$ (c) $1 - \dfrac{2}{2x^2 - x} = \dfrac{3}{(2x^2 - x)^2}$ (e) $\sqrt{2x + 3} - \sqrt{4 - x} = 2$

(b) $\dfrac{21}{x + 2} - \dfrac{1}{x - 4} = 2$ (d) $\sqrt{4x + 1} - \sqrt{3x - 2} = 5$ (f) $\sqrt{3x - 2} - \sqrt{x - 2} = 2$

Ans. (a) $\pm 2, \pm 5$ (b) $5, 7$ (c) $-1, \frac{3}{2}, \frac{1}{4}(1 \pm i\sqrt{7})$ (d) 342 (e) 3 (f) $2, 6$

7.28 Form the quadratic equation whose roots are

(a) The negatives of the roots of $3x^2 + 5x - 8 = 0$.

(b) Twice the roots of $2x^2 - 5x + 2 = 0$.

(c) One-half the roots of $2x^2 - 5x - 3 = 0$.

Ans. (a) $3x^2 - 5x - 8 = 0$ (b) $x^2 - 5x + 4 = 0$ (c) $8x^2 - 10x - 3 = 0$

7.29 The length of a rectangle is 7 cm more than its width; its area is 228 cm^2. What are its dimensions?

Ans. 12 cm × 19 cm

7.30 A rectangular garden plot 16 m × 24 m is to be bordered by a strip of uniform width x meters so as to double the area. Find x.

Ans. 4 *m*

7.31 The interior of a cubical box is lined with insulating material $\frac{1}{2}$ cm thick. Find the original interior dimensions if the volume is thereby decreased by 271 cm^3.

Ans. 10 cm

7.32 What are the dimensions of the largest rectangular field which can be enclosed by 1200 m of fencing?

Ans. 300 m × 300 m

Chapter 8

Inequalities

AN INEQUALITY is a statement that one (real) number is greater than or less than another; for example $3 > -2$, $-10 < -5$.

Two inequalities are said to have the *same sense* if their signs of inequality point in the same direction. Thus $3 > -2$ and $-5 > -10$ have the same sense; $3 > -2$ and $-10 < -5$ have opposite senses.

The sense of an equality is *not* changed:

(a) If the same number is added to or subtracted from both sides

(b) If both sides are multiplied or divided by the same *positive* number

The sense of an equality *is* changed if both sides are multiplied or divided by the same negative number. (See Problems 8.1–8.3.)

AN ABSOLUTE INEQUALITY is one which is true for all real values of the letters involved; for example, $x^2 + 1 > 0$ is an absolute inequality.

A CONDITIONAL INEQUALITY is one which is true for certain values of the letters involved; for example, $x + 2 > 5$ is a conditional inequality since it is true for $x = 4$ but not for $x = 1$.

SOLUTION OF CONDITIONAL INEQUALITIES.   The solution of a conditional inequality in one letter, say x, consists of all values of x for which the inequality is true. These values lie on one or more intervals of the real number scale as illustrated in the examples below.

To solve a linear inequality, proceed as in solving a linear equality keeping in mind the rules for keeping or reversing the sense.

EXAMPLE 1.   Solve the inequality $5x + 4 > 2x + 6$.

Subtract $2x$ from each member $\left.\begin{array}{l} \\ \\ \end{array}\right\}$
Subtract 4 from each member $3x > 2$

Divide by 3: $x > \frac{2}{3}$

Graphical representation: (See Fig. 8-1.)

Fig. 8-1

(See Problems 8.5–8.6.)

To solve a quadratic inequality, $f(x) = ax^2 + bx + c > 0$, solve the equality $f(x) = 0$, locate the roots r_1 and r_2 on a number scale, and determine the sign of $f(x)$ on each of the resulting intervals.

EXAMPLE 2.   Solve the inequality $3x^2 - 8x + 7 > 2x^2 - 3x + 1$.

Subtract $2x - 3x + 1$ from each member:

$$x^2 - 5x + 6 > 0$$

51

Solve the equality $x^2 - 5x + 6 = 0$:

$$x = 2, \qquad x = 3$$

Locate the roots on a number scale (see Fig. 8-2).

Fig. 8-2

Determine the sign of $f(x) = x^2 - 5x + 6$:

On the interval $x < 2$: $f(0) = 6 > 0$
On the interval $2 < x < 3$: $f(\frac{5}{2}) = \frac{25}{4} - \frac{25}{2} + 6 < 0$
On the interval $x > 3$: $f(4) = 16 - 20 + 6 > 0$

The given inequality is satisfied (see darkened portions of the scale) when $x < 2$ and $x > 3$. (See Problems 8.7–8.11.)

Solved Problems

8.1 Given the inequality $-3 < 4$, write the result when (a) 5 is added to both sides, (b) 5 is subtracted from both sides, (c) -6 is subtracted from both sides, (d) both sides are doubled, (e) both sides are divided by -2.

   Ans.   (a) $2 < 9$,   (b) $-8 < -1$,   (c) $3 < 10$,   (d) $-6 < 8$,   (e) $\frac{3}{2} > -2$

8.2 Square each of the inequalities: (a) $-3 < 4$,   (b) $-5 < 4$.

   Ans.   (a) $9 < 16$,   (b) $25 > 16$

8.3 If $a > 0$, $b > 0$ prove that $a^2 > b^2$ if and only if $a > b$.

   Suppose $a > b$. Since $a > 0$, $a^2 > ab$ and, since $b > 0$, $ab > b^2$. Hence, $a^2 > ab > b^2$ and $a^2 > b^2$.
   Suppose $a^2 > b^2$. Then $a^2 - b^2 = (a - b)(a + b) > 0$. Dividing by $a + b > 0$, we have $a - b > 0$ and $a > b$.

8.4 Prove $\dfrac{a}{b^2} + \dfrac{b}{a^2} > \dfrac{1}{a} + \dfrac{1}{b}$ if $a > 0$, $b > 0$, and $a \neq b$.

   Suppose $a > b$; then $a^2 > b^2$ and $a - b > 0$. Now $a^2(a - b) > b^2(a - b)$ or $a^3 - a^2b > ab^2 - b^3$ and $a^3 + b^3 > ab^2 + a^2b$. Since $a^2b^2 > 0$,

$$\frac{a^3 + b^3}{a^2b^2} > \frac{ab^2 + a^2b}{a^2b^2} \qquad \text{and} \qquad \frac{a}{b^2} + \frac{b}{a^2} > \frac{1}{a} + \frac{1}{b}$$

   Why is it necessary that $a > 0$ and $b > 0$? **Hint:** See Problem 8.3.

8.5   Solve $3x + 4 > 5x + 2$.

Subtract $5x + 4$ from each member:

$$-2x > -2$$

Divide by -2:

$$x < 1$$

See Fig. 8-3.

Graphical representation

Fig. 8-3

8.6   Solve $2x - 8 < 7x + 12$.

Subtract $7x - 8$ from each member:

$$-5x < 20$$

Divide by -5:

$$x > -4$$

See Fig. 8-4.

Graphical representation

Fig. 8-4

8.7   Solve $x^2 > 4x + 5$.

Subtract $4x + 5$ from each member:

$$x^2 - 4x - 5 > 0$$

Solve the equality $f(x) = x^2 - 4x - 5 = 0$:

$$x = -1, 5$$

Locate the roots on a number scale.

Determine the sign of $f(x)$

On the interval $x < -1$: $f(-2) = 4 + 8 - 5 > 0$
On the interval $-1 < x < 5$: $f(0) = -5 < 0$
On the interval $x > 5$: $f(6) = 36 - 24 - 5 > 0$

The inequality is satisfied when $x < -1$ and $x > 5$. See Fig. 8-5.

Fig. 8-5

8.8 Solve $3x^2 + 2x + 2 < 2x^2 + x + 4$.

Subtract $2x^2 + x + 4$ from each member:

$$x^2 + x - 2 < 0$$

Solve $f(x) = x^2 + x - 2 = 0$:

$$x = -2, 1$$

Locate the roots on a number scale.

Determine the sign of $f(x)$

On the interval $x < -2$: $\quad f(-3) = 9 - 3 - 2 > 0$
On the interval $-2 < x < 1$: $\quad f(0) = -2 < 0$
On the interval $x > 1$: $\quad f(2) = 4 + 2 - 2 > 0$

The inequality is satisfied when $-2 < x < 1$. See Fig. 8-6.

Fig. 8-6

8.9 Solve $(x + 5)(x - 1)(x - 2) < 0$.

Solve the equality $f(x) = (x + 5)(x - 1)(x - 2) = 0$. $\quad x = 1, 2, -5$

Locate the roots on a number scale.

Determine the sign of $f(x)$

On the interval $x < -5$: $\quad f(-6) = (-1)(-7)(-8) < 0$
On the interval $-5 < x < 1$: $\quad f(0) = 5(-1)(-2) > 0$
On the interval $1 < x < 2$: $\quad f(\frac{3}{2}) = (\frac{13}{2})(\frac{1}{2})(-\frac{1}{2}) < 0$
On the interval $x > 2$: $\quad f(3) = 8 \cdot 2 \cdot 1 > 0$

The inequality is satisfied when $x < -5$ and $1 < x < 2$. See Fig. 8-7.

Fig. 8-7

8.10 Solve $(x - 2)^2(x - 5) > 0$.

Solve the equality $f(x) = (x - 2)^2(x - 5) = 0$: $x = 2, 2, 5$

Locate the roots on a number scale.

Determine the sign of $f(x)$

On the interval $x < 2$: $\quad f(0) = (+)(-) < 0$
On the interval $2 < x < 5$: $\quad f(3) = (+)(-) < 0$
On the interval $x > 5$: $\quad f(6) = (+)(+) > 0$

The inequality is satisfied when $x > 5$. See Fig. 8-8.

Fig. 8-8

(NOTE.   The inequality $(x - 2)^2(x - 5) < 0$ is satisfied when $x < 2$ and $2 < x < 5$.
The inequality $(x - 2)^2(x - 5) \geq 0$ is satisfied when $x \geq 5$ and $x = 2$.)

8.11 Determine the values of k so that $3x^2 + kx + 4 = 0$ will have real roots.

The discriminant $b^2 - 4ac = k^2 - 48 = (k - 4\sqrt{3})(k + 4\sqrt{3}) \geq 0$. The roots will be real when $k \geq 4\sqrt{3}$ and when $k \leq -4\sqrt{3}$, that is, when $|k| \geq 4\sqrt{3}$.

Supplementary Problems

8.12 If $2y^2 + 4xy - 3x = 0$, determine the range of values of x for which the corresponding y roots are real.

Ans. Here

$$y = \frac{-4x \pm \sqrt{16x^2 + 24x}}{4} = \frac{-2x \pm \sqrt{4x^2 + 6x}}{2}$$

will be real provided $4x^2 + 6x \geq 0$. Thus y will be real for $x \leq -\frac{3}{2}$ and for $x \geq 0$.

8.13 Prove: If $a > b$ and $c > d$, then $a + c > b + d$.

Hint: $(a - b) + (c - d) = (a + c) - (b + d) > 0$.

8.14 Prove: If $a \neq b$ are real numbers, then $a^2 + b^2 > 2ab$.

Hint: $(a - b)^2 > 0$.

8.15 Prove: If $a \neq b \neq c$ are real numbers, then $a^2 + b^2 + c^2 > ab + bc + ca$.

8.16 Prove: If $a > 0$, $b > 0$, and $a \neq b$, then $a/b + b/a > 2$.

8.17 Prove: If $a^2 + b^2 = 1$ and $c^2 + d^2 = 1$, then $ac + bd \leq 1$.

8.18 Solve: (*a*) $x - 4 > -2x + 5$ (*c*) $x^2 - 16 > 0$ (*e*) $x^2 - 6x > -5$
           (*b*) $4 + 3x < 2x + 24$ (*d*) $x^2 - 4x < 5$ (*f*) $5x^2 + 5x - 8 \leq 3x^2 + 4$

    Ans.   (*a*) $x > 3$ (*b*) $x < 20$ (*c*) $|x| > 4$ (*d*) $-1 < x < 5$
          (*e*) $x < 1, x > 5$,   (*f*) $-4 \leq x \leq \frac{3}{2}$

8.19 Solve: (*a*) $(x + 1)(x - 2)(x + 4) > 0$ (*b*) $(x - 1)^3(x - 3)(x + 2) < 0$
          (*c*) $(x + 3)(x - 2)^2(x - 5)^3 < 0$

    Ans.   (*a*) $-4 < x < -1, x > 2$ (*b*) $x < -2, 1 < x < 3$ (*c*) $-3 < x < 2, 2 < x < 5$

8.20 In each of the following determine the range of x for which y will be real:

  (*a*) $y = \sqrt{2x^2 - 7x + 3}$ (*c*) $y = \sqrt{61x^2 + 144x + 144}$ (*e*) $xy^2 + 3xy + 3x - 4y - 4 = 0$
  (*b*) $y = \sqrt{6 - 5x - 4x^2}$ (*d*) $y^2 + 2xy + 4y + x + 14 = 0$ (*f*) $6x^2 + 5xy - 6y^2 + x + 8y - 2 = 0$

    Ans.   (*a*) $x \leq \frac{1}{2}, x \geq 3$ (*c*) all values of x (*e*) $-4 \leq x \leq \frac{4}{3}$
         (*b*) $-2 \leq x \leq \frac{3}{4}$ (*d*) $x \leq -5, x \geq 2$ (*f*) all values of x

Chapter 9

The Locus of an Equation

DEGENERATE LOCI. The locus of an equation $f(x, y) = 0$ is called *degenerate* if $f(x, y)$ is the product of two or more real factors $g(x, y)$, $h(x, y)$, ... The locus of $f(x, y) = 0$ then consists of the loci of $g(x, y) = 0$, $h(x, y) = 0$, ... (See Problem 9.1.)

INTERCEPTS. The intercepts on the coordinate axes of a locus are the directed distances from the origin to the points of intersection of the locus and the coordinate axes.

To find the x intercepts, set $y = 0$ in the equation of the locus and solve for x; to find the y intercepts, set $x = 0$ and solve for y. (See Problem 9.2.)

SYMMETRY. Two points P and Q are said to be symmetric with respect to a point R if R is the midpoint of the segment PQ (see Fig. 9-1). Each of the points is called the symmetric point of the other with respect to the point R, the *center of symmetry*.

Two points P and Q are said to be symmetric with respect to a line l if l is the perpendicular bisector of the segment PQ (see Fig. 9-2). Each of the points P, Q is called the symmetric point of the other with respect to l, the *axis of symmetry*.

A locus is said to be symmetric with respect to a point R or to a line l if the symmetric point with respect to R or l of every point of the locus is also a point of the locus (see Figs. 9-3 and 9-4).

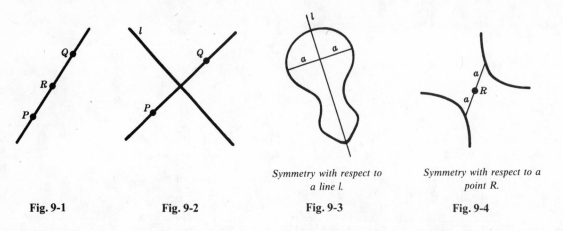

Symmetry with respect to a line l.

Symmetry with respect to a point R.

| Fig. 9-1 | Fig. 9-2 | Fig. 9-3 | Fig. 9-4 |

SYMMETRY OF A LOCUS. The locus of a given equation $f(x, y) = 0$ is symmetric with respect to the x axis if an equivalent equation is obtained when y is replaced by $-y$, is symmetric with respect to the y axis if an equivalent equation is obtained when x is replaced by $-x$, and is symmetric with respect to the origin if an equivalent equation is obtained when x is replaced by $-x$ and y is replaced by $-y$ simultaneously.

An equation whose graph is symmetric with respect to the y axis is called *even*; one whose graph is symmetric with respect to the x axis is *odd*.

EXAMPLE 1. Examine $x^2 + 2y^2 + x = 0$ for symmetry with respect to the coordinate axes and the origin.

When y is replaced by $-y$, we have $x^2 + 2y^2 + x = 0$; the locus is symmetric with respect to the x axis.

When x is replaced by $-x$, we have $x^2 + 2y^2 - x = 0$; the locus is not symmetric with respect to the y axis.

When x is replaced by $-x$ and y by $-y$, we have $x^2 + 2y^2 - x = 0$; the locus is not symmetric with respect to the origin. (See Problem 9.3.)

EXTENT OF A LOCUS. The extent of the locus of $f(x, y) = 0$ is indicated by listing either (a) the real values of x (usually intervals) which yield real values of y and the real values of y which yield real values of x or (b) the real values of x for which the values of y are imaginary or for which no value of y is defined and the real values of y for which the values of x are imaginary or for which no value of x is defined.

EXAMPLE 2. Examine for extent $x^2 + 4y^2 = 4$.

Solve for x: $x^2 = 4 - 4y^2$ and $x = \pm\sqrt{4 - 4y^2} = \pm 2\sqrt{1 - y^2}$. Now x is real provided $1 - y^2 \geq 0$ or $y^2 \leq 1$, i.e., provided $|y| \leq 1$ or $-1 \leq y \leq 1$.

Solve for y: $4y^2 = 4 - x^2$ and $y = \pm\frac{1}{2}\sqrt{4 - x^2}$. Now y is real provided $4 - x^2 \geq 0$ or $x^2 \leq 4$, i.e., provided $|x| \leq 2$ or $-2 \leq x \leq 2$.

The entire curve is confined to the ranges $-2 \leq x \leq 2$ and $-1 \leq y \leq 1$.

EXAMPLE 3. Examine for extent $xy + 3x - y - 6 = 0$.

Solve for x:

$$x = \frac{y + 6}{y + 3}$$

There is a value of x for every real value of y, except $y = -3$, since there the denominator is zero. The locus does not exist when $y = -3$. Note that this is not because the corresponding value of x is imaginary as in Example 2 but is due to the fact that no corresponding value of x is defined.

Solve for y:

$$y = \frac{6 - 3x}{x - 1}$$

There is a value of y for every real value of x, except $x = 1$. Thus, the curve exists for $x \neq 1$ and $y \neq -3$. (See Problem 9.4.)

ASYMPTOTES. The line $x = a$ is a vertical asymptote to the graph of an equation (locus) if, as x gets arbitrarily close to a, the locus gets arbitrarily large.

The line $y = b$ is a horizontal asymptote to a locus if the locus gets arbitrarily close to b as x gets arbitrarily large. See Fig. 9-5.

Fig. 9-5

EXAMPLE 4. Show that the line $y - 2 = 0$ is a horizontal asymptote of the curve $xy - 2x - 1 = 0$.

The locus exists for all $x \neq 0$ and for all $y \neq 2$. Since $y > 0$ for $x > 0$, there is a branch of the locus in the first quadrant. On this branch choose a point, say $A(1, 3)$, and let a point $P(x, y)$, moving to the right from A, trace the locus.

From Table 9.1 which lists a few of the positions assumed by P, it is clear that as P moves to the right, (a) its ordinate y remains greater than 2 and (b) the difference $y - 2$ may be made as small as we please by taking x sufficiently large. For example, if we wish $y - 2 < 1/10^{12}$, we take $x > 10^{12}$; if we wish $y - 2 < 1/10^{999}$, we take $x > 10^{999}$, and so on. The line $y - 2 = 0$ is therefore a horizontal asymptote.

Table 9.1

x	y
1	3
10	2.1
100	2.01
1000	2.001
10 000	2.0001

To find the horizontal asymptotes: Solve the equation of the locus for x and set each *real* linear factor of the denominator, if any, equal to zero.

To find the vertical asymptotes: Solve the equation of the locus for y and set each *real* linear factor of the denominator, if any, equal to zero.

EXAMPLE 5.   It is clear that the locus of Example 2 has neither horizontal nor vertical asymptotes. The locus of Example 3 has $y + 3 = 0$ as a horizontal asymptote and $x - 1 = 0$ as a vertical asymptote.

Solved Problems

9.1 (a) The locus $xy + x^2 = x(y + x) = 0$ consists of the lines $x = 0$ and $y + x = 0$.

(b) The locus $y^3 + xy^2 - xy - x^2 = (x + y)(y^2 - x) = 0$ consists of the line $x + y = 0$ and the parabola $y^2 - x = 0$.

(c) The locus $y^4 + y^2 - x^2 - x = (y^2 - x)(y^2 + x + 1) = 0$ consists of the parabolas $y^2 - x = 0$ and $y^2 + x + 1 = 0$.

9.2   Examine for intercepts.

(a) $4x^2 - 9y^2 = 36$.

Set $y = 0$: $4x^2 = 36$, $x^2 = 9$; the x intercepts are ± 3.
Set $x = 0$: $-9y^2 = 36$, $y^2 = -4$; the y intercepts are imaginary.

(b) $x^2 - 2x = y^2 - 4y + 3$.

Set $y = 0$: $x^2 - 2x - 3 = (x - 3)(x + 1) = 0$; the x intercepts are -1 and 3.
Set $x = 0$: $y^2 - 4y + 3 = (y - 1)(y - 3) = 0$; the y intercepts are 1 and 3.

9.3   Examine for symmetry with respect to the coordinate axis and the origin.

(a) $4x^2 - 9y^2 = 36$.

Replacing y by $-y$, we have $4x^2 - 9y^2 = 36$; the locus is symmetric with respect to the x axis.
Replacing x by $-x$, we have $4x^2 - 9y^2 = 36$; the locus is symmetric with respect to the y axis.
Replacing x by $-x$ and y by $-y$, we have $4x^2 - 9y^2 = 36$; the locus is symmetric with respect to the origin. Note that if a locus is symmetric with respect to the coordinate axes it is automatically symmetric with respect to the origin. It will be shown in the next problem that the converse is not true.

(b) $x^3 - x^2y + y^3 = 0$.

Replacing y by $-y$, we have $x^3 + x^2y - y^3 = 0$; the locus is not symmetric with respect to the x axis.
Replacing x by $-x$, we have $-x^3 - x^2y + y^3 = 0$; the locus is not symmetric with respect to the y axis.
Replacing x by $-x$ and y by $-y$, we have $-x^3 + x^2y - y^3 = -(x^3 - x^2y + y^3) = 0$; the locus is symmetric with respect to the origin.

(c) $x^2 - 4y^2 - 2x + 8y = 0$.

Replacing y by $-y$, we have $x^2 - 4y^2 - 2x - 8y = 0$, replacing x by $-x$, we have $x^2 - 4y^2 + 2x + 8y = 0$, replacing x by $-x$ and y by $-y$, we have $x^2 - 4y^2 + 2x - 8y = 0$; the locus is not symmetric with respect to either axis or the origin.

9.4   Discuss for extent.

(a) $3x + 4y - 12 = 0$.

Solve for y: $y = 3 - 3x/4$. Solve for x: $x = 4 - 4y/3$. The locus exists for all values of x and for all values of y.

(b) $4x^2 + 5y^2 = 20$.

Solve for y: $y^2 = \frac{4}{5}(5 - x^2)$ or $y = \pm\dfrac{2\sqrt{5}}{5}\sqrt{5 - x^2}$

Now y is real when $5 - x^2 \geq 0$ or $x^2 \leq 5$, that is, when $|x| \leq \sqrt{5}$ or $-\sqrt{5} \leq x \leq \sqrt{5}$.

Solve for x: $x^2 = \frac{5}{4}(4 - y^2)$ or $x = \dfrac{\sqrt{5}}{2}\sqrt{4 - y^2}$

Now x is real when $4 - y^2 \geq 0$ or $y^2 \leq 4$, that is, when $|y| \leq 2$ or $-2 \leq y \leq 2$.

(c) $x^2 + y^2 - 4x = 0$.

Solve for y: $y = \pm\sqrt{4x - x^2}$. The locus exists when $4x - x^2 \geq 0$ or $x(4 - x) \geq 0$, that is, when $0 \leq x \leq 4$.
Solve for x: $x^2 - 4x + 4 = 4 - y^2$, $x - 2 = \pm\sqrt{4 - y^2}$, $x = 2 \pm \sqrt{4 - y^2}$. The locus exists when $4 - y^2 \geq 0$ or $y^2 \leq 4$, that is, when $|y| \leq 2$ or $-2 \leq y \leq 2$.

(d) $y^2 - 4x + 8y - 12 = 0$.

Solve for y: $y = -4 \pm 2\sqrt{x + 7}$. Solve for x: $x = \frac{1}{4}(y^2 + 8y - 12)$. The locus exists when $x + 7 \geq 0$, that is, when $x \geq -7$ and for all values of y.

(e) $xy = 4$.

Solve for y: $y = 4/x$. Solve for x: $x = 4/y$. The locus exists for all $x \neq 0$ and for all $y \neq 0$.

(f) $xy - y - x - 2 = 0$.

Solve for y: $y = \dfrac{x + 2}{x - 1}$. Solve for x: $x = \dfrac{y + 2}{y - 1}$. The locus exists for all $x \neq 1$ and for all $y \neq 1$.

(g) $x^2y - x - 4y = 0$.

Solve for y: $y = \dfrac{x}{x^2 - 4}$. Solve for x: $x = \dfrac{1 \pm \sqrt{1 + 16y^2}}{2y}$.

The locus exists for all $x \neq \pm 2$ and for all values of y. At first glance it would seem that $y = 0$ must be excluded because of the denominator in the solution for x. This solution of the given equation as a quadratic in x assumes, however, that $y \neq 0$. When $y = 0$, the given equation becomes $-x = 0$, a linear equation, and $x = 0$.

9.5   Investigate for horizontal and vertical asymptotes.

(a) $3x + 4y - 12 = 0$.

Solve for y: $y = \dfrac{12 - 3x}{4}$. Solve for x: $x = \dfrac{12 - 4y}{3}$.

Since the denominators do not involve the variables, there are neither horizontal nor vertical asymptotes.

(b) $xy = 8$.

Solve for y: $y = 8/x$. Solve for x: $x = 8/y$. Set each denominator equal to zero:

$x = 0$ is the vertical asymptote,

$y = 0$ is the horizontal asymptote.

(c) $xy - y - x - 2 = 0$. [See Prob. 9.4(f).]

Solve for y: $y = \dfrac{x+2}{x-1}$. Solve for x: $x = \dfrac{y+2}{y-1}$.

Set each denominator equal to zero:

$x = 1$ is the vertical asymptote,
$y = 1$ is the horizontal asymptote.

(d) $x^2y - x - 4y = 0$. [See Prob. 9.4(g).]

Solve for y: $y = \dfrac{x}{x^2-4}$. Solve for x: $x = \dfrac{1 \pm \sqrt{1+16y^2}}{2y}$.

Then $x = 2$ and $x = -2$ are vertical asymptotes and $y = 0$ is the horizontal asymptote.

(e) $x^2y - x^2 + 4y = 0$.

Solve for y: $y = \dfrac{x^2}{x^2+4}$. Solve for x: $\pm 2\sqrt{\dfrac{y}{1-y}}$.

There are no vertical asymptotes since when $x^2 + 4 = 0$, x is imaginary. The horizontal asymptote is $y = 1$.

Discuss the following equations and sketch their loci.

9.6 $y^2 = -8x$.

Intercepts: When $y = 0$, $x = 0$ (x intercept); when $x = 0$, $y = 0$ (y intercept).
Symmetry: When y is replaced by $-y$ the equation is unchanged; the locus is symmetric with respect to the x axis.
Extent: Solve for y: $y = \pm 2\sqrt{-2x}$. Solve for x: $x = -y^2/8$. The locus exists when $-2x \geq 0$, i.e., when $x \leq 0$, and for all values of y.

The locus is a parabola with vertex at $(0, 0)$. It may be sketched after locating the following points: $(-1, \pm 2\sqrt{2})$, $(-2, \pm 4)$, and $(-3, \pm 2\sqrt{6})$. See Fig. 9-6.

9.7 $x^2 - 4x + 4y + 8 = 0$.

Intercepts: When $y = 0$, x is imaginary; when $x = 0$, $y = -2$ (y intercept).
Symmetry: The locus is not symmetric with respect to the coordinate axes or the origin.
Extent: Solve for y: $y = \frac{1}{4}(4x - 8 - x^2)$. Solve for x: $x = 2 \pm 2\sqrt{-y-1}$. The locus exists when $-y - 1 \geq 0$, i.e., when $y \leq -1$ and for all values of x.

The locus is a parabola with vertex at $(2, -1)$. Other points on the locus are: $(-2, -5)$, $(4, -2)$, and $(6, -5)$. See Fig. 9-7.

Fig. 9-6   Fig. 9-7

9.8 $x^2 + y^2 - 4x + 6y - 23 = 0$.

Intercepts: When $y = 0$, $x = \dfrac{4 \pm \sqrt{16 + 92}}{2} = 2 \pm 3\sqrt{3}$ (x intercepts); when $x = 0$, $y = \dfrac{-6 \pm \sqrt{36 + 92}}{2} = -3 \pm 4\sqrt{2}$ (y intercepts).

Symmetry: There is no symmetry with respect to the coordinate axes or the origin.

Extent: Solve for y: $y = -3 \pm \sqrt{32 + 4x - x^2}$. Solve for x: $x = 2 \pm \sqrt{27 - 6y - y^2}$. The locus exists when $32 + 4x - x^2 \geq 0$, i.e., when $x^2 - 4x - 32 = (x - 8)(x + 4) \leq 0$ or $-4 \leq x \leq 8$, and when $27 - 6y - y^2 \geq 0$, i.e., when $y^2 + 6y - 27 = (y - 3)(y + 9) \leq 0$ or $-9 \leq y \leq 3$.

Completing the squares, we have

$$(x^2 - 4x + 4) + (y^2 + 6y + 9) = 23 + 4 + 9 = 36 \qquad \text{or} \qquad (x - 2)^2 + (y + 3)^2 = 36$$

the equation of a circle having center at $C(2, -3)$ and radius 6. See Fig. 9-8.

Fig. 9-8   Fig. 9-9

9.9 $4x^2 + 9y^2 = 36$.

Intercepts: When $y = 0$, $x = \pm 3$ (x intercepts); when $x = 0$, $y = \pm 2$ (y intercepts).

Symmetry: The locus is symmetric with respect to the coordinate axes and the origin.

Extent: Solve for y: $y = \pm \frac{2}{3}\sqrt{9 - x^2}$. Solve for x: $x = \pm \frac{3}{2}\sqrt{4 - y^2}$. The locus exists when $9 - x^2 \geq 0$ or $x^2 \leq 9$, i.e., when $-3 \leq x \leq 3$, and when $4 - y^2 \geq 0$ or $y^2 \leq 4$, i.e., when $-2 \leq y \leq 2$.

Since the locus is symmetric with respect to both the axes, only sufficient points to sketch the portion of the locus in the first quadrant are needed. Two such points are $(1, 4\sqrt{2}/3)$ and $(2, 2\sqrt{5}/3)$. The locus is called an *ellipse*. See Fig. 9-9.

9.10 $9x^2 - 4y^2 = 36$.

Intercepts: When $y = 0$, $x = \pm 2$ (x intercepts); when $x = 0$, y is imaginary.

Symmetry: The locus is symmetric with respect to the coordinate axes and the origin.

Extent: Solve for y: $y = \pm \frac{3}{2}\sqrt{x^2 - 4}$. Solve for x: $x = \pm \frac{2}{3}\sqrt{y^2 + 9}$. The locus exists when $y^2 + 9 \geq 0$, that is, for all values of y and when $x^2 - 4 \geq 0$ or $x^2 \geq 4$, that is, when $x \leq -2$ and $x \geq 2$.

The locus consists of two separate pieces and is not closed. The portion in the first quadrant has been sketched using the points $(3, 3\sqrt{5}/2)$, $(4, 3\sqrt{3})$, and $(5, 3\sqrt{21}/2)$. The locus is called a *hyperbola*. See Fig. 9-10.

Fig. 9-10   Fig. 9-11

9.11 $xy - y - x - 2 = 0$.

     Intercepts: The x intercept is -2; the y intercept is -2.

     Symmetry: There is no symmetry with respect to the coordinate axes or the origin.

     Extent: $y = \dfrac{x+2}{x-1}$, $x = \dfrac{y+2}{y-1}$. The locus exists for $x \neq 1$ and for $y \neq 1$.

     Asymptotes: $x = 1$, $y = 1$.

To sketch the locus, first draw in the asymptotes $x = 1$ and $y = 1$ (dotted lines). While the asymptotes are *not* a part of the locus, they serve as very convenient guide lines. Since the locus does not exist for $x = 1$ and $y = 1$, it does not cross the asymptotes. Since there is one value of y for each value of $x \neq 1$, that is, since y is single-valued, the locus appears in only two of the four regions into which the plane is separated by the asymptotes.

From Table 9.2 it is evident that the locus lies in the region to the right of the vertical asymptote and above the horizontal asymptote (see the portion of the table to the right of the double line) and in the region to the left of the vertical asymptote and below the horizontal asymptote (see the portion of the table to the left of the double line). The locus is shown in Fig. 9-11; note that it is symmetric with respect to $(1, 1)$ the point of intersection of the asymptotes.

Table 9.2

x	-10	-4	-3	-2	-1	0	$\frac{1}{2}$	$\frac{3}{4}$	$\frac{5}{4}$	$\frac{3}{2}$	2	10
y	$\frac{8}{11}$	$\frac{2}{5}$	$\frac{1}{4}$	0	$-\frac{1}{2}$	-2	-5	-11	13	7	4	$\frac{4}{3}$

9.12 $x^2 y - x^2 - 4y = 0$.

     Intercepts: The x intercept is 0, the y intercept is 0.

     Symmetry: The locus is symmetric with respect to the y axis.

     Extent:

$$y = \frac{x^2}{x^2 - 4} \qquad x = \pm 2 \sqrt{\frac{y}{y-1}}$$

The locus exists for $x \neq \pm 2$ and for all y such that $\dfrac{y}{y-1} \geq 0$. To solve this inequality, locate on the number scale the value of y for which the numerator is equal to zero, that is, $y = 0$, and the value of y for which the denominator is zero, that is, $y = 1$. See Fig. 9-12.

Fig. 9-12

For $y < 0$, $\dfrac{y}{y-1} > 0$; for $0 < y < 1$, $\dfrac{y}{y-1} < 0$; for $y > 1$, $\dfrac{y}{y-1} > 0$. Thus the locus exists for $y \leq 0$ and for $y > 1$.

     Asymptotes: $x = \pm 2$, $y = 1$.

The asymptotes divide the plane into six regions. Since the locus does not exist when $x = \pm 2$ and when $y = 1$ (that is, does not cross an asymptote) and since y is single-valued, the locus appears in only three of these regions. By means of Table 9.3 the locus is sketched in Fig. 9-13. Note that only half of the table is necessary since the locus is symmetric with respect to the y axis.

Table 9.3

x	-10	-5	-4	-3	$-\frac{5}{2}$	$-\frac{7}{4}$	$-\frac{3}{2}$	-1	0	1	$\frac{3}{2}$	$\frac{7}{4}$
y	$\frac{25}{24}$	$\frac{25}{21}$	$\frac{4}{3}$	$\frac{9}{5}$	$\frac{25}{9}$	$-\frac{49}{15}$	$-\frac{9}{7}$	$-\frac{1}{3}$	0	$-\frac{1}{3}$	$-\frac{9}{7}$	$-\frac{49}{15}$

x	$\frac{5}{2}$	3	4	5	10
y	$\frac{25}{9}$	$\frac{9}{5}$	$\frac{4}{3}$	$\frac{25}{21}$	$\frac{25}{24}$

Fig. 9-13   Fig. 9-14

9.13 $x^2y - x^2 + 4y = 0.$

Intercepts: The x intercept is 0, the y intercept is 0.
Symmetry: The locus is symmetric with respect to the y axis.

Extent:
$$y = \frac{x^2}{x^2 + 4} \qquad x = \pm 2\sqrt{\frac{y}{1-y}}$$

The locus exists for all values of x, and for y such that $\dfrac{y}{1-y} \geq 0$. Using Fig. 9-15 (see

Problem 9.12), the locus exists for $0 \leq y < 1$.
Asymptotes: $y = 1.$

$$\frac{y}{1-y} < 0 \qquad \frac{y}{1-y} > 0 \qquad \frac{y}{1-y} < 0$$

Fig. 9-15

Since the locus exists only for $0 \leq y < 1$, it lies entirely below its asymptote. The locus is sketched in Fig. 9-14 using Table 9.4.

Table 9.4

x	-10	-5	-3	-2	-1	0	1	2	3	5	10
y	$\frac{25}{26}$	$\frac{25}{29}$	$\frac{9}{13}$	$\frac{1}{2}$	$\frac{1}{5}$	0	$\frac{1}{5}$	$\frac{1}{2}$	$\frac{9}{13}$	$\frac{25}{29}$	$\frac{25}{26}$

Only half of the table is necessary since the locus is symmetric with respect to the y axis.

Supplementary Problems

9.14   Discuss and sketch.

(a) $x^2 - 4y^2 = 0$

(b) $x^2 + 2xy + y^2 = 4$

(c) $y = 9x^2$

(d) $y^2 = 6x - 3$

(e) $y^2 = 4 - 2x$

(f) $x^2 + y^2 = 16$

(g) $x^2 + y^2 = 0$

(h) $4x^2 + 9y^2 = 36$

(i) $9x^2 - 4y^2 = 36$

(j) $9x^2 - 4y^2 + 36 = 0$

(k) $xy = -4$

(l) $x^2y = 4$

(m) $xy^2 = -9$

(n) $y^3 + xy^2 = 2xy - 2x^2$

(o) $y = x^3$

(p) $xy - x + 4 = 0$

(q) $xy - 3x - y = 0$

(r) $x^2 + xy + y - 2 = 0$

(s) $x^2y - x - 4y = 0$

(t) $x^2y - 4xy + 3y - x - 2 = 0$

(u) $x^2y - x^2 + xy + 3x - 2 = 0$

(v) $x^3 + xy^2 - y^2 = 0$

Chapter 10

Simultaneous Equations Involving Quadratics

ONE LINEAR AND ONE QUADRATIC EQUATION

Procedure: Solve the linear equation for one of the two unknowns (your choice) and substitute in the quadratic equation. Since this results in a quadratic equation in one unknown, the system can always be solved.

EXAMPLE 1. Solve the system $\begin{cases} 4x^2 + 3y^2 = 16 \\ 5x + y = 7 \end{cases}$

Solve the linear equation for y: $y = 7 - 5x$. Substitute in the quadratic equation:

$$4x^2 + 3(7 - 5x)^2 = 16$$

$$4x^2 + 3(49 - 70x + 25x^2) = 16$$

$$79x^2 - 210x + 131 = (x - 1)(79x - 131) = 0$$

and $x = 1, \frac{131}{79}$.

When $x = 1$, $y = 7 - 5x = 2$; when $x = \frac{131}{79}$, $y = -\frac{102}{79}$. The solutions are $x = 1$, $y = 2$ and $x = \frac{131}{79}$, $y = -\frac{102}{79}$.

The locus of the linear equation is the straight line and the locus of the quadratic equation is the ellipse in Fig. 10-1. (See Problems 10.1–10.2.)

(1, 2)

$\left(\frac{131}{79}, -\frac{102}{79}\right)$

Fig. 10-1

TWO QUADRATIC EQUATIONS. In general, solving a system of two quadratic equations in two unknowns involves solving an equation of the fourth degree in one of the unknowns. Since the solution of the general equation of the fourth degree in one unknown is beyond the scope of this book, only those systems which require the solution of a quadratic equation in one unknown will be treated here.

TWO QUADRATIC EQUATIONS OF THE FORM $ax^2 + by^2 = c$

Procedure: Eliminate one of the unknowns by the method of addition of Chapter 6.

EXAMPLE 2. Solve the system $\begin{cases} 4x^2 + 9y^2 = 72 & (10.1) \\ 3x^2 - 2y^2 = 19 & (10.2) \end{cases}$

Multiply (10.1) by 2: $8x^2 + 18y^2 = 144$
Multiply (10.2) by 9: $27x^2 - 18y^2 = 171$
                    Add: $\overline{35x^2 \qquad\quad = 315}$

Then $x^2 = 9$ and $x = \pm 3$.
  When $x = 3$, (10.1) gives $9y^2 = 72 - 4x^2 = 72 - 36 = 36$, $y^2 = 4$, and $y = \pm 2$.
  When $x = -3$, (10.1) gives $9y^2 = 72 - 36 = 36$, $y^2 = 4$, and $y = \pm 2$.
  The four solutions $x = 3$, $y = 2$; $x = 3$, $y = -2$; $x = -3$, $y = 2$; $x = -3$, $y = -2$ may also be written as $x = \pm 3$, $y = \pm 2$; $x = \pm 3$, $y = \mp 2$. By convention, we read the two upper signs and the two lower signs in the latter form.
  The ellipse and the hyperbola intersect in the points $(3, 2)$, $(3, -2)$, $(-3, 2)$, $(-3, -2)$. See Fig. 10-2. (See Problems 10.3–10.4.)

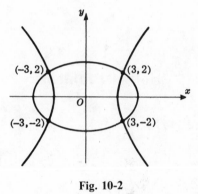

Fig. 10-2

TWO QUADRATIC EQUATIONS, ONE HOMOGENEOUS.

TWO QUADRATIC EQUATIONS, ONE HOMOGENEOUS. An expression, as $2x^2 - 3xy + y^2$, whose terms are all of the same degree in the variables, is called *homogeneous*. A homogeneous expression equated to zero is called a *homogeneous equation*. A homogeneous quadratic equation in two unknowns can always be solved for one of the unknowns in terms of the other.

EXAMPLE 3. Solve the system $\begin{cases} x^2 - 3xy + 2y^2 = 0 & (10.3) \\ 2x^2 + 3xy - y^2 = 13 & (10.4) \end{cases}$

Solve (10.3) for x in terms of y: $(x - y)(x - 2y) = 0$ and $x = y$, $x = 2y$.
Solve the systems (see Example 1):

$$\begin{cases} 2x^2 + 3xy - y^2 = 13 \\ \qquad\qquad\quad x = y \end{cases} \qquad \begin{cases} 2x^2 + 3xy - y^2 = 13 \\ \qquad\qquad\quad x = 2y \end{cases}$$

$$2y^2 + 3y^2 - y^2 = 4y^2 = 13 \qquad\qquad 8y^2 + 6y^2 - y^2 = 13y^2 = 13$$

$$y^2 = \tfrac{13}{4}, \; y = \pm \frac{\sqrt{13}}{2} \qquad\qquad y^2 = 1, \; y = \pm 1$$

$$\text{Then } x = y = \pm\sqrt{13}/2. \qquad\qquad \text{Then } x = 2y = \pm 2.$$

The solutions are $x = \sqrt{13}/2$, $y = \sqrt{13}/2$; $x = -\sqrt{13}/2$, $y = -\sqrt{13}/2$; $x = 2$, $y = 1$; $x = -2$, $y = -1$ or $x = \pm\sqrt{13}/2$, $y = \pm\sqrt{13}/2$; $x = \pm 2$, $y = \pm 1$. (See Problem 10.5.)

TWO QUADRATIC EQUATIONS OF THE FORM $ax^2 + bxy + cy^2 = d$

Procedure: Combine the two given equations to obtain a homogeneous equation. Solve, as in Example 3, the system consisting of this homogeneous equation and either of the given equations. (See Problems 10.6–10.7.)

TWO QUADRATIC EQUATIONS, EACH SYMMETRICAL IN x **AND** y. An equation, as $2x^2 - 3xy + 2y^2 + 5x + 5y = 1$, which is unchanged when the two unknowns are interchanged is called a *symmetrical equation.*

Procedure: Substitute $x = u + v$ and $y = u - v$ and then eliminate v^2 from the resulting equations. (See Problem 10.8.)

Frequently a careful study of a given system will reveal some special device for solving it. (See Problems 10.9–10.13.)

Solved Problems

10.1   Solve the system $\begin{cases} 2y^2 - 3x = 0 & (1) \\ 4y - x = 6 & (2) \end{cases}$

Solve (2) for x: $x = 4y - 6$. Substitute in (1):

$$2y^2 - 3(4y - 6) = 2(y - 3)^2 = 0 \qquad y = 3, 3$$

When $y = 3$: $x = 4y - 6 = 12 - 6 = 6$. The solutions are $x = 6$, $y = 3$; $x = 6$, $y = 3$. The straight line is tangent to the parabola (see Fig. 10-3) at $(6, 3)$.

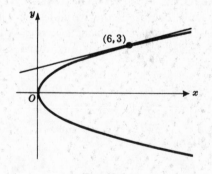

Fig. 10-3

10.2   Solve the system $\begin{cases} y^2 - 4y - 3x + 1 = 0 & (1) \\ 3y - 4x = 7 & (2) \end{cases}$

Solve (2) for x: $x = \frac{1}{4}(3y - 7)$. Substitute in (1):

$$y^2 - 4y - \tfrac{3}{4}(3y - 7) + 1 = 0$$

$$4y^2 - 16y - 9y + 21 + 4 = 4y^2 - 25y + 25 = (y - 5)(4y - 5) = 0 \qquad \text{or} \qquad y = 5 \qquad \text{and} \qquad y = \tfrac{5}{4}$$

When $y = 5$, $x = \frac{1}{4}(3y - 7) = 2$; when $y = \frac{5}{4}$, $x = \frac{1}{4}(3y - 7) = -\frac{13}{16}$. The solutions are $x = 2$, $y = 5$; $x = -\frac{13}{16}$, $y = \frac{5}{4}$. The straight line intersects the parabola in the points $(2, 5)$ and $(-\frac{13}{16}, \frac{5}{4})$.

10.3   Solve the system $\begin{cases} 3x^2 - y^2 = 27 & (1) \\ x^2 - y^2 = -45 & (2) \end{cases}$

Subtract: $2x^2 = 72$, $x^2 = 36$, and $x = \pm 6$.

When $x = 6$, $y^2 = x^2 + 45 = 36 + 45 = 81$, and $y = \pm 9$.

When $x = -6$, $y^2 = x^2 + 45 = 36 + 45 = 81$, and $y = \pm 9$. The solutions are $x = \pm 6$, $y = \pm 9$; $x = \pm 6$, $y = \mp 9$. The two hyperbolas intersect in the points $(6, 9)$, $(-6, 9)$, $(-6, -9)$, and $(6, -9)$.

10.4   Solve the system $\begin{cases} 5x^2 + 3y^2 = 92 & (1) \\ 2x^2 + 5y^2 = 52 & (2) \end{cases}$

Multiply (1) by 5: $25x^2 + 15y^2 = 460$
Multiply (2) by -3: $-6x^2 - 15y^2 = -156$
Add: $\overline{19x^2 = 304}$ $x^2 = 16$ and $x = \pm 4$

When $x = \pm 4$: $3y^2 = 92 - 5x^2 = 92 - 80 = 12$; $y^2 = 4$ and $y = \pm 2$. The solutions are $x = \pm 4$, $y = \pm 2$; $x = \pm 4$, $y = \mp 2$. See Fig. 10-4.

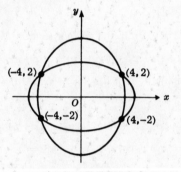

Fig. 10-4

10.5   Solve the system $\begin{cases} x^2 + 4xy = 0 & (1) \\ x^2 - xy + y^2 = 21 & (2) \end{cases}$

Solve (1) for x: $x(x + 4y) = 0$ and $x = 0$, $x = -4y$.

Solve the systems

$\begin{cases} x^2 - xy + y^2 = 21 \\ x = 0 \end{cases}$ $\begin{cases} x^2 - xy + y^2 = 21 \\ x = -4y \end{cases}$

$y^2 = 21, \qquad y = \pm\sqrt{21}$ $y^2 = 1, \qquad y = \pm 1; \qquad x = -4y = \mp 4$

The solutions are $x = 0$, $y = \pm\sqrt{21}$; $x = \pm 4$, $y = \mp 1$.

10.6   Solve the system $\begin{cases} 3x^2 + 8y^2 = 140 & (1) \\ 5x^2 + 8xy = 84 & (2) \end{cases}$

Multiply (1) by -3: $-9x^2 - 24y^2 = -420$
Multiply (2) by 5: $25x^2 + 40xy = 420$
Add: $\overline{16x^2 + 40xy - 24y^2 = 0}$

Then

$$8(2x^2 + 5xy - 3y^2) = 8(2x - y)(x + 3y) = 0 \qquad \text{and} \qquad x = \tfrac{1}{2}y, \ x = -3y.$$

Solve the systems

$\begin{cases} 3x^2 + 8y^2 = 140 \\ x = \tfrac{1}{2}y \end{cases}$ $\begin{cases} 3x^2 + 8y^2 = 140 \\ x = -3y \end{cases}$

$\tfrac{3}{4}y^2 + 8y^2 = \tfrac{35}{4}y^2 = 140$ $27y^2 + 8y^2 = 35y^2 = 140$

$y^2 = 16, \qquad y = \pm 4, \qquad x = \tfrac{1}{2}y = \pm 2$ $y^2 = 4, \ y = \pm 2; \ x = -3y = \mp 6$

The solutions are $x = \pm 2$, $y = \pm 4$; $x = \mp 6$, $y = \pm 2$.

10.7   Solve the system $\begin{cases} x^2 - 3xy + 2y^2 = 15 & (1) \\ 2x^2 + y^2 = 6 & (2) \end{cases}$

Multiply (1) by -2: $-2x^2 + 6xy - 4y^2 = -30$
Multiply (2) by 5: $\underline{10x^2 \qquad + 5y^2 = 30}$
Add: $8x^2 + 6xy + y^2 = (4x + y)(2x + y) = 0.$ Then $y = -4x$ and $y = -2x.$

Solve the systems

$$\begin{cases} 2x^2 + y^2 = 6 \\ y = -4x \end{cases} \qquad\qquad \begin{cases} 2x^2 + y^2 = 6 \\ y = -2x \end{cases}$$

$2x^2 + 16x^2 = 18x^2 = 6,\ x^2 = 1/3;$ $2x^2 + 4x^2 = 6x^2 = 6,\ x^2 = 1;$

$x = \pm\sqrt{3}/3$ and $y = -4x = \mp 4\sqrt{3}/3$ $x = \pm 1$ and $y = -2x = \mp 2$

The solutions are $x = \pm\sqrt{3}/3,\ y = \mp 4\sqrt{3}/3;\ x = \pm 1,\ y = \mp 2.$

10.8   Solve the system $\begin{cases} x^2 + y^2 + 3x + 3y = 8 \\ xy + 4x + 4y = 2 \end{cases}$

Substitute $x = u + v,\ y = u - v$ in the given system:

$$(u + v)^2 + (u - v)^2 + 3(u + v) + 3(u - v) = 2u^2 + 2v^2 + 6u = 8 \quad (1)$$

$$(u + v)(u - v) + 4(u + v) + 4(u - v) = u^2 - v^2 + 8u = 2 \quad (2)$$

Add (1) and $2(2)$:

$$4u^2 + 22u - 12 = 2(2u - 1)(u + 6) = 0; \qquad u = \tfrac{1}{2}, -6.$$

For $u = \tfrac{1}{2}$, (2) yields $v^2 = u^2 + 8u - 2 = \tfrac{1}{4} + 4 - 2 = \tfrac{9}{4};\ v = \pm\tfrac{3}{2}.$

When $u = \tfrac{1}{2},\ v = \tfrac{3}{2}$: $x = u + v = 2,\ y = u - v = -1.$
When $u = \tfrac{1}{2},\ v = -\tfrac{3}{2}$: $x = u + v = -1,\ y = u - v = 2.$

For $u = -6$, (2) yields $v^2 = u^2 + 8u - 2 = 36 - 48 - 2 = -14;\ v = \pm i\sqrt{14}.$

When $u = -6,\ v = i\sqrt{14}$: $x = u + v = -6 + i\sqrt{14},\ y = u - v = -6 - i\sqrt{14}.$
When $u = -6,\ v = -i\sqrt{14}$: $x = u + v = -6 - i\sqrt{14},\ y = u - v = -6 + i\sqrt{14}.$

The solutions are $x = 2,\ y = -1;\ x = -1,\ y = 2;\ x = -6 \pm i\sqrt{14},\ y = -6 \mp i\sqrt{14}.$

10.9   Solve the system $\begin{cases} x^2 + y^2 = 25 & (1) \\ xy = 12 & (2) \end{cases}$

Multiply (2) by 2 and add to (1): $x^2 + 2xy + y^2 = 49$ or $x + y = \pm 7.$
Multiply (2) by -2 and add to (1): $x^2 - 2xy + y^2 = 1$ or $x - y = \pm 1.$
Solve the systems

$$\begin{cases} x + y = 7 \\ \underline{x - y = 1} \\ 2x = 8; \quad x = 4 \end{cases} \qquad\qquad \begin{cases} x + y = 7 \\ \underline{x - y = -1} \\ 2x = 6; \quad x = 3 \end{cases}$$

$$y = 7 - x = 3 \qquad\qquad\qquad y = 7 - x = 4$$

$$\begin{cases} x + y = -7 \\ \underline{x - y = 1} \\ 2x = -6; \quad x = -3 \end{cases} \qquad\qquad \begin{cases} x + y = -7 \\ \underline{x - y = -1} \\ 2x = -8; \quad x = -4 \end{cases}$$

$$y = -7 - x = -4 \qquad\qquad\qquad y = -7 - x = -3$$

The solutions are $x = \pm 4,\ y = \pm 3;\ x = \pm 3,\ y = \pm 4.$ See Fig. 10-5.

Alternate Solution.   Solve (2) for $y = 12/x$ and substitute in (1). The resulting quartic $x^4 - 25x^2 + 144 = 0$ can be factored readily.

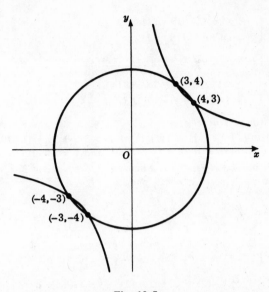

Fig. 10-5

10.10   Solve the system $\begin{cases} x^2 - xy - 12y^2 = 8 & (1) \\ x^2 + xy - 10y^2 = 20 & (2) \end{cases}$

This system may be solved by the procedure used in Problems 10.6 and 10.7. Here we give an alternate solution.

Procedure: Substitute $y = mx$ in the given equations to obtain a system in the unknowns m and x: then eliminate x to obtain a quadratic in m.

Put $y = mx$ in (1) and (2):

$$x^2 - mx^2 - 12m^2x^2 = x^2(1 - m - 12m^2) = 8$$
$$x^2 + mx^2 - 10m^2x^2 = x^2(1 + m - 10m^2) = 20$$

Now

$$x^2 = \frac{8}{1 - m - 12m^2} \qquad \text{and} \qquad x^2 = \frac{20}{1 + m - 10m^2}$$

so that

$$\frac{8}{1 - m - 12m^2} = \frac{20}{1 + m - 10m^2}$$

$$8 + 8m - 80m^2 = 20 - 20m - 240m^2$$

$$160m^2 + 28m - 12 = 4(5m - 1)(8m + 3) = 0 \qquad \text{and} \qquad m = \tfrac{1}{5}, \; m = -\tfrac{3}{8}$$

When $m = \tfrac{1}{5}$: $x^2 = \dfrac{8}{1 - m - 12m^2} = 25$; $x = \pm 5$, $y = mx = \dfrac{1}{5}(\pm 5) = \pm 1$.

When $m = -\tfrac{3}{8}$:

$$x^2 = \frac{8}{1 - m - 12m^2} = -\frac{128}{5}; \qquad x = \pm\frac{8i\sqrt{10}}{5}, \qquad y = mx = \mp\frac{3i\sqrt{10}}{5}.$$

The solutions are $x = \pm 5$, $y = \pm 1$; $x = \pm\dfrac{8i\sqrt{10}}{5}$, $y = \mp\dfrac{3i\sqrt{10}}{5}$.

10.11 Solve the system $\begin{cases} x^3 - y^3 = 19 & (1) \\ x^2 + xy + y^2 = 19 & (2) \end{cases}$

Divide (1) by (2): $x - y = 1$.

Solve the system $\begin{cases} x^2 + xy + y^2 = 19 & (2) \\ x - y = 1 & (3) \end{cases}$

Solve (3) for x: $x = y + 1$.
Substitute in (2): $(y + 1)^2 + (y + 1)y + y^2 = 3y^2 + 3y + 1 = 19$.
Then $3y^2 + 3y - 18 = 3(y + 3)(y - 2) = 0$ and $y = -3, 2$.
When $y = -3$, $x = y + 1 = -2$; when $y = 2$, $x = y + 1 = 3$.
The solutions are $x = -2$, $y = -3$; $x = 3$, $y = 2$.

10.12 Solve the system $\begin{cases} (2x - y)^2 - 4(2x - y) = 5 & (1) \\ x^2 - y^2 = 3 & (2) \end{cases}$

Factor (1): $(2x - y)^2 - 4(2x - y) - 5 = (2x - y - 5)(2x - y + 1) = 0$. Then $2x - y = 5$ and $2x - y = -1$.
Solve the systems

$\begin{cases} x^2 - y^2 = 3 \\ 2x - y = 5 \end{cases}$ $\begin{cases} x^2 - y^2 = 3 \\ 2x - y = -1 \end{cases}$

$y = 2x - 5$ $y = 2x + 1$
$x^2 - (2x - 5)^2 = 3$ $x^2 - (2x + 1)^2 = 3$
$3x^2 - 20x + 28 = (x - 2)(3x - 14) = 0$ $3x^2 + 4x + 4 = 0$

$x = 2, \tfrac{14}{3}$ $x = \dfrac{-4 \pm \sqrt{16 - 48}}{6} = \dfrac{-2 \pm 2i\sqrt{2}}{3}$

When $x = 2$, $y = 2x - 5 = -1$. $y = 2x + 1 = \dfrac{-1 \pm 4i\sqrt{2}}{3}$

When $x = \tfrac{14}{3}$, $y = 2x - 5 = \tfrac{13}{3}$.

The solutions are $x = 2$, $y = -1$; $x = \tfrac{14}{3}$, $y = \tfrac{13}{3}$; $x = \dfrac{-2 \pm 2i\sqrt{2}}{3}$, $y = \dfrac{-1 \pm 4i\sqrt{2}}{3}$.

10.13 Solve the system $\begin{cases} 5/x^2 + 3/y^2 = 32 & (1) \\ 4xy = 1 & (2) \end{cases}$

Write (1) as $3x^2 + 5y^2 = 32x^2y^2 = 2(4xy)^2$. Substitute ($2$): $3x^2 + 5y^2 = 2(1)^2 = 2$ (3)
Subtract 2(2) from (3): $3x^2 - 8xy + 5y^2 = 0$. Then $(x - y)(3x - 5y) = 0$ and $x = y$, $x = 5y/3$.
Solve the systems

$\begin{cases} 4xy = 1 \\ x = y \end{cases}$ $\begin{cases} 4xy = 1 \\ x = 5y/3 \end{cases}$

$4y^2 = 1$; $y = \pm\tfrac{1}{2}$ $\dfrac{20}{3}y^2 = 1$, $y^2 = \dfrac{3}{20} = \dfrac{15}{100}$;

and $x = y = \pm\tfrac{1}{2}$. $y = \pm\dfrac{\sqrt{15}}{10}$ and $x = \dfrac{5}{3}y = \pm\dfrac{\sqrt{15}}{6}$.

The solutions are $x = \pm\tfrac{1}{2}$, $y = \pm\tfrac{1}{2}$; $x = \pm\dfrac{\sqrt{15}}{6}$, $y = \pm\dfrac{\sqrt{15}}{10}$.

Supplementary Problems

Solve.

10.14 $\begin{cases} xy + y^2 = 5 \\ 2x + 3y = 7 \end{cases}$ *Ans.* $x = -4, y = 5$
$x = \frac{1}{2}, y = 2$

10.15 $\begin{cases} y = x^2 - x - 1 \\ y = 2x + 3 \end{cases}$ *Ans.* $x = -1, y = 1$
$x = 4, y = 11$

10.16 $\begin{cases} 3x^2 - 7y^2 = 12 \\ x - 3y = -2 \end{cases}$ *Ans.* $x = -2, y = 0$
$x = \frac{17}{5}, y = \frac{9}{5}$

10.17 $\begin{cases} x^2 + 3y^2 = 43 \\ 3x^2 + y^2 = 57 \end{cases}$ *Ans.* $x = \pm 4, y = \pm 3$
$x = \pm 4, y = \mp 3$

10.18 $\begin{cases} 9x^2 + y^2 = 90 \\ x^2 + 9y^2 = 90 \end{cases}$ *Ans.* $x = \pm 3, y = \pm 3$
$x = \pm 3, y = \mp 3$

10.19 $\begin{cases} 2/x^2 - 3/y^2 = 5 \\ 1/x^2 + 2/y^2 = 6 \end{cases}$ *Ans.* $x = \pm \frac{1}{2}, y = \pm 1$
$x = \pm \frac{1}{2}, y = \mp 1$

10.20 $\begin{cases} x^2 - xy + y^2 = 28 \\ 2x^2 + 3xy - 2y^2 = 0 \end{cases}$ *Ans.* $x = \pm 4, y = \mp 2$
$x = \pm 2\sqrt{21}/3, y = \pm 4\sqrt{21}/3$

10.21 $\begin{cases} x^2 - xy - 12y^2 = 0 \\ x^2 + xy - 10y^2 = 20 \end{cases}$ *Ans.* $x = \pm 4\sqrt{2}, y = \pm\sqrt{2}$
$x = \mp 3i\sqrt{5}, y = \pm i\sqrt{5}$

10.22 $\begin{cases} 6x^2 + 3xy + 2y^2 = 24 \\ 3x^2 + 2xy + 2y^2 = 18 \end{cases}$ *Ans.* $x = \pm 2, y = \mp 3$
$x = \pm\sqrt{30}/5, y = \pm 2\sqrt{30}/5$

10.23 $\begin{cases} y^2 = 4x - 8 \\ y^2 = -6x + 32 \end{cases}$ *Ans.* $x = 4, y = \pm 2\sqrt{2}$

10.24 $\begin{cases} x^2 - y^2 = 16 \\ y^2 = 2x - 1 \end{cases}$ *Ans.* $x = 5, y = \pm 3$
$x = -3, y = \pm i\sqrt{7}$

10.25 $\begin{cases} 2x^2 + y^2 = 6 \\ x^2 + y^2 + 2x = 3 \end{cases}$ *Ans.* $x = -1, y = \pm 2$
$x = 3, y = \pm 2i\sqrt{3}$

10.26 $\begin{cases} x^2 + y^2 - 2x - 2y = 12 \\ xy = 6 \end{cases}$ *Ans.* $x = 3 \pm \sqrt{3}, y = 3 \mp \sqrt{3}$
$x = -2 \pm i\sqrt{2}, y = -2 \mp i\sqrt{2}$

10.27 $\begin{cases} x^3 - y^3 = 28 \\ x - y = 4 \end{cases}$ *Ans.* $x = 1, y = -3$
$x = 3, y = -1$

10.28 $\begin{cases} x + y + 3\sqrt{x + y} = 18 \\ x - y - 2\sqrt{x - y} = 15 \end{cases}$ **Hint:** Let $\sqrt{x + y} = u,\ \sqrt{x - y} = v.$ *Ans.* $x = 17,\ y = -8$

10.29 Two numbers differ by 2 and their squares differ by 48. Find the numbers.   *Ans.*   11, 13.

10.30 The sum of the circumference of two circles is 88 cm and the sum of their areas is $\frac{2200}{7}$ cm^2, when $\frac{22}{7}$ is used for π. Find the radius of each circle.   *Ans.*   6 cm, 8 cm

10.31 A party costing \$30 is planned. It is found that by adding three more to the group, the cost per person would be reduced by 50 cents. For how many was the party originally plannned?   *Ans.*   12

Chapter 11

Logarithms

THE LOGARITHM OF A POSITIVE NUMBER N to a given base b (written $\log_b N$) is the exponent of the power to which b must be raised to produce N. It will be understood throughout this chapter that b is positive and different from 1.

EXAMPLE 1.
(a)   Since $9 = 3^2$, $\log_3 9 = 2$.
(b)   Since $64 = 4^3$, $\log_4 64 = 3$.
(c)   Since $64 = 2^6$, $\log_2 64 = 6$.
(d)   Since $1000 = 10^3$, $\log_{10} 1000 = 3$.
(e)   Since $0.01 = 10^{-2}$, $\log_{10} 0.01 = -2$.

(See Problems 11.1–11.3.)

FUNDAMENTAL LAWS OF LOGARITHMS

1.   The logarithm of the product of two or more positive numbers is equal to the sum of the logarithms of the several numbers. For example,

$$\log_b(P \cdot Q \cdot R) = \log_b P + \log_b Q + \log_b R$$

2.   The logarithm of the quotient of two positive numbers is equal to the logarithm of the dividend minus the logarithm of the divisor. For example,

$$\log_b \frac{P}{Q} = \log_b P - \log_b Q$$

3.   The logarithm of a power of a positive number is equal to the logarithm of the number, multiplied by the exponent of the power. For example,

$$\log_b P^n = n \log_b P$$

4.   The logarithm of a root of a positive number is equal to the logarithm of the number, divided by the index of the root. For example,

$$\log_b \sqrt[n]{P} = \frac{1}{n} \log_b P$$

(See Problems 11.4–11.7.)

IN NUMERICAL COMPUTATIONS the most useful base for a system of logarithms is 10. Such logarithms are called *common logarithms*.

The common logarithm of a positive number N (hereafter written $\log N$ instead of $\log_{10} N$) consists of two parts: an integer (positive, negative, or zero) called the *characteristic* and a positive decimal fraction called the *mantissa*.

The characteristic of the common logarithm of any number equal to or greater than 1 is one less than the number of digits to the left of the decimal point in the given number. For example, the characteristics of the common logarithms of 132, 54, and 2380.6 are 2, 1, 3 respectively.

The characteristic of the common logarithm of any positive number smaller than 1 is negative. However, because of greater convenience in computing, we shall write $-1 = 9 - 10$, $-2 = 8 - 10$, and so on. This latter form of the characteristic is obtained directly by subtracting the number of zeros immediately following the decimal point from 9 and affixing -10. For example, the characteristics of the common logarithms of 0.000436, 0.5, and 0.07086 are $6 - 10$, $9 - 10$, $8 - 10$, respectively. (See Problem 11.8.)

The mantissa of the common logarithm of a positive number is usually a continuous decimal fraction. The discussion below refers to a four-place table of mantissas.

TO FIND THE COMMON LOGARITHM OF A GIVEN POSITIVE NUMBER

(a)   Write down the characteristic.

(b_1)   When the given number contains three or fewer significant digits, read the mantissa directly from the table.

> **EXAMPLE 2.**   Find (a) log 2.6, (b) 32.8.
> (a)   The characteristic is 0. To find the mantissa, consider the number as 2.60 and locate in the row opposite 26 the entry 4150 in the column headed 0. Thus, log 2.6 = 0.4150.
>
> (b)   The characteristic is 1. To find the mantissa, locate in the row opposite 32 the entry 5159 in the column headed 8. Thus, log 32.8 = 1.5159.

(b_2)   When the given number contains four digits interpolate, using the method of proportional parts.

> **EXAMPLE 3.**   Find (a) log 5462, (b) log 0.08367.
> (a)   The characteristic is 3. For the mantissa, we have
>
> $$\begin{aligned} \text{Mantissa of log 5460} &= .7372 \\ \text{Mantissa of log 5470} &= \underline{.7380} \\ \text{Tabular difference} &= .0008 \\ 0.2 \text{ (tabular difference)} &= .00016 \\ \text{Mantissa of log 5462} &= .7372 + .00016 = .73736 \end{aligned}$$
>
> or .7374 to four decimal places. Hence log 5462 = 3.7374.
> The essential calculation here is $7372 + 0.2(8) = 7373.6$ or 7374.
>
> (b)   The characteristic is $8 - 10$. For the mantissa, we have
>
> $$\begin{aligned} \text{Mantissa of log 8360} &= .9222 \\ \text{Mantissa of log 8370} &= \underline{.9227} \\ \text{Tabular difference} &= .0005 \\ 0.7 \text{ (tabular difference)} &= .00035 \\ \text{Mantissa of log 8367} &= .9222 + .00035 = .92255 \end{aligned}$$
>
> or .9226 to four decimal places. Hence log 0.08367 = 8.9226 - 10.
> The essential calculation here is $9222 + 0.7(5) = 9225.5$ or 9226.

The mantissas in Example 3 were found to five digits and rounded off to four. In general, a number N is said to be rounded off to p digits when it is the nearest number to N that can be written with p digits; for example, .52687 is rounded off to .5269, .38233 is rounded off to .3823.

When the rounding off process can lead to two numbers, we shall agree to use that one which ends in an even digit; for example, .52685 is rounded off to .5268, .38235 is rounded off to .3824. (See Problems 11.9–11.10.)

TO FIND THE NUMBER CORRESPONDING TO A GIVEN COMMON LOGARITHM

(a)   When the given mantissa is found in the table, write down the row number followed by the column heading and then point off in accordance with the rule for characteristics. The resulting number is called the *antilogarithm* (antilog) of the given logarithm.

EXAMPLE 4.   Find (*a*) antilog 1.6551, (*b*) antilog 8.9090 − 10.

(*a*) The mantissa .6551 is found in the row opposite 45 and in the column headed 2; thus, we have 452. Since the characteristic is 1, there are two digits to the left of the decimal point. Then antilog 1.6551 = 45.20.

(*b*) The mantissa .9090 is found in the row opposite 81 and under the column headed 1; thus, we have 811. Since the characteristic is 8 − 10, the number is smaller than 1 with one zero immediately following the decimal point. Hence, antilog 8.9090 − 10 = 0.0811.

(*b*) When the given mantissa is not found in the table, interpolation must be used.

EXAMPLE 5.   Find (*a*)   antilog 2.7949, (*b*)   antilog 9.6271–10.

(*a*)   Mantissa of log 6230 = .7945   Given mantissa = .7949
       Mantissa of log 6240 = .7952   Next smaller mantissa = .7945
          Tabular difference = .0007   Difference = .0004

$$\text{Correction} = \frac{.0004}{.0007}(10) = 5.7. \text{ Add correction to smaller sequence:}$$

$$6230 + 5.7 = 6235.7 \quad \text{or} \quad 6236 \text{ to four digits}$$

Then antilog 2.7949 = 623.6. The essential calculation here is $\frac{4}{7}(10) = 5.7$.

(*b*)   Mantissa of log 4230 = .6263   Given mantissa = .6271
       Mantissa of log 4240 = .6274   Next smaller mantissa = .6263
          Tabular difference = .0011   Difference = .0008

$$\text{Correction} = \frac{.0008}{.0011}(10) = 7.3. \text{ Add correction to smaller sequence:}$$

$$4230 + 7.3 = 4237.3 \quad \text{or} \quad 4237 \text{ to four digits}$$

Then antilog 9.6271 − 10 = 0.4237. The essential calculation here is $\frac{8}{11}(10) = 7.3$.

(See Problem 11.11.)

AN EXPONENTIAL EQUATION is an equation involving one or more unknowns in an exponent. For example, $2^x = 7$ and $(1.03)^{-x} = 2.5$ are exponential equations. Such equations are solved by means of logarithms.

EXAMPLE 6.   Solve the exponential equation $2^x = 7$.
Take logarithms of both sides: $x \log 2 = \log 7$

Solve for x: $x = \dfrac{\log 7}{\log 2} = \dfrac{0.8451}{0.3010}$

Evaluate, using logarithms: log 0.8451 = 9.9270 − 10
                        −log 0.3010 = 9.4786 − 10
                             log x = 0.4484
                                  x = 2.808

(See Problem 11.16.)

IN THE CALCULUS the most useful system of logarithms is the *natural system* in which the base is a certain irrational number $e = 2.71828$, approximately.

The natural logarithm of N, ln N, and the common logarithm of N, log N, are related by the formula

$$\ln N = 2.3026 \log N$$

THE CALCULATOR can be used to do logarithmic calculation with extreme ease.

EXAMPLE 7. Evaluate log 82,734 rounded to six decimal places.

Press: $\boxed{8}\ \boxed{2}\ \boxed{7}\ \boxed{3}\ \boxed{4}\ \boxed{\log}$. On screen: 4.917684.

EXAMPLE 8. Solve for x to four significant digits: $\ln x = -0.3916$. Press: .3916 $\boxed{\pm}\ \boxed{e^x}$. On screen: 0.6759745.
Then $x \approx 0.6760$.

Solved Problems

11.1 Change the following from exponential to logarithmic form:

   (a) $7^2 = 49$,   (b) $6^{-1} = \frac{1}{6}$,   (c) $10^0 = 1$,   (d) $4^0 = 1$,   (e) $\sqrt[3]{8} = 2$.

   Ans. (a) $\log_7 49 = 2$,   (b) $\log_6 \frac{1}{6} = -1$,   (c) $\log_{10} 1 = 0$,   (d) $\log_4 1 = 0$,

   (e) $\log_8 2 = \frac{1}{3}$.

11.2 Change the following from logarithmic to exponential form:

   (a) $\log_3 81 = 4$,   (b) $\log_5 \frac{1}{625} = -4$,   (c) $\log_{10} 10 = 1$,   (d) $\log_9 27 = \frac{3}{2}$.

   Ans. (a) $3^4 = 81$,   (b) $5^{-4} = \frac{1}{625}$,   (c) $10^1 = 10$,   (d) $9^{3/2} = 27$.

11.3 Evaluate x, given:

   (a) $x = \log_5 125$ (d) $x = \log_2 \frac{1}{16}$ (g) $\log_x \frac{1}{16} = -2$

   (b) $x = \log_{10} 0.001$ (e) $x = \log_{1/2} 32$ (h) $\log_6 x = 2$

   (c) $x = \log_8 2$ (f) $\log_x 243 = 5$ (i) $\log_a x = 0$

   Ans. (a) 3, since $5^3 = 125$ (d) -4, since $2^{-4} = \frac{1}{16}$ (g) 4, since $4^{-2} = \frac{1}{16}$

   (b) -3, since $10^{-3} = 0.001$ (e) -5, since $(\frac{1}{2})^{-5} = 32$ (h) 36, since $6^2 = 36$

   (c) $\frac{1}{3}$, since $8^{1/3} = 2$ (f) 3, since $3^5 = 243$ (i) 1, since $a^0 = 1$

11.4 Prove the four laws of logarithms.

   Let $P = b^p$ and $Q = b^q$; then $\log_b P = p$ and $\log_b Q = q$.

   1. Since $P \cdot Q = b^p \cdot b^q = b^{p+q}$, $\log_b PQ = p + q = \log_b P + \log_b Q$; that is, the logarithm of the product of two positive numbers is equal to the sum of the logarithms of the numbers.

   2. Since $P/Q = b^p/b^q = b^{p-q}$, $\log_b (P/Q) = p - q = \log_b P - \log_b Q$; that is, the logarithm of the quotient of two positive numbers is the logarithm of the numerator minus the logarithm of the denominator.

   3. Since $P^n = (b^p)^n = b^{np}$, $\log_b P^n = np = n \log_b P$; that is, the logarithm of a power of a positive number is equal to the product of the exponent and the logarithm of the number.

   4. Since $\sqrt[n]{P} = P^{1/n} = b^{p/n}$, $\log_b \sqrt[n]{P} = \dfrac{p}{n} = \dfrac{1}{n} \log_b P$; that is, the logarithm of a root of a positive number

       is equal to the logarithm of the number divided by the index of the root.

11.5 Express the logarithms of the given expressions in terms of the logarithms of the individual letters involved.

   (a) $\log_b \dfrac{P \cdot Q}{R} = \log_b (P \cdot Q) - \log_b R = \log_b P + \log_b Q - \log_b R$

   (b) $\log_b \dfrac{P}{Q \cdot R} = \log_b P - \log_b (Q \cdot R) = \log_b P - (\log_b Q + \log_b R) = \log_b P - \log_b Q - \log_b R$

(c) $\log_b P^2 \cdot \sqrt[3]{Q} = \log_b P^2 + \log_b \sqrt[3]{Q} = 2 \log_b P + \frac{1}{3} \log_b Q$

(d) $\log_b \sqrt{\dfrac{P \cdot Q^3}{R^{1/2} \cdot S}} = \dfrac{1}{2} \log_b \dfrac{P \cdot Q^3}{R^{1/2} \cdot S} = \dfrac{1}{2} \left[\log_b (P \cdot Q^3) - \log_b (R^{1/2} \cdot S) \right]$

$$= \tfrac{1}{2}(\log_b P + 3 \log_b Q - \tfrac{1}{2} \log_b R - \log_b S)$$

11.6 Express each of the following as a single logarithm:

(a) $\log_b x - 2 \log_b y + \log_b z = (\log_b x + \log_b z) - 2 \log_b y = \log_b xz - \log_b y^2 = \log_b \dfrac{xz}{y^2}$

(b) $\log_b 2 + \log_b \pi + \frac{1}{2} \log_b l - \frac{1}{2} \log_b g = (\log_b 2 + \log_b \pi) + \frac{1}{2}(\log_b l - \log_b g)$

$$= \log_b (2\pi) + \tfrac{1}{2} \log_b \dfrac{l}{g} = \log_b \left(2\pi \sqrt{\dfrac{l}{g}} \right)$$

11.7 Show that $b^{3 \log_b x} = x^3$.

Let $3 \log_b x = t$. Then $\log_b x^3 = t$ and $x^3 = b^t = b^{3 \log_b x}$.

11.8 Determine the characteristic of the common logarithm of

(a) 3860	(c) 7.84	(e) 5.463	(g) 0.02345	(i) 2.866	(k) 77.62
(b) 52.6	(d) 728 000	(f) 0.3748	(h) 0.0001234	(j) 0.00005	(l) 0.002945

Ans. (a) 3   (c) 0   (e) 0   (g) $8 - 10$ (i) 0   (k) 1

(b) 1   (d) 5   (f) $9 - 10$ (h) $6 - 10$ (j) $5 - 10$ (l) $7 - 10$

11.9 Find

(a) $\log 3860 = 3.5866$ (c) $\log 7.84 = 0.8943$

(b) $\log 52.6 = 1.7210$ (d) $\log 728\,000 = 5.8621$

(e) $\log 5.463 = 0.7374$ $[7372 + .3(8)]$

(f) $\log 0.3748 = 9.5738 - 10$ $[5729 + .8(11)]$

(g) $\log 0.02345 = 8.3702 - 10$ $[3692 + .5(19)]$

(h) $\log 0.0001234 = 6.0913 - 10$ $[0899 + .4(35)]$

(i) $\log 2.866 = 0.4573$ $[4564 + .6(15)]$

(j) $\log 0.00005 = 5.6990 - 10$

(k) $\log 77.62 = 1.8900$ $[8899 + .2(5)]$

(l) $\log 0.002945 = 7.4690 - 10$ $[4683 + .5(15)]$

11.10 Find (a) $\log 2.864^3$, (b) $\log \sqrt{2.864}$, (c) $\log 0.007463^2$, (d) $\log \sqrt[3]{0.007463}$.

Since $\log 2.864 = 0.4570$,

(a) $\log 2.864^3 = 3 \log 2.864 = 3(0.4570) = 1.3710$

(b) $\log \sqrt{2.864} = \frac{1}{2} \log 2.864 = \frac{1}{2}(0.4570) = 0.2285$

Since $\log 0.007463 = 7.8729 - 10$,

(c) $\log 0.007463^2 = 2 \log 0.007463 = 2(7.8729 - 10) = 15.7458 - 20 = 5.7458 - 10$

(d) $\log \sqrt[3]{0.007463} = \frac{1}{3} \log 0.007463 = \frac{1}{3}(7.8729 - 10) = \frac{1}{3}(27.8729 - 30) = 9.2910 - 10$

11.11   Find

    (*a*)   antilog $1.4232 = 26.50$

    (*b*)   antilog $7.9217 - 10 = 0.008350$

    (*c*)   antilog $2.7514 = 564.1$; $5640 + \frac{1}{7}(10)$

    (*d*)   antilog $8.6362 - 10 = 0.04327$; $4320 + \frac{7}{10}(10)$

    (*e*)   antilog $0.9702 = 9.338$; $9330 + \frac{3}{4}(10)$

    (*f*)   antilog $9.5884 - 10 = 0.3876$; $3870 + \frac{7}{11}(10)$

    (*g*)   antilog $4.8353 = 68430$; $6840 + \frac{2}{6}(10)$

    (*h*)   antilog $1.5829 = 38.27$; $3820 + \frac{8}{11}(10)$

    (*i*)   antilog $7.9231 - 10 = 0.008378$; $8370 + \frac{4}{5}(10)$

    (*j*)   antilog $0.9150 = 8.222$; $8220 + \frac{1}{5}(10)$

    (*k*)   antilog $9.5306 - 10 = 0.3393$; $3390 + \frac{4}{13}(10)$.

11.12   Evaluate, using logarithms.

(*a*) $N = 3.268 \times 0.8794$

$$\begin{aligned}
\log 3.268 &= 0.5142 \\
+\log 0.8794 &= \underline{9.9442 - 10} \\
\log N &= 0.4584 \\
N &= 2.873
\end{aligned}$$

(*b*) $N = \dfrac{0.8183}{0.0544}$

$$\begin{aligned}
\log 0.8183 &= 9.9130 - 10 \\
-\log 0.0544 &= \underline{8.7356 - 10} \\
\log N &= 1.1774 \\
N &= 15.04
\end{aligned}$$

11.13   Evaluate, using five-place tables of logarithms.

(*a*) $N = 36.234 \times 2.6748 \times 0.0071756$

$$\begin{aligned}
\log 36.234 &= 1.55912 \\
+\log 2.6748 &= 0.42729 \\
+\log 0.0071756 &= \underline{7.85586 - 10} \\
\log N &= 9.84227 - 10 \\
N &= 0.69546
\end{aligned}$$

(*b*) $N = \dfrac{1.7834}{0.62315}$

$$\begin{aligned}
\log 1.7834 &= 10.25125 - 10 \\
-\log 0.62315 &= \underline{9.79460} \\
\log N &= 0.45665 \\
N &= 2.8619
\end{aligned}$$

(*c*) $N = \sqrt{6.3794 \times 0.95327}$

$$\begin{aligned}
\log 6.3794 &= 0.80478 \\
+\log 0.95327 &= \underline{9.97922 - 10} \\
2 \log N &= 0.78400 \\
\log N &= 0.39200 \\
N &= 2.4661
\end{aligned}$$

11.14   Solve.

(*a*) $(1.06)^x = 3$.

    Taking logarithms, $x \log 1.06 = \log 3$.

$$x = \frac{\log 3}{\log 1.06} = \frac{0.4771}{0.0253}$$

$$\begin{aligned}
\log 0.4771 &= 9.6786 - 10 \\
-\log 0.0253 &= \underline{8.4031 - 10} \\
\log x &= 1.2755 \\
x &= 18.86
\end{aligned}$$

(b) $12^{2x+5} = 55(7^{3x})$.

Taking logarithms, $(2x + 5) \log 12 = \log 55 + 3x \log 7$.

$$2x \log 12 - 3x \log 7 = \log 55 - 5 \log 12$$

$$x = \frac{\log 55 - 5 \log 12}{2 \log 12 - 3 \log 7} = \frac{1.7404 - 5(1.0792)}{2(1.0792) - 3(0.8451)} = \frac{3.6556}{0.3769}$$

Using a four-place table, we solve $x = \dfrac{3.656}{0.3769}$.

$$\log 3.656 = 0.5630$$
$$-\log 0.3769 = \underline{9.5762 - 10}$$
$$\log x = 0.9868 - 10$$
$$x = 9.700$$

(c) $41.2^x = 12.6^{x-1}$.

Taking logarithms, $x \log 41.2 = (x - 1) \log 12.6$.

$$x \log 41.2 - x \log 12.6 = -\log 12.6 \qquad \text{or} \qquad x = \frac{-\log 12.6}{\log 41.2 - \log 12.6}$$

$$y = -x = \frac{\log 12.6}{\log 41.2 - \log 12.6} = \frac{1.1004}{0.5145}$$

$$\log 1.100 = 0.0414$$
$$-\log 0.5145 = \underline{9.7114 - 10}$$
$$\log y = 0.3300$$
$$y = 2.138$$
$$x = -2.138$$

(d) $(0.8)^{2x-3} = 1.5^x$.

Taking logarithms, $(2x - 3) \log 0.8 = x \log 1.5$.

$$2x \log 0.8 - x \log 1.5 = 3 \log 0.8$$

$$x = \frac{3 \log 0.8}{2 \log 0.8 - \log 1.5} = \frac{-3 \log 0.8}{\log 1.5 - 2 \log 0.8} = \frac{3 \operatorname{colog} 0.8}{\log 1.5 - 2 \log 0.8} = \frac{0.2907}{0.3699}$$

$$\log 0.2907 = 9.4634 - 10$$
$$-\log 0.3699 = \underline{9.5681 - 10}$$
$$\log x = 9.8953 - 10$$
$$x = 0.7858$$

Supplementary Problems

11.15 Evaluate, using a calculator. Round your answer to four digits.

(a) $3.141 \times 0.9856 \times 58.44$ (b) $\dfrac{42.25}{386.2}$ (c) $\sqrt{76.94}$ (d) $\dfrac{222.6 \times 0.8988}{(5.344)^2}$

  Ans. (a) 180.9   (b) 0.1094   (c) 8.770   (d) 7.005

11.16 Evaluate, using a calculator. Round your answer to four places.

(a) $36.234 \times 2.6748 \times 0.0071756$ (b) $\sqrt[3]{0.48476}$ (c) $\dfrac{47.75 \times 8.643}{6467}$ (c) $\dfrac{(3.2486)^{2/3}}{\sqrt[5]{316.48}}$

  Ans. (a) 0.69546   (b) 0.78554   (c) 0.063816   (d) 0.69351

11.17 Solve for x using a calculator.

(a) $3^x = 30$ (b) $1.07^x = 3$ (c) $5.72^x = 8.469$ (d) $38.5^x = 6.5^{x-2}$

  Ans. (a) 3.096   (b) 16.23   (c) 1.225   (d) −2.104

Chapter 12

Power, Exponential, and Logarithmic Curves

POWER FUNCTIONS in x are of the form x^n. If $n > 0$, the graph of $y = x^n$ is said to be of the *parabolic* type (the curve is a parabola for $n = 2$). If $n < 0$, the graph of $y = x^n$ is said to be of the *hyperbolic* type (the curve is a hyperbola for $n = -1$).

EXAMPLE 1. Sketch the graphs of (a) $y = x^{3/2}$, (b) $y = -x^{-3/2}$.

Table 12.1 has been computed for selected values of x. We shall assume that the points corresponding to intermediate values of x lie on a smooth curve joining the points given in the table. See Figs. 12-1 and 12-2. (See Problems 12.1–12.3.)

Table 12.1

x	$y = x^{3/2}$	$y = -x^{-3/2}$
9	27	$-\frac{1}{27}$
4	8	$-\frac{1}{8}$
1	1	-1
$\frac{1}{4}$	$\frac{1}{8}$	-8
$\frac{1}{9}$	$\frac{1}{27}$	-27
0	0	—

$$y = x^{3/2}$$
Fig. 12-1

$$y = -x^{-3/2}$$
Fig. 12-2

EXPONENTIAL FUNCTIONS in x are of the form b^x where b is a constant. The discussion will be limited here to the case $b > 1$.

The curve whose equation is $y = b^x$ is called an *exponential curve*. The general properties of such curves are

(a) The curve passes through the point $(0, 1)$.

(b) The curve lies above the x axis and has that axis as an asymptote.

EXAMPLE 2. Sketch the graphs of (a) $y = 2^x$, (b) $y = 3^x$. (See Problem 12.4.)

Table 12.2

x	$y = 2^x$	$y = 3^x$
3	8	27
2	4	9
1	2	3
0	1	1
-1	$\frac{1}{2}$	$\frac{1}{3}$
-2	$\frac{1}{4}$	$\frac{1}{9}$
-3	$\frac{1}{8}$	$\frac{1}{27}$

$(0, 1)$

$$y = 2^x$$
Fig. 12-3

$(0, 1)$

$$y = 3^x$$
Fig. 12-4

The exponential equation appears frequently in the form $y = c e^{kx}$ where c and k are nonzero constants and $e = 2.71828\ldots$ is the natural logarithmic base. See Table 12.2 and Figs. 12-3 and 12-4. (See Problems 12.5–12.6.)

80

THE CURVE WHOSE EQUATION IS $y = \log_b x$, $b > 1$, is called a *logarithmic curve*. The general properties are

(*a*) The curve passes through the point $(1, 0)$.

(*b*) The curve lies to the right of the y axis and has that axis as an asymptote.

EXAMPLE 3. Sketch the graph of $y = \log_2 x$.

Table 12.3

x	8	4	2	1	$\frac{1}{2}$	$\frac{1}{4}$	$\frac{1}{8}$
y	3	2	1	0	-1	-2	-3

$y = \log_2 x$

Fig. 12-5

Since $x = 2^y$, the table of values in Table 12.3 may be obtained from the table for $y = 2^x$ of Example 2 by interchanging x and y. See Fig. 12-5. (See Problem 12.7.)

Solved Problems

12.1 Sketch the graph of the *semicubic parabola* $y^2 = x^3$.

Since the given equation is equivalent to $y = \pm x^{3/2}$, the graph consists of the curve of Example 1(*a*) together with its reflection in the x axis. See Fig. 12-6.

12.2 Sketch the graph of $y^3 = x^2$.

Refer to Fig. 12-7 and Table 12.4.

Table 12.4

x	± 3	± 2	± 1	0
y	2.1	1.6	1	0

$y^2 = x^3$

Fig. 12-6

$y^3 = x^2$

Fig. 12-7

12.3   Sketch the graph of $y = x^{-2}$.

See Fig. 12-8 and Table 12.5.

Table 12.5

x	± 3	± 2	± 1	$\pm\frac{1}{2}$	$\pm\frac{1}{4}$	0
y	$\frac{1}{9}$	$\frac{1}{4}$	1	4	16	—

$y = x^{-2}$

Fig. 12-8

12.4   Sketch the graph of $y = 3^{-x}$.

See Fig. 12-9 and Table 12.6.
Note that the graph of $y = b^{-x}$ is a reflection in the y axis of the graph of $y = b^x$.

Table 12.6

x	3	2	1	0	-1	-2	-3
y	$\frac{1}{27}$	$\frac{1}{9}$	$\frac{1}{3}$	1	3	9	27

$y = 3^{-x}$

Fig. 12-9

12.5   Sketch the graph of $y = e^{2x}$.

See Fig. 12-10 and Table 12.7.

Table 12.7

x	2	1	$\frac{1}{2}$	0	$-\frac{1}{2}$	-1	-2
$y = e^{2x}$	54.6	7.4	2.7	1	0.4	0.14	0.02

$y = e^{2x}$

Fig. 12-10

12.6   Sketch the graph of $y = e^{-x^2}$.

Refer to Fig. 12-11 and Table 12.8.

Table 12.8

x	± 2	$\pm\frac{3}{2}$	± 1	$\pm\frac{1}{2}$	0
y	0.02	0.1	0.4	0.8	1

$y = e^{-x^2}$

Fig. 12-11

This is a simple form of the *normal probability curve* used in statistics.

12.7.   Sketch the graphs of (*a*) $y = \log x$, (*b*) $y = \log x^2 = 2 \log x$.

See Table 12.9 and Figs. 12-12 and 12-13.

Table 12.9

x	10	5	4	3	2	1	0.5	0.25	0.1	0.01
$y = \log x$	1	0.7	0.6	0.5	0.3	0	−0.3	−0.6	−1	−2
$y = \log x^2$	2	1.4	1.2	1	0.6	0	−0.6	−1.2	−2	−4

$y = \log x$ $y = \log x^2$

Fig. 12-12   **Fig. 12-13**

Supplementary Problems

12.8   Sketch the graphs of (*a*) $y^2 = x^{-3}$, (*b*) $y^3 = x^{-2}$, (*c*) $y^2 = \frac{1}{x}$, (*d*) the cubical parabola $y = x^3$.

12.9   Sketch the graphs of

(*a*) $y = (2.5)^x$ (*c*) $y = 2^{-1/x}$ (*e*) $y = e^{x/2}$ (*g*) $y = e^{x+2}$

(*b*) $y = 2^{x+1}$ (*d*) $y = \frac{1}{2}e^x$ (*f*) $y = e^{-x/2}$ (*h*) $y = xe^{-x}$

12.10 Sketch the graphs of (*a*) $y = \frac{1}{2} \log x$,   (*b*) $y = \log (3x + 2)$,   (*c*) $y = \log (x^2 + 1)$.

12.11 Show that the curve $y^q = x^p$, where p and q are positive integers, lies entirely in

(*a*) Quadrants I and III if p and q are both odd

(*b*) Quadrants I and IV if p is odd and q is even

(*c*) Quadrants I and II if p is even and q is odd

12.12 Show that the curve $y^q = x^{-p}$, where p and q are positive integers, lies entirely in

(*a*) Quadrants I and III if p and q are both odd

(*b*) Quadrants I and II if p is even and q is odd

(*c*) Quadrants I and IV if p is odd and q is even

Chapter 13

Polynomial Equations, Rational Roots

A POLYNOMIAL EQUATION (or rational integral equation) is obtained when any polynomial in one variable is set equal to zero. As in Chapter 15, we shall work with polynomials having integral coefficients although many of the theorems will be stated for polynomial equations with weaker restrictions on the coefficients.

A polynomial equation is said to be in *standard form* when written as

$$a_0x^n + a_1x^{n-1} + a_2x^{n-2} + \cdots + a_{n-2}x^2 + a_{n-1}x + a_n = 0 \qquad (13.1)$$

where the terms are arranged in descending powers of x, a zero has been inserted as coefficient of each missing term, the coefficients have no common factor except ± 1, and $a_0 \neq 0$. (See Problem 13.1.)

A NUMBER r IS CALLED A ROOT of $f(x) = 0$ if and only if $f(r) = 0$. It follows that the abscissas of the points of intersection of the graph of $y = f(x)$ and the x axis are roots of $f(x) = 0$.

THE FUNDAMENTAL THEOREM OF ALGEBRA. Every polynomial equation $f(x) = 0$ has at least one root, real or complex.

A polynomial equation of degree n has exactly n roots. These n roots may not all be distinct. If r is one of the roots and occurs just once, it is called a *simple root*; if r occurs exactly $m > 1$ times among the roots, it is called a *root of multiplicity m* or an *m-fold root*. If $m = 2$, r is called a *double root*; if $m = 3$, a *triple root*; and so on. (See Problems 13.2, 13.3.)

COMPLEX ROOTS. If the polynomial equation $f(x) = 0$ has real coefficients and if the complex $a + bi$ is a root of $f(x) = 0$, then the *complex conjugate $a - bi$* is also a root. (For a proof, see Problem 13.11.)

IRRATIONAL ROOTS. Given the polynomial equation $f(x) = 0$, if the irrational number $a + \sqrt{b}$, where a and b are rational, is a root of $f(x) = 0$, then the conjugate irrational $a - \sqrt{b}$ is also a root. (See Problem 13.4.)

LIMITS TO THE REAL ROOTS. A real number L is called an *upper limit* of the real roots of $f(x) = 0$ if no (real) root is greater than L; a real number l is called a *lower limit* if no (real) root is smaller than l.

If $L > 0$ and if, when $f(x)$ is divided by $x - L$ by synthetic division, every number in the third line is nonnegative, then L is an upper limit of the real roots of $f(x) = 0$.

If $l < 0$ and if, when $f(x)$ is divided by $x - l$ by synthetic division, the numbers in the third line alternate in sign, then l is a lower limit of the real roots of $f(x) = 0$.

RATIONAL ROOTS. A polynomial equation has 0 as a root if and only if the constant term of the equation is zero.

EXAMPLE 1. The roots of $x^5 - 2x^4 + 6x^3 - 5x^2 = x^2(x^3 - 2x^2 + 6x - 5) = 0$ are $0, 0$, and the three roots of $x^3 - 2x^2 + 6x - 5 = 0$.

If a rational fraction p/q , expressed in lowest terms, is a root of (13.1) in which $a_n \neq 0$, then p is a divisor of the constant term a_n and q is a divisor of the leading coefficient of a_0 of (13.1) . (For a proof, see Problem 13.12.)

EXAMPLE 2. The theorem permits us to say that $\frac{2}{3}$ is a *possible* root of the equation $9x^4 - 5x^2 + 8x + 4 = 0$ since the numerator 2 divides the constant term 4 and the denominator 3 divides the leading coefficient 9. It does *not* assure that $\frac{2}{3}$ is a root. However, the theorem does assure that neither $\frac{1}{2}$ nor $-\frac{4}{5}$ is a root. In each case the denominator does not divide the leading coefficient.

If p , an integer, is a root of (13.1) then p is a divisor of its constant term.

EXAMPLE 3. The *possible* rational roots of the equation

$$12x^4 - 40x^3 - 5x^2 + 45x + 18 = 0$$

are all numbers $\pm p/q$ in which the values of p are the positive divisors 1, 2, 3, 6, 9, 18 of the constant term 18 and the values of q are the positive divisors 1, 2, 3, 4, 6, 12 of the leading coefficient 12. Thus the rational roots, if any, of the equation are among the numbers

$$\pm 1, \pm 2, \pm 3, \pm 6, \pm 9, \pm 18, \pm \tfrac{1}{2}, \pm \tfrac{3}{2}, \pm \tfrac{9}{2}, \pm \tfrac{1}{3}, \pm \tfrac{2}{3}, \pm \tfrac{1}{4}, \pm \tfrac{3}{4}, \pm \tfrac{1}{6}, \pm \tfrac{1}{12}$$

THE PRINCIPAL PROBLEM OF THIS CHAPTER is to find the rational roots of a given polynomial equation. The general procedure is: Test the possible rational roots by synthetic division, accepting as roots all those for which the last number in the third line *is* zero and rejecting all those for which it is not. Certain refinements, which help to shorten the work, are pointed out in the examples and solved problems below.

EXAMPLE 4. Find the rational roots of $x^5 + 2x^4 - 18x^3 - 8x^2 + 41x + 30 = 0$.
Since the leading coefficient is 1, all rational roots p/q are integers. The possible integral roots, the divisors (both positive and negative) of the constant term 30, are

$$\pm 1, \pm 2, \pm 3, \pm 5, \pm 6, \pm 10, \pm 15, \pm 30$$

$$
\begin{array}{r}
1 + 2 - 18 - \;\,8 + 41 + 30 \quad \underline{|1} \\
\text{Try 1:} \quad \underline{1 + \;\,3 - 15 - 23 + 18} \\
1 + 3 - 15 - 23 + 18 + 48
\end{array}
$$

Then 1 is not a root. This number ($+1$) should be removed from the list of possible roots lest we forget and try it again later on.

$$
\begin{array}{r}
1 + 2 - 18 - \;\,8 + 41 + 30 \quad \underline{|2} \\
\text{Try 2:} \quad \underline{2 + \;\,8 - 20 - 56 - 30} \\
1 + 4 - 10 - 28 - 15 + \;\,0
\end{array}
$$

Then 2 is a root and the remaining rational roots of the given equation are the rational roots of the *depressed equation*

$$x^4 + 4x^3 - 10x^2 - 28x - 15 = 0$$

Now $\pm 2, \pm 6, \pm 10$, and ± 30 cannot be roots of this equation (they are not divisors of 15) and should be removed from the list of possibilities. We return to the depressed equation.

$$
\begin{array}{r}
1 + 4 - 10 - 28 - 15 \quad \underline{|3} \\
\text{Try 3:} \quad \underline{3 + 21 + 33 + 15} \\
1 + 7 + 11 + \;\,5 + \;\,0
\end{array}
$$

Then 3 is a root and the new depressed equation is

$$x^3 + 7x^2 + 11x + 5 = 0$$

Since the coefficients of this equation are nonnegative, it has no positive roots. We now remove $+3$, $+5$, $+15$ from the original list of possible roots and return to the new depressed equation.

$$\begin{array}{r} 1+7+11+5 \quad \underline{|-1} \\ \text{Try } -1: \quad \underline{-1-\ 6-5} \\ 1+6+\ 5+0 \end{array}$$

Then -1 is a root and the depressed equation

$$x^2 + 6x + 5 = (x+1)(x+5) = 0$$

has -1 and -5 as roots.

The necessary computations may be neatly displayed as follows:

$$\begin{array}{r} 1+2-18-\ 8+41+30 \quad \underline{|2} \\ \underline{2+\ 8-20-56-30} \\ 1+4-10-28-15 \quad\quad\ \underline{|3} \\ \underline{3+21+33+15} \\ 1+7+11+\ 5 \quad\quad\quad\quad\ \underline{|-1} \\ \underline{-1-\ 6-\ 5} \\ 1+6+\ 5 \end{array}$$

$$x^2 + 6x + 5 = (x+1)(x+5) = 0 \qquad x = -1, -5$$

The roots are 2, 3, -1, -1, -5.

Note that the roots here are numerically small numbers, that is, 3 is a root but 30 is not, -1 is a root but -15 is not. Hereafter we shall not list integers which are large numerically or fractions with large numerator or denominator among the possible roots. (See Problems 13.5–13.9.)

Solved Problems

13.1 Write each of the following in standard form.

(a) $4x^2 + 2x^3 - 6 + 5x = 0$ *Ans.* $2x^3 + 4x^2 + 5x - 6 = 0$

(b) $-3x^3 + 6x - 4x^2 + 2 = 0$ *Ans.* $3x^3 + 4x^2 - 6x - 2 = 0$

(c) $2x^5 + x^3 + 4 = 0$ *Ans.* $2x^5 + 0 \cdot x^4 + x^3 + 0 \cdot x^2 + 0 \cdot x + 4 = 0$

(d) $x^3 + \frac{1}{2}x^2 - x + 2 = 0$ *Ans.* $2x^3 + x^2 - 2x + 4 = 0$

(e) $4x^4 + 6x^3 - 8x^2 + 12x - 10 = 0$ *Ans.* $2x^4 + 3x^3 - 4x^2 + 6x - 5 = 0$

13.2 (a) Show that -1 and 2 are roots of $x^4 - 9x^2 + 4x + 12 = 0$.

Using synthetic division,

$$\begin{array}{r} 1+0-9+\ 4+12 \quad \underline{|-1} \\ \underline{-1+1+\ 8-12} \\ 1-1-8+12+\ 0 \end{array} \qquad \begin{array}{r} 1+0-9+\ 4+12 \quad \underline{|2} \\ \underline{2+4-10-12} \\ 1+2-5-\ 6+\ 0 \end{array}$$

Since $f(-1) = 0$ and $f(2) = 0$, both -1 and 2 are roots.

(b) Show that the equation in (a) has at least two other roots by finding them.

From the synthetic division in (a) and the Factor Theorem (Chapter 15),

$$x^4 - 9x^2 + 4x + 12 = (x+1)(x^3 - x^2 - 8x + 12)$$

Since 2 is also a root of the given equation, 2 is a root of $x^3 - x^2 - 8x + 12 = 0$.

Using synthetic division

$$\begin{array}{r} 1 - 1 - 8 + 12 \ \underline{\lvert 2} \\ 2 + 2 - 12 \\ \hline 1 + 1 - 6 + \ \ 0 \end{array}$$

we obtain $x^3 - x^2 - 8x + 12 = (x - 2)(x^2 + x - 6)$. Then

$$x^4 - 9x^2 + 4x + 12 = (x + 1)(x - 2)(x^2 + x - 6) = (x + 1)(x - 2)(x + 3)(x - 2)$$

Thus, the roots of the given equation are $-1, 2, -3, 2$.

(NOTE.   Since $x - 2$ appears twice among the factors of $f(x)$, 2 appears twice among the roots of $f(x) = 0$ and is a double root of the equation.)

13.3   (a)   Find all of the roots of $(x + 1)(x - 2)^3(x + 4)^2 = 0$.

The roots are $-1, 2, 2, 2, -4, -4$; thus, -1 is a simple root, 2 is a root of multiplicity three or a triple root, and -4 is a root of multiplicity two or a double root.

(b)   Find all the roots of $x^2(x - 2)(x - 5) = 0$.

The roots are $0, 0, 2, 5$; 2 and 5 are simple roots and 0 is a double root.

13.4   Form the equation of lowest degree with integral coefficients having the roots $\sqrt{5}$ and $2 - 3i$.

To assure integral coefficients, the conjugates $-\sqrt{5}$ and $2 + 3i$ must also be roots. Thus the required equation is

$$(x - \sqrt{5})(x + \sqrt{5})[x - (2 - 3i)][x - (2 + 3i)] = x^4 - 4x^3 + 8x^2 + 20x - 65 = 0$$

In Problems 13.5–13.9 find the rational roots; when possible, find all the roots.

13.5 $2x^4 - x^3 - 11x^2 + 4x + 12 = 0$

The possible rational roots are $\pm 1, \pm 2, \pm 3, \pm 4, \pm\frac{1}{2}, \pm\frac{3}{2}, \ldots$. Discarding all false trials, we find

$$\begin{array}{r} 2 - 1 - 11 + \ \ 4 + 12 \ \underline{\lvert 2} \\ 4 + \ \ 6 - 10 - 12 \\ \hline 2 + 3 - \ \ 5 - \ \ 6 \qquad \underline{\lvert -1} \\ -2 - \ \ 1 + \ \ 6 \\ \hline 2 + 1 - \ \ 6 \end{array}$$

$2x^2 + x - 6 = (2x - 3)(x + 2) = 0; \qquad x = \frac{3}{2}, -2$

The roots are $2, -1, \frac{3}{2}, -2$.

13.6 $4x^4 - 3x^3 - 4x + 3 = 0$

The possible rational roots are: $\pm 1, \pm 3, \pm\frac{1}{2}, \pm\frac{3}{2}, \pm\frac{1}{4}, \pm\frac{3}{4}$. By inspection the sum of the coefficients is 0; then $+1$ is a root. Discarding all false trials, we find

$$\begin{array}{r} 4 - 3 + 0 - 4 + 3 \ \underline{\lvert 1} \\ 4 + 1 + 1 - 3 \\ \hline 4 + 1 + 1 - 3 \qquad \underline{\lvert \frac{3}{4}} \\ 3 + 3 + 3 \\ \hline 4 + 4 + 4 \\ 1 + 1 + 1 \end{array}$$

$x^2 + x + 1 = 0; \qquad x = \dfrac{-1 \pm i\sqrt{3}}{2}$

(Factor out 4.)

The roots are $1, \dfrac{3}{4}, -\dfrac{1}{2} \pm \dfrac{i\sqrt{3}}{2}$.

13.7　$24x^6 - 20x^5 - 6x^4 + 9x^3 - 2x^2 = 0$

Since $24x^6 - 20x^5 - 6x^4 + 9x^3 - 2x^2 = x^2(24x^4 - 20x^3 - 6x^2 + 9x - 2)$ the roots of the given equation are $0, 0$ and the roots of $24x^4 - 20x^3 - 6x^2 + 9x - 2 = 0$. Possible rational roots are: $\pm 1, \pm 2, \pm\frac{1}{2}, \pm\frac{1}{3}, \pm\frac{2}{3}, \pm\frac{1}{4}, \dots$. Discarding all false trials, we find

$$
\begin{array}{l}
24 - 20 - 6 + 9 - 2 \quad \lfloor\tfrac{1}{2} \qquad 6x^2 + x - 2 = (2x - 1)(3x + 2) = 0; \qquad x = \tfrac{1}{2}, -\tfrac{2}{3} \\
\underline{12 - 4 - 5 + 2} \\
24 - 8 - 10 + 4 \qquad \text{(Factor out 2.)} \\
12 - 4 - 5 + 2 \qquad \lfloor\tfrac{1}{2} \\
\underline{6 + 1 - 2} \\
12 + 2 - 4 \qquad \text{(Factor out 2.)} \\
6 + 1 - 2
\end{array}
$$

The roots are $0, 0, \frac{1}{2}, \frac{1}{2}, \frac{1}{2}, -\frac{2}{3}$.

13.8　$4x^5 - 32x^4 + 93x^3 - 119x^2 + 70x - 25 = 0$

Since the signs of the coefficients alternate, the rational roots (if any) are positive. Possible rational roots are $1, 5, \frac{1}{2}, \frac{5}{2}, \frac{1}{4}, \dots$. Discarding all false trials, we find

$$
\begin{array}{l}
4 - 32 + 93 - 119 + 70 - 25 \quad \lfloor\tfrac{5}{2} \\
\underline{10 - 55 + 95 - 60 + 25} \\
4 - 22 + 38 - 24 + 10 \qquad \text{(Factor out 2.)} \\
2 - 11 + 19 - 12 + 5 \qquad \lfloor\tfrac{5}{2} \\
\underline{5 - 15 + 10 - 5} \\
2 - 6 + 4 - 2 \qquad \text{(Factor out 2.)} \\
1 - 3 + 2 - 1
\end{array}
$$

The rational roots are $\frac{5}{2}, \frac{5}{2}$.

The equation $f(x) = x^3 - 3x^2 + 2x - 1 = 0$ has at least one real (irrational) root since $f(0) < 0$ while for sufficiently large x $(x > 3)$, $f(x) > 0$.

The only possible rational root of $x^3 - 3x^2 + 2x - 1 = 0$ is 1; it is not a root.

13.9　$6x^4 + 13x^3 - 11x^2 + 5x + 1 = 0$

The possible rational roots are $\pm 1, \pm\frac{1}{2}, \pm\frac{1}{3}, \pm\frac{1}{6}$. After testing each possibility, we conclude that the equation has no rational roots.

13.10　If $ax^3 + bx^2 + cx + d = 0$, with integral coefficients, has a rational root r and if $c + ar^2 = 0$, then all of the roots are rational.

Using synthetic division to remove the known root, the depressed equation is

$$ax^2 + (b + ar)x + c + br + ar^2 = 0$$

which reduces to $ax^2 + (b + ar)x + br = 0$ when $c + ar^2 = 0$. Since its discriminant is $(b - ar)^2$, a perfect square, its roots are rational. Thus, all the roots of the given equation are rational.

13.11　Prove: If the polynomial $f(x)$ has real coefficients and if the imaginary $a + bi$, $b \neq 0$, is a root of $f(x) = 0$, then the conjugate imaginary $a - bi$ is also a root.

Since $a + bi$ is a root of $f(x) = 0$, $x - (a + bi)$ is a factor of $f(x)$. Similarly, if $a - bi$ is to be a root of $f(x) = 0$, $x - (a - bi)$ must be a factor of $f(x)$. We need to show then that when $a + bi$ is a root of $f(x) = 0$, it follows that

$$[x - (a + bi)][x - (a - bi)] = x^2 - 2ax + a^2 + b^2$$

is a factor of $f(x)$. By division we find

$$f(x) = [x^2 - 2ax + a^2 + b^2] \cdot Q(x) + Mx + N$$

$$= [x - (a + bi)][x - (a - bi)] \cdot Q(x) + Mx + N$$

$$(1)$$

where $Q(x)$ is a polynomial of degree 2 less than that of $f(x)$ and the remainder $Mx + N$ is of degree at most one in x, that is, M and N are constants.

Since $a + bi$ is a root of $f(x) = 0$, we have from (1),

$$f(a + bi) = 0 \cdot Q(a + bi) + M(a + bi) + N = (aM + N) + bMi = 0$$

Then $aM + N = 0$ and $bM = 0$. Now $b \neq 0$; hence, $bM = 0$ requires $M = 0$ and then $aM + N = 0$ requires $N = 0$. Since $M = N = 0$, (1) becomes

$$f(x) = (x^2 - 2ax + a^2 + b^2) \cdot Q(x)$$

Then $x^2 - 2ax + a^2 + b^2$ is a factor of $f(x)$ as was to be proved.

13.12   Prove: If a rational fraction p/q, expressed in lowest terms, is a root of the polynomial equation (13.1) whose constant terms $a_n \neq 0$, then p is a divisor of the constant term a_n and q is a divisor of the leading coefficient a_0.

Let the given equation be

$$a_0 x^n + a_1 x^{n-1} + a_2 x^{n-2} + \cdots + a_{n-2} x^2 + a_{n-1} x + a_n = 0, \qquad a_0 a_n \neq 0$$

If p/q is a root, then

$$a_0 (p/q)^n + a_1 (p/q)^{n-1} + a_2 (p/q)^{n-2} + \cdots + a_{n-2} (p/q)^2 + a_{n-1} (p/q) + a_n = 0$$

Multiplying both members by q^n, this becomes

$$a_0 p^n + a_1 p^{n-1} q + a_2 p^{n-2} q^2 + \cdots + a_{n-2} p^2 q^{n-2} + a_{n-1} p q^{n-1} + a_n q^n = 0 \qquad (2)$$

When (2) is written as

$$a_0 p^n + a_1 p^{n-1} q + a_2 p^{n-2} q^2 + \cdots + a_{n-2} p^2 q^{n-2} + a_{n-1} p q^{n-1} = -a_n q^n$$

it is clear that p, being a factor of every term of the left member of the equality, must divide $a_n q^n$. Since p/q is expressed in lowest terms, no factor of p will divide q. Hence, p must divide a_n as was to be shown.

Similarly, when (2) is written as

$$a_1 p^{n-1} q + a_2 p^{n-2} q^2 + \cdots + a_{n-2} p^2 q^{n-2} + a_{n-1} p q^{n-1} + a_n q^n = -a_0 p^n$$

it follows that q must divide a_0.

Supplementary Problems

13.13   Find all the roots.

(a) $x^5 - 2x^4 - 9x^3 + 22x^2 + 4x - 24 = 0$ (d) $6x^4 + 5x^3 - 16x^2 - 9x - 10 = 0$

(b) $18x^4 - 27x^3 + x^2 + 12x - 4 = 0$ (e) $9x^4 - 19x^2 - 6x + 4 = 0$

(c) $12x^4 - 40x^3 - 5x^2 + 45x + 18 = 0$ (f) $2x^5 + 3x^4 - 3x^3 - 2x^2 = 0$

Ans.   (a) $2, 2, 2, -1, -3$ (c) $-\frac{2}{3}, -\frac{1}{2}, \frac{3}{2}, 3$ (e) $-1, \frac{1}{3}, (1 \pm \sqrt{13})/3$

(b) $1, \frac{1}{2}, \frac{2}{3}, -\frac{2}{3}$ (d) $-2, \frac{5}{3}, (-1 \pm i\sqrt{7})/4$ (f) $0, 0, 1, -2, -\frac{1}{2}$

13.14   Solve the inequalities.

(a) $x^3 - 5x^2 + 2x + 8 > 0$ (c) $x^4 - 3x^3 + x^2 + 4 > 0$

(b) $6x^3 - 17x^2 - 5x + 6 < 0$ (d) $x^5 - x^4 - 2x^3 + 2x^2 + x - 1 > 0$

Ans.   (a) $x > 4, -1 < x < 2$ (b) $x < -\frac{2}{3}, \frac{1}{2} < x < 3$ (c)   all $x \neq 2$ (d) $x > 1$

13.15   Prove: if $r \neq 1$ is a root of $f(x) = 0$ then $r - 1$ divides $f(1)$.

Chapter 14

Irrational Roots of Polynomial Equations

IF $f(x) = 0$ IS A POLYNOMIAL EQUATION, the equation $f(-x) = 0$ has as roots the negatives of the roots of $f(x) = 0$. When $f(x) = 0$ is written in standard form, the equation whose roots are the negatives of the roots of $f(x) = 0$ may be obtained by changing the signs of alternate terms, beginning with the second.

EXAMPLE 1
(a) The roots of $x^3 + 3x^2 - 4x - 12 = 0$ are $2, -2, -3$; the roots of $x^3 - 3x^2 - 4x + 12 = 0$ are $-2, 2, 3$.
(b) The equation $6x^4 + 13x^3 - 13x - 6 = 0$ has roots $1, -1, -\frac{2}{3}, -\frac{3}{2}$; the equation $6x^4 - 13x^3 + 13x - 6 = 0$ has roots $-1, 1, \frac{2}{3}, \frac{3}{2}$.

(See Problem 14.1.)

VARIATION OF SIGN.   If, when a polynomial is arranged in descending powers of the variable, two successive terms differ in sign, the polynomial is said to have a *variation of sign*.

EXAMPLE 2
(a) The polynomial $x^3 - 3x^2 - 4x + 12$ has two variations of sign, one from $+x^3$ to $-3x^2$ and one from $-4x$ to $+12$; the polynomial $x^3 + 3x^2 - 4x - 12$ has one variation of sign.
(b) The polynomial $6x^4 + 13x^3 - 13x - 6$ has one variation of sign; the polynomial $6x^4 - 13x^3 + 13x - 6$ has three. Note that here the term with zero coefficient has not been considered.

DESCARTES' RULE OF SIGNS.   The number of positive roots of a polynomial equation $f(x) = 0$, with real coefficients, is equal either to the number of variations of sign in $f(x)$ or to that number diminished by an even number.

The number of negative roots of $f(x) = 0$ is equal to the number of positive roots of $f(-x) = 0$.

EXAMPLE 3.   Since $f(x) = x^3 - 3x^2 - 4x + 12$ of Example 2(a) has two variations of sign, $f(x) = 0$ has either two or no positive roots.

Since $x^3 + 3x^2 - 4x - 12$ has one variation of sign, $f(-x) = 0$ has one positive root and $f(x) = 0$ has one negative root. (See Problem 14.2.)

DIMINISHING THE ROOTS OF AN EQUATION.   Let

$$f(x) = a_0 x^n + a_1 x^{n-1} + \cdots + a_{n-1}x + a_n = 0 \tag{14.1}$$

be a polynomial equation of degree n. Let

$$f(x) = (x - h) \cdot q_1(x) + R_1,$$
$$q_1(x) = (x - h) \cdot q_2(x) + R_2,$$
$$\vdots \tag{14.2}$$
$$q_{n-2}(x) = (x - h) \cdot q_{n-1}(x) + R_{n-1},$$
$$q_{n-1}(x) = (x - h) \cdot q_n(x) + R_n,$$

where each R is a constant and $q_n(x) = a_0$. Then the roots of

$$g(y) = a_0 y^n + R_n y^{n-1} + R_{n-1} y^{n-2} + \cdots + R_2 y + R_1 = 0 \qquad (14.3)$$

are the roots of $f(x) = 0$ diminished by h.

We shall show that if r is any root of $f(x) = 0$ then $r - h$ is a root of $g(y) = 0$. Since $f(r) = 0$,

$$R_1 = -(r - h) \cdot q_1(r),$$

$$R_2 = q_1(r) - (r - h) \cdot q_2(r),$$

$$\vdots$$

$$R_{n-1} = q_{n-2}(r) - (r - h) \cdot q_{n-1}(r),$$

$$R_n = q_{n-1}(r) - (r - h)a_0.$$

When these replacements are made in (14.3), we have

$$a_0 y^n + [q_{n-1}(r) - (r - h)a_0]y^{n-1} + [q_{n-2}(r) - (r - h) \cdot q_{n-1}(r)]y^{n-2} + \cdots$$

$$+ [q_1(r) - (r - h) \cdot q_2(r)]y - (r - h) \cdot q_1(r)$$

$$= a_0[y - (r - h)]y^{n-1} + q_{n-1}(r)[y - (r - h)]y^{n-2} + \cdots + q_1(r)[y - (r - h)] = 0.$$

It is clear that $r - h$ is a root of (14.3) as was to be proved.

EXAMPLE 4.   Find the equation each of whose roots is 4 less than the roots of $x^3 + 3x^2 - 4x - 12 = 0$.

$$
\begin{array}{ll}
\begin{array}{l}
1 +\ 3 -\ 4 - 12\ \underline{\lvert 4} \\
\quad\ \ \underline{4 + 28 + 96} \\
1 +\ 7 + 24 + \circled{84} \\
\quad\ \ \underline{4 + 44} \\
1 + 11 + \circled{68} \\
\quad\ \ \underline{4} \\
1 + \circled{15}
\end{array}
&
\begin{array}{l}
1 + 15 + 68 + 84\ \underline{\lvert -2} \\
\ \ \underline{-\ 2 - 26 - 84} \\
1 + 13 + 42 \qquad \underline{\lvert -6} \\
\ \ \underline{-\ 6 - 42} \\
1 +\ 7 \qquad\qquad \underline{\lvert -7} \\
\ \ \underline{-\ 7} \\
1
\end{array}
\end{array}
$$

On the left the successive remainders have been found and circled. The resulting equation is $y^3 + 15y^2 + 68y + 84 = 0$. The given equation has roots $x = 2, -2, -3$; on the right, it is shown that $2 - 4 = -2$, $-2 - 4 = -6$, $-3 - 4 = -7$ are roots of the newly formed equation. (See Problem 14.3.)

APPROXIMATION OF IRRATIONAL ROOTS

LET $f(x) = 0$ **BE A POLYNOMIAL EQUATION** having no rational roots. If the given equation had rational roots, we suppose that they have been found using synthetic division and that $f(x)$ is then the last third line in this process. (See Chapter 13.)

THE METHOD OF SUCCESSIVE LINEAR APPROXIMATIONS will be explained by means of examples.

EXAMPLE 5.   The equation $f(x) = x^3 + x - 4 = 0$ has no rational roots. By Descartes' rule of signs, it has one positive (real) root and two imaginary roots. To approximate the real root we shall first isolate it as lying between two consecutive integers. Since $f(1) = -2$ and $f(2) = 6$, the root lies between $x = 1$ and $x = 2$. Figure 14-1 exhibits the portion of the graph of $f(x)$ between $(1, -2)$ and $(2, 6)$.

Fig. 14-1   Fig. 14-2   Fig. 14-3

In Fig. 14-2, the curve joining the two points has been replaced by a straight line which meets the x axis at R. We shall take OR, measured to the nearest tenth of a unit, as the first approximation of the required root and use it to isolate the root between successive tenths. From the similar triangles RSQ and PTQ,

$$\frac{RS}{PT} = \frac{SQ}{TQ} \qquad \text{or} \qquad RS = \frac{SQ}{TQ}(PT) = \frac{6}{8}(1) = \frac{3}{4} = 0.7$$

The first approximation of the root is given by $OR = OS - RS = 2 - 0.7 = 1.3$. Since $f(1.3) = -0.50$ and $f(1.4) = 0.14$, the required root lies between $x = 1.3$ and $x = 1.4$.

We now repeat the above process using the points $(1.3, -0.50)$ and $(1.4, 0.14)$ and isolate the root between successive hundredths. From Fig. 14-3,

$$RS = \frac{SQ}{TQ}(PT) = \frac{0.14}{0.64}(0.1) = 0.02 \qquad \text{and} \qquad OR = OS - RS = 1.4 - 0.02 = 1.38$$

is the next approximation. Since $f(1.38) = 0.008$ (hence, too large) and $f(1.37) = -0.059$, the root lies between $x = 1.37$ and $x = 1.38$.

Using the points $(1.37, -0.059)$ and $(1.38, 0.008)$, we isolate the root between successive thousandths. We find (no diagram needed)

$$RS = \frac{0.008}{0.067}(0.01) = 0.001 \qquad \text{and} \qquad OR = 1.38 - 0.001 = 1.379$$

Since $f(1.379) = 0.0012$ and $f(1.378) = -0.0054$, the root lies between $x = 1.378$ and $x = 1.379$.

For the next approximation

$$RS = \frac{0.0012}{0.0066}(0.001) = 0.0001 \qquad \text{and} \quad OR = 1.379 - 0.0001 = 1.3789$$

The root correct to three decimal places is 1.379. (See Problem 14.4.)

HORNER'S METHOD OF APPROXIMATION.

This method will be explained by means of examples.

EXAMPLE 6. The equation $x^3 + x^2 + x - 4 = 0$ has no rational roots. By Descartes' rule of signs, it has one positive root. Since $f(1) = -1$ and $f(2) = 10$, this root is between $x = 1$ and $x = 2$. We first diminish the roots of the given equation by 1.

$$
\begin{array}{rrrr|l}
1 & 1 & 1 & -4 & \underline{1} \\
  & 1 & 2 & 3 & \\
\hline
1 & 2 & 3 & -1 & \\
  & 1 & 3 & & \\
\hline
1 & 3 & 6 & & \\
  & 1 & & & \\
\hline
1 & 4 & & &
\end{array}
$$

and obtain the equation $g(y) = y^3 + 4y^2 + 6y - 1 = 0$ having a root between $y = 0$ and $y = 1$. To approximate it, we disregard the first two terms of the equation and solve $6y - 1 = 0$ for $y = 0.1$. Since $g(0.1) = -0.359$ and $g(0.2) = 0.368$, the root of $g(y) = 0$ lies between $y = 0.1$ and $y = 0.2$, and we diminish the roots of $g(y) = 0$ by 0.1.

$$
\begin{array}{rrrr|l}
1 & 4 & 6 & -1 & \underline{0.1} \\
  & 0.1 & 0.41 & 0.641 & \\
\hline
1 & 4.1 & 6.41 & -0.359 & \\
  & 0.1 & 0.42 & & \\
\hline
1 & 4.2 & 6.83 & & \\
  & 0.1 & & & \\
\hline
1 & 4.3 & & &
\end{array}
$$

We obtain the equation $h(z) = z^3 + 4.3z^2 + 6.83z - 0.359 = 0$ having a root between 0 and 0.01. Disregarding the first two terms of this equation and solving $6.83z - 0.359 = 0$, we obtain $z = 0.05$ as an approximation of the root. Since $h(0.05) = -0.007$ and $h(0.06) = 0.07$, the root of $h(z) = 0$ lies between $z = 0.05$ and $z = 0.06$ and we diminish the roots by 0.05

$$
\begin{array}{rrrr|l}
1 & 4.3 & 6.83 & -0.359 & \underline{0.05} \\
  & 0.05 & 0.2175 & 0.352375 & \\
\hline
1 & 4.35 & 7.0475 & -0.006625 & \\
  & 0.05 & 0.2200 & & \\
\hline
1 & 4.40 & 7.2675 & & \\
  & 0.05 & & & \\
\hline
1 & 4.45 & & &
\end{array}
$$

and obtain the equation $k(w) = w^3 + 4.45w^2 + 7.2675w - 0.006625 = 0$ having a root between $w = 0$ and $w = 0.001$. An approximation of this root, obtained by solving $7.2675w - 0.006625 = 0$ is $w = 0.0009$.

Without further computation, we are safe in stating the root of the given equation to be

$$x = 1 + 0.1 + 0.05 + 0.0009 = 1.1509$$

The complete solution may be exhibited more compactly, as follows:

$$
\begin{array}{rrrr|l}
1 & 1 & 1 & -4 & \underline{1} \\
  & 1 & 2 & 3 & \\
\hline
1 & 2 & 3 & -1 & \\
  & 1 & 3 & & \\
\hline
1 & 3 & 6 & & \\
  & 1 & & & \\
\hline
1 & 4 & 6 & -1 & \underline{0.1} \\
  & 0.1 & 0.41 & 0.641 & \\
\hline
1 & 4.1 & 6.41 & -0.359 & \\
  & 0.1 & 0.42 & & \\
\hline
1 & 4.2 & 6.83 & & \\
  & 0.1 & & & \\
\hline
1 & 4.3 & 6.83 & -0.359 & \underline{0.05} \\
  & 0.05 & 0.2175 & 0.352375 & \\
\hline
1 & 4.35 & 7.0475 & -0.006625 & \\
  & 0.05 & 0.2200 & & \\
\hline
1 & 4.40 & 7.2675 & & \\
  & 0.05 & & & \\
\hline
1 & 4.45 & 7.2675 & -0.006625 &
\end{array}
$$

$$y = \tfrac{1}{6} = 0.1$$

$$z = \frac{0.359}{6.83} = 0.05$$

$$w = \frac{0.006625}{7.2675} = 0.0009$$

(See Problems 14.5–14.6.)

Solved Problems

14.1   For each of the equations $f(x) = 0$, write the equation whose roots are the negatives of those of $f(x) = 0$.

    (a) $x^3 - 8x^2 + x - 1 = 0$ *Ans.* $x^3 + 8x^2 + x + 1 = 0$

    (b) $x^4 + 3x^2 + 2x + 1 = 0$ *Ans.* $x^4 + 3x^2 - 2x + 1 = 0$

    (c) $2x^4 - 5x^2 + 8x - 3 = 0$ *Ans.* $2x^4 - 5x^2 - 8x - 3 = 0$

    (d) $x^5 + x + 2 = 0$ *Ans.* $x^5 + x - 2 = 0$

14.2   Give all the information obtainable from Descartes' rule of signs about the roots of the following equations:

    (a) $f(x) = x^3 - 8x^2 + x - 1 = 0$ [Problem 14.1(a)].

        Since there are three variations of sign in $f(x) = 0$ and no variation of sign in $f(-x) = 0$, the given equation has either three positive roots or one positive root and two imaginary roots.

    (b) $f(x) = 2x^4 - 5x^2 + 8x - 3 = 0$ [Problem 14.1(c)].

        Since there are three variations of sign in $f(x) = 0$ and one variation of sign in $f(-x) = 0$, the given equation has either three positive and one negative root or one positive, one negative, and two imaginary roots.

    (c) $f(x) = x^5 + x + 2 = 0$ [Problem 14.1(d)].

        Since there is no variation of sign in $f(x) = 0$ and one in $f(-x) = 0$, the given equation has one negative and four imaginary roots.

14.3   Form the equation whose roots are equal to the roots of the given equation diminished by the indicated number.

    (a) $x^3 - 4x^2 + 8x - 5 = 0; 2.$ (b) $2x^3 + 9x^2 - 5x - 8 = 0; -3.$ (c) $x^4 - 8x^3 + 5x^2 + x + 8 = 0; 2.$

```
1 - 4 + 8 - 5  |2          2 + 9 -  5 -  8  |-3         1 - 8 +  5 +  1 +  8  |2
    2 - 4 + 8                  - 6 -  9 + 42               2 - 12 - 14 - 26
1 - 2 + 4 + 3              2 + 3 - 14 + 34             1 - 6 -  7 - 13 - 18
    2 + 0                     - 6 +  9                     2 -  8 - 30
1 + 0 + 4                 2 - 3 -  5                   1 - 4 - 15 - 43
    2                        - 6                           2 -  4
1 + 2                     2 - 9                        1 - 2 - 19
                                                           2
                                                      1 + 0
```

    The required equation is   The required equation is   The required equation is

    $y^3 + 2y^2 + 4y + 3 = 0.$ $2y^3 - 9y^2 - 5y + 34 = 0.$ $y^4 - 19y^2 - 43y - 18 = 0.$

14.4   Use the method of successive linear approximation to approximate the irrational roots of

$$f(x) = x^3 + 3x^2 - 2x - 5 = 0$$

    By Descartes' rule of signs the equation has either one positive and two negative roots or one positive and two imaginary roots. By the location principle (see Table 14.1), there are roots between $x = 1$ and $x = 2$, $x = -1$ and $x = -2$, and $x = -3$ and $x = -4$.

Table 14.1

x	2	1	0	-1	-2	-3	-4
$f(x)$	11	-3	-5	-1	3	1	-13

(*a*) To approximate the positive root, use Fig. 14-4. Then

$$RS = \frac{SQ}{TQ}(PT) = \frac{11}{14}(1) = 0.7 \qquad \text{and} \qquad OR = 2 - 0.7 = 1.3$$

Fig. 14-4

Since $f(1.3) = -0.33$ and $f(1.4) = 0.82$, the root lies between $x = 1.3$ and $x = 1.4$.
    For the next approximation,

$$RS = \frac{0.82}{1.15}(0.1) = 0.07 \qquad \text{and} \qquad OR = 1.4 - 0.07 = 1.33$$

Since $f(1.33) = -0.0006$ and $f(1.34) = 0.113$, the root lies between $x = 1.33$ and $x = 1.34$.
    For the next approximation,

$$RS = \frac{0.113}{0.1136}(0.01) = 0.009 \qquad \text{and} \qquad OR = 1.34 - 0.009 = 1.331$$

Now $f(1.331) > 0$ so that this approximation is too large; in fact, $f(1.3301) > 0$. Thus, the root to three decimal places is $x = 1.330$.

In approximating a negative root of $f(x) = 0$, it is more convenient to approximate the equally positive root of $f(-x) = 0$.

(*b*) To approximate the root of $f(x) = 0$ between $x = -1$ and $x = -2$, we shall approximate the positive root between $x = 1$ and $x = 2$ of $g(x) = x^3 - 3x^2 - 2x + 5 = 0$. Since $g(1) = 1$ and $g(2) = -3$, we obtain from Fig. 14-5

$$SR = \frac{SP}{TP}(TQ) = \frac{1}{4}(1) = 0.2 \qquad \text{and} \qquad OR = OS + SR = 1.2$$

Fig. 14-5

Since $g(1.2) = 0.01$ and $g(1.3) = -0.47$, the root is between $x = 1.2$ and $x = 1.3$.

For the next approximation,

$$SR = \frac{0.01}{0.48}(0.1) = 0.002 \quad \text{and} \quad OR = 1.2 + 0.002 = 1.202$$

Since $g(1.202) = -0.0018$ (hence, too large) and $g(1.201) = 0.0031$, the root is between $x = 1.201$ and $x = 1.202$.

For the next approximation,

$$SR = \frac{0.0031}{0.0049}(0.001) = 0.0006 \quad \text{and} \quad OR = 1.201 + 0.0006 = 1.2016$$

Since $g(1.2016) = -0.00281$ and $g(1.2015) = 0.00007$, the root of $g(x) = 0$ to three decimal places is $x = 1.202$. The corresponding root of the given equation is $x = -1.202$.

(c)   The approximation of the root -3.128 between $x = -3$ and $x = -4$ is left as an exercise.

14.5   Use Horner's method to approximate the irrational roots of $x^3 + 2x^2 - 4 = 0$.

By Descartes' rule of signs the equation has either one positive and two negative roots or one positive and two imaginary roots. By the location principle there is one root, between $x = 1$ and $x = 2$.

Arranged in compact form, the computation is as follows:

$$
\begin{array}{llll}
1 + 2 & +0 & -4 & \underline{|1} \\
\quad 1 & +3 & +3 \\
\overline{1 + 3} & +3 & -1 \\
\quad 1 & +4 & & \qquad y = \tfrac{1}{7} = 0.1 \\
\overline{1 + 4} & +7 \\
\quad 1 \\
\overline{1 + 5} & +7 & -1 & \underline{|0.1} \\
\quad 0.1 & +0.51 & +0.751 \\
\overline{1 + 5.1} & +7.51 & -0.249 \\
\quad 0.1 & +0.52 & & \qquad z = \dfrac{0.249}{8.03} = 0.03 \\
\overline{1 + 5.2} & +8.03 \\
\quad 0.1 \\
\overline{1 + 5.3} & +8.03 & -0.249 & \underline{|0.03} \\
\quad 0.03 & +0.1599 & +0.245697 \\
\overline{1 + 5.33} & +8.1899 & -0.003303 \\
\quad 0.03 & +0.1608 & & \qquad w = \dfrac{0.003303}{8.3507} = 0.00039 \\
\overline{1 + 5.36} & +8.3507 \\
\quad 0.03 \\
\overline{1 + 5.39} & +8.3507 & -0.003303
\end{array}
$$

The root, correct to four decimal places, is $1 + 0.1 + 0.03 + 0.00039 = 1.1304$.

14.6   Use Horner's method to approximate the irrational roots of $x^3 + 3x^2 - 2x - 5 = 0$.

By the location principle there are roots between $x = 1$ and $x = 2$, $x = -1$ and $x = -2$, $x = -3$ and $x = -4$.

(a)   The computation for the root between $x = 1$ and $x = 2$ is as follows:

$$
\begin{array}{llll}
1+3 & -\ 2 & -5 & \underline{|1} \\
\underline{\quad 1} & \underline{+\ 4} & \underline{+2} & \\
1+4 & +\ 2 & -3 & \\
\underline{\quad 1} & \underline{+\ 5} & & \\
1+5 & +\ 7 & & \\
\underline{\quad 1} & & & \\
1+6 & +\ 7 & -3 & \underline{|0.3} \\
\underline{\ 0.3} & \underline{+\ 1.89} & \underline{+2.667} & \\
1+6.3 & +\ 8.89 & -0.333 & \\
\underline{\ 0.3} & \underline{+\ 1.98} & & \\
1+6.6 & +10.87 & & \\
\underline{\ 0.3} & & & \\
1+6.9 & +10.87 & -0.333 & \underline{|0.03} \\
\underline{\ 0.03} & \underline{+\ 0.2079} & \underline{+0.332337} & \\
1+6.93 & +11.0779 & -0.000663 & \\
\underline{\ 0.03} & \underline{+\ 0.2088} & & \\
1+6.96 & +11.2867 & & \\
\underline{\ 0.03} & & & \\
1+6.99 & +11.2867 & -0.000663 & \\
\end{array}
$$

$y = \frac{3}{7} = 0.4$
but is too large since, when used,
the last number in the third line
is positive.

$z = \dfrac{0.333}{10.87} = 0.03$

$w = \dfrac{0.000663}{11.2867} = 0.000058$

The root is 1.330.

When approximating a negative root of $f(x) = 0$ using Horner's method, it is more convenient to approximate the equally positive root of $f(-x) = 0$.

(b)   To approximate the root between $x = -1$ and $x = -2$ of the given equation, we approximate the root between $x = 1$ and $x = 2$ of the equation $x^3 - 3x^2 - 2x + 5 = 0$. The computation is as follows:

$$
\begin{array}{llll}
1-3 & -\ 2 & +5 & \underline{|1} \\
\underline{\quad 1} & \underline{-\ 2} & \underline{-4} & \\
1-2 & -\ 4 & +1 & \\
\underline{\quad 1} & \underline{-\ 1} & & \\
1-1 & -\ 5 & & \\
\underline{\quad 1} & & & \\
1+0 & -\ 5 & +1 & \underline{|0.2} \\
\underline{\ 0.2} & \underline{+\ 0.04} & \underline{-0.992} & \\
1+0.2 & -\ 4.96 & +0.008 & \\
\underline{\ 0.2} & \underline{+\ 0.08} & & \\
1+0.4 & -\ 4.88 & & \\
\underline{\ 0.2} & & & \\
1+0.6 & -\ 4.88 & +0.008 & \underline{|0.001} \\
\underline{\ 0.001} & \underline{+\ 0.000601} & \underline{-0.004879399} & \\
1+0.601 & -\ 4.879399 & +0.003120601 & \\
\underline{\ 0.001} & \underline{+\ 0.000602} & & \\
1+0.602 & -\ 4.878797 & & \\
\underline{\ 0.001} & & & \\
1+0.603 & -\ 4.878797 & +0.003120601 & \\
\end{array}
$$

$y = \frac{1}{5} = 0.2$

$z = \dfrac{0.008}{4.88} = 0.001$

$w = \dfrac{0.003120601}{4.878797} = 0.00063$

To four decimal places the root is $x = 1.2016$; thus, the root of the given equation is $x = -1.2016$.

(c)   The approximation of the root $x = -3.1284$ between $x = -3$ and $x = -4$ is left as an exercise.

Supplementary Problems

14.7 Use Descartes' rule of signs to show

(a) $x^4 + 5x^2 + 24 = 0$ has only complex roots.

(b) $x^n - 1 = 0$ has exactly two real roots if n is even and only one real root if n is odd.

(c) $x^3 + 3x + 2 = 0$ has exactly one real root.

(d) $x^7 - x^5 + 2x^4 + 3x^2 + 5 = 0$ has at least four complex roots.

(e) $x^7 - 2x^4 + 3x^3 - 5 = 0$ has at most three real roots.

14.8 Find all the irrational roots of the following equations:

(a) $x^3 + x - 3 = 0$ (c) $x^3 - 9x + 3 = 0$ (e) $x^4 + 4x^3 + 6x^2 - 51x - 40 = 0$

(b) $x^3 - 3x + 1 = 0$ (d) $x^3 + 6x^2 + 7x - 3 = 0$

> Ans. (a) 1.2134   (c) 0.3376, 2.8169, −3.1546   (e) 2.7325, −0.7325
>
> (b) 0.3473, 1.5321, −1.8794   (d) 0.3301, −2.2016, −4.1284

14.9 Show that

(a)   The equation $f(x) = x^n + p_1 x^{n-1} + \cdots + p_{n-1} x + p_n = 0$, with integral coefficients, has no rational root if $f(0)$ and $f(1)$ are odd integers.

Hint: Suppose r is an integral root; then r is odd and $r - 1$ does not divide $f(1)$.

(b)   The equation $x^4 - 301x - 1275 = 0$ has no rational roots.

14.10 In a polynomial equation of the form

$$x^n + p_1 x^{n-1} + \cdots + p_{n-1} x + p_n = 0$$

the following relations exist between the coefficients and roots:

(1)   The sum of the roots is $-p_1$.
(2)   The sum of the products of the roots taken two at a time is p_2.
(3)   The sum of the products of the roots taken three at a time is $-p_3$.
$\vdots$
(n)   The product of the roots is $(-1)^n p_n$.

If a, b, c are the roots of $x^3 - 3x^2 + 4x + 2 = 0$, find

(a) $a + b + c$

(b) $ab + bc + ca$

(c) abc

(d) $a^2 + b^2 + c^2 = (a + b + c)^2 - 2(ab + bc + ca)$

(e) $a^3 + b^3 + c^3$

(f) $\dfrac{1}{a} + \dfrac{1}{b} + \dfrac{1}{c} = \dfrac{ab + bc + ca}{abc}$

(g) $\dfrac{1}{ab} + \dfrac{1}{bc} + \dfrac{1}{ca}$

> Ans. (a) 3,   (b) 4,   (c) −2,   (d) 1,   (e) −15,   (f) −2,   (g) $-\frac{3}{2}$

14.11 For a, b, c defined as in Problem 14.10, find an equation whose roots are $1/a, 1/b, 1/c$.

Hint: Write $x = 1/y$.   Ans. $2y^3 + 4y^2 - 3y + 1 = 0$

14.12 Two of the roots of $2x^3 - 11x^2 + \cdots = 0$ are $2, 3$. Find the third root and complete the equation.

> Ans. $2x^3 - 11x^2 + 17x - 6 = 0$

Chapter 15

Graphs of Polynomials

THE GENERAL POLYNOMIAL (or rational integral function) of the *n*th degree in *x* has the form

$$f(x) = a_0 x^n + a_1 x^{n-1} + a_2 x^{n-2} + \cdots + a_{n-2} x^2 + a_{n-1} x + a_n \qquad (15.1)$$

in which *n* is a positive integer and the *a*'s are constants, real or complex, with $a_0 \neq 0$. The term $a_0 x^n$ is called the *leading term*, a_n the *constant term*, and a_0 the *leading coefficient*.

Although most of the theorems and statements below apply to the general polynomial, attention in this chapter will be restricted to polynomials whose coefficients (the *a*'s) are integers.

REMAINDER THEOREM. If a polynomial $f(x)$ is divided by $x - h$ until a remainder free of *x* is obtained, this remainder is $f(h)$. (For a proof, see Problem 15.1.)

EXAMPLE 1. Let $f(x) = x^3 + 2x^2 - 3x - 4$ and $x - h = x - 2$; then $h = 2$. By actual division

$$\frac{x^3 + 2x^2 - 3x - 4}{x - 2} = x^2 + 4x + 5 + \frac{6}{x - 2}$$

or $x^3 + 2x^2 - 3x - 4 = (x^2 + 4x + 5)(x - 2) + 6$, and the remainder is 6.

By the remainder theorem, the remainder is

$$f(2) = 2^3 + 2 \cdot 2^2 - 3 \cdot 2 - 4 = 6$$

FACTOR THEOREM. If $x - h$ is a factor of $f(x)$ then $f(h) = 0$, and conversely. (For a proof, see Problem 15.2.)

SYNTHETIC DIVISION. By a process known as synthetic division, the necessary work in dividing a polynomial $f(x)$ by $x - h$ may be displayed in three lines, as follows:

(*1*) Arrange the dividend $f(x)$ in descending powers of *x* (as usual in division) and set down in the first line the coefficients, supplying zero as coefficient whenever a term is missing.

(*2*) Place *h*, the synthetic divisor, in the first line to the right of the coefficients.

(*3*) Recopy the leading coefficient a_0 directly below it in the third line.

(*4*) Multiply a_0 by *h*; place the product $a_0 h$ in the second line under a_1 (in the first line), add to a_1, and place the sum $a_0 h + a_1$ in the third line under a_1.

(*5*) Multiply the sum in Step 4 by *h*; place the product in the second line under a_2, add to a_2, and place the sum in the third line under a_2.

(*6*) Repeat the process of Step 5 until a product has been added to the constant term a_n.

The first *n* numbers in the third line are the coefficients of the quotient, a polynomial of degree $n - 1$, and the last number of the third line is the remainder $f(h)$.

EXAMPLE 2. Divide $5x^4 - 8x^2 - 15x - 6$ by $x - 2$, using synthetic division.

Following the procedure outlined above, we have

$$
\begin{array}{r}
5 + 0 - 8 - 15 - 6 \;\underline{\lfloor 2} \\
\underline{10 + 20 + 24 + 18} \\
5 + 10 + 12 + 9 + 12
\end{array}
$$

The quotient is $Q(x) = 5x^3 + 10x^2 + 12x + 9$ and the remainder is $f(2) = 12$.

(See Problem 15.4.)

DESCARTES' RULE OF SIGNS can be used to discuss the number of positive and negative zeros of a given polynomial (see Chapter 14).

EXAMPLE 3.   Consider the polynomial $P(x) = 2x^2 + x - 4$. $P(x)$ has *one* variation of sign (from + to −), $P(-x) = 2x^2 - x - 4$ has *one* variation of sign (from + to −). Descartes' rule of signs tells us that there will be one positive zero [one variation in sign for $P(x)$] and one negative zero [one variation in sign for $P(-x)$].

EXAMPLE 4.   Suppose $S(x) = x^5 + x^4 + x^3 - x^2 + 1$. $S(x)$ has two variations; $S(-x)$ has three. Thus $S(x)$ has 2 or 0 positive and 3 or 1 negative zeros.

THE GRAPH OF A POLYNOMIAL $y = f(x)$ may be obtained by computing a table of values, locating the several points (x, y), and joining them by a smooth curve. In order to avoid unnecessary labor in constructing the table, the following systematic procedure is suggested:

(*1*)   When $x = 0$, $y = f(0)$ is the constant term of the polynomial.
(*2*)   Use synthetic division to find $f(1), f(2), f(3), \ldots$ stopping as soon as the numbers in the third line of the synthetic division have the same sign.
(*3*)   Use synthetic division to find $f(-1), f(-2), f(-3), \ldots$ stopping as soon as the numbers in the third line of the synthetic division have alternating signs.

In advanced mathematics it is proved:

(*a*)   The graph of a polynomial in x with integral coefficients is always a smooth curve without breaks or sharp corners.
(*b*)   The number of *real* intersections of the graph of a polynomial of degree n with the x axis is *never* greater than n.
(*c*)   If a and b are real numbers such that $f(a)$ and $f(b)$ have opposite signs, the graph has an *odd* number of real intersections with the x axis between $x = a$ and $x = b$.
(*d*)   If a and b are real numbers such that $f(a)$ and $f(b)$ have the same signs, the graph either does not intersect the x axis or intersects it an *even* number of times between $x = a$ and $x = b$. See Fig. 15-1.

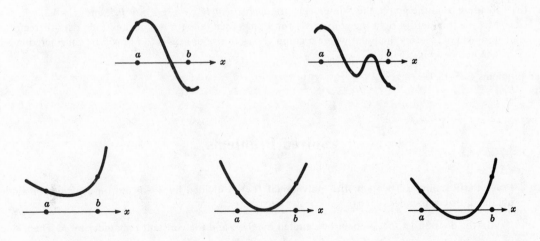

Fig. 15-1

EXAMPLE 5.   Construct the graph of $y = 2x^3 - 7x^2 - 7x + 5$. Form the table in Table 15.1.

Table 15.1

x	-2	-1	0	1	2	3	4	5
y	-25	3	5	-7	-21	-25	-7	45

$$
\begin{array}{r|}
2 - 7 - \;7 + \;5 \quad \underline{|1} \\
\underline{\quad\;\; 2 - \;\;5 - 12\quad\;} \\
2 - 5 - 12 - \;7
\end{array}
$$

$$
\begin{array}{r|}
2 - 7 - \;7 + \;5 \quad \underline{|2} \\
\underline{\quad\;\; 4 - \;\;6 - 26\quad\;} \\
2 - 3 - 13 - 21
\end{array}
$$

$$
\begin{array}{r|}
2 - 7 - \;7 + \;5 \quad \underline{|3} \\
\underline{\quad\;\; 6 - \;\;3 - 30\quad\;} \\
2 - 1 - 10 - 25
\end{array}
$$

$$
\begin{array}{r|}
2 - 7 - 7 + \;5 \quad \underline{|4} \\
\underline{\quad\;\; 8 + 4 - 12\quad\;} \\
2 + 1 - 3 - \;7
\end{array}
$$

$$
\begin{array}{r|}
2 - \;7 - \;7 + \;5 \quad \underline{|5} \\
\underline{\quad\; 10 + 15 + 40\quad\;} \\
2 + \;3 + \;8 + 45
\end{array}
$$

$$
\begin{array}{r|}
2 - 7 - 7 + 5 \quad \underline{|-1} \\
\underline{\quad\; -2 + 9 - 2\quad\;} \\
2 - 9 + 2 + 3
\end{array}
$$

$$
\begin{array}{r|}
2 - \;7 - \;7 + \;\;5 \quad \underline{|-2} \\
\underline{\quad\;\; - \;\;4 + 22 - 30\quad\;} \\
2 - 11 + 15 - 25
\end{array}
$$

Fig. 15-2

It is to be noted that

(a) The numbers in the third line are all nonnegative for the first time in finding $f(5)$; they alternate for the first time in finding $f(-2)$.

(b) The graph intersects the x axis between $x = -2$ and $x = -1$ since $f(-2)$ and $f(-1)$ have opposite signs, between $x = 0$ and $x = 1$, and between $x = 4$ and $x = 5$. Since the polynomial is of degree three, there are no other intersections.

(c) Reading from the graph, the x intercepts are *approximately* $x = -1.2$, $x = 0.5$, and $x = 4.2$.

(d) Moving from left to right the graph rises for a time, then falls for a time, and then rises thereafter. The problem of locating the point where a graph ceases to rise or ceases to fall will be considered in a later chapter. See Fig. 15-2.

(See Problems 15.5–15.9.)

Solved Problems

15.1   Prove the Remainder Theorem: If a polynomial $f(x)$ is divided by $x - h$ until a constant remainder is obtained, that remainder is $f(h)$.

In the division let the quotient be denoted by $Q(x)$ and the constant remainder by R. Then, since

Dividend = divisor times quotient + remainder,

$$f(x) = (x - h)Q(x) + R$$

is true for *all* values of x. When $x = h$, we have

$$f(h) = (h - h)Q(h) + R = R$$

15.2 Prove the Factor Theorem: If $x - h$ is a factor of $f(x)$ then $f(h) = 0$, and conversely.

By the Remainder Theorem, $f(x) = (x - h)Q(x) + f(h)$.
If $x - h$ is a factor of $f(x)$, the remainder must be zero when $f(x)$ is divided by $x - h$. Thus, $f(h) = 0$.
Conversely, if $f(h) = 0$, then $f(x) = (x - h)Q(x)$ and $x - h$ is a factor of $f(x)$.

15.3 Without performing the division, show that

(a) $x - 2$ is a factor of $f(x) = x^3 - x^2 - 14x + 24$.

(b) $x + a$ is not a factor of $f(x) = x^n + a^n$ for n an even positive integer and $a \neq 0$.

(a) $f(2) = 2^3 - 2^2 - 14 \cdot 2 + 24 = 0$ (b) $f(-a) = (-a)^n + a^n = 2a^n \neq 0$

15.4 Use synthetic division to divide $4x^4 + 12x^3 - 21x^2 - 65x + 9$ by (a) $2x - 1$, (b) $2x + 3$.

(a) Write the divisor as $2(x - \frac{1}{2})$. By synthetic division with synthetic divisor $h = \frac{1}{2}$, we find

$$\begin{array}{r} 4 + 12 - 21 - 65 + 9 \quad \lfloor \tfrac{1}{2} \\ \underline{2 + 7 - 7 - 36} \\ 4 + 14 - 14 - 72 - 27 \end{array}$$

Now

$$4x^4 + 12x^3 - 21x^2 - 65x + 9 = (4x^3 + 14x^2 - 14x - 72)(x - \tfrac{1}{2}) - 27$$
$$= (2x^3 + 7x^2 - 7x - 36)(2x - 1) - 27$$

Thus, when dividing $f(x)$ by $h = m/n$, the coefficients of the quotient have n as a common factor.

(b) Here $h = -\frac{3}{2}$. Then from

$$\begin{array}{r} 4 + 12 - 21 - 65 + 9 \quad \lfloor -\tfrac{3}{2} \\ \underline{- 6 - 9 + 45 + 30} \\ 4 + 6 - 30 - 20 + 39 \end{array}$$

we have

$$4x^4 + 12x^3 - 21x^2 - 65x + 9 = (2x^3 + 3x^2 - 15x - 10)(2x + 3) + 39$$

15.5 For the polynomial $y = f(x) = 4x^4 + 12x^3 - 31x^2 - 72x + 42$ form a table of values for integral values of x from $x = -5$ to $x = 3$.

Table 15.2 is the required table.

Table 15.2

x	-5	-4	-3	-2	-1	0	1	2	3
y	627	90	-21	30	75	42	-45	-66	195

$$\begin{array}{r} 4 + 12 - 31 - 72 + 42 \quad \lfloor -5 \\ \underline{- 20 + 40 - 45 + 585} \\ 4 - 8 + 9 - 117 + 627 \end{array}$$

$$\begin{array}{r} 4 + 12 - 31 - 72 + 42 \quad \lfloor -4 \\ \underline{- 16 + 16 + 60 + 48} \\ 4 - 4 - 15 - 12 + 90 \end{array}$$

$$\begin{array}{r} 4 + 12 - 31 - 72 + 42 \quad \lfloor -3 \\ \underline{- 12 + 0 + 93 - 63} \\ 4 + 0 - 31 + 21 - 21 \end{array}$$

$$\begin{array}{r} 4 + 12 - 31 - 72 + 42 \quad \lfloor -2 \\ \underline{- 8 - 8 + 78 - 12} \\ 4 + 4 - 39 + 6 + 30 \end{array}$$

$$\begin{array}{r} 4 + 12 - 31 - 72 + 42 \quad \lfloor -1 \\ \underline{- 4 - 8 + 39 + 33} \\ 4 + 8 - 39 - 33 + 75 \end{array}$$

$f(0) = 42$, the constant term of the polynomial.

$$\begin{array}{r} 4 + 12 - 31 - 72 + 42 \quad \lfloor 1 \\ \underline{4 + 16 - 15 - 87} \\ 4 + 16 - 15 - 87 - 45 \end{array}$$

$$\begin{array}{r} 4 + 12 - 31 - 72 + 42 \quad \lfloor 2 \\ \underline{8 + 40 + 18 - 108} \\ 4 + 20 + 9 - 54 - 66 \end{array}$$

$$\begin{array}{r} 4 + 12 - 31 - 72 + 42 \quad \lfloor 3 \\ \underline{12 + 72 + 123 + 153} \\ 4 + 24 + 41 + 51 + 195 \end{array}$$

15.6   Sketch the graph of $y = f(x) = x^4 - 9x^2 + 7x + 4$.

From Table 15.3 it is seen that the graph crosses the x axis between $x = -4$ and $x = -3$, $x = -1$ and $x = 0$, $x = 1$ and $x = 2$, and $x = 2$ and $x = 3$. From the graph in Fig. 15-3, the points of crossing are approximately $x = -3.1$, -0.4, 1.4, and 2.1.

Table 15.3

x	-4	-3	-2	-1	0	1	2	3
y	88	-17	-30	-11	4	3	-2	25

Fig. 15-3

15.7   Sketch the graph of $y = f(x) = x^3 - 5x^2 + 8x - 3$.

From Table 15.4 (formed for integral values of x from $x = -1$ to $x = 5$) it is seen that the graph crosses the x axis between $x = 0$ and $x = 1$. If there are two other real intersections, they are both between $x = 1$ and $x = 3$ since on this interval the graph rises to the right of $x = 1$, then falls, and then begins to rise to the left of $x = 3$.

By computing additional points on the interval (some are shown in the table) we are led to suspect that there are no further intersections (see Fig. 15-4). Note that were we able to locate the exact points at which the graph ceases to rise or fall these additional points would not have been necessary.

Table 15.4

x	-1	0	1	$\frac{5}{4}$	$\frac{3}{2}$	$\frac{7}{4}$	2	$\frac{9}{4}$	$\frac{5}{2}$	$\frac{11}{4}$	3	4	5
y	-17	-3	1	$\frac{73}{64}$	$\frac{9}{8}$	$\frac{67}{64}$	1	$\frac{69}{64}$	$\frac{11}{8}$	$\frac{127}{64}$	3	13	37

Fig. 15-4

15.8   Sketch the graph of $y = f(x) = x^3 - 4x^2 - 3x + 18$.

From Table 15.5 it is evident that the graph crosses the x axis at $x = -2$ and meets it again at $x = 3$. If there is a third distinct intersection, it must be between $x = 2$ and $x = 3$ or between $x = 3$ and $x = 4$. By computing additional points for x on these intervals, we are led to suspect that no such third intersection exists.

This function has been selected so that the question of intersections can be definitely settled. When $f(x)$ is divided by $x + 2$, the quotient is $x^2 - 6x + 9 = (x - 3)^2$ and the remainder is zero. Thus, $f(x) = (x + 2)(x - 3)^2$ in factored form. It is now clear that the function is positive for $x > -2$, that is, the graph never falls below the x axis on this interval. Thus, the graph is tangent to the x axis at $x = 3$, the point of tangency accounting for two of its intersections with the x axis. (See Fig. 15-5.)

Table 15.5

x	-3	-2	-1	0	1	2	3	4	5
y	-36	0	16	18	12	4	0	6	28

Fig. 15-5

15.9   Sketch the graph of $y = f(x) = x^3 - 6x^2 + 12x - 8$.

From Table 15.6 it is evident that the graph crosses the x axis at $x = 2$ and is symmetrical with respect to that point [i.e., $f(2 + h) = -f(2 - h)$]. Suppose the graph intersects the x axis to the right of $x = 2$. Then, since $f(x)$ is positive for $x = 3, 4, 5$ and $x \geq 6$, the graph is either tangent at the point or crosses the axis twice between some two consecutive values of x shown in the table. But, by symmetry, the graph would then have $2 + 2 + 1 = 5$ intersections with the x axis and this is impossible.

When $f(x)$ is divided by $x - 2$, the quotient is $x^2 - 4x + 4 = (x - 2)^2$ and the remainder is zero. Thus, $f(x) = (x - 2)^3$. Now it is clear that the graph lies above the x axis when $x > 2$ and below the axis when $x < 2$. The graph crosses the x axis at $x = 2$ and is also tangent there. (See Fig. 15-6.) In determining the intersections of the graph and the x axis, the point $x = 2$ is to be counted three times.

Table 15.6

x	-2	-1	0	1	2	3	4	5	6
y	-64	-27	-8	-1	0	1	8	27	64

Fig. 15-6

Supplementary Problems

15.10   Given $f(x) = x^4 + 5x^3 - 7x^2 - 13x - 2$, use synthetic division to find

   (a) $f(2) = 0$ (b) $f(4) = 410$ (c) $f(-2) = -28$ (d) $f(-3) = -80$

15.11   Using synthetic division, find the quotient and remainder when

   (a) $2x^4 - 3x^3 - 6x^2 + 11x - 10$ is divided by $x - 2$;   (b) $3x^4 + 11x^3 + 7x^2 - 8$ is divided by $x + 2$.

   Ans.   (a) $2x^3 + x^2 - 4x + 3$; -4 (b) $3x^3 + 5x^2 - 3x + 6$; -20

15.12   Use synthetic division to show

   (a) $x + 2$ and $3x - 2$ are factors of $3x^4 - 20x^3 + 80x - 48$.

   (b) $x - 7$ and $3x + 5$ are not factors of $6x^4 - x^3 - 94x^2 + 74x + 35$.

15.13   Use synthetic division to form a table of values and sketch the graph of

   (a) $y = x^3 - 13x + 12$ (c) $y = 3x^3 + 5x^2 - 4x - 3$ (e) $y = -x^3 + 13x + 12$
   (b) $y = 2x^3 + x^2 - 12x - 5$ (d) $y = x^4 - x^3 - 7x^2 + 13x - 6$

15.14   Sketch the graph of

   (a) $y = x(x^2 - 4)$ (b) $y = x(4 - x^2)$ (c) $y = x(x - 2)^2$ (d) $y = x(2 - x)^2$.

15.15   Use Descartes' rule of signs to discuss positive and negative zeros for

   (a) $P(x) = x^4 + x^2 + 1$ (none)

   (b) $M(x) = 7x^2 + 2x + 4$ (0 positive zeros, 0 or 2 negative)

Part II

TOPICS IN
DISCRETE MATHEMATICS

Chapter 16

Arithmetic and Geometric Progressions

A SEQUENCE IS A SET OF NUMBERS, called *terms*, arranged in a definite order; that is, there is a rule by which the terms after the first may be formed. Sequences may be finite or infinite. Only finite sequences will be treated in this chapter.

EXAMPLE 1

(*a*) Sequence: 3, 7, 11, 15, 19, 23, 27.
    Type: Finite of 7 terms.
    Rule: Add 4 to a given term to produce the next.

(*b*) Sequence: 3, 6, 12, 24, 48, 96.
    Type: Finite of 6 terms.
    Rule: Multiply a given term by 2 to produce the next.

AN ARITHMETIC PROGRESSION is a sequence in which each term after the first is formed by adding a fixed amount, called the *common difference*, to the preceding term. The sequence of Example 1(*a*) is an arithmetic progression whose common difference is 4. (See Problems 16.1–16.2.)

If a is the first term, d is the common difference, and n is the number of terms of an arithmetic progression, the successive terms are

$$a, a + d, a + 2d, a + 3d, \ldots, a + (n - 1)d. \qquad (16.1)$$

Thus, the *last* term (or *n*th term) l is given by

$$l = a + (n - 1)d \qquad (16.2)$$

The *sum S* of the n terms of this progression is given by

$$S = \frac{n}{2}(a + l) \qquad \text{or} \qquad S = \frac{n}{2}[2a + (n - 1)d] \qquad (16.3)$$

(For a proof, see Problem 16.3.)

EXAMPLE 2. Find the twentieth term and the sum of the first 20 terms of the arithmetic progression 4, 9, 14, 19, ...

For this progression $a = 4$, $d = 5$, and $n = 20$; then the twentieth term is $l = a + (n - 1)d = 4 + 19 \cdot 5 = 99$ and the sum of the first 20 terms is

$$S = \frac{n}{2}(a + l) = \frac{20}{2}(4 + 99) = 1030$$

(See Problems 16.4–16.8.)

THE TERMS BETWEEN THE FIRST AND LAST TERMS of an arithmetic progression are called *arithmetic means* between these two terms. Thus, to insert k arithmetic means between two numbers is to form an arithmetic progression of $(k + 2)$ terms having the two given numbers as first and last terms.

EXAMPLE 3. Insert five arithmetic means between 4 and 22.

We have $a = 4$, $l = 22$, and $n = 5 + 2 = 7$. Then $22 = 4 + 6d$ and $d = 3$. The first mean is $4 + 3 = 7$, the second is $7 + 3 = 10$, and so on. The required means are 7, 10, 13, 16, 19 and the resulting progression is 4, 7, 10, 13, 16, 19, 22.

108

When just one mean is to be inserted between two numbers, it is called *the arithmetic mean* (also, the average) of the two numbers.

EXAMPLE 4.   Find the arithmetic mean of the two numbers a and l.

We seek the middle term of an arithmetic progression of three terms having a and l as first and third terms, respectively. If d is the common difference, then $a + d = l - d$ and $d = \frac{1}{2}(l - a)$. The arithmetic mean is $a + d = a + \frac{1}{2}(l - a) = \frac{1}{2}(a + l)$. (See Problem 16.9.)

A GEOMETRIC PROGRESSION is a sequence in which each term after the first is formed by multiplying the preceding term by a fixed number, called the *common ratio*. The sequence $3, 6, 12, 24, 48, 96$ of Example 1(b) is a geometric progression whose common ratio is 2. (See Problems 16.10–16.11.)

If a is the first term, r is the common ratio, and n is the number of terms, the geometric progression is

$$a, ar, ar^2, \ldots, ar^{n-1}. \tag{16.4}$$

Thus, the last (or nth) term l is given by

$$l = ar^{n-1} \tag{16.5}$$

The *sum S* of the first n terms of the geometric progression (*16.4*) is given by

$$S = \frac{a - rl}{1 - r} \quad \text{or} \quad S = \frac{a(1 - r^n)}{1 - r} \tag{16.6}$$

(For a proof, see Problem 16.12.)

EXAMPLE 5.   Find the ninth term and the sum of the first nine terms of the geometric progression $8, 4, 2, 1, \ldots$.

Here $a = 8$, $r = \frac{1}{2}$, and $n = 9$; the ninth term is $l = ar^{n-1} = 8(\frac{1}{2})^8 = (\frac{1}{2})^5 = \frac{1}{32}$ and the sum of the first nine terms is

$$S = \frac{a - rl}{1 - r} = \frac{8 - \frac{1}{2}(\frac{1}{32})}{1 - \frac{1}{2}} = 16 - \frac{1}{32} = \frac{511}{32}$$

(See Problems 16.13–16.18.)

THE TERMS BETWEEN THE FIRST AND LAST TERMS of a geometric progression are called *geometric means* between the two terms. Thus, to insert k geometric means between two numbers is to form a geometric progression of $(k + 2)$ terms having the two given numbers as first and last terms.

EXAMPLE 6.   Insert four geometric means between 25 and $\frac{1}{125}$.

We have $a = 25$, $l = \frac{1}{125}$, and $n = 4 + 2 = 6$. Using $l = ar^{n-1}$, $\frac{1}{125} = 25r^5$; then $r^5 = (\frac{1}{5})^5$ and $r = \frac{1}{5}$. The first mean is $25(\frac{1}{5}) = 5$, the second is $5(\frac{1}{5}) = 1$, and so on. The required means are $5, 1, \frac{1}{5}, \frac{1}{25}$ and the geometric progression is $25, 5, 1, \frac{1}{5}, \frac{1}{25}, \frac{1}{125}$.

When one mean is to be inserted between two numbers, it is called *the geometric mean* of the two numbers.

The geometric mean of two numbers a and l, having like signs, is $(\pm)\sqrt{a \cdot l}$. The sign to be used is the common sign of a and l. (See Problem 16.19.)

Solved Problems

ARITHMETIC PROGRESSIONS

16.1 Determine which of the following sequences are arithmetic progressions (A.P.). In the case of an A.P., write three more terms.

(a) 3, 6, 9, 12, 15, 18.

Since $6 - 3 = 9 - 6 = 12 - 9 = 15 - 12 = 18 - 15 = 3$, the sequence is an A.P. with common difference 3. The next three terms are $18 + 3 = 21$, $21 + 3 = 24$, and $24 + 3 = 27$.

(b) 25, 19, 13, 7, 1, −5.

Since $19 - 25 = 13 - 19 = 7 - 13 = 1 - 7 = -5 - 1 = -6$, the sequence is an A.P. with $d = -6$. The next three terms are $-5 + (-6) = -11$, $-11 + (-6) = -17$, and $-17 + (-6) = -23$.

(c) 5, 10, 14, 20, 25.

Since $10 - 5 \neq 14 - 10$, the sequence is not an A.P.

(d) $3a - 2b$, $4a - b$, $5a$, $6a + b$.

Since $(4a - b) - (3a - 2b) = 5a - (4a - b) = (6a + b) - 5a = a + b$, the sequence is an A.P. with $d = a + b$. The next three terms are $(6a + b) + (a + b) = 7a + 2b$, $8a + 3b$, and $9a + 4b$.

16.2 Find the value of k such that each sequence is an A.P.

(a) $k - 1$, $k + 3$, $3k - 1$.

If the sequence is to form an A.P., $(k + 3) - (k - 1) = (3k - 1) - (k + 3)$. Then $k = 4$ and the A.P. is 3, 7, 11.

(b) $3k^2 + k + 1$, $2k^2 + k$; $4k^2 - 6k + 1$.

Setting $(2k^2 + k) - (3k^2 + k + 1) = (4k^2 - 6k + 1) - (2k^2 + k)$, we have $3k^2 - 7k + 2 = 0$ and $k = 2, \frac{1}{3}$. The progressions are 15, 10, 5 when $k = 2$ and $\frac{5}{3}, \frac{5}{9}, -\frac{5}{9}$ when $k = \frac{1}{3}$.

16.3 Obtain the formula $S = \dfrac{n}{2}[2a + (n - 1)d]$ for an arithmetic progression.

Write the indicated sum of the n terms in the order given by (16.1), then write this sum in reverse order, and sum term by term. Thus,

$$S = \{a \qquad\qquad\} + \{ \ a + d \qquad\quad\} + \cdots + \{a + (n - 2)d\} + \{a + (n - 1)d\}$$

$$S = \{a + (n - 1)d\} + \{ \ a + (n - 2)d\} + \cdots + \{a + d \qquad\quad\} + \{a \qquad\qquad\}$$

$$2S = \{2a + (n - 1)d\} + \{2a + (n - 1)d\} + \cdots + \{2a + (n - 1)d\} + \{2a + (n - 1)d\}$$

$$= n[2a + (n - 1)d] \qquad \text{and} \qquad S = \frac{n}{2}[2a + (n - 1)d]$$

16.4 (a) Find the eighteenth term and the sum of the first 18 terms of the A.P. $2, 6, 10, 14, \ldots$.

Here $a = 2$, $d = 4$, $n = 18$. Then $l = a + (n - 1)d = 2 + 17 \cdot 4 = 70$ and $S = \dfrac{n}{2}(a + l) = \dfrac{18}{2}(2 + 70) = 648$.

(b) Find the forty-ninth term and the sum of the first 49 terms of the A.P. $10, 4, -2, -8, \ldots$.

Here $a = 10$, $d = -6$, $n = 49$. Then $l = 10 + 48(-6) = -278$ and $S = \frac{49}{2}(10 - 278) = -6566$.

(c) Find the twelfth term and the sum of the first 15 terms of the A.P. $8, \frac{19}{3}, \frac{14}{3}, 3, \ldots$.

Since $a = 8$ and $d = -\frac{5}{3}$, the twelfth term is $l = 8 + 11(-\frac{5}{3}) = -\frac{31}{3}$ and the sum of the first 15 terms is

$$S = \frac{n}{2}[2a + (n - 1)d] = \frac{15}{2}\left[16 + 14\left(-\frac{5}{3}\right)\right] = \frac{15}{2}\left(-\frac{22}{3}\right) = -55$$

(d) Find the tenth term, the sum of the first 10 terms, and the sum of the first 13 terms of the A.P. $2x + 3y$, $x + y$, $-y$, ...

Here $a = 2x + 3y$ and $d = -x - 2y$. The tenth term is $l = (2x + 3y) + 9(-x - 2y) = -7x - 15y$.
Sum of first 10 terms is $S = 5[(2x + 3y) + (-7x - 15y)] = -25x - 60y$.
Sum of first 13 terms is $S = \frac{13}{2}[2(2x + 3y) + 12(-x - 2y)] = \frac{13}{2}(-8x - 18y) = -52x - 117y$.

16.5 The seventh term of an A.P. is 41 and the thirteenth term is 77. Find the twentieth term.
If a is the first term and d is the common difference, then for

the seventh term $\qquad\qquad\qquad\qquad a + 6d = 41$

and for the thirteenth term $\qquad\qquad a + 12d = 77$

Subtracting, $6d = 36$; then $d = 6$ and $a = 41 - 6 \cdot 6 = 5$. The twentieth term is $l = 5 + 19 \cdot 6 = 119$.

16.6 The sixth term of an A.P. is 21 and the sum of the first 17 terms is 0. Write the first three terms.

If a is the first term and d is the common difference,

$$a + 5d = 21 \qquad \text{and} \qquad 0 = \frac{17}{2}(2a + 16d) \qquad \text{or} \qquad a + 8d = 0$$

Then $d = -7$ and $a = -8d = 56$. The first three terms are $56, 49, 42$.

16.7 Obtain formulas for (a) l in terms of a, n, S; (b) a in terms of d, n, S.

(a) From $S = \dfrac{n}{2}(a + l)$, $\quad a + l = \dfrac{2S}{n}$ and $l = \dfrac{2S}{n} - a$.

(b) From $S = \dfrac{n}{2}[2a + (n - 1)d]$, $2a + (n - 1)d = \dfrac{2S}{n}$, $2a = \dfrac{2S}{n} - (n - 1)d$ and $a = \dfrac{S}{n} - \dfrac{1}{2}(n - 1)d$.

16.8 If a body is dropped, the distance (s meters) through which it falls freely in t seconds, is approximately $16t^2$. (a) Show that the distances through which it falls during the first, second, third, . . . seconds form an A.P. (b) How far will the body fall in the tenth second? (c) How far will it fall in the first 20 seconds?

(a) The distance through which the body falls during the first second is 16 m, during the second second is $16(2)^2 - 16 = 48$ m, during the third second is $16(3)^2 - 16(2)^2 = 80$ m, during the fourth second is $16(4)^2 - 16(3)^2 = 112$ m, and so on. These are the first four terms of an A.P. whose common difference is 32.

(b) When $n = 10$, $l = 16 + 9(32) = 304$ m.

(c) In the first 20 s, the body falls $16(20)^2 = 6400$ m.

16.9 (a) Insert six arithmetic means between 7 and 77.
For the A.P. having $a = 7$, $l = 77$ and $n = 6 + 2 = 8$, $77 = 7 + 7d$ and $d = 10$. The required means are $17, 27, 37, 47, 57, 67$ and the A.P. is $7, 17, 27, 37, 47, 57, 67, 77$.

(b) Find the arithmetic mean of 8 and -56.
From Example 4, the arithmetic mean is $\frac{1}{2}(a + l) = \frac{1}{2}[8 + (-56)] = -24$.

GEOMETRIC PROGRESSIONS

16.10 Determine which of the following sequences are geometric progressions (G.P.). In the case of a G.P., write the next three terms.

(a) $4, 8, 16, 32, 64$.
Since $\frac{8}{4} = \frac{16}{8} = \frac{32}{16} = \frac{64}{32} = 2$, the sequence is a G.P. with common ratio 2. The next three terms are $128, 256, 512$.

(b) $1, \frac{1}{4}, \frac{1}{16}, \frac{1}{48}$.
Since $\frac{1}{16}/\frac{1}{4} \neq \frac{1}{48}/\frac{1}{16}$, the sequence is not a G.P.

(c) $12, -4, \frac{4}{3}, -\frac{4}{9}$.
Since $-\frac{4}{12} = \frac{4}{3}/-4 = -\frac{4}{9}/\frac{4}{3} = -\frac{1}{3}$, the sequence is a G.P. with common ratio $-\frac{1}{3}$. The next three terms are $\frac{4}{27}, -\frac{4}{81}, \frac{4}{243}$.

16.11   Find the value of k so that the sequence $2k - 5$, $k - 4$, $10 - 3k$ forms a G.P.

If the sequence is to form a G.P.,

$$\frac{k-4}{2k-5} = \frac{10-3k}{k-4} \quad \text{or} \quad k^2 - 8k + 16 = -6k^2 + 35k - 50$$

Then $7k^2 - 43k + 66 = (k - 3)(7k - 22) = 0$ and $k = 3, \frac{22}{7}$.

The sequences are $1, -1, 1$ when $k = 3$ and $\frac{9}{7}, -\frac{6}{7}, \frac{4}{7}$ when $k = \frac{22}{7}$.

16.12   Obtain the formula $S = \dfrac{a(1 - r^n)}{1 - r}$ for a geometric progression.

Write the indicated sum of the n terms given by (16.4), then multiply this sum by r, and subtract term by term. Thus,

$$S = a + ar + ar^2 + \cdots + ar^{n-1}$$

$$rS = \qquad ar + ar^2 + \cdots + ar^{n-1} + ar^n$$

$$S - rS = a \qquad\qquad\qquad\qquad\qquad - ar^n$$

Then $S(1 - r) = a - ar^n = a(1 - r^n)$ and $S = \dfrac{a(1 - r^n)}{1 - r}$.

16.13   (a)   Find the seventh term and the sum of the first seven terms of the G.P. $12, 16, \frac{64}{3}, \ldots$.

Here $a = 12$, $r = \frac{4}{3}$, $n = 7$. The seventh term is $l = ar^{n-1} = 12(\frac{4}{3})^6 = 4^7/3^5 = 16384/243$ and the sum of the first seven terms is

$$S = \frac{a - rl}{1 - r} = \frac{12 - \frac{4}{3}(4^7/3^5)}{1 - \frac{4}{3}} = \frac{4^8}{3^5} - 36 = \frac{65536 - 8748}{243} = \frac{56788}{243}$$

(b)   Find the sixth term and the sum of the first nine terms of the G.P. $4, -6, 9, \ldots$.

Since $a = 4$ and $r = -\frac{3}{2}$, the sixth term is $l = 4(-\frac{3}{2})^5 = -3^5/2^3 = -\frac{243}{8}$ and the sum of the first nine terms is

$$S = \frac{a(1 - r^n)}{1 - r} = \frac{4[1 - (-\frac{3}{2})^9]}{1 - (-\frac{3}{2})} = \frac{8(1 + 3^9/2^9)}{5} = \frac{2^9 + 3^9}{5 \cdot 2^6} = \frac{4039}{64}$$

(c)   Find the sum of the G.P. $8, -4, 2, \ldots, \frac{1}{128}$.

$$S = \frac{a - rl}{1 - r} = \frac{8 - (-\frac{1}{2})(\frac{1}{128})}{1 - (-\frac{1}{2})} = \frac{2^4 + (\frac{1}{2})^7}{3} = \frac{2^{11} + 1}{3 \cdot 2^7} = \frac{683}{128}$$

16.14   The fourth term of a G.P. is 1 and the eighth term is $\frac{1}{256}$. Find the tenth term.

Since the fourth term is 1, $ar^3 = 1$; since the eighth term is $\frac{1}{256}$, $ar^7 = \frac{1}{256}$. Then $ar^7/ar^3 = \frac{1}{256}$, $r^4 = \frac{1}{256}$, and $r = \pm\frac{1}{4}$. From $ar^3 = 1$, we have $a = \pm 64$. In each case, the tenth term is $\frac{1}{4096}$.

16.15   Given $S = \frac{3367}{64}$, $r = \frac{3}{4}$, $l = \frac{243}{64}$. Find a and n.

Since $S = \dfrac{3367}{64} = \dfrac{a - (\frac{3}{4})(\frac{243}{64})}{1 - \frac{3}{4}} = 4a - \dfrac{729}{64}$, $4a = \dfrac{4096}{64}$ and $a = 16.$

Now $l = \dfrac{243}{64} = 16\left(\dfrac{3}{4}\right)^{n-1}$, $\left(\dfrac{3}{4}\right)^{n-1} = \dfrac{243}{16 \cdot 64} = \left(\dfrac{3}{4}\right)^5$, $n - 1 = 5$ and $n = 6.$

16.16 Given $a = 8$, $r = \frac{3}{2}$, $S = \frac{2059}{8}$. Find l and n.

Since $\qquad S = \frac{2059}{8} = \frac{8 - \left(\frac{3}{2}\right)l}{1 - \frac{3}{2}} = 3l - 16, \qquad 3l = \frac{2059}{8} + 16 = \frac{2187}{8} \qquad$ and $\qquad l = \frac{729}{8}.$

Now $\qquad l = \frac{729}{8} = 8\left(\frac{3}{2}\right)^{n-1}, \qquad \left(\frac{3}{2}\right)^{n-1} = \frac{729}{64} = \left(\frac{3}{2}\right)^{6}, \qquad n - 1 = 6 \qquad$ and $\qquad n = 7.$

16.17 If a boy undertakes to deposit 1¢ on Sept. 1, 2¢ on Sept. 2, 4¢ on Sept. 3, 8¢ on Sept. 4, and so on, (*a*) how much will he deposit from Sept. 1 to Sept. 15 inclusive, (*b*) how much would he deposit on Sept. 30?

Here, $a = 0.01$ and $r = 2$.
(*a*) When $n = 15$,

$$S = \frac{0.01(1 - 2^{15})}{1 - 2} = 0.01(2^{15} - 1) = \$327.67.$$

(*b*) When $n = 30$, $l = 0.01(2)^{29} = \$5\,368\,709.12.$

16.18 A rubber ball is dropped from a height of 81 m. Each time it strikes the ground it rebounds two-thirds of the distance through which it last fell. (*a*) Through what distance did the ball fall when it struck the ground for the sixth time? (*b*) Through what distance had it traveled from the time it was dropped until it struck the ground for the sixth time?

(*a*) The successive distances through which the ball falls form a G.P. in which $a = 81$, $r = \frac{2}{3}$.
When $n = 6$,
$$l = 81\left(\frac{2}{3}\right)^5 = \frac{32}{3}\text{ m}$$

(*b*) The required distance is the sum of the distances for the first six falls and the first five rebounds.
For the falls: $a = 81$, $r = \frac{2}{3}$, $n = 6$, and

$$S = \frac{81\left[1 - \left(\frac{2}{3}\right)^6\right]}{1 - \frac{2}{3}} = 81\left(3 - \frac{2^6}{3^5}\right) = \frac{3^6 - 2^6}{3} = \frac{665}{3}\text{ m}$$

For the rebounds: $a = 54$, $r = \frac{2}{3}$, $n = 5$, and

$$S = \frac{54\left[1 - \left(\frac{2}{3}\right)^5\right]}{1 - \frac{2}{3}} = 54\left(3 - \frac{2^5}{3^4}\right) = \frac{2}{3}(3^5 - 2^5) = \frac{422}{3}\text{ m}$$

Thus, the total distance is $\frac{665}{3} + \frac{422}{3} = 362\frac{1}{3}$ m.

16.19 (*a*) Insert five geometric means between 8 and $\frac{1}{8}$.
We have $a = 8$, $l = \frac{1}{8}$, $n = 5 + 2 = 7$. Since $l = ar^{n-1}$, $\frac{1}{8} = 8r^6$ and $r = \pm\frac{1}{2}$.
When $r = \frac{1}{2}$, the first mean is $8\left(\frac{1}{2}\right) = 4$, the second is $4\left(\frac{1}{2}\right) = 2$, and so on. The required means are $4, 2, 1, \frac{1}{2}, \frac{1}{4}$ and the G.P. is $8, 4, 2, 1, \frac{1}{2}, \frac{1}{4}, \frac{1}{8}$.
When $r = -\frac{1}{2}$, the means are $-4, 2, -1, \frac{1}{2}, -\frac{1}{4}$ and the G.P. is $8, -4, 2, -1, \frac{1}{2}, -\frac{1}{4}, \frac{1}{8}$.

(*b*) Insert four geometric means between $\frac{81}{2}$ and $-\frac{16}{3}$.
Here $a = \frac{81}{2}$, $l = -\frac{16}{3}$, $n = 6$. Then $-\frac{16}{3} = \frac{81}{2}r^5$, $r^5 = -\frac{32}{243}$ and $r = -\frac{2}{3}$. The required means are $-27, 18, -12, 8$ and the G.P. is $\frac{81}{2}, -27, 18, -12, 8, -\frac{16}{3}$.

(*c*) Find the geometric mean of $\frac{1}{3}$ and 243.
The required mean is $\sqrt{\frac{1}{3}(243)} = \sqrt{81} = 9$.

(*d*) Find the geometric mean of $-\frac{2}{3}$ and $-\frac{32}{27}$.
The required mean is $-\sqrt{\left(-\frac{2}{3}\right)\left(-\frac{32}{27}\right)} = -\sqrt{\frac{64}{81}} = -\frac{8}{9}$.

Supplementary Problems

ARITHMETIC PROGRESSION

16.20   Find

(a)   The fifteenth term and the sum of the first 15 terms of the A.P. 3, 8, 13, 18, ...
(b)   The twelfth term and the sum of the first 20 terms of the A.P. 11, 8, 5, 2, ...
(c)   The sum of the A.P. for which $a = 6\frac{3}{4}$, $l = -3\frac{1}{4}$, $n = 17$.
(d)   The sum of all the integers from 1 to 200 which are divisible by 3.

> Ans.   (a)   73 570   (b)   −22, −350   (c) $29\frac{3}{4}$ (d)   6633

16.21   The fourth term of an A.P. is 14 and the ninth term is 34. Find the thirteenth term.

> Ans.   50

16.22   The sum of the first 7 terms of an A.P. is 98 and the sum of the first 12 terms is 288. Find the sum of the first 20 terms.

> Ans.   800

16.23   Find the sum of (a) the first n positive integers, (b) the first n odd positive integers.

> Ans.   (a) $\frac{1}{2}n(n+1)$ (b) n^2

16.24   (a)   Sum all the integers between 200 and 1000 that are divisible by 3.

> Ans.   160 200

(b)   Sum all the even positive integers less than 200 which are not divisible by 6.

> Ans.   6534

16.25   In a potato race, 10 potatoes are placed 8 ft apart in a straight line. If the potatoes are to be picked up singly and returned to the basket, and if the first potato is 20 ft in front of the basket, find the total distance covered by a contestant who finishes the race.

> Ans.   1120 ft

16.26   In a lottery, tickets are numbered consecutively from 1 to 100. Customers draw a ticket at random and pay an amount in cents corresponding to the number on the ticket except for those tickets with numbers divisible by 5, which are free. How much is realized if 100 tickets are sold?

> Ans. $40

16.27   Find the arithmetic mean between (a) 6 and 60, (b) $a - 2d$ and $a + 6d$.

> Ans.   (a)   33   (b) $a + 2d$

16.28   Insert five arithmetic means between 12 and 42.

> Ans.   17, 22, 27, 32, 37

16.29   After inserting x arithmetic means between 2 and 38, the sum of the resulting progression is 200. Find x.

> Ans.   8

GEOMETRIC PROGRESSION

16.30 Find

(a) The eighth term and the sum of the first eight terms of the G.P. $4, 12, 36, \ldots$.

(b) The tenth term and the sum of the first 12 terms of the G.P. $8, 4, 2, \ldots$.

(c) The sum of the G.P. for which $a = 64$, $l = 729$, and $n = 7$.

   Ans.   (a) 8748, 13 120   (b) 1/64, 15 255/256   (c) 2059, 463

16.31 The third term of a G.P. is 36 and the fifth term is 16. Find the tenth term.

   Ans. $\pm\frac{512}{243}$

16.32 The sum of the first three terms of a G.P. is 21 and the sum of the first six terms is $20\frac{2}{3}$. Find the sum of the first nine terms.

   Ans. $20\frac{61}{243}$

16.33 Given $S = \frac{255}{192}$, $l = -\frac{1}{64}$, $r = -\frac{1}{2}$; find a and n.

   Ans. $a = 2$, $n = 8$

16.34 Find three numbers in geometric progression such that their sum is 14 and the sum of their squares is 84.

   Ans.   2, 4, 8

16.35 Prove: $x^n - y^n = (x - y)(x^{n-1} + x^{n-2}y + \cdots + xy^{n-2} + y^{n-1})$, n being a positive integer.

16.36 In a certain colony of bacteria each divides into two every hour. How many will be produced from a single bacillus if the rate of division continues for 12 hr?

   Ans.   4096

16.37 Find the geometric mean between (a) 2 and 32, (b) -4 and -25.

   Ans.   (a) 8   (b) -10

16.38 Insert five geometric means between 6 and 384.

   Ans.   12, 24, 48, 96, 192

16.39 Show that for $p > q$, positive integers, their arithmetic mean A is greater than their geometric mean G.

   Hint: Consider $A - G$.

16.40 The sum of three numbers in A.P. is 24. If the first is decreased by 1 and the second is decreased by 2, the three numbers are in G.P. Find the A.P.

   Ans.   4, 8, 12 or 13, 8, 3

Chapter 17

Infinite Geometric Series

THE INDICATED SUM of the terms of a finite or infinite sequence is called a finite or infinite *series*. The sums of arithmetic and geometric progressions in the preceding chapter are examples of finite series.

Of course it is impossible to add up all the terms of an infinite series; that is, in the usual meaning of the word *sum*, there is no such thing as the sum of such a series. However, it is possible to associate with certain infinite series a well-defined number which, for convenience, will be called the sum of the series.

Infinite series will be treated in some detail in a later chapter. For the study of the infinite geometric series here, we shall need only to examine the behavior of r^n, where $|r| < 1$, as n increases indefinitely.

EXAMPLE 1. From the table of values of $(\frac{1}{2})^n$ in Table 17.1, it appears that, as n increases indefinitely, $(\frac{1}{2})^n$ decreases indefinitely while remaining positive. Moreover, it can be made to have a value as near 0 as we please by choosing n sufficiently large. We describe this state of affairs by saying: The limit of $(\frac{1}{2})^n$, as n increases indefinitely, is 0.

Table 17.1

n	1	3	5	10
$(\frac{1}{2})^n$	0.5	0.125	0.03125	0.0009765625

By examining the behavior of r^n for other values of r, it becomes tolerably clear that

The limit of r^n, as n increases indefinitely, is 0 when $|r| < 1$.

THE SUM S of the infinite geometric series

$$a + ar + ar^2 + \cdots + ar^{n-1} + \cdots, \qquad |r| < 1, \qquad \text{is } S = \frac{a}{1 - r}.$$

(For a proof, see Problem 17.1.)

EXAMPLE 2. For the infinite geometric series $12 + 4 + \frac{4}{3} + \frac{4}{9} + \cdots$, $a = 12$ and $r = \frac{1}{3}$. The sum of the series is

$$S = \frac{a}{1 - r} = \frac{12}{1 - \frac{1}{3}} = 18. \text{ (See Problems 17.2–17.5.)}$$

EVERY REPEATING DECIMAL approximates a rational number. This rational number is also called the *limiting value of the decimal*.

EXAMPLE 3. Find the limiting value of the repeating decimal .727272.... We write .727272... = .72 + .0072 + .000072 + $\cdots$ and note that for this infinite geometric series $a = .72$ and $r = .01$.

Then

$$S = \frac{a}{1 - r} = \frac{.72}{1 - .01} = \frac{.72}{.99} = \frac{72}{99} = \frac{8}{11}$$

(See Problem 17.5.)

116

SIGMA NOTATION is a convenient notation for expressing sums. $\sum_{i=a_1}^{a_m} x_i$ simply means: add $x_{a_1} + x_{a_2} + \cdots + x_{a_{m-1}} + x_m$. For example,

$$\sum_{i=1}^{4} 2i = 2(1) + 2(2) + 2(3) + 2(4)$$

$$= 2 + 4 + 6 + 8 = 20$$

$$\sum_{j=2}^{6} j^2 = 2^2 + 3^2 + 4^2 + 5^2 + 6^2$$

Thus, $$a + ar + ar^2 + \cdots + ar^{n-1} + \cdots = \frac{a}{1-r} = \sum_{i=0}^{\infty} ar^i, \qquad |r| < 1$$

Solved Problems

17.1   Prove: The sum of the infinite geometric series $a + ar + ar^2 + \cdots + ar^{n-1} + \cdots$, where $|r| < 1$, is $S = \dfrac{a}{1-r}$.

The sum of the first n terms of the series is

$$S_n = \frac{a(1-r^n)}{1-r} = \frac{a}{1-r} - \frac{a}{1-r} r^n$$

As n increases indefinitely, the first term $\dfrac{a}{1-r}$ remains fixed while r^n, and hence $\dfrac{a}{1-r} r^n$, approaches zero in value. Thus, $S = \dfrac{a}{1-r}$.

17.2   Determine the sum of each of the following infinite geometric series:

(a) $18 + 12 + 8 + \cdots$

Here $a = 18$, $r = \frac{2}{3}$, and $S = \dfrac{a}{1-r} = \dfrac{18}{1-\frac{2}{3}} = 54.$

(b) $25 - 20 + 16 - \cdots$

Here $a = 25$, $r = -\frac{4}{5}$, and $S = \dfrac{a}{1-r} = \dfrac{25}{1-(-\frac{4}{5})} = \dfrac{125}{9}.$

(c) $.6 + .06 + .006 + \cdots$

Here $a = .6$, $r = .1$, and $S = \dfrac{.6}{1-.1} = \dfrac{.6}{.9} = \dfrac{6}{9} = \dfrac{2}{3}.$

17.3   An equilateral triangle has a perimeter of 30 cm. Another triangle is formed by joining the midpoints of the sides of the given triangle, another is formed by joining the midpoints of the sides of the second triangle, and so on. Find the sum of the perimeters of the triangles thus formed.

Since the side of each new triangle is $\frac{1}{2}$ the side of the triangle from which it is formed, the perimeters of the triangles are $30, 15, \frac{15}{2}, \ldots$.

Then $$30 + 15 + \frac{15}{2} + \cdots = \frac{30}{1-\frac{1}{2}} = 60 \text{ cm.}$$

17.4 A rubber ball is dropped from a height of 81 m. Each time it strikes the ground it rebounds two-thirds of the distance through which it last fell. Find the total distance it travels in coming to rest.

For the falls: $a = 81$ and $r = \frac{2}{3}$; $S = \dfrac{a}{1-r} = \dfrac{81}{1-\frac{2}{3}} = 243$ m.

For the rebounds: $a = 54$ and $r = \frac{2}{3}$; $S = \dfrac{54}{1-\frac{2}{3}} = 162$ m.

Thus, the total distance traveled is $243 + 162 = 405$ m.

17.5 For what values of x does $\dfrac{1}{x+1} + \dfrac{1}{(x+1)^2} + \dfrac{1}{(x+1)^3} + \cdots$ have a sum? Find the sum.

There will be a sum provided $|r| = \left|\dfrac{1}{x+1}\right| < 1$.

When $\left|\dfrac{1}{x+1}\right| = 1$, $|x+1| = 1$, $(x+1)^2 = 1$, $x^2 + 2x = 0$, and $x = -2, 0$. By examining the intervals $x < -2$, $-2 < x < 0$, and $x > 0$, we find that $\left|\dfrac{1}{x+1}\right| < 1$ when $x < -2$ and $x > 0$.

Thus, the series has a sum $S = \dfrac{1/(x+1)}{1 - 1/(x+1)} = \dfrac{1}{x}$ when $x < -2$ and $x > 0$.

17.6 Find the limiting value of each of the repeating decimals.

(a) .0123123123 ...

Since $.0123123123\ldots = .0123 + .0000123 + .0000000123 + \cdots$ in which $a = .0123$ and $r = .001$,

$$S = \frac{a}{1-r} = \frac{.0123}{1-.001} = \frac{.0123}{.999} = \frac{123}{9990} = \frac{41}{3330}$$

(b) 2.373737 ...

The given number may be written as $2 + [.37 + .0037 + .000037 + \cdots]$. For the infinite geometric series in the brackets, $a = .37$ and $r = .01$; hence, $S = \dfrac{.37}{1-.01} = \dfrac{.37}{.99} = \dfrac{37}{99}$.

The limiting value is $2 + \frac{37}{99} = \frac{235}{99}$.

(c) 23.1454545 ...

Write $23.1454545\ldots = 23.1 + [.045 + .00045 + .0000045 + \cdots]$

$$= 23.1 + \frac{.045}{1-.01} = 23.1 + \frac{45}{990} = \frac{231}{10} + \frac{1}{22} = \frac{1273}{55}$$

Supplementary Problems

17.7 Sum the following infinite geometric series:

(a) $36 + 12 + 4 + \cdots$ (d) $5.6 - 2.24 + 0.896 - \cdots$ (f) $3 - \dfrac{3}{\sqrt{2}-1} + \dfrac{3}{3-2\sqrt{2}} - \cdots$

(b) $18 - 12 + 8 - \cdots$ (e) $1 + \frac{1}{2}\sqrt{2} + \frac{1}{2} + \cdots$

(c) $5 + 3 + 1.8 + \cdots$

Ans. (a) 54   (b) $\frac{54}{5}$ (c) $12\frac{1}{2}$ (d) 4   (e) $2 + \sqrt{2}$ (f) $\frac{3}{2}(2 - \sqrt{2})$

17.8 A swinging pendulum bob traverses the following distances: 40, 30, $22\frac{1}{2}$, ... cm. Find the distance which it travels before coming to rest.

    Ans. 160 cm.

17.9 An unlimited sequence of squares are inscribed one within another by joining the midpoints of the sides of each preceding square. If the initial square is 8 cm on a side, find the sum of the perimeters of these squares.

    Ans. $32(2 + \sqrt{2})$ cm.

17.10 Express each repeating decimal as a rational fraction.

    (*a*) 0.272727...   (*b*) 1.702702...   (*c*) 2.4242...   (*d*) 0.076923076923...

    Ans. (*a*) $\frac{3}{11}$ (*b*) $\frac{189}{111}$ (*c*) $\frac{80}{33}$ (*d*) $\frac{1}{13}$

17.11 Find the values of x for which each of the following geometric series may be *summed*:

    (*a*) $3 + 3x + 3x^2 + \cdots$ **Hint:** $|r| = |x|$

    (*b*) $1 + (x - 1) + (x - 1)^2 + \cdots$ **Hint:** $|r| = |x - 1|$

    (*c*) $5 + 5(x - 3) + 5(x - 3)^2 + \cdots$

    Ans. (*a*) $-1 < x < 1$ (*b*) $0 < x < 2$ (*c*) $2 < x < 4$

17.12 Find the exact error when $\frac{1}{6}$ is approximated as 0.1667.

    Ans. $0.000033... = 1/30\,000$

17.13 Evaluate

    (*a*) $\sum\limits_{j=1}^{2} 3j^2$

    Ans. 15

    (*b*) $\sum\limits_{i=0}^{\infty} 3 \cdot (0.5)^i$

    Ans. 6

Chapter 18

Mathematical Induction

EVERYONE IS FAMILIAR with the process of reasoning, called *ordinary* or *incomplete induction*, in which a generalization is made on the basis of a number of simple observations.

EXAMPLE 1. We observe that $1 = 1^2$, $1 + 3 = 4 = 2^2$, $1 + 3 + 5 = 9 = 3^2$, $1 + 3 + 5 + 7 = 16 = 4^2$, and conclude that

$$1 + 3 + 5 + \cdots + (2n - 1) = n^2$$

or, in words, the sum of the first n odd integers is n^2.

EXAMPLE 2. We observe that 2 points determine $1 = \frac{1}{2} \cdot 2(2 - 1)$ line; that 3 points, not on a line, determine $3 = \frac{1}{2} \cdot 3(3 - 1)$ lines; that 4 points, no 3 on a line, determine $6 = \frac{1}{2} \cdot 4(4 - 1)$ lines; that 5 points, no 3 on a line, determine $10 = \frac{1}{2} \cdot 5(5 - 1)$ lines; and conclude that n points, no 3 on a line, determine $\frac{1}{2}n(n - 1)$ lines.

EXAMPLE 3. We observe that for $n = 1, 2, 3, 4, 5$ the values of

$$f(n) = \frac{n^4}{8} - \frac{17n^3}{12} + \frac{47n^2}{8} - \frac{103n}{12} + 6$$

are $2, 3, 5, 7, 11$, respectively, and conclude that $f(n)$ is a prime number for every positive integral value of n.

The conclusions in Examples 1 and 2 are valid as we shall prove later. The conclusion in Example 3 is false since $f(6) = 22$ is not a prime number.

MATHEMATICAL INDUCTION or complete induction is a type of reasoning by which such conclusions as were drawn in the above examples may be proved or disproved.

The steps are

(*1*) The verification of the proposed formula or theorem for some positive integral value of n, usually the smallest. (Of course, we would not attempt to prove an unknown theorem by mathematical induction without first verifying it for several values of n.)

(*2*) The proof that if the proposed formula or theorem is true for $n = k$, some positive integer, it is true also for $n = k + 1$.

(*3*) The conclusion that the proposed formula or theorem is true for all values of n greater than the one for which verification was made in Step *1*.

EXAMPLE 4. Prove: $1 + 3 + 5 + \cdots + (2n - 1) = n^2$.

(*1*) The formula is true for $n = 1$ since $1 = 1^2$.

(*2*) Let us assume the formula true for $n = k$, any positive integer; that is, let us assume that

$$1 + 3 + 5 + \cdots + (2k - 1) = k^2 \qquad (18.1)$$

We wish to show that, when (*18.1*) is true, the proposed formula is then true for $n = k + 1$; that is, that

$$1 + 3 + 5 + \cdots + (2k - 1) + (2k + 1) = (k + 1)^2 \qquad (18.2)$$

[NOTE: Statements (*18.1*) and (*18.2*) are obtained by replacing n in the proposed formula by k and $k + 1$, respectively. Now it is clear that the left member of (*18.2*) can be obtained from the left member of (*18.1*) by adding $(2k + 1)$. At this point the proposed formula is true or false according as we do or do not obtain the right member of (*18.2*) when $(2k + 1)$ is added to the right member of (*18.1*).]

Adding $(2k + 1)$ to both members of (18.1), we have

$$1 + 3 + 5 + \cdots + (2k - 1) + (2k + 1) = k^2 + (2k + 1) = (k + 1)^2 \qquad (18.3)$$

Now (18.3) is identical with (18.2); thus, if the proposed formula is true for any positive integer $n = k$ it is true for the next positive integer $n = k + 1$.

(3) Since the formula is true for $n = k = 1$ (Step 1), it is true for $n = k + 1 = 2$; being true for $n = k = 2$, it is true for $n = k + 1 = 3$; and so on. Hence, the formula is true for all positive integral values of n.

Solved Problems

Prove by mathematical induction.

18.1 $1 + 7 + 13 + \cdots + (6n - 5) = n(3n - 2)$.

(1) The proposed formula is true for $n = 1$, since $1 = 1(3 - 2)$.

(2) Assume the formula to be true for $n = k$, a positive integer; that is, assume

$$1 + 7 + 13 + \cdots + (6k - 5) = k(3k - 2) \qquad (1)$$

Under this assumption we wish to show that

$$1 + 7 + 13 + \cdots + (6k - 5) + (6k + 1) = (k + 1)(3k + 1) \qquad (2)$$

When $(6k + 1)$ is added to both members of (1), we have on the right

$$k(3k - 2) + (6k + 1) = 3k^2 + 4k + 1 = (k + 1)(3k + 1)$$

hence, if the formula is true for $n = k$ it is true for $n = k + 1$.

(3) Since the formula is true for $n = k = 1$ (Step 1), it is true for $n = k + 1 = 2$; being true for $n = k = 2$ it is true for $n = k + 1 = 3$; and so on, for every positive integral value of n.

18.2 $1 + 5 + 5^2 + \cdots + 5^{n-1} = \frac{1}{4}(5^n - 1)$.

(1) The proposed formula is true for $n = 1$, since $1 = \frac{1}{4}(5 - 1)$.

(2) Assume the formula to be true for $n = k$, a positive integer; that is, assume

$$1 + 5 + 5^2 + \cdots + 5^{k-1} = \frac{1}{4}(5^k - 1) \qquad (1)$$

Under this assumption we wish to show that

$$1 + 5 + 5^2 + \cdots + 5^{k-1} + 5^k = \frac{1}{4}(5^{k+1} - 1) \qquad (2)$$

When 5^k is added to both members of (1), we have on the right

$$\tfrac{1}{4}(5^k - 1) + 5^k = \tfrac{5}{4}(5^k) - \tfrac{1}{4} = \tfrac{1}{4}(5 \cdot 5^k - 1) = \tfrac{1}{4}(5^{k+1} - 1)$$

hence, if the formula is true for $n = k$ it is true for $n = k + 1$.

(3) Since the formula is true for $n = k = 1$ (Step 1), it is true for $n = k + 1 = 2$; being true for $n = k = 2$ it is true for $n = k + 1 = 3$; and so on, for every positive integral value of n.

18.3 $\dfrac{5}{1 \cdot 2 \cdot 3} + \dfrac{6}{2 \cdot 3 \cdot 4} + \dfrac{7}{3 \cdot 4 \cdot 5} + \cdots + \dfrac{n + 4}{n(n + 1)(n + 2)} = \dfrac{n(3n + 7)}{2(n + 1)(n + 2)}$.

(1) The formula is true for $n = 1$, since $\dfrac{5}{1 \cdot 2 \cdot 3} = \dfrac{1(3 + 7)}{2 \cdot 2 \cdot 3} = \dfrac{5}{6}$.

(2)   Assume the formula to be true for $n = k$, a positive integer; that is, assume

$$\frac{5}{1 \cdot 2 \cdot 3} + \frac{6}{2 \cdot 3 \cdot 4} + \cdots + \frac{k+4}{k(k+1)(k+2)} = \frac{k(3k+7)}{2(k+1)(k+2)} \qquad (1)$$

Under this assumption we wish to show that

$$\frac{5}{1 \cdot 2 \cdot 3} + \frac{6}{2 \cdot 3 \cdot 4} + \cdots + \frac{k+4}{k(k+1)(k+2)} + \frac{k+5}{(k+1)(k+2)(k+3)} = \frac{(k+1)(3k+10)}{2(k+2)(k+3)} \qquad (2)$$

When $\dfrac{k+5}{(k+1)(k+2)(k+3)}$ is added to both members of (1), we have on the right

$$\frac{k(3k+7)}{2(k+1)(k+2)} + \frac{k+5}{(k+1)(k+2)(k+3)} = \frac{1}{(k+1)(k+2)} \left[\frac{k(3k+7)}{2} + \frac{k+5}{k+3} \right]$$

$$= \frac{1}{(k+1)(k+2)} \frac{k(3k+7)(k+3) + 2(k+5)}{2(k+3)}$$

$$= \frac{1}{(k+1)(k+2)} \frac{3k^3 + 16k^2 + 23k + 10}{2(k+3)}$$

$$= \frac{1}{(k+1)(k+2)} \frac{(k+1)^2(3k+10)}{2(k+3)}$$

$$= \frac{(k+1)(3k+10)}{2(k+2)(k+3)};$$

hence, if the formula is true for $n = k$ it is true for $n = k + 1$.

(3)   Since the formula is true for $n = k = 1$ (Step 1), it is true for $n = k + 1 = 2$; being true for $n = k = 2$, it is true for $n = k + 1 = 3$; and so on, for all positive integral values of n.

18.4 $x^{2n} - y^{2n}$ is divisible by $x + y$.

(1)   The theorem is true for $n = 1$, since $x^2 - y^2 = (x - y)(x + y)$ is divisible by $x + y$.

(2)   Let us assume the theorem true for $n = k$, a positive integer; that is, let us assume

$$x^{2k} - y^{2k} \text{ is divisible by } x + y. \qquad (1)$$

We wish to show that, when (1) is true.

$$x^{2k+2} - y^{2k+2} \text{ is divisible by } x + y. \qquad (2)$$

Now $x^{2k+2} - y^{2k+2} = (x^{2k+2} - x^2 y^{2k}) + (x^2 y^{2k} - y^{2k+2}) - x^2(x^{2k} - y^{2k}) + y^{2k}(x^2 - y^2)$. In the first term $(x^{2k} - y^{2k})$ is divisible by $(x + y)$ by assumption, and in the second term $(x^2 - y^2)$ is divisible by $(x + y)$ by Step 1; hence, if the theorem is true for $n = k$, a positive integer, it is true for the next one $n = k + 1$.

(3)   Since the theorem is true for $n = k = 1$, it is true for $n = k + 1 = 2$; being true for $n = k = 2$, it is true for $n = k + 1 = 3$; and so on, for every positive integral value of n.

18.5   The number of straight lines determined by $n > 1$ points, no three on the same straight line, is $\frac{1}{2}n(n - 1)$.

(1)   The theorem is true when $n = 2$, since $\frac{1}{2} \cdot 2(2 - 1) = 1$ and two points determine one line.

(2)   Let us assume that k points, no three on the same straight line, determine $\frac{1}{2}k(k - 1)$ lines.

When an additional point is added (not on any of the lines already determined) and is joined to each of the original k points, k new lines are determined. Thus, altogether we have $\frac{1}{2}k(k - 1) + k = \frac{1}{2}k(k - 1 + 2) = \frac{1}{2}k(k + 1)$ lines and this agrees with the theorem when $n = k + 1$.

Hence, if the theorem is true for $n = k$, a positive integer greater than 1, it is true for the next one $n = k + 1$.

(3)   Since the theorem is true for $n = k = 2$ (Step 1), it is true for $n = k + 1 = 3$; being true for $n = k = 3$, it is true for $n = k + 1 = 4$; and so on, for every possible integral value > 1 of n.

Supplementary Problems

Prove by mathematical induction, n being a positive integer.

18.6 $\quad 1 + 2 + 3 + \cdots + n = \frac{1}{2}n(n+1)$

18.7 $\quad 1 + 4 + 7 + \cdots + (3n-2) = \frac{1}{2}n(3n-1)$

18.8 $\quad 1^2 + 3^2 + 5^2 + \cdots + (2n-1)^2 = \frac{1}{3}n(4n^2-1)$

18.9 $\quad 1^2 + 2^2 + 3^2 + \cdots + n^2 = \frac{1}{6}n(n+1)(2n+1)$

18.10 $\quad 1^3 + 2^3 + 3^3 + \cdots + n^3 = \frac{1}{4}n^2(n+1)^2$

18.11 $\quad 1^4 + 2^4 + 3^4 + \cdots + n^4 = \frac{1}{30}n(n+1)(2n+1)(3n^2+3n-1)$

18.12 $\quad 1 \cdot 2 + 2 \cdot 3 + 3 \cdot 4 + \cdots + n(n+1) = \frac{1}{3}n(n+1)(n+2)$

18.13 $\quad \dfrac{1}{1 \cdot 3} + \dfrac{1}{3 \cdot 5} + \dfrac{1}{5 \cdot 7} + \cdots + \dfrac{1}{(2n-1)(2n+1)} = \dfrac{n}{2n+1}$

18.14 $\quad 1 \cdot 3 + 2 \cdot 3^2 + 3 \cdot 3^3 + \cdots + n \cdot 3^n = \frac{3}{4}[(2n-1)3^n + 1]$

18.15 $\quad \dfrac{3}{1 \cdot 2 \cdot 2} + \dfrac{4}{2 \cdot 3 \cdot 2^2} + \dfrac{5}{3 \cdot 4 \cdot 2^3} + \cdots + \dfrac{n+2}{n(n+1)2^n} = 1 - \dfrac{1}{(n+1)2^n}$

18.16 A convex polygon of n sides has $\frac{1}{2}n(n-3)$ diagonals.

18.17 The sum of the interior angles of a regular polygon of n sides is $(n-2)180°$.

Chapter 19

The Binomial Theorem

By actual multiplication

$$(a + b)^1 = a + b, \quad (a + b)^2 = a^2 + 2ab + b^2, \quad (a + b)^3 = a^3 + 3a^2b + 3ab^2 + b^3,$$

$$(a + b)^4 = a^4 + 4a^3b + 6a^2b^2 + 4ab^3 + b^4,$$

$$(a + b)^5 = a^5 + 5a^4b + 10a^3b^2 + 10a^2b^3 + 5ab^4 + b^5, \text{etc.}$$

From these cases we conclude that, when n is a positive integer,

$$(a + b)^n = a^n + na^{n-1}b + \frac{n(n - 1)}{1 \cdot 2} a^{n-2}b^2 + \frac{n(n - 1)(n - 2)}{1 \cdot 2 \cdot 3} a^{n-3}b^3 + \cdots + nab^{n-1} + b^n$$

and note the following properties:

 (1)   The number of terms in the expansion is $(n + 1)$.
 (2)   The first term a of the binomial enters the first term of the expansion with exponent n, the second term with exponent $(n - 1)$, the third term with exponent $(n - 2)$, and so on.
 (3)   The second term b of the binomial enters the second term of the expansion with exponent 1, the third term with exponent 2, the fourth term with exponent 3, and so on.
 (4)   The sum of the exponents of a and b in any term is n.
 (5)   The coefficient of the first term in the expansion is 1, of the second term is $n/1$, of the third term is $\dfrac{n(n - 1)}{1 \cdot 2}$, of the fourth term is $\dfrac{n(n - 1)(n - 2)}{1 \cdot 2 \cdot 3}$, etc.
 (6)   The coefficients of terms equidistant from the ends of the expansion are the same. Note that the number of factors in the numerator and denominator of any coefficient except the first and last is then either the exponent of a or of b, whichever is the smaller.

The above properties may be proved by mathematical induction.

EXAMPLE 1.   Expand $(3x + 2y^2)^5$ and simplify term by term.
   We put the several powers of $(3x)$ in first, then the powers of $(2y^2)$, and finally the coefficients, recalling Property 6.

$$(3x + 2y^2)^5 = (3x)^5 + \frac{5}{1}(3x)^4(2y^2) + \frac{5 \cdot 4}{1 \cdot 2}(3x)^3(2y^2)^2 + \frac{5 \cdot 4}{1 \cdot 2}(3x)^2(2y^2)^3 + \frac{5}{1}(3x)(2y^2)^4 + (2y^2)^5$$

$$= 3^5x^5 + 5 \cdot 3^4x^4 \cdot 2y^2 + 10 \cdot 3^3x^3 \cdot 2^2y^4 + 10 \cdot 3^2x^2 \cdot 2^3y^6 + 5 \cdot 3x \cdot 2^4y^8 + 2^5y^{10}$$

$$= 243x^5 + 810x^4y^2 + 1080x^3y^4 + 720x^2y^6 + 240xy^8 + 32y^{10}$$

(See Problems 19.1–19.2.)

THE rth TERM $(r \le n + 1)$ in the expansion of $(a + b)^n$ is

$$\frac{n(n - 1)(n - 2) \cdots (n - r + 2)}{1 \cdot 2 \cdot 3 \cdots (r - 1)} a^{n-r+1}b^{r-1}.$$

(See Problem 19.3.)

WHEN THE LAWS ABOVE are used to expand $(a + b)^n$, where n is real but not a positive integer, an endless succession of terms is obtained. Such expansions are valid [see Problem 19.7(a) for a verification] when $|b| < |a|$.

EXAMPLE 2.   Write the first five terms in the expansion of $(a + b)^{-3}$, $|b| < |a|$.

$$(a + b)^{-3} = a^{-3} + (-3)a^{-4}b + \frac{(-3)(-4)}{1 \cdot 2} a^{-5}b^2 + \frac{(-3)(-4)(-5)}{1 \cdot 2 \cdot 3} a^{-6}b^3$$

$$+ \frac{(-3)(-4)(-5)(-6)}{1 \cdot 2 \cdot 3 \cdot 4} a^{-7}b^4 + \cdots$$

$$= \frac{1}{a^3} - \frac{3b}{a^4} + \frac{6b^2}{a^5} - \frac{10b^3}{a^6} + \frac{15b^4}{a^7} - \cdots$$

(See Problems 19.5–19.8.)

Solved Problems

19.1   Expand and simplify term by term.

(a) $\left(x^2 + \frac{1}{2}y\right)^6 = (x^2)^6 + \frac{6}{1}(x^2)^5\left(\frac{1}{2}y\right) + \frac{6 \cdot 5}{1 \cdot 2}(x^2)^4\left(\frac{1}{2}y\right)^2 + \frac{6 \cdot 5 \cdot 4}{1 \cdot 2 \cdot 3}(x^2)^3\left(\frac{1}{2}y\right)^3$

$$+ \frac{6 \cdot 5}{1 \cdot 2}(x^2)^2\left(\frac{1}{2}y\right)^4 + \frac{6}{1}(x^2)\left(\frac{1}{2}y\right)^5 + \left(\frac{1}{2}y\right)^6$$

$$= x^{12} + 6(x^{10})\tfrac{1}{2}y + 15(x^8)\tfrac{1}{4}y^2 + 20(x^6)\tfrac{1}{8}y^3 + 15(x^4)\tfrac{1}{16}y^4 + 6(x^2)\tfrac{1}{32}y^5 + \tfrac{1}{64}y^6$$

$$= x^{12} + 3x^{10}y + \tfrac{15}{4}x^8y^2 + \tfrac{5}{2}x^6y^3 + \tfrac{15}{16}x^4y^4 + \tfrac{3}{16}x^2y^5 + \tfrac{1}{64}y^6$$

(b) $(x^{1/2} + 2y^{1/3})^4 = (x^{1/2})^4 + \frac{4}{1}(x^{1/2})^3(2y^{1/3}) + \frac{4 \cdot 3}{1 \cdot 2}(x^{1/2})^2(2y^{1/3})^2 + \frac{4}{1}(x^{1/2})(2y^{1/3})^3 + (2y^{1/3})^4$

$$= x^2 + 4(x^{3/2})2y^{1/3} + 6(x)4y^{2/3} + 4(x^{1/2})8y + 16y^{4/3}$$

$$= x^2 + 8x^{3/2}y^{1/3} + 24xy^{2/3} + 32x^{1/2}y + 16y^{4/3}$$

(c) $\left(\frac{2}{3}x^{1/2} - \frac{1}{2x}\right)^6 = \left(\frac{2}{3}x^{1/2}\right)^6 + \frac{6}{1}\left(\frac{2}{3}x^{1/2}\right)^5\left(-\frac{1}{2x}\right)$

$$+ \frac{6 \cdot 5}{1 \cdot 2}\left(\frac{2}{3}x^{1/2}\right)^4\left(-\frac{1}{2x}\right)^2 + \frac{6 \cdot 5 \cdot 4}{1 \cdot 2 \cdot 3}\left(\frac{2}{3}x^{1/2}\right)^3\left(-\frac{1}{2x}\right)^3$$

$$+ \frac{6 \cdot 5}{1 \cdot 2}\left(\frac{2}{3}x^{1/2}\right)^2\left(-\frac{1}{2x}\right)^4 + \frac{6}{1}\left(\frac{2}{3}x^{1/2}\right)\left(-\frac{1}{2x}\right)^5 + \left(-\frac{1}{2x}\right)^6$$

$$= \frac{64}{729}x^3 - \frac{32}{81}x^{3/2} + \frac{20}{27} - \frac{20x^{1/2}}{27x^2} + \frac{5}{12x^3} - \frac{x^{1/2}}{8x^5} + \frac{1}{64x^6}$$

19.2   Write the first five terms in each expansion and simplify term by term.

(a) $\left(\frac{2}{3}m^{1/2} + \frac{3}{2m^{3/2}}\right)^{12} = \left(\frac{2}{3}m^{1/2}\right)^{12} + \frac{12}{1}\left(\frac{2}{3}m^{1/2}\right)^{11}\left(\frac{3}{2m^{3/2}}\right) + \frac{12 \cdot 11}{1 \cdot 2}\left(\frac{2}{3}m^{1/2}\right)^{10}\left(\frac{3}{2m^{3/2}}\right)^2$

$$+ \frac{12 \cdot 11 \cdot 10}{1 \cdot 2 \cdot 3}\left(\frac{2}{3}m^{1/2}\right)^9\left(\frac{3}{2m^{3/2}}\right)^3 + \frac{12 \cdot 11 \cdot 10 \cdot 9}{1 \cdot 2 \cdot 3 \cdot 4}\left(\frac{2}{3}m^{1/2}\right)^8\left(\frac{3}{2m^{3/2}}\right)^4 + \cdots$$

$$= \frac{2^{12}}{3^{12}}m^6 + 12\left(\frac{2^{11}}{3^{11}}m^{11/2}\right)\frac{3}{2m^{3/2}} + 66\left(\frac{2^{10}}{3^{10}}m^5\right)\frac{3^2}{2^2m^3}$$

$$+ 220\left(\frac{2^9}{3^9}m^{9/2}\right)\frac{3^3}{2^3m^{9/2}} + 495\left(\frac{2^8}{3^8}m^4\right)\frac{3^4}{2^4m^6} + \cdots$$

$$= \frac{2^{12}}{3^{12}}m^6 + \frac{2^{12}}{3^9}m^4 + 11\frac{2^9}{3^7}m^2 + 55\frac{2^8}{3^6} + 55\frac{2^4}{3^2m^2} + \cdots$$

(b) $\left(\dfrac{x^{1/2}}{y^{2/3}z} - \dfrac{yz^2}{2x}\right)^{11} = \left(\dfrac{x^{1/2}}{y^{2/3}z}\right)^{11} + \dfrac{11}{1}\left(\dfrac{x^{1/2}}{y^{2/3}z}\right)^{10}\left(-\dfrac{yz^2}{2x}\right) + \dfrac{11\cdot 10}{1\cdot 2}\left(\dfrac{x^{1/2}}{y^{2/3}z}\right)^{9}\left(-\dfrac{yz^2}{2x}\right)^2$

$\qquad + \dfrac{11\cdot 10\cdot 9}{1\cdot 2\cdot 3}\left(\dfrac{x^{1/2}}{y^{2/3}z}\right)^{8}\left(-\dfrac{yz^2}{2x}\right)^3 + \dfrac{11\cdot 10\cdot 9\cdot 8}{1\cdot 2\cdot 3\cdot 4}\left(\dfrac{x^{1/2}}{y^{2/3}z}\right)^{7}\left(-\dfrac{yz^2}{2x}\right)^4 + \cdots$

$\qquad = \dfrac{x^{11/2}}{y^{22/3}z^{11}} - 11\left(\dfrac{x^5}{y^{20/3}z^{10}}\right)\dfrac{yz^2}{2x} + 55\left(\dfrac{x^{9/2}}{y^6z^9}\right)\dfrac{y^2z^4}{2^2x^2}$

$\qquad - 165\left(\dfrac{x^4}{y^{16/3}z^8}\right)\dfrac{y^3z^6}{2^3x^3} + 330\left(\dfrac{x^{7/2}}{y^{14/3}z^7}\right)\dfrac{y^4z^8}{2^4x^4} - \cdots$

$\qquad = \dfrac{x^{11/2}y^{2/3}}{y^8z^{11}} - \dfrac{11x^4y^{1/3}}{2y^6z^8} + \dfrac{55x^{5/2}}{4y^4z^5} - \dfrac{165xy^{2/3}}{8y^3z^2} + \dfrac{165x^{1/2}y^{1/3}z}{8xy} - \cdots$

19.3 Find the indicated term and simplify.

(a) The seventh term of $(a + b)^{15}$.

In the seventh term the exponent of b is $7 - 1 = 6$, the exponent of a is $15 - 6 = 9$, and the coefficient has six factors in the numerator and denominator. Hence, the term is

$$\dfrac{15 \cdot 14 \cdot 13 \cdot 12 \cdot 11 \cdot 10}{1 \cdot 2 \cdot 3 \cdot 4 \cdot 5 \cdot 6} a^9 b^6 = 5005 a^9 b^6$$

(b) The ninth term of $\left(x - \dfrac{1}{x^{1/2}}\right)^{12}$.

In the ninth term the exponent of $b = -\dfrac{1}{x^{1/2}}$ is $9 - 1 = 8$, the exponent of $a = x$ is $12 - 8 = 4$, and the coefficient has four factors in numerator and denominator. Hence the required term is

$$\dfrac{12 \cdot 11 \cdot 10 \cdot 9}{1 \cdot 2 \cdot 3 \cdot 4} (x)^4 \left(-\dfrac{1}{x^{1/2}}\right)^8 = 495(x^4)\dfrac{1}{x^4} = 495$$

(c) The twelfth term of $\left(\dfrac{x^{1/2}}{4} - \dfrac{2y}{x^{3/2}}\right)^{18}$.

The required term is

$$\dfrac{18 \cdot 17 \cdot 16 \cdot 15 \cdot 14 \cdot 13 \cdot 12}{1 \cdot 2 \cdot 3 \cdot 4 \cdot 5 \cdot 6 \cdot 7}\left(\dfrac{x^{1/2}}{4}\right)^7\left(-\dfrac{2y}{x^{3/2}}\right)^{11} = -9 \cdot 17 \cdot 16 \cdot 13\left(\dfrac{x^{7/2}}{2^{14}}\right)\dfrac{2^{11}y^{11}}{x^{33/2}} = -3978\dfrac{y^{11}}{x^{13}}$$

(d) The middle term in the expansion of $\left(x^{2/3} + \dfrac{1}{x^{1/2}}\right)^{10}$.

Since there are 11 terms in all, the middle term is the sixth. This term is

$$\dfrac{10 \cdot 9 \cdot 8 \cdot 7 \cdot 6}{1 \cdot 2 \cdot 3 \cdot 4 \cdot 5}(x^{2/3})^5\left(\dfrac{1}{x^{1/2}}\right)^5 = 252\dfrac{x^{10/3}}{x^{5/2}} = 252x^{5/6}$$

(e) The term involving y^{12} in the expansion of $\left(y^3 - \dfrac{x}{3}\right)^9$.

The first term of the binomial must be raised to the fourth power to produce y^{12}; hence, the second term must be raised to the fifth power and we are to write the six term. This term is

$$\dfrac{9 \cdot 8 \cdot 7 \cdot 6}{1 \cdot 2 \cdot 3 \cdot 4}(y^3)^4\left(-\dfrac{x}{3}\right)^5 = -3^2 \cdot 14y^{12}\dfrac{x^5}{3^5} = -\dfrac{14}{27}x^5y^{12}$$

(f) The term involving x^4 in the expansion of $\left(\dfrac{2}{x} + \dfrac{x^2}{4}\right)^{14}$.

Let p and q be positive integers so that $p + q = 14$. We are required to determine p and q so that $\left(\dfrac{2}{x}\right)^p\left(\dfrac{x^2}{4}\right)^q$ yields a term in x^4. Then $2q - p = 4$ or $2q - (14 - q) = 3q - 14 = 4$ and $q = 6$. The required term, the seventh in the expansion, is

$$\frac{14 \cdot 13 \cdot 12 \cdot 11 \cdot 10 \cdot 9}{1 \cdot 2 \cdot 3 \cdot 4 \cdot 5 \cdot 6} \left(\frac{2}{x}\right)^8 \left(\frac{x^2}{4}\right)^6 = 3003 \frac{2^8}{x^8} \frac{x^{12}}{2^{12}} = \frac{3003}{16} x^4$$

19.4   Evaluate $(1.02)^{12}$ correct to four decimal places.

$$(1.02)^{12} = (1 + .02)^{12} = 1 + 12(.02) + \frac{12 \cdot 11}{1 \cdot 2}(.02)^2 + \frac{12 \cdot 11 \cdot 10}{1 \cdot 2 \cdot 3}(.02)^3 + \frac{12 \cdot 11 \cdot 10 \cdot 9}{1 \cdot 2 \cdot 3 \cdot 4}(.02)^4 + \cdots$$

$$= 1 + .24 + .0264 + .00176 + .00008 + \cdots = 1.26824 \text{ (approx.)}$$

Thus, $(1.02)^{12} = 1.2682$ correct to four decimal places.

19.5   Write the first five terms and simplify term by term.

(a) $$\left(x^2 - \frac{2}{x^4}\right)^{1/2} = (x^2)^{1/2} + \frac{1}{2}(x^2)^{-1/2}\left(-\frac{2}{x^4}\right) + \frac{(\frac{1}{2})(-\frac{1}{2})}{1 \cdot 2}(x^2)^{-3/2}\left(-\frac{2}{x^4}\right)^2$$

$$+ \frac{(\frac{1}{2})(-\frac{1}{2})(-\frac{3}{2})}{1 \cdot 2 \cdot 3}(x^2)^{-5/2}\left(-\frac{2}{x^4}\right)^3 + \frac{(\frac{1}{2})(-\frac{1}{2})(-\frac{3}{2})(-\frac{5}{2})}{1 \cdot 2 \cdot 3 \cdot 4}(x^2)^{-7/2}\left(-\frac{2}{x^4}\right)^4 + \cdots$$

$$= x - \frac{1}{2} \cdot \frac{1}{x} \cdot \frac{2}{x^4} - \frac{1}{2^3} \cdot \frac{1}{x^3} \cdot \frac{2^2}{x^8} - \frac{1}{2^4} \cdot \frac{1}{x^5} \cdot \frac{2^3}{x^{12}} - \frac{5}{2^7} \cdot \frac{1}{x^7} \cdot \frac{2^4}{x^{16}} - \cdots$$

$$= x - \frac{1}{x^5} - \frac{1}{2x^{11}} - \frac{1}{2x^{17}} - \frac{5}{8x^{23}} - \cdots, \qquad (|x| > \sqrt[6]{2})$$

(b) $$(1 - x^3)^{-2/3} = 1^{-2/3} + \left(-\frac{2}{3}\right)1^{-5/3}(-x^3) + \frac{(-\frac{2}{3})(-\frac{5}{3})}{1 \cdot 2}1^{-8/3}(-x^3)^2$$

$$+ \frac{(-\frac{2}{3})(-\frac{5}{3})(-\frac{8}{3})}{1 \cdot 2 \cdot 3}1^{-11/3}(-x^3)^3 + \frac{(-\frac{2}{3})(-\frac{5}{3})(-\frac{8}{3})(-\frac{11}{3})}{1 \cdot 2 \cdot 3 \cdot 4}1^{-14/3}(-x^3)^4 + \cdots$$

$$= 1 + \frac{2}{3}x^3 + \frac{5}{9}x^6 + \frac{40}{81}x^9 + \frac{110}{243}x^{12} + \cdots, \qquad (|x| < 1)$$

19.6   Find and simplify the sixth term in the expansion of $\left(x^2 - \frac{3}{2x}\right)^{5/3}$, $|x| > \sqrt[3]{\frac{3}{2}}$.

The required term is

$$\frac{(\frac{5}{3})(\frac{2}{3})(-\frac{1}{3})(-\frac{4}{3})(-\frac{7}{3})}{1 \cdot 2 \cdot 3 \cdot 4 \cdot 5}(x^2)^{-10/3}\left(-\frac{3}{2x}\right)^5 = \frac{7}{3^6}x^{-20/3}\frac{3^5}{2^5 x^5} = \frac{7x^{1/3}}{96x^{12}}$$

19.7   (a)   Evaluate $\sqrt{26}$ correct to four decimal places.

$$\sqrt{26} = (5^2 + 1)^{1/2} = (5^2)^{1/2} + \frac{1}{2}(5^2)^{-1/2}(1) + \frac{(\frac{1}{2})(-\frac{1}{2})}{1 \cdot 2}(5^2)^{-3/2}(1)^2$$

$$+ \frac{(\frac{1}{2})(-\frac{1}{2})(-\frac{3}{2})}{1 \cdot 2 \cdot 3}(5^2)^{-5/2}(1)^3 + \frac{(\frac{1}{2})(-\frac{1}{2})(-\frac{3}{2})(-\frac{5}{2})}{1 \cdot 2 \cdot 3 \cdot 4}(5^2)^{-7/2}(1)^4 + \cdots$$

$$= 5 + \frac{1}{2} \cdot \frac{1}{5} - \frac{1}{2^3} \cdot \frac{1}{5^3} + \frac{1}{2^4} \cdot \frac{1}{5^5} - \frac{1}{2^7} \cdot \frac{1}{5^6} + \cdots$$

$$= 5.00000 + .10000 - .00100 + .00002 - \cdots$$

$$= 5.09902 \quad \text{(approximately)}.$$

Thus $\sqrt{26} = 5.0990$, correct to four decimal places.

(NOTE.   If we write $\sqrt{26} = (1 + 5^2)^{1/2} = 1 + \frac{1}{2}(5^2) - \frac{1}{8}(5^2)^2 + \frac{1}{16}(5^2)^3 - \cdots$

$$= 1 + 12.5 - 78.125 + 976.5625 - \cdots$$

it is clear that the expansion is not valid.)

(b) Evaluate $\sqrt{23}$ correct to four decimal places.

$$\sqrt{23} = (5^2 - 2)^{1/2} = (5^2)^{1/2} + \frac{1}{2}(5^2)^{-1/2}(-2) + \frac{(\frac{1}{2})(-\frac{1}{2})}{1 \cdot 2}(5^2)^{-3/2}(-2)^2$$

$$+ \frac{(\frac{1}{2})(-\frac{1}{2})(-\frac{3}{2})}{1 \cdot 2 \cdot 3}(5^2)^{-5/2}(-2)^3 + \frac{(\frac{1}{2})(-\frac{1}{2})(-\frac{3}{2})(-\frac{5}{2})}{1 \cdot 2 \cdot 3 \cdot 4}(5^2)^{-7/2}(-2)^4 + \cdots$$

$$= 5 - \frac{1}{5} - \frac{1}{2} \cdot \frac{1}{5^3} - \frac{1}{2} \cdot \frac{1}{5^5} - \frac{1}{8} \cdot \frac{1}{5^6} - \cdots$$

$$= 4.79583 \quad \text{(approximately)}.$$

Thus $\sqrt{23} = 4.7958$, correct to four decimal places.

19.8   (a)   Show that the sum of the coefficients in the expansion of $(a + b)^n$, n a positive integer, is 2^n.

       (b)   Show that the sum of the coefficients in the expansion of $(a - b)^n$, n a positive integer, is 0.

   (a)   In $(a + b)^n = a^n + na^{n-1}b + \frac{n(n-1)}{1 \cdot 2}a^{n-2}b^2 + \cdots + \frac{n(n-1)}{1 \cdot 2}a^2b^{n-2} + nab^{n-1} + b^n$, let $a = b = 1$; then $(1 + 1)^n = 2^n = 1 + n + \frac{1}{2}n(n-1) + \cdots + \frac{1}{2}n(n-1) + n + 1$, as was to be proved.

   (b)   Similarly, let $a = b = 1$ in the expansion of $(a - b)^n$ and obtain

$$1 - n + \frac{1}{2}n(n-1) - \cdots + (-1)^{n-2}[\tfrac{1}{2}n(n-1)] + (-1)^{n-1}n + (-1)^n = (1 - 1)^n = 0$$

Supplementary Problems

19.9   Expand by the binomial theorem and simplify term by term.

   (a) $(a + \frac{1}{2}b)^6 = a^6 + 3a^5b + \frac{15}{4}a^4b^2 + \frac{5}{2}a^3b^3 + \frac{15}{16}a^2b^4 + \frac{3}{16}ab^5 + \frac{1}{64}b^6$

   (b) $(4x + \frac{1}{4}y)^5 = 1024x^5 + 320x^4y + 40x^3y^2 + \frac{5}{2}x^2y^3 + \frac{5}{64}xy^4 + \frac{1}{1024}y^5$

   (c) $\left(\dfrac{x}{4y^3} - \dfrac{2y}{x^2}\right)^5 = \dfrac{x^5}{1024y^{15}} - \dfrac{5x^2}{128y^{11}} + \dfrac{5}{8xy^7} - \dfrac{5}{x^4y^3} + \dfrac{20y}{x^7} - \dfrac{32y^5}{x^{10}}$

19.10   Find the indicated term and simplify.

   (a)   Fifth term of $(\frac{1}{2} + x)^{10}$ (d)   Seventh term of $\left(x^{1/3} - \dfrac{1}{2x^{2/3}}\right)^{10}$

   (b)   Sixth term of $\left(\dfrac{2}{x^{1/2}} - \dfrac{x^{1/4}}{4}\right)^9$

         (e)   Middle term of $\left(\dfrac{1}{x} - x^2\right)^{12}$

   (c)   Tenth term of $\left(\dfrac{27a^2}{b^3} + \dfrac{b^2}{6a^4}\right)^{12}$

         (f)   Middle term of $\left(a^{1/2}b^{1/2} - \dfrac{a}{2b^{3/2}}\right)^9$

   (g)   The term involving x^{14} in the expansion of $(2/x - x^2)^{10}$

   (h)   The term free of y in the expansion of $(xy^{1/6} - y^{-2/3})^{15}$

     Ans.   (a) $\dfrac{105}{32}x^4$ (b) $-\dfrac{63x^{1/4}}{32x}$ (c) $\dfrac{55b^9}{128a^{30}}$ (d) $\dfrac{105x^{1/3}}{32x^3}$ (e) $924x^6$

          (f) $\dfrac{63a^{13/2}b^{1/2}}{8b^4}, -\dfrac{63a^7b^{1/2}}{16b^6}$ (g) $180x^{14}$ (h) $-455x^{12}$

19.11 Expand $(a + b - c)^3$. **Hint:** Write $(a + b - c)^3 = [(a + b) - c]^3$.

  Ans. $a^3 + b^3 - c^3 + 3a^2b - 3a^2c + 3ab^2 - 3b^2c + 3ac^2 + 3bc^2 - 6abc$

19.12 Find the value of n if the coefficients of the sixth and sixteenth terms in the expansion of $(a + b)^n$ are equal.

  Ans. $n = 20$

19.13 Find the first five terms in the expansion and simplify term by term.

  (a) $\left(\dfrac{x}{a} + \dfrac{a^3}{x^2}\right)^{-3} = \dfrac{a^3}{x^3} - \dfrac{3a^7}{x^6} + \dfrac{6a^{11}}{x^9} - \dfrac{10a^{15}}{x^{12}} + \dfrac{15a^{19}}{x^{15}} - \cdots, (|x| > a^{4/3})$

  (b) $\left(x^2 - \dfrac{2y^2}{x}\right)^{3/2} = x^3 - 3y^2 + \dfrac{3y^4}{2x^3} + \dfrac{y^6}{2x^6} + \dfrac{3y^8}{8x^9} + \cdots, (|x| < \sqrt[3]{2y^2})$

  (c) $\left(1 - \dfrac{3y^2}{x}\right)^{1/3} = 1 - \dfrac{y^2}{x} - \dfrac{y^4}{x^2} - \dfrac{5y^6}{3x^3} - \dfrac{10y^8}{3x^4} - \cdots, (|x| > 3y^2)$

19.14 Find and simplify the indicated term.

  (a) The sixth term in $(a^2 - 4b^2)^{1/2}$, $(|a| > 2|b|)$.

    Ans. $-28\dfrac{b^{10}}{a^9}$

  (b) The seventh term in $\left(x^{1/4} - \dfrac{3}{x^{1/2}}\right)^{-1}$, $(|x| > 3^{4/3})$.

    Ans. $729\dfrac{x^{1/4}}{x^5}$

  (c) The term involving x^{20} in $\left(\dfrac{2}{x} + x^2\right)^{-2}$, $(|x| < \sqrt[3]{2})$.

    Ans. $\dfrac{7x^{20}}{256}$

Chapter 20

Permutations

ANY ARRANGEMENT OF A SET OF OBJECTS in a definite order is called a *permutation* of the set taken all at a time. For example, *abcd, acbd, bdca* are permutations of the set of letters *a, b, c, d* taken all at a time.

If a set contains *n* objects, any ordered arrangement of any $r \leq n$ of the objects is called a permutation of the *n* objects taken *r* at a time. For example, *ab, ba, ca, db* are permutations of the $n = 4$ letters *a, b, c, d* taken $r = 2$ at a time, while *abc, adb, bad, cad* are permutations of the $n = 4$ letters taken $r = 3$ at a time.

THE NUMBER OF PERMUTATIONS which may be formed in each situation can be found by means of the

> **FUNDAMENTAL PRINCIPLE:** If one thing can be done in *u* different ways, if after it has been done in any one of these a second thing can be done in *v* different ways, if after it has been done in any one of these a third thing can be done in *w* different ways, ..., the several things can be done in the order stated in $u \cdot v \cdot w \cdots$ different ways.

EXAMPLE 1. In how many ways can 6 students be assigned to (*a*) a row of 6 seats, (*b*) a row of 8 seats?

(*a*) Let the seats be denoted xxxxxx. The seat on the left may be assigned to any one of the 6 students; that is, it may be assigned in 6 different ways. After the assignment has been made, the next seat may be assigned to any one of the 5 remaining students. After the assignment has been made, the next seat may be assigned to any one of the 4 remaining students, and so on. Placing the number of ways in which each seat may be assigned under the x marking the seat, we have

$$\begin{matrix} x & x & x & x & x & x \\ 6 & 5 & 4 & 3 & 2 & 1 \end{matrix}$$

By the fundamental principle, the seats may be assigned in

$$6 \cdot 5 \cdot 4 \cdot 3 \cdot 2 \cdot 1 = 720 \text{ ways}$$

The reader should assure himself that the seats might have been assigned to the students with the same result.

(*b*) Here each student must be assigned a seat. The first student may be assigned any one of the 8 seats, the second student any one of the 7 remaining seats, and so on. Letting x represent a student, we have

$$\begin{matrix} x & x & x & x & x & x \\ 8 & 7 & 6 & 5 & 4 & 3 \end{matrix}$$

and the assignment may be made in $8 \cdot 7 \cdot 6 \cdot 5 \cdot 4 \cdot 3 = 20160$ ways.

(NOTE: If we attempt to assign students to seats, we must first select the six seats to be used. The problem of selections will be considered under Combinations.) (See Problems 20.1–20.5.)

Define *n*! (*n* factorial) to be

$$n \cdot (n - 1) \cdot (n - 2) \cdot \cdots \cdot (2)(1)$$

(where $0! = 1! = 1$). Then, if we define $\binom{n}{k}$ to be $\dfrac{n!}{k!(n-k)!}$, we call $\binom{n}{k}$ a binomial coefficient and note that

$$(a + b)^n = \sum_{k=0}^{n} \binom{n}{k} a^{n-k} b^k, \qquad n \geq 1.$$

130

PERMUTATIONS OF OBJECTS NOT ALL DIFFERENT. If there are n objects of which k are alike while the remaining $(n - k)$ objects are different from them and from each other, it is clear that the number of different permutations of the n objects taken all together is not $_nP_n$.

EXAMPLE 2. How many different permutations of four letters can be formed using the letters of the word bass? For the moment, think of the given letters as b, a, s_1, s_2 so that they are all different. Then

$$bas_1s_2 \quad as_1bs_2 \quad s_2s_1ba \quad s_1as_2b \quad bs_1s_2a$$

$$bas_2s_1 \quad as_2bs_1 \quad s_1s_2ba \quad s_2as_1b \quad bs_2s_1a$$

are 10 of the 24 permutations of the four letters taken all together. However, when the subscripts are removed, it is seen that the two permutations in each column are alike.

Thus, there are $\dfrac{1 \cdot 2 \cdot 3 \cdot 4}{1 \cdot 2} = 12$ different permutations.

In general, given n objects of which k_1 are of one sort, k_2 of another, k_3 of another, ...; then the number of different permutations that can be made from the n objects taken all together is

$$\frac{n!}{k_1! \, k_2! \, k_3! \ldots}$$

(See Problem 20.10.)

Solved Problems

20.1 Using the letters of the word MARKING and calling any arrangement a word, (a) how many different 7-letter words can be formed, (b) how many different 3-letter words can be formed?

(a) We must fill each of the positions xxxxxxx with a different letter. The first position may be filled in 7 ways, the second in 6 ways, and so on.

Thus, we have $\begin{array}{ccccccc} x & x & x & x & x & x & x \\ 7 & 6 & 5 & 4 & 3 & 2 & 1 \end{array}$ and there are $7 \cdot 6 \cdot 5 \cdot 4 \cdot 3 \cdot 2 \cdot 1 = 5040$ words.

(b) We must fill each of the positions xxx with a different letter. The first position can be filled in 7 ways, the second in 6 ways, and the third in 5 ways. Thus, there are $7 \cdot 6 \cdot 5 = 210$ words.

20.2 In forming 5-letter words using the letters of the word EQUATIONS, (a) how many consist only of vowels, (b) how many contain all of the consonants, (c) how many begin with E and end in S, (d) how many begin with a consonant, (e) how many contain N, (f) how many in which the vowels and consonants alternate, (g) how many in which Q is immediately followed by U?

There are 9 letters, consisting of 5 vowels and 4 consonants.

(a) There are five places to be filled and 5 vowels at our disposal. Hence, we can form $5 \cdot 4 \cdot 3 \cdot 2 \cdot 1 = 120$ words.

(b) Each word is to contain the 4 consonants and one of the 5 vowels. There are now six things to do: first pick the vowel to be used (in 5 ways), next place the vowel (in 5 ways), and fill the remaining four positions with consonants.

We have $\begin{array}{ccccc} x & x & x & x & x \\ 5 & 5 & 4 & 3 & 2 & 1 \end{array}$; hence there are $5 \cdot 5 \cdot 4 \cdot 3 \cdot 2 \cdot 1 = 600$ words.

(c) Indicate the fact that the position of certain letters is fixed by writing $\begin{array}{ccccc} \overset{E}{x} & x & x & x & \overset{S}{x} \end{array}$.

Now there are just three positions to be filled and 7 letters at our disposal. Thus, there are $7 \cdot 6 \cdot 5 = 210$ words.

(*d*) Here we have $\begin{matrix} & c \\ x & x & x & x & x \\ 4 & 8 & 7 & 6 & 5 \end{matrix}$ since, after filling the first position with any one of the 4 consonants, there are 8 letters remaining. Hence, there are $4 \cdot 8 \cdot 7 \cdot 6 \cdot 5 = 6720$ words.

(*e*) There are five things to do: first, place the letter N in any one of the five positions and then fill the other four positions from among the 8 letters remaining.

We have $\begin{matrix} x & x & x & x & x \\ 5 & 8 & 7 & 6 & 5 \end{matrix}$; hence, there are $5 \cdot 8 \cdot 7 \cdot 6 \cdot 5 = 8400$ words.

(*f*) We may have $\begin{matrix} v & c & v & c & v \\ x & x & x & x & x \\ 5 & 4 & 4 & 3 & 3 \end{matrix}$ or $\begin{matrix} c & v & c & v & c \\ x & x & x & x & x \\ 4 & 5 & 3 & 4 & 2 \end{matrix}$. Hence, there are $5 \cdot 4 \cdot 4 \cdot 3 \cdot 3 + 4 \cdot 5 \cdot 3 \cdot 4 \cdot 2 = 1200$ words.

(*g*) First we place Q so that U may follow it (Q may occupy any one of the first four positions but not the last), next we place U (in only 1 way), then we fill the three other positions from among the 7 letters remaining.

Thus, we have $\begin{matrix} x & x & x & x & x \\ 4 & 1 & 7 & 6 & 5 \end{matrix}$ and there are $4 \cdot 1 \cdot 7 \cdot 6 \cdot 5 = 840$ words.

20.3 If repetitions are not allowed, (*a*) how many three-digit numbers can be formed with the digits 0, 1, 2, 3, 4, 5, 6, 7, 8, 9? (*b*) How many of these are odd numbers? (*c*) How many are even numbers? (*d*) How many are divisible by 5? (*e*) How many are greater than 600?

In each case we have $\begin{matrix} \neq 0 \\ x & x & x \end{matrix}$.

(*a*) The position on the left can be filled in 9 ways (0 cannot be used), the middle position can be filled in 9 ways (0 can be used), and the position on the right can be filled in 8 ways. Thus, there are $9 \cdot 9 \cdot 8 = 648$ numbers.

(*b*) We have $\begin{matrix} \neq 0 & & odd \\ x & x & x \end{matrix}$. Care must be exercised here in choosing the order in which to fill the positions. If the position on the left is filled first (in 9 ways), we cannot determine the number of ways in which the position on the right can be filled since, if the former is filled with an odd digit there are 4 ways of filling the latter but if the former is filled with an even digit there are 5 ways of filling the latter.

We fill first the position on the right (in 5 ways), then the position on the left (in 8 ways, since one odd digit and 0 are excluded), and the middle position (in 8 ways, since two digits are now excluded). Thus, there are $8 \cdot 8 \cdot 5 = 320$ numbers.

(*c*) We have $\begin{matrix} \neq 0 & & even \\ x & x & x \end{matrix}$. We note that the argument above excludes the possibility of first filling the position on the left. But if we fill the first position on the right (in 5 ways), we are unable to determine the number of ways the position on the left can be filled (9 ways if 0 was used on the right, 8 ways if 2, 4, 6, or 8 was used). Thus, we must separate the two cases.

First, we form all numbers ending in 0; there are $9 \cdot 8 \cdot 1$ of them. Next, we form all numbers ending in 2, 4, 6, or 8; there are $8 \cdot 8 \cdot 4$ of them. Thus, in all, there are $9 \cdot 8 \cdot 1 + 8 \cdot 8 \cdot 4 = 328$ numbers.

As a check, we have 320 odd and 328 even numbers for a total of 648 as found in (*a*) above.

(*d*) A number is divisible by 5 if and only if it ends in 0 or 5. There are $9 \cdot 8 \cdot 1$ numbers ending in 0 and $8 \cdot 8 \cdot 1$ numbers ending in 5. Hence, in all, there are $9 \cdot 8 \cdot 1 + 8 \cdot 8 \cdot 1 = 136$ numbers divisible by 5.

(*e*) The position on the left can be filled in 4 ways (with 6, 7, 8, or 9) and the remaining positions in $9 \cdot 8$ ways. Thus, there are $4 \cdot 9 \cdot 8 = 288$ numbers.

20.4 Solve Problem 20.3 (a), (b), (c), (d) if any digit may be used once, twice, or three times in forming the three-digit number.

(a) The position on the left can be filled in 9 ways and each of the other positions can be filled in 10 ways. Thus, there are $9 \cdot 10 \cdot 10 = 900$ numbers.

(b) The position on the right can be filled in 5 ways, the middle position in 10 ways, and the position on the left in 9 ways. Thus, there are $9 \cdot 10 \cdot 5 = 450$ numbers.

(c) There are $9 \cdot 10 \cdot 5 = 450$ even numbers.

(d) There are $9 \cdot 10 \cdot 1 = 90$ numbers ending in 0 and the same number ending in 5. Thus, there are 180 numbers divisible by 5.

20.5 In how many ways can 10 boys be arranged (a) in a straight line, (b) in a circle?

(a) The boys may be arranged in a straight line in $10 \cdot 9 \cdot 8 \cdot 7 \cdot 6 \cdot 5 \cdot 4 \cdot 3 \cdot 2 \cdot 1$ ways.

(b) We first place a boy at any point on the circle. The other 9 boys may then be arranged in $9 \cdot 8 \cdot 7 \cdot 6 \cdot 5 \cdot 4 \cdot 3 \cdot 2 \cdot 1$ ways.

This is an example of a *circular permutation*. In general, n objects may be arranged in a circle in $(n-1)(n-2) \cdots 2 \cdot 1$ ways.

20.6 Evaluate.

(a) $\dfrac{8!}{3!} = \dfrac{1 \cdot 2 \cdot 3 \cdot 4 \cdot 5 \cdot 6 \cdot 7 \cdot 8}{1 \cdot 2 \cdot 3} = 6720$ (c) $\dfrac{10!}{3!\,3!\,4!} = \dfrac{1 \cdot 2 \cdot 3 \cdot 4 \cdot 5 \cdot 6 \cdot 7 \cdot 8 \cdot 9 \cdot 10}{1 \cdot 2 \cdot 3 \cdot 1 \cdot 2 \cdot 3 \cdot 1 \cdot 2 \cdot 3 \cdot 4} = 4200$

(b) $\dfrac{7!}{6!} = \dfrac{1 \cdot 2 \cdot 3 \cdot 4 \cdot 5 \cdot 6 \cdot 7}{1 \cdot 2 \cdot 3 \cdot 4 \cdot 5 \cdot 6} = 7$ (d) $\dfrac{(r+1)!}{(r-1)!} = \dfrac{1 \cdot 2 \cdots (r-1)r(r+1)}{1 \cdot 2 \cdots (r-1)} = r(r+1)$

20.7 The number of permutations $_nP_r$ of n different objects taken $r < n$ at a time is given by

$$_nP_r = n(n-1) \cdots (n-r+1) = \frac{n(n-1) \cdots (n-r+1) \cdot (n-r) \cdots 2 \cdot 1}{(n-r) \cdots 2 \cdot 1} = \frac{n!}{(n-r)!}$$

and the number of permutations $_nP_n$ of n different objects taken n at a time is given by

$$_nP_n = n(n-1) \cdots 2 \cdot 1 = n!$$

20.8 Solve for n, given (a) $_nP_2 = 110$, (b) $_nP_4 = 30\,_nP_2$.

(a) $_nP_2 = n(n-1) = n^2 - n = 110$. Then $n^2 - n - 110 = (n-11)(n+10) = 0$ and, since n is positive, $n = 11$.

(b) We have $n(n-1)(n-2)(n-3) = 30n(n-1)$ or $n(n-1)(n-2)(n-3) - 30n(n-1) = 0$.

Then $n(n-1)[(n-2)(n-3) - 30] = n(n-1)(n^2 - 5n - 24) = n(n-1)(n-8)(n+3) = 0$. Since $n \geq 4$, the required solution is $n = 8$.

20.9 (a) How many permutations can be made of the letters, taken all together, of the "word" MASSESS?
(b) In how many ways will the four S's be together? (c) How many will end in SS?

(a) There are seven letters of which four are S's. The number of permutations is $7!/4! = 210$.

(b) First, permute the non-S's in $1 \cdot 2 \cdot 3 = 6$ ways and then place the four S's at the ends or between any two letters in each of the six permutations. Thus, there will be $4 \cdot 6 = 24$ permutations.

(c) After filling the last two places with S, we have to fill five places with 5 letters of which 2 are S's. Thus, there are $5!/2! = 60$ permutations.

Supplementary Problems

20.10 In how many different ways can 5 persons be seated on a bench?

   Ans. 120

20.11 In how many ways can the offices of chairman, vice-chairman, secretary, and treasurer be filled from a committee of seven?

   Ans. 840

20.12 How many 3-digit numbers can be formed with the digits $1, 2, \ldots, 9$, if no digit is repeated in any number?

   Ans. 504

20.13 How many 3-digit odd numbers can be formed with the digits $1, 2, 3, \ldots, 9$, if no digit is repeated in any number?

   Ans. 280

20.14 How many 3-digit numbers >300 can be formed with the digits $1, 2, 3, 4, 5, 6$, if no digit is repeated in any number?

   Ans. 80

20.15 How many 4-digit numbers >3000 can be formed with the digits $2, 3, 4, 5$, if repetitions of digits (*a*) are not allowed, (*b*) are allowed?

   Ans. (*a*) 18   (*b*) 192

20.16 In how many ways can 3 girls and 3 boys be seated in a row if boys and girls alternate?

   Ans. 72

20.17 In how many ways can 2 letters be mailed if 5 letter boxes are available?

   Ans. 25

20.18 Seven-letter words are formed using the letters of the word BLACKER. (*a*) How many can be formed? (*b*) How many which end in R? (*c*) How many in which E immediately follows K? (*d*) How many do not begin with B? (*e*) How many in which the vowels are separated by exactly two letters? (*f*) How many in which the vowels are separated by two or more letters?

   Ans. (*a*) 5040   (*b*) 720   (*c*) 720   (*d*) 4320   (*e*) 960   (*f*) 2400

20.19 Eight books are to be arranged on a shelf. (*a*) In how many ways can this be done? (*b*) In how many ways if two of the books are to be placed together? (*c*) In how many ways if five of the books have red binding and three have blue binding, and the books of the same color are to be kept together? (*d*) In how many ways if four of the books belong to a numbered set and are to be kept together and in order?

   Ans. (*a*) 40 320   (*b*) 10080   (*c*) 1440   (*d*) 120

20.20 How many six-letter words can be formed using the letters of the word ASSIST (*a*) in which the S's alternate with other letters? (*b*) in which the three S's are together? (*c*) which begin and end with S? (*d*) which neither begin nor end with S?

     Ans. (*a*) 12   (*b*) 24   (*c*) 24   (*d*) 24

20.21 (*a*) In how many ways can 8 persons be seated about a round table? (*b*) With 8 beads of different colors, how many bracelets can be formed by stringing them all together?

     Ans. (*a*) 5040   (*b*) 2520

20.22 How many signals can be made with 3 white, 3 green, and 2 blue flags by arranging them on a mast (*a*) all at a time? (*b*) three at a time? (*c*) five at a time?

     Ans. (*a*) 560   (*b*) 26   (*c*) 170

20.23 A car will hold 2 in the front seat and 1 in the rear seat. If among 6 persons only 2 can drive, in how many ways can the car be filled?

     Ans. 40

20.24 A chorus consists of 6 boys and 6 girls. How many arrangements of them can be made (*a*) in a row, facing front, the boys and girls alternating? (*b*) in two rows, facing front, with a boy behind each girl? (*c*) in a ring with the boys facing the center and the girls facing away from the center? (*d*) in two concentric rings, both facing the center, with a boy behind each girl?

     Ans. (*a*) 1 036 800   (*b*) 518 400   (*c*) 39 916 800   (*d*) 86 400

20.25 (*a*) In how many ways can 10 boys take positions in a straight line if two particular boys must not stand side by side? (*b*) In how many ways can 10 boys take positions about a round table if two particular boys must not be seated side by side?

     Ans. (*a*) $8 \cdot 9!$ (*b*) $7 \cdot 8!$

20.26 A man has 5 large books, 7 medium-sized books, and 3 small books. In how many different ways can they be arranged on a shelf if all books of the same size are to be kept together?

     Ans. 21 772 800

20.27 (*a*) How many words can be made from the letters of the word MASSACHUSETTS taken all together? (*b*) Of the words in (*a*), how many begin and end with SS? (*c*) Of the words in (*a*), how many begin and end with S? (*d*) Show that there are as many words having H as middle letter as there are circular permutations, using all letters.

     Ans. (*a*) 64 864 800   (*b*) 90 720   (*c*) 4 989 600

Chapter 21

Combinations

THE COMBINATIONS of n objects taken r at a time consist of all possible sets of r of the objects, without regard to the order of arrangement. The number of combinations of n objects taken r at a time will be denoted by $_nC_r$.

For example, the combinations of the $n = 4$ letters a, b, c, d taken $r = 3$ at a time, are

$$abc, \qquad abd, \qquad acd, \qquad bcd$$

Thus, $_4C_3 = 4$. When the letters of each combination are rearranged (in 3! ways), we obtain the $_4P_3$ permutations of the 4 letters taken 3 at a time. Hence, $_4P_3 = 3!(_4C_3)$ and $_4C_3 = _4P_3/3!$.

The number of combinations of n different objects taken r at a time is equal to the number of permutations of the n objects taken r at a time divided by factorial r, or

$$_nC_r = \frac{_nP_r}{r!} = \frac{n(n-1)\cdots(n-r+1)}{1 \cdot 2 \cdots r}$$

(For a proof, see Problem 21.1.)

EXAMPLE. From a shelf containing 12 different toys, a child is permitted to select 3. In how many ways can this be done?

The required number is

$$_{12}C_3 = \frac{_{12}P_3}{3!} = \frac{12 \cdot 11 \cdot 10}{1 \cdot 2 \cdot 3} = 220.$$

Notice that $_nC_r$ is the rth term's coefficient in the binomial theorem. See Chapter 19.

Solved Problems

21.1 Derive the formula $_nC_r = \dfrac{_nP_r}{r!}$.

From each of the $_nC_r$ combinations of n objects taken r at a time, $r!$ permutations can be formed. Since two combinations differ at least in one element, the $_nC_r \cdot r!$ permutations thus formed are precisely the number of permutations $_nP_r$ of n objects taken r at a time. Thus

$$_nC_r \cdot r! = _nP_r \qquad \text{and} \qquad _nC_r = \frac{_nP_r}{r!}$$

21.2 Show (a) $_nC_r = \dfrac{n!}{r!(n-r)!}$, (b) $_nC_r = _nC_{n-r}$.

(a) $\quad _nC_r = \dfrac{n(n-1)\cdots(n-r+1)}{1 \cdot 2 \cdots r} = \dfrac{n(n-1)\cdots(n-r+1)}{1 \cdot 2 \cdots r} \cdot \dfrac{(n-r)\cdots 2 \cdot 1}{1 \cdot 2 \cdots (n-r)} = \dfrac{n!}{r!(n-r)!}$

(b) $\quad _nC_r = \dfrac{n!}{r!(n-r)!} = \dfrac{n(n-1)\cdots(r+1)r(r-1)\cdots 2 \cdot 1}{1 \cdot 2 \cdots r \cdot (n-r)!} = \dfrac{n(n-1)\cdots(r+1)}{(n-r)!} = \dfrac{_nP_{n-r}}{(n-r)!} = _nC_{n-r}$

21.3   Compute (a) $_{10}C_2$, (b) $_{12}C_5$, (c) $_{15}C_{12}$, (d) $_{25}C_{21}$.

(a) $_{10}C_2 = \dfrac{10 \cdot 9}{1 \cdot 2} = 45$ (c) $_{15}C_{12} = {}_{15}C_3 = \dfrac{15 \cdot 14 \cdot 13}{1 \cdot 2 \cdot 3} = 455$

(b) $_{12}C_5 = \dfrac{12 \cdot 11 \cdot 10 \cdot 9 \cdot 8}{1 \cdot 2 \cdot 3 \cdot 4 \cdot 5} = 792$ (d) $_{25}C_{21} = {}_{25}C_4 = \dfrac{25 \cdot 24 \cdot 23 \cdot 22}{1 \cdot 2 \cdot 3 \cdot 4} = 12650$

21.4   A lady gives a dinner party for six guests. (a) In how many ways may they be selected from among 10 friends? (b) In how many ways if two of the friends will not attend the party together?

(a)   The six guests may be selected in $_{10}C_6 = {}_{10}C_4 = \dfrac{10 \cdot 9 \cdot 8 \cdot 7}{1 \cdot 2 \cdot 3 \cdot 4} = 210$ ways.

(b)   Let A and B denote the two who will not attend together. If neither A nor B is included, the guests may be selected in $_8C_6 = {}_8C_2 = \dfrac{8 \cdot 7}{1 \cdot 2} = 28$ ways. If one of A and B is included, the guests may be selected in $2 \cdot {}_8C_5 = 2 \cdot {}_8C_3 = 2\left(\dfrac{8 \cdot 7 \cdot 6}{1 \cdot 2 \cdot 3}\right) = 112$ ways. Thus, the six guests may be selected in $28 + 112 = 140$ ways.

21.5   A committee of 5 is to be selected from 12 seniors and 8 juniors. In how many ways can this be done (a) if the committee is to consist of 3 seniors and 2 juniors, (b) if the committee is to contain at least 3 seniors and 1 junior?

(a)   With each of the $_{12}C_3$ selections of 3 seniors, we may associate any one of the $_8C_2$ selections of 2 juniors. Thus, a committee can be selected in $_{12}C_3 \cdot {}_8C_2 = \dfrac{12 \cdot 11 \cdot 10}{1 \cdot 2 \cdot 3} \cdot \dfrac{8 \cdot 7}{1 \cdot 2} = 6160$ ways.

(b)   The committee may consist of 3 seniors and 2 juniors or of 4 seniors and 1 junior. A committee of 3 seniors and 2 juniors can be selected in 6160 ways, and a committee of 4 seniors and 1 junior can be selected in $_{12}C_4 \cdot {}_8C_1 = 3960$ ways. In all, a committee may be selected in $6160 + 3960 = 10120$ ways.

21.6   There are ten points $A, B, \ldots$, in a plane, no three on the same straight line. (a) How many lines are determined by the points? (b) How many of the lines pass through A? (c) How many triangles are determined by the points? (d) How many of the triangles have A as a vertex? (e) How many of the triangles have AB as a side?

(a)   Since any two points determine a line, there are $_{10}C_2 = 45$ lines.

(b)   To determine a line through A, one other point must be selected. Thus there are nine lines through A.

(c)   Since any three of the points determine a triangle, there are $_{10}C_3 = 120$ triangles.

(d)   Two additional points are needed to form a triangle. These points may be selected from the nine points in $_9C_2 = 36$ ways.

(e)   One additional point is needed; there are eight triangles having AB as a side.

21.7   In how many ways may 12 persons be divided into three groups (a) of 2, 4, and 6 persons, (b) of 4 persons each?

(a)   The group of two can be selected in $_{12}C_2$ ways, then the group of four in $_{10}C_4$ ways, and the group of six in $_6C_6 = 1$ way. Thus, the division may be made in $_{12}C_2 \cdot {}_{10}C_4 \cdot 1 = 13860$ ways.

(b)   One group of four can be selected in $_{12}C_4$ ways, then another in $_8C_4$ ways, and the third in 1 way. Since the order in which the groups are formed is now immaterial, the division may be made in $_{12}C_4 \cdot {}_8C_4 \cdot 1 \div 3! = 5775$ ways.

21.8 The English alphabet consists of 21 consonants and 5 vowels.
  (*a*) In how many ways can 4 consonants and 2 vowels be selected?
  (*b*) How many words consisting of 4 consonants and 2 vowels can be formed?
  (*c*) How many of the words in (*b*) begin with *R*?
  (*d*) How many of the words in (*c*) contain *E*?

  (*a*) The 4 consonants can be selected in $_{21}C_4$ ways and the 2 vowels can be selected in $_5C_2$ ways. Thus, the selections may be made in $_{21}C_4 \cdot _5C_2 = 59850$ ways.

  (*b*) From each of the selections in (*a*), 6! words may be formed by permuting the letters. Therefore, $59850 \cdot 6! = 43092000$ words can be formed.

  (*c*) Since the position of the consonant *R* is fixed, we must select 3 other consonants (in $_{20}C_3$ ways) and 2 vowels (in $_5C_2$ ways), and arrange each selection of 5 letters in all possible ways. Thus, there are $_{20}C_3 \cdot _5C_2 \cdot 5! = 1368000$ words.

  (*d*) Since the position of the consonant *R* is fixed but the position of the vowel *E* is not, we must select 3 other consonants (in $_{20}C_3$ ways) and 1 other vowel (in four ways), and arrange each set of 5 letters in all possible ways. Thus, there are $_{20}C_3 \cdot 4 \cdot 5! = 547200$ words.

21.9 From an ordinary deck of playing cards, in how many different ways can five cards be dealt (*a*) consisting of spades only, (*b*) consisting of black cards only, (*c*) containing the four aces, (*d*) consisting of three cards of one suit and two of another, (*e*) consisting of three kings and a pair, (*f*) consisting of three of one kind and two of another?

  (*a*) From the 13 spades, 5 can be selected in $_{13}C_5 = 1287$ ways.

  (*b*) From the 26 black cards, 5 can be selected in $_{26}C_5 = 65780$ ways.

  (*c*) One card must be selected from the 48 remaining cards. This can be done in 48 different ways.

  (*d*) A suit can be selected in four ways and three cards from the suit can be selected in $_{13}C_3$ ways; a second suit can now be selected in three ways and two cards of this suit in $_{13}C_2$ ways. Thus three cards of one suit and two of another can be selected in $4 \cdot _{13}C_3 \cdot 3 \cdot _{13}C_2 = 267696$ ways.

  (*e*) Three kings can be selected from the four kings in $_4C_3$ ways, another kind can be selected in 12 ways, and two cards of this kind can be selected in $_4C_2$ ways. Thus, three kings and another pair can be dealt in $_4C_3 \cdot 12 \cdot _4C_2 = 288$ ways.

  (*f*) A kind can be selected in 13 ways and three of this kind in $_4C_3$ ways; another kind can be selected in 12 ways and two of this kind can be selected in $_4C_2$ ways. Thus, 3 of one kind and 2 of another can be dealt in $13 \cdot _4C_3 \cdot 12 \cdot _4C_2 = 3744$ ways.

21.10 (*a*) Prove: The total number of combinations of *n* objects taken successively $1, 2, 3, \ldots, n$ at a time is $2^n - 1$. (*b*) In how many different ways can one invite one or more of five friends to the movies.

  (*a*) The total number of combinations is

$$_nC_1 + _nC_2 + _nC_3 + \cdots + _nC_n = 2^n - 1$$

  since, from Problem 14.8(*a*),

$$_nC_0 + _nC_1 + _nC_2 + \cdots + _nC_n = 2^n$$

  (*b*) The number is $_5C_1 + _5C_2 + _5C_3 + _5C_4 + _5C_5 = 2^5 - 1 = 31$.

Supplementary Problems

21.11 Evaluate (*a*) $_6C_2$ (*b*) $_8C_6$ (*c*) $_nC_3$.

  Ans. (*a*) 15   (*b*) 28   (*c*) $\dfrac{n(n-1)(n-2)}{1 \cdot 2 \cdot 3}$

21.12 Find n if (a) $\ _nC_2 = 55$, $\quad$ (b) $\ _nC_3 = 84$, $\quad$ (c) $\ _{2n}C_3 = 11 \cdot\ _nC_3$.

$\quad$ *Ans.* $\quad (a)$ $\ \ 11$ $\quad (b)$ $\ \ 9$ $\quad (c)$ $\ \ 6$

21.13 Two dice can be tossed in 36 ways. In how many of these is the sum equal to (a) 4, (b) 7, (c) 11?

$\quad$ *Ans.* $\quad (a)$ $\ \ 3$ $\quad (b)$ $\ \ 6$ $\quad (c)$ $\ \ 2$

21.14 Four delegates are to be chosen from eight members of a club. (a) How many choices are possible? (b) How many contain member A? (c) How many contain A or B but not both?

$\quad$ *Ans.* $\quad (a)$ $\ \ 70$ $\quad (b)$ $\ \ 35$ $\quad (c)$ $\ \ 40$

21.15 A party of 8 boys and 8 girls are going on a picnic. Six of the party go in one automobile, four go in another, and the rest walk. (a) In how many ways can the party be distributed for the trip? (b) In how many ways if no girl walks?

$\quad$ *Ans.* $\quad (a)$ $\ \ 1681680$ $\quad (b)$ $\ \ 5880$

21.16 Solve Problem 21.15 if the owner of each car (a boy) drives his own car.

$\quad$ *Ans.* $\quad (a)$ 168168 $\quad (b)$ $\ 56$

21.17 How many selections of five letters each can be made from the letters of the word CANADIANS?

$\quad$ *Ans.* $\ \ 41$

21.18 A bag contains nine balls numbered $1, 2, \ldots, 9$. In how many ways can two balls be drawn so that (a) both are odd? (b) their sum is odd?

$\quad$ *Ans.* $\quad (a)$ $\ \ 10$ $\quad (b)$ $\ \ 20$

21.19 How many diagonals has (a) a hexagon, (b) an octagon, (c) an n-gon?

$\quad$ *Ans.* $\quad (a)$ $\ \ 9$ $\quad (b)$ $\ \ 20$ $\quad (c)$ $\ \ \frac{1}{2}n(n-3)$

21.20 (a) How many words consisting of 3 consonants and 2 vowels can be formed from 10 consonants and 5 vowels? (b) In how many of these will the consonants occupy the odd places?

$\quad$ *Ans.* $\quad (a)$ $\ \ 144000$ $\quad (b)$ $\ \ 14400$

21.21 Three balls are drawn from a bag containing five red, four white, and three black balls. In how many ways can this can be done if (a) each is of a different color? (b) they are of the same color? (c) exactly two are red? (d) at least two are red?

$\quad$ *Ans.* $\quad (a)$ $\ \ 60$ $\quad (b)$ $\ \ 15$ $\quad (c)$ $\ \ 70$ $\quad (d)$ $\ \ 80$

21.22 A squad is made up of 10 privates and 5 privates first class. (a) In how many ways can a detail of 4 privates and 2 privates first class be formed? (b) On how many of these details will private X serve? (c) On how many will private X but not private first class Y serve?

$\quad$ *Ans.* $\quad (a)$ $\ \ 2100$ $\quad (b)$ $\ \ 840$ $\quad (c)$ $\ \ 504$

21.23 A civic club has 60 members including 2 bankers, 4 lawyers, and 5 doctors. In how many ways can a committee of 10 be formed to contain 1 banker, 2 lawyers, and 2 doctors?

  Ans.   228 826 080

21.24 How many committees of two or more can be selected from 10 people?

  Ans. $2^{10} - 11$

21.25 Hands consisting of three cards are dealt from an ordinary deck. Show that a hand consisting of three different kinds should show 352 times as often as a hand consisting of three cards of the same kind.

21.26 Prove (*a*) $_nC_r + {_n}C_{r+1} = {_{n+1}}C_{r+1}$ (*b*) $_{2n}C_n = 2 \cdot {_{2n-1}}C_{n-1}$

Chapter 22

Probability

IN ESTIMATING THE PROBABILITY that a given event will or will not happen, we may, as in the case of drawing a face card from an ordinary deck, count the different ways in which this event may or may not happen. On the other hand, in the case of estimating the probability that a person who is now 25 years old will live to receive a bequest at age 30, we are forced to depend upon such knowledge of what has happened on similar occasions in the past as is available. In the first case, the result is called *mathematical* or *theoretical probability*; in the latter case, the result is called *statistical* or *empirical probability*.

MATHEMATICAL PROBABILITY. If an event must result in some one of n, $(n \neq 0)$ different but *equally likely ways* and if a certain s of these ways are considered successes and the other $f = n - s$ ways are considered failures, then the probability of success in a given trial is $p = s/n$ and the probability of failure is $q = f/n$. Since $p + q = \dfrac{s+f}{n} = \dfrac{n}{n} = 1$, $p = 1 - q$ and $q = 1 - p$.

EXAMPLE 1. One card is drawn from an ordinary deck. What is the probability (a) that it is a red card, (b) that it is a spade, (c) that it is a king, (d) that it is not the ace of hearts?
  One card can be drawn from the deck in $n = 52$ different ways.

(a) A red card can be drawn from the deck in $s = 26$ different ways. Thus, the probability of drawing a red card is $s/n = \frac{26}{52} = \frac{1}{2}$.
(b) A spade can be drawn from the deck in 13 different ways. The probability of drawing a spade is $\frac{13}{52} = \frac{1}{4}$.
(c) A king can be drawn in 4 ways. The required probability is $\frac{4}{52} = \frac{1}{13}$.
(d) The ace of hearts can be drawn in 1 way; the probability of drawing the ace of hearts is $\frac{1}{52}$. Thus, the probability of *not* drawing the ace of hearts is $1 - \frac{1}{52} = \frac{51}{52}$.

(See Problems 22.1–22.4.)

[NOTE: We write $P(A)$ to denote the "probability that A occurs."]

Two or more events are called *mutually exclusive* if not more than one of them can occur in a single trial. Thus, the drawing of a jack and the drawing of a queen on a single draw from an ordinary deck are mutually exclusive events; however, the drawing of a jack and the drawing of a spade are not mutually exclusive.

EXAMPLE 2. Find the probability of drawing a jack or a queen from an ordinary deck of cards.
  Since there are four jacks and four queens, $s = 8$ and $p = \frac{8}{52} = \frac{2}{13}$. Now the probability of drawing a jack is $\frac{1}{13}$, the probability of drawing a queen is $\frac{1}{13}$, and the required probability is $\frac{2}{13} = \frac{1}{13} + \frac{1}{13}$.

We have verified

THEOREM A. The probability that some one of a set of mutually exclusive events will happen at a single trial is the sum of their separate probabilities of happening.

THEOREM A′. The probability that A will occur or B will occur, $P(A \cup B) = P(A) + P(B) - P(A \cap B)$.

Two events A and B are called *independent* if the happening of one does not affect the happening of the other. Thus, in a toss of two dice, the fall of either does not affect the fall of the other. However, in drawing two cards from a deck, the probability of obtaining a red card on the second draw depends upon whether or not a red card was obtained on the draw of the first card. Two such events are called *dependent*. More explicitly, if $P(A \cap B) = P(A) \cdot P(B)$, then A and B are independent.

EXAMPLE 3.   One bag contains 4 white and 4 black balls, a second bag contains 3 white and 6 black balls, and a third contains 1 white and 5 black balls. If one ball is drawn from each bag, find the probability that all are white.

A ball can be drawn from the first bag in any one of 8 ways, from the second in any one of 9 ways, and from the third in any one of 6 ways; hence, three balls can be drawn one from each bag in $8 \cdot 9 \cdot 6$ ways. A white ball can be drawn from the first bag in 4 ways, from the second in 3 ways, and from the third in 1 way; hence, three white balls can be drawn one from each bag in $4 \cdot 3 \cdot 1$ ways. Thus the required probability is

$$\frac{4 \cdot 3 \cdot 1}{8 \cdot 9 \cdot 6} = \frac{1}{36}$$

Now drawing a white ball from one bag does not affect the drawing of a white ball from another so that here we are concerned with three independent events. The probability of drawing a white ball from the first bag is $\frac{4}{8}$, from the second is $\frac{3}{9}$, and from the third bag is $\frac{1}{6}$.

Since the probability of drawing three white balls, one from each bag, is $\frac{4}{8} \cdot \frac{3}{9} \cdot \frac{1}{6}$, we have verified

THEOREM B.   The probability that all of a set of independent events will happen in a single trial is the product of their separate probabilities.

THEOREM C.   (concerning dependent events).   If the probability that an event will happen is p_1, and if after it has happened the probability that a second event will happen is p_2, the probability that the two events will happen in that order is $p_1 p_2$.

EXAMPLE 4.   Two cards are drawn from an ordinary deck. Find the probability that both are face cards (king, queen, jack) if (a) the first card drawn is replaced before the second is drawn, (b) the first card drawn is not replaced before the second is drawn.

(a) Since each drawing is made from a complete deck, we have the case of two independent events. The probability of drawing a face card in a single draw is $\frac{12}{52}$; thus, the probability of drawing two face cards, under the conditions imposed, is $\frac{12}{52} \cdot \frac{12}{52} = \frac{9}{169}$.

(b) Here the two events are dependent. The probability that the first drawing results in a face card is $\frac{12}{52}$. Now, of the 51 cards remaining in the deck, there are 11 face cards; the probability that the second drawing results in a face card is $\frac{11}{51}$. Hence, the probability of drawing two face cards is $\frac{12}{52} \cdot \frac{11}{51} = \frac{11}{221}$.

(See Problems 22.6–22.10.)

EXAMPLE 5.   Two dice are tossed six times. Find the probability (a) that 7 will show on the first four tosses and will not show on the other two, (b) that 7 will show on exactly four of the tosses.

The probability that 7 will show on a single toss is $p = \frac{1}{6}$ and the probability that 7 will not show is $q = 1 - p = \frac{5}{6}$.

(a) The probability that 7 will show on the first four tosses and will not show on the other two is

$$\frac{1}{6} \cdot \frac{1}{6} \cdot \frac{1}{6} \cdot \frac{1}{6} \cdot \frac{5}{6} \cdot \frac{5}{6} = \frac{25}{46\,656}$$

(b) The four tosses on which 7 is to show may be selected in $_6C_4 = 15$ ways. Since these 15 ways constitute mutually exclusive events and the probability of any one of them is $(\frac{1}{6})^4(\frac{5}{6})^2$, the probability that 7 will show exactly four times in six tosses is $_6C_4\left(\frac{1}{6}\right)^4\left(\frac{5}{6}\right)^2 = \frac{125}{15\,552}$.

We have verified

THEOREM D.   If p is the probability that an event will happen and q is the probability that it will fail to happen at a given trial, the probability that it will happen exactly r times in n trials is $_nC_r p^r q^{n-r}$. (See Problems 22.11–22.12.)

EMPIRICAL PROBABILITY.   If an event has been observed to happen s times in n trials, the ratio $p = s/n$ is defined as the *empirical probability* that the event will happen at any future trial. The confidence which can be placed in such probabilities depends in a large measure on the number of observations used. Life insurance companies, for example, base their premium rate on empirical probabilities. For this purpose they use a mortality table based on an enormous number of observations over the years.

The American Experience Table of Mortality begins with 100 000 persons all of age 10 years and indicates the number of the group who die each year thereafter. In using this table, it will be assumed that the laws stated above for mathematical probability hold also for empirical probability.

EXAMPLE 6. Find the probability that a person 20 years old (a) will die during the year, (b) will die during the next 10 years, (c) will reach age 75.

(a)   Of the 100 000 persons alive at age 10 years, 92 637 are alive at age 20 years. Of these 92 637 a total of 723 will die during the year. The probability that a person 20 years of age will die during the year is $\dfrac{723}{92\,637} = 0.0078$.

(b)   Of the 92 637 who reach age 20 years, 85 441 reach age 30 years; thus, 92 637 − 85 441 = 7196 die during the 10-year period. The required probability is $\dfrac{7196}{92\,637} = 0.0777$.

(c)   Of the 92 637 alive at age 20 years, 26 237 will reach age 75 years. The required probability is $\dfrac{26\,237}{92\,637} = 0.2832$.

(See Problem 22.13.)

Solved Problems

22.1   One ball is drawn from a bag containing 3 white, 4 red, and 5 black balls. What is the probability (a) that it is white, (b) that it is white or red, (c) that it is not red?

A ball can be drawn from the bag in $n = 3 + 4 + 5 = 12$ different ways.

(a)   A white ball can be drawn in $s = 3$ different ways. Thus, the probability of drawing a white ball is $p = s/n = \frac{3}{12} = \frac{1}{4}$.

(b)   Here success consists in drawing either a white or a red ball; hence, $s = 3 + 4 = 7$ and the required probability is $\frac{7}{12}$.

(c)   The probability of drawing a red ball is $p = \frac{4}{12} = \frac{1}{3}$. The probability that the ball drawn is *not* red is $1 - p = 1 - \frac{1}{3} = \frac{2}{3}$. This problem may also be solved as in (b).

22.2   If two dice are tossed, what is the probability (a) of throwing a total of 7, (b) of throwing a total of 8, (c) of throwing a total of 10 or more, (d) of both dice showing the same number?

Two dice may turn up in $6 \times 6 = 36$ ways.

(a)   A total of 7 may result in 6 ways (1, 6; 6, 1; 2, 5; 5, 2; 3, 4; 4, 3). The probability of a throw of 7 is then $\frac{6}{36} = \frac{1}{6}$.

(b)   A total of 8 may result in 5 ways (2, 6; 6, 2; 3, 5; 5, 3; 4, 4). The probability of a throw of 8 is $\frac{5}{36}$.

(c)   Here, success consists in throwing a total of 10, 11, or 12. Since a total of 10 may result in 3 ways, a total of 11 in 2 ways, and a total of 12 in 1 way, the probability is $(3 + 2 + 1)/36 = \frac{1}{6}$.

(d)   The probability that the second die will show the same number as the first is $\frac{1}{6}$. This problem may also be solved by counting the number of successes 1, 1; 2, 2; etc.

22.3   If five coins are tossed, what is the probability (*a*) that all will show heads, (*b*) that exactly three will show heads, (*c*) that at least three will show heads?

Each coin can turn up in 2 ways; hence, the five coins can turn up in $2^5 = 32$ ways. (The assumption here is that HHHTT and THHHT are different results.)

(*a*)   Five heads can turn up in only 1 way; hence, the probability of a toss of five heads is $\frac{1}{32}$.

(*b*)   Exactly three heads can turn up in $_5C_3 = 10$ ways; thus, the probability of a toss of exactly three heads is $\frac{10}{32} = \frac{5}{16}$.

(*c*)   Success here consists of a throw of exactly three, exactly four, or all heads. Exactly three heads can turn up in 10 ways, exactly four heads in 5 ways, and all heads in 1 way. Thus, the probability of throwing at least three heads is $(10 + 5 + 1)/32 = \frac{1}{2}$.

22.4   If three cards are drawn from an ordinary deck, find the probability (*a*) that all are red cards, (*b*) that all are of the same suit, (*c*) that all are aces.

Three cards may be drawn from a deck in $_{52}C_3$ ways.

(*a*)   Three red cards may be drawn from the 26 red cards in $_{26}C_3$ ways. Hence, the probability of drawing three red cards in a single draw is $\dfrac{_{26}C_3}{_{52}C_3} = \dfrac{26 \cdot 25 \cdot 24}{1 \cdot 2 \cdot 3} \cdot \dfrac{1 \cdot 2 \cdot 3}{52 \cdot 51 \cdot 50} = \dfrac{2}{17}$.

(*b*)   There are 4 ways of choosing a suit and $_{13}C_3$ ways of selecting three cards of that suit. Thus, the probability of drawing three cards of the same suit is $4 \cdot {_{13}C_3}/{_{52}C_3} = \frac{22}{425}$.

This problem may also be solved as follows: The first card drawn determines a suit. The deck now contains 51 cards of which 12 are of that suit; hence, the probability that the next two cards drawn will be of that suit is $_{12}C_2/{_{51}C_2} = \frac{22}{425}$, as before.

(*c*)   Three aces may be selected from the four aces in 4 ways; hence, the required probability is $\dfrac{4}{_{52}C_3} = \dfrac{1}{5525}$.

22.5   One bag contains 8 black balls and a second bag contains 1 white and 6 black balls. One of the bags is selected and then a ball is drawn from that bag. What is the probability that it is the white ball?

The probability that the second bag is chosen is $\frac{1}{2}$ and the probability that the white ball is drawn from this bag is $\frac{1}{7}$. Thus, the required probability is $\frac{1}{2}(\frac{1}{7}) = \frac{1}{14}$.

22.6   Two cards are drawn in succession from an ordinary deck. What is the probability (*a*) that the first will be the jack of diamonds and the second will be the queen of spades, (*b*) that the first will be a diamond and the second a spade, (*c*) that both cards are diamonds or both are spades?

(*a*)   The probability that the first card is the jack of diamonds is $\frac{1}{52}$ and the probability that the second is the queen of spades is $\frac{1}{51}$. The required probability is $\left(\frac{1}{52}\right)\left(\frac{1}{51}\right) = \frac{1}{2652}$.

(*b*)   The probability that the first card is a diamond is $\frac{13}{52}$ and the probability that the second card is a spade is $\frac{13}{51}$. The probability of the required draw is $\frac{13}{52} \cdot \frac{13}{51} = \frac{13}{204}$.

(*c*)   The probability that both cards are of a specified suit is $\frac{13}{52} \cdot \frac{12}{51}$. Thus, the probability that both are diamonds or both are spades is $\frac{13}{52} \cdot \frac{12}{51} + \frac{13}{52} \cdot \frac{12}{51} = \frac{2}{17}$.

22.7   *A*, *B*, and *C* work independently on a problem. If the respective probabilities that they will solve it are $\frac{1}{2}$, $\frac{1}{3}$, $\frac{2}{5}$, find the probability that the problem will be solved.

The problem will be solved unless all three fail; the probability that this will happen is $\frac{1}{2} \cdot \frac{2}{3} \cdot \frac{3}{5} = \frac{1}{5}$. Thus, the probability that the problem will be solved is $1 - \frac{1}{5} = \frac{4}{5}$.

22.8 A tosses a coin and if a head appears he wins the game; if a tail appears, B tosses the coin under the same conditions, and so on. If the stakes are $15, find the expectation of each.

We first compute the probability that A will win. The probability that he will win on the first toss is $\frac{1}{2}$; the probability that he will win on his second toss (that is, that A first tosses a tail, B tosses a tail, and A then tosses a head) is $\frac{1}{2} \cdot \frac{1}{2} \cdot \frac{1}{2} = (\frac{1}{2})^3$; the probability that he will win on his third toss (that is, that A first tosses a tail, B tosses a tail, A tosses a tail, B tosses a tail, and A then tosses a head) is $\frac{1}{2} \cdot \frac{1}{2} \cdot \frac{1}{2} \cdot \frac{1}{2} \cdot \frac{1}{2} = (\frac{1}{2})^5$, and so on. Thus, the probability that A will win is

$$\frac{1}{2} + \frac{1}{2^3} + \frac{1}{2^5} + \cdots = \frac{\frac{1}{2}}{1 - 1/2^2} = \frac{2}{3}$$

and his expectation is $\frac{2}{3}$ ($15) = $10. Then B's expectation is $5.

22.9 On a toss of two dice, X throws a total of 5. Find the probability that he will throw another 5 before he throws 7.

X will succeed should he throw a total of 5 on the next toss, or should he not throw 5 or 7 on this toss but throw 5 on the next, or should he not throw 5 or 7 on either of these tosses but throw 5 on the next, and so on. The respective probabilities are $\frac{4}{36}$, $\frac{26}{36} \cdot \frac{4}{36}$, $\frac{26}{36} \cdot \frac{26}{36} \cdot \frac{4}{36}$, and so on. Thus, the probability that he throws 5 before 7 is

$$\frac{4}{36} + \frac{26}{36} \cdot \frac{4}{36} + \frac{26}{36} \cdot \frac{26}{36} \cdot \frac{4}{36} + \cdots = \frac{\frac{4}{36}}{1 - \frac{26}{36}} = \frac{2}{5}$$

22.10 A bag contains 2 white and 3 black balls. A ball is drawn 5 times, each being replaced before another is drawn. Find the probability that (a) the first 4 balls drawn are white and the last is black, (b) exactly 4 of the balls drawn are white, (c) at least 4 of the balls drawn are white, (d) at least 1 ball is white.

The probability of drawing a white ball is $p = \frac{2}{5}$ and the probability of drawing a black ball is $\frac{3}{5}$.

(a) The probability that the first 4 are white and the last black is $\frac{2}{5} \cdot \frac{2}{5} \cdot \frac{2}{5} \cdot \frac{2}{5} \cdot \frac{3}{5} = \frac{48}{3125}$.

(b) Here $n = 5$, $r = 4$; the probability that exactly 4 of the balls drawn are white is

$$nC_r p^r q^{n-r} = {}_5C_4(\tfrac{2}{5})^4(\tfrac{3}{5}) = \frac{48}{625}$$

(c) Since success consists of drawing either 4 white and 1 black ball or 5 white balls, the probability is

$$_5C_4(\tfrac{2}{5})^4(\tfrac{3}{5}) + {}_5C_5(\tfrac{2}{5})^5 = \frac{272}{3125}$$

(d) Here failure consists of drawing 5 black balls. Since the probability of failure is $(\tfrac{3}{5})^5 = \frac{243}{3125}$, the probability of success is $1 - \frac{243}{3125} = \frac{2882}{3125}$. The problem may also be solved as in (c).

22.11 One bag contains 2 white balls and 2 black balls, and another contains 3 white balls and 5 black balls. At five different trials, a bag is chosen at random and 1 ball is drawn from that bag and replaced. Find the probability (a) that exactly 3 white balls are drawn, (b) that at least 3 white balls are drawn.

At any trial the probability that a white ball is drawn from the first bag is $\frac{1}{2} \cdot \frac{1}{2}$ and the probability that a white ball is drawn from the second bag is $\frac{1}{2}(\frac{3}{8})$. Thus, the probability that a white ball is drawn at any trial is $p = \frac{1}{2} \cdot \frac{1}{2} + \frac{1}{2}(\frac{3}{8}) = \frac{7}{16}$, and the probability that a black ball is drawn is $q = \frac{9}{16}$.

(a) The probability of drawing exactly 3 white balls in 5 trials is $_5C_3\left(\dfrac{7}{16}\right)^3\left(\dfrac{9}{16}\right)^2 = \dfrac{138\,915}{524\,288}$.

(b) The probability of drawing at least 3 white balls in 5 trials is

$$_5C_3\left(\frac{7}{16}\right)^3\left(\frac{9}{16}\right)^2 + {}_5C_4\left(\frac{7}{16}\right)^4\left(\frac{9}{16}\right) + {}_5C_5\left(\frac{7}{16}\right)^5 = \frac{201\,341}{524\,288}$$

22.12 A husband is 35 years old and his wife is 28. Find the probability that at the end of 20 years (a) the husband will be alive, (b) the wife will be alive, (c) both will be alive, (d) both will be dead, (e) the wife will be alive and the husband will not, (f) one will be alive but not the other.

(a) Of the 81 822 alive at age 35, 64 563 will reach age 55. The probability that the husband will be alive at the end of 20 years is $\dfrac{64\,563}{81\,822} = 0.7890$.

(b) The probability that the wife will be alive at the end of 20 years is $\dfrac{71\,627}{86\,878} = 0.8245$.

(c) Since the survival of the husband and of the wife are independent events, the probability that both are alive after 20 years is $\dfrac{64\,563}{81\,822} \cdot \dfrac{71\,627}{86\,878} = 0.6506$.

(d) From (a) 17 259 of the 81 822 alive at age 35 will not reach age 55; thus, the probability that the husband will not live for 20 years is $\dfrac{17\,259}{81\,822}$. Similarly, the probability that the wife will not live for 20 years is $\dfrac{15\,251}{86\,878}$. Hence, the probability that after 20 years both will be dead is $\dfrac{17\,259}{81\,822} \cdot \dfrac{15\,251}{86\,878} = 0.0370$.

(e) The probability that the husband will be dead and the wife will be alive after 20 years is $\dfrac{17\,259}{81\,822} \cdot \dfrac{71\,627}{86\,878} = 0.1739$.

(f) The probability that the wife will survive but the husband will not is found in (e). The probability that the husband will survive but the wife will not is $\dfrac{64\,563}{81\,822} \cdot \dfrac{15\,251}{86\,878}$. Thus, the probability that just one will survive is $\dfrac{17\,259}{81\,822} \cdot \dfrac{71\,627}{86\,878} + \dfrac{64\,563}{81\,822} \cdot \dfrac{15\,251}{86\,878} = 0.3116$.

Supplementary Problems

22.13 One ball is drawn from a bag containing 4 white and 6 black balls. Find the probability that it is (a) white, (b) black.

  Ans. (a) $\frac{2}{5}$ (b) $\frac{3}{5}$

22.14 Three balls are drawn together from a bag containing 8 white and 12 black balls. Find the probability that (a) all are white, (b) just two are white, (c) just one is white, (d) all are black.

  Ans. (a) $\frac{14}{285}$ (b) $\frac{28}{95}$ (c) $\frac{44}{95}$ (d) $\frac{11}{57}$

22.15 Ten students are seated at random in a row. Find the probability that two particular students are not seated side by side.

  Ans. $\frac{4}{5}$

22.16 If a die is cast three times, find the probability (a) that an even number will be thrown each time, (b) that an odd number will appear just once, (c) that the sum of the three numbers will be even.

  Ans. (a) $\frac{1}{8}$ (b) $\frac{3}{8}$ (c) $\frac{1}{2}$

22.17   From a box containing 10 cards numbered 1, 2, 3, ..., 10, four cards are drawn. Find the probability that their sum will be even (a) if the cards are drawn together, (b) if each card drawn is replaced before the next is drawn.

  Ans. (a) $\frac{11}{21}$ (b) $\frac{1}{2}$

22.18   A and B, having equal skill, are playing a game of three points. After A has won 2 points and B has won 1 point, what is the probability that A will win the game?

  Ans. $\frac{3}{4}$

22.19   One bag contains 3 white and 2 black balls, and another contains 2 white and 3 black balls. A ball is drawn from the second bag and placed in the first; then a ball is drawn from the first bag and placed in the second. When the pair of operations is repeated, what is the probability that the first bag will contain 5 white balls?

  Ans. $\frac{1}{225}$

22.20   Three bags contain respectively 2 white and 1 black ball, 3 white and 3 black balls, 6 white and 2 black balls. Two bags are selected and a ball is drawn from each. Find the probability (a) that both balls are white, (b) that both balls are of the same color.

  Ans. (a) $\frac{29}{72}$ (b) $\frac{19}{36}$

22.21   If four trials are made in Problem 22.20, find the probability (a) that the first two will result in pairs of white balls and the other two in pairs of black balls, (b) that a pair of black balls will be obtained at least three times.

  Ans. (a) $\frac{841}{331\,776}$ (b) $\frac{125}{36\,864}$

22.22   Five cards numbered 1, 2, 3, 4, 5 respectively are placed in a revolving box. If the cards are drawn one at a time from the box, what is the probability that they will be drawn in their natural order?

  Ans. $\frac{1}{120}$

22.23   Brown, Jones, and Smith shoot at a target in alphabetical order with probabilities $\frac{1}{4}, \frac{1}{3}, \frac{1}{2}$ respectively of hitting the bull's eye. (a) Find the probability that on his first shot each will be the first to hit the bull's-eye, (b) find the probability that the bull's-eye is not hit on the first round, (c) find the probability that the first to hit the bull's-eye is Jones on his second shot.

  Ans. (a) $\frac{1}{4}, \frac{1}{4}, \frac{1}{4}$ (b) $\frac{1}{4}$ (c) $\frac{1}{16}$

22.24   The probability that X will win a game of checkers is $\frac{2}{5}$. In a five-game match, what is the probability (a) that X will win the first, third, and fifth games, and lose the others? (b) that he will win exactly three games? (c) that he will win at least three games?

  Ans. (a) $\frac{72}{3125}$ (b) $\frac{144}{625}$ (c) $\frac{992}{3125}$

22.25   Three pennies are tossed at the same time. Find the probability that two are heads and one is a tail.

  Ans. $\frac{3}{8}$

22.26   Prove: $_nC_r + {_nC_{r+1}} = {_{n+1}C_{r+1}}$.

22.27   Prove: $_{2n}C_n = 2 \cdot {_{2n-1}C_{n-1}}$.

22.28   Find the probability that in a random arrangement of 52 bridge cards, no two aces are adjacent.

Chapter 23

Determinants of Order Two and Three

DETERMINANTS OF ORDER TWO. The symbol $\begin{vmatrix} a_1 & b_1 \\ a_2 & b_2 \end{vmatrix}$, consisting of 2^2 numbers called *elements* arranged in two rows and two columns, is called a *determinant of order two*. The elements a_1 and b_2 are said to lie along the *principal diagonal*; the elements a_2 and b_1 are said to lie along the *secondary diagonal*.

The *value* of the determinant is obtained by forming the product of the elements along the principal diagonal and subtracting from it the product of the elements along the secondary diagonal; thus,

$$\begin{vmatrix} a_1 & b_1 \\ a_2 & b_2 \end{vmatrix} = a_1 b_2 - a_2 b_1$$

(See Problem 23.1.)

THE SOLUTION of the consistent and independent equations

$$\begin{cases} a_1 x + b_1 y = c_1 \\ a_2 x + b_2 y = c_2 \end{cases} \tag{23.1}$$

may be expressed as quotients of determinants of order two:

$$x = \frac{c_1 b_2 - c_2 b_1}{a_1 b_2 - a_2 b_1} = \frac{\begin{vmatrix} c_1 & b_1 \\ c_2 & b_2 \end{vmatrix}}{\begin{vmatrix} a_1 & b_1 \\ a_2 & b_2 \end{vmatrix}}, \qquad y = \frac{a_1 c_2 - a_2 c_1}{a_1 b_2 - a_2 b_1} = \frac{\begin{vmatrix} a_1 & c_1 \\ a_2 & c_2 \end{vmatrix}}{\begin{vmatrix} a_1 & b_1 \\ a_2 & b_2 \end{vmatrix}}$$

These equations are consistent and independent if and only if $\begin{vmatrix} a_1 & b_1 \\ a_2 & b_2 \end{vmatrix} \neq 0$.

EXAMPLE 1. Solve $\begin{cases} y = 3x + 1 \\ 4x + 2y - 7 = 0 \end{cases}$ using determinants.

Arrange the equations in the form (*23.1*): $\begin{cases} 3x - y = -1 \\ 4x + 2y = 7 \end{cases}$. The solution requires the values of three determinants:

The denominator, D, formed by writing the coefficients of x and y in order

$$D = \begin{vmatrix} 3 & -1 \\ 4 & 2 \end{vmatrix} = 3 \cdot 2 - 4(-1) = 6 + 4 = 10$$

The numerator of x, N_x, formed from D by replacing the coefficients of x by the constant terms

$$N_x = \begin{vmatrix} -1 & -1 \\ 7 & 2 \end{vmatrix} = -1 \cdot 2 - 7(-1) = -2 + 7 = 5$$

The numerator of y, N_y, formed from D by replacing the coefficients of y by the constant terms

$$N_y = \begin{vmatrix} 3 & -1 \\ 4 & 7 \end{vmatrix} = 3 \cdot 7 - 4(-1) = 21 + 4 = 25$$

Then $x = \dfrac{N_x}{D} = \dfrac{5}{10} = \dfrac{1}{2}$ and $y = \dfrac{N_y}{D} = \dfrac{25}{10} = \dfrac{5}{2}$. (See Problem 23.2.)

DETERMINANTS OF ORDER THREE.   The symbol

$$\begin{vmatrix} a_1 & b_1 & c_1 \\ a_2 & b_2 & c_2 \\ a_3 & b_3 & c_3 \end{vmatrix},$$

consisting of 3^2 elements arranged in three rows and three columns, is called a *determinant of order three*. Its value is

$$a_1b_2c_3 + a_2b_3c_1 + a_3b_1c_2 - a_1b_3c_2 - a_2b_1c_3 - a_3b_2c_1$$

This may be written as

$$a_1(b_2c_3 - b_3c_2) - b_1(a_2c_3 - a_3c_2) + c_1(a_2b_3 - a_3b_2)$$

or
$$a_1\begin{vmatrix} b_2 & c_2 \\ b_3 & c_3 \end{vmatrix} - b_1\begin{vmatrix} a_2 & c_2 \\ a_3 & c_3 \end{vmatrix} + c_1\begin{vmatrix} a_2 & b_2 \\ a_3 & b_3 \end{vmatrix} \tag{23.2}$$

to involve three determinants of order two. Note that the elements which multiply the determinants of order two are the elements of the first row of the given determinant. In all, six such representations using the elements of each of the three rows and each of the three columns may be worked out. We shall use only the representation (*23.2*) in this chapter. (See Problem 23.3.)

Another procedure for evaluating determinants of order three is indicated in Problem 23.4.

THE SOLUTION of the system of consistent and independent equations

$$\begin{cases} a_1x + b_1y + c_1z = d_1 \\ a_2x + b_2y + c_2z = d_2 \\ a_3x + b_3y + c_3z = d_3 \end{cases}$$

in determinant form is given by

$$x = \frac{N_x}{D} = \frac{\begin{vmatrix} d_1 & b_1 & c_1 \\ d_2 & b_2 & c_2 \\ d_3 & b_3 & c_3 \end{vmatrix}}{\begin{vmatrix} a_1 & b_1 & c_1 \\ a_2 & b_2 & c_2 \\ a_3 & b_3 & c_3 \end{vmatrix}}, \qquad y = \frac{N_y}{D} = \frac{\begin{vmatrix} a_1 & d_1 & c_1 \\ a_2 & d_2 & c_2 \\ a_3 & d_3 & c_3 \end{vmatrix}}{D}, \qquad z = \frac{N_z}{D} = \frac{\begin{vmatrix} a_1 & b_1 & d_1 \\ a_2 & b_2 & d_2 \\ a_3 & b_3 & d_3 \end{vmatrix}}{D}.$$

The determinant D is formed by writing the coefficients of x, y, z in order while the determinant appearing in the numerator for any unknown is obtained from D by replacing the column of coefficients of that unknown by the column of constants.

The system is consistent and independent if and only if $D \neq 0$.

EXAMPLE 2.   Solve, using determinants: $\begin{cases} x + 3y + 2z = -13 \\ 2x - 6y + 3z = 32 \\ 3x - 4y - z = 12 \end{cases}$

The solution requires the values of four determinants:
The denominator,

$$D = \begin{vmatrix} 1 & 3 & 2 \\ 2 & -6 & 3 \\ 3 & -4 & -1 \end{vmatrix} = 1(6 + 12) - 3(-2 - 9) + 2(-8 + 18)$$

$$= 18 + 33 + 20 = 71$$

The numerator of x,

$$N_x = \begin{vmatrix} -13 & 3 & 2 \\ 32 & -6 & 3 \\ 12 & -4 & -1 \end{vmatrix} = -13(6+12) - 3(-32-36) + 2(-128+72)$$

$$= -234 + 204 - 112 = -142$$

The numerator of y,

$$N_y = \begin{vmatrix} 1 & -13 & 2 \\ 2 & 32 & 3 \\ 3 & 12 & -1 \end{vmatrix} = 1(-32-36) - (-13)(-2-9) + 2(24-96)$$

$$= -68 - 143 - 144 = -355$$

The numerator of z,

$$N_z = \begin{vmatrix} 1 & 3 & -13 \\ 2 & -6 & 32 \\ 3 & -4 & 12 \end{vmatrix} = 1(-72+128) - 3(24-96) + (-13)(-8+18)$$

$$= 56 + 216 - 130 = 142$$

Then

$$x = \frac{N_x}{D} = \frac{-142}{71} = -2, \qquad y = \frac{N_y}{D} = \frac{-355}{71} = -5, \qquad z = \frac{N_z}{D} = \frac{142}{71} = 2$$

(See Problems 23.5–23.6.)

Solved Problems

23.1   Evaluate each of the following determinants:

(a) $\begin{vmatrix} 2 & 3 \\ 4 & 5 \end{vmatrix} = 2 \cdot 5 - 4 \cdot 3 = -2$ (b) $\begin{vmatrix} 5 & -2 \\ 3 & 1 \end{vmatrix} = 5 \cdot 1 - 3(-2) = 11$

23.2   Solve for x and y, using determinants: (a) $\begin{cases} x + 2y = -4 \\ 5x + 3y = 1 \end{cases}$ (b) $\begin{cases} ax - 2by = c \\ 2ax - 3by = 4c \end{cases}$

(a) $D = \begin{vmatrix} 1 & 2 \\ 5 & 3 \end{vmatrix} = 3 - 10 = -7, \qquad N_x = \begin{vmatrix} -4 & 2 \\ 1 & 3 \end{vmatrix} = -12 - 2 = -14, \qquad N_y = \begin{vmatrix} 1 & -4 \\ 5 & 1 \end{vmatrix} = 1 + 20 = 21$

$$x = \frac{N_x}{D} = \frac{-14}{-7} = 2, \qquad y = \frac{N_y}{D} = \frac{21}{-7} = -3$$

(b) $D = \begin{vmatrix} a & -2b \\ 2a & -3b \end{vmatrix} = ab, \qquad N_x = \begin{vmatrix} c & -2b \\ 4c & -3b \end{vmatrix} = 5bc, \qquad N_y = \begin{vmatrix} a & c \\ 2a & 4c \end{vmatrix} = 2ac$

$$x = \frac{N_x}{D} = \frac{5bc}{ab} = \frac{5c}{a}, \qquad y = \frac{N_y}{D} = \frac{2ac}{ab} = \frac{2c}{b}$$

23.3   Evaluate the following determinants:

(a) $\begin{vmatrix} 2 & -2 & -1 \\ 6 & 1 & -1 \\ 4 & 3 & 5 \end{vmatrix} = 2 \begin{vmatrix} 1 & -1 \\ 3 & 5 \end{vmatrix} - (-2) \begin{vmatrix} 6 & -1 \\ 4 & 5 \end{vmatrix} + (-1) \begin{vmatrix} 6 & 1 \\ 4 & 3 \end{vmatrix}$

$$= 2(5+3) + 2(30+4) - (18-4) = 2 \cdot 8 + 2 \cdot 34 - 14 = 70$$

(b) $\begin{vmatrix} 2 & 5 & 0 \\ 0 & 3 & 4 \\ -5 & 3 & 6 \end{vmatrix} = 2\begin{vmatrix} 3 & 4 \\ 3 & 6 \end{vmatrix} - 5\begin{vmatrix} 0 & 4 \\ -5 & 6 \end{vmatrix} + 0\begin{vmatrix} 0 & 3 \\ -5 & 3 \end{vmatrix} = 2(18 - 12) - 5(0 + 20) + 0 = -88$

(c) $\begin{vmatrix} 3 & 2 & 1 \\ 1 & -2 & 4 \\ 4 & 2 & 3 \end{vmatrix} = 3(-6 - 8) - 2(3 - 16) + 1(2 + 8) = -6$

(d) $\begin{vmatrix} 4 & -3 & 2 \\ 5 & 9 & -7 \\ 4 & -1 & 7 \end{vmatrix} = 4(36 - 7) + 3(20 + 28) + 2(-5 - 36) = 178$

23.4 The following procedure for evaluating a determinant of order three makes use of six lines each of which joins the three elements whose product is to be formed.

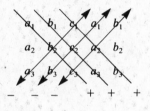

Here the first two columns are rewritten to the right of the determinant. The products formed by following the lines running down from left to right have a plus sign attached, and those formed by following the lines running up from left to right have a negative sign attached.

The algebraic sum of the products thus formed is the value of the determinant.

23.5 Solve, using determinants: $\begin{cases} 2x - 3y + 2z = 6 \\ x + 8y + 3z = -31 \\ 3x - 2y + z = -5 \end{cases}$

We evaluate $D = \begin{vmatrix} 2 & -3 & 2 \\ 1 & 8 & 3 \\ 3 & -2 & 1 \end{vmatrix} = 2(8 + 6) + 3(1 - 9) + 2(-2 - 24) = -48$

$N_x = \begin{vmatrix} 6 & -3 & 2 \\ -31 & 8 & 3 \\ -5 & -2 & 1 \end{vmatrix} = 6(8 + 6) + 3(-31 + 15) + 2(62 + 40) = 240$

$N_y = \begin{vmatrix} 2 & 6 & 2 \\ 1 & -31 & 3 \\ 3 & -5 & 1 \end{vmatrix} = 2(-31 + 15) - 6(1 - 9) + 2(-5 + 93) = 192$

$N_z = \begin{vmatrix} 2 & -3 & 6 \\ 1 & 8 & -31 \\ 3 & -2 & -5 \end{vmatrix} = 2(-40 - 62) + 3(-5 + 93) + 6(-2 - 24) = -96.$

Then $x = \dfrac{N_x}{D} = \dfrac{240}{-48} = -5, \qquad y = \dfrac{N_y}{D} = \dfrac{192}{-48} = -4, \qquad z = \dfrac{N_z}{D} = \dfrac{-96}{-48} = 2.$

23.6 Solve, using determinants:
$$\begin{cases} 2x + y = 2 \\ z - 4y = 0 \\ 4x + z = 6 \end{cases}$$

$$D = \begin{vmatrix} 2 & 1 & 0 \\ 0 & -4 & 1 \\ 4 & 0 & 1 \end{vmatrix} = -4, \quad N_x = \begin{vmatrix} 2 & 1 & 0 \\ 0 & -4 & 1 \\ 6 & 0 & 1 \end{vmatrix} = -2, \quad N_y = \begin{vmatrix} 2 & 2 & 0 \\ 0 & 0 & 1 \\ 4 & 6 & 1 \end{vmatrix} = -4, \quad N_z = \begin{vmatrix} 2 & 1 & 2 \\ 0 & -4 & 0 \\ 4 & 0 & 6 \end{vmatrix} = -16$$

Then
$$x = \frac{-2}{-4} = \frac{1}{2}, \quad y = \frac{-4}{-4} = 1, \quad \text{and} \quad z = \frac{-16}{-4} = 4.$$

Supplementary Problems

23.7 Evaluate. (a) $\begin{vmatrix} 1 & 2 \\ -3 & 1 \end{vmatrix}$ (b) $\begin{vmatrix} 2 & 0 \\ -5 & 1 \end{vmatrix}$ (c) $\begin{vmatrix} 3 & 2 \\ 1 & -4 \end{vmatrix}$ (d) $\begin{vmatrix} 2 & -10 \\ 3 & -15 \end{vmatrix}$

  Ans. (a) 7 (b) 2 (c) −14 (d) 0

23.8 Solve, using determinants. (a) $\begin{cases} 2x + y = 4 \\ 3x + 4y = 1 \end{cases}$ (b) $\begin{cases} 5x + 2y = 2 \\ 3x - 5y = 26 \end{cases}$ (c) $\begin{cases} 5x + 3y = -6 \\ 3x + 5y = -18 \end{cases}$

  Ans. (a) $x = 3, y = -2$ (b) $x = 2, y = -4$ (c) $x = \frac{3}{2}, y = -\frac{9}{2}$

23.9 Evaluate. (a) $\begin{vmatrix} 1 & 2 & 3 \\ 2 & -3 & 4 \\ -3 & 4 & 5 \end{vmatrix}$ (b) $\begin{vmatrix} 2 & -1 & 3 \\ 2 & 1 & 5 \\ -2 & 3 & -2 \end{vmatrix}$ (c) $\begin{vmatrix} -2 & 3 & 1 \\ 3 & 5 & 1 \\ 4 & -2 & 1 \end{vmatrix}$ (d) $\begin{vmatrix} 2 & 6 & 1 \\ 4 & -4 & 1 \\ 3 & 1 & 1 \end{vmatrix}$

  Ans. (a) −78 (b) −4 (c) −37 (d) 0

23.10 Solve, using determinants.

(a) $\begin{cases} x + 2y + 2z = 4 \\ 3x - y + 4z = 25 \\ 3x + 2y - z = -4 \end{cases}$ (b) $\begin{cases} 2x - 3y + 5z = 4 \\ 3x + 2y + 2z = 3 \\ 4x + y - 4z = -6 \end{cases}$ (c) $\begin{cases} \dfrac{1}{x} + \dfrac{2}{y} + \dfrac{1}{z} = 2 \\ \dfrac{3}{x} - \dfrac{4}{y} - \dfrac{2}{z} = 1 \\ \dfrac{2}{x} + \dfrac{5}{y} - \dfrac{2}{z} = 3 \end{cases}$

Hint: In (c) solve first for $1/x$, $1/y$, $1/z$.

  Ans. (a) $x = 2, y = -3, z = 4$ (b) $x = -\frac{1}{3}, y = \frac{2}{3}, z = \frac{4}{3}$ (c) $x = 1, y = z = 3$

23.11 Verify, by evaluating the determinants.

(a) $\begin{vmatrix} 1 & 2 & 3 \\ a & 2a & 3a \\ 8 & 9 & 10 \end{vmatrix} = 0$ (c) $\begin{vmatrix} 3 & 4 & 5 \\ 7 & -2 & 3 \\ 2 & 5 & -1 \end{vmatrix} = - \begin{vmatrix} 4 & -2 & 5 \\ 3 & 7 & 2 \\ 5 & 3 & -1 \end{vmatrix}$

(b) $\begin{vmatrix} 2 & 1 & -1 \\ 3 & 4 & 2 \\ -2 & -5 & 3 \end{vmatrix} = \begin{vmatrix} 0 & 1 & 0 \\ -5 & 4 & 6 \\ 8 & -5 & -2 \end{vmatrix}$ (d) $\begin{vmatrix} 4 & 2 & 5 \\ -7 & -3 & 1 \\ 9 & 4 & 8 \end{vmatrix} + \begin{vmatrix} -1 & 2 & 5 \\ 5 & -3 & 1 \\ -3 & 4 & 8 \end{vmatrix} = \begin{vmatrix} 3 & 2 & 5 \\ -2 & -3 & 1 \\ 6 & 4 & 8 \end{vmatrix}$

23.12 Solve, using determinants. $\begin{vmatrix} x & y & 1 \\ 1 & -1 & 1 \\ 13 & 2 & 1 \end{vmatrix} = 0,$ $\begin{vmatrix} x & y & 1 \\ 3 & 2 & 1 \\ -6 & -4 & 1 \end{vmatrix} = 0.$

Ans. $x = -3, y = -2$

Chapter 24

Determinants of Order n

A DETERMINANT of order n consists of n^2 numbers called elements arranged in n rows and columns, and enclosed by two vertical lines. For example,

$$\Delta_1 = |a_1| \qquad \Delta_2 = \begin{vmatrix} a_1 & b_1 \\ a_2 & b_2 \end{vmatrix} \qquad \Delta_3 = \begin{vmatrix} a_1 & b_1 & c_1 \\ a_2 & b_2 & c_2 \\ a_3 & b_3 & c_3 \end{vmatrix} \qquad \Delta_4 = \begin{vmatrix} a_1 & b_1 & c_1 & d_1 \\ a_2 & b_2 & c_2 & d_2 \\ a_3 & b_3 & c_3 & d_3 \\ a_4 & b_4 & c_4 & d_4 \end{vmatrix}$$

are determinants of orders one, two, three, and four, respectively. In this notation the letters designate columns and the subscripts designate rows. Thus, all elements with basal letter c are in the third column and all elements with subscript 2 are in the second row.

THE MINOR OF A GIVEN ELEMENT of a determinant is the determinant of the elements which remain after deleting the row and the column in which the given element stands. For example, the minor of a_1 in Δ_4 is

$$\begin{vmatrix} b_2 & c_2 & d_2 \\ b_3 & c_3 & d_3 \\ b_4 & c_4 & d_4 \end{vmatrix}$$

and the minor of b_3 is

$$\begin{vmatrix} a_1 & c_1 & d_1 \\ a_2 & c_2 & d_2 \\ a_4 & c_4 & d_4 \end{vmatrix}$$

Note that the minor of a given element contains no element having either the letter or the subscript of the given element. (See Problem 24.1.)

THE VALUE OF A DETERMINANT of order one is the single element of the determinant. A determinant of order $n > 1$ may be expressed as the sum of n products formed by multiplying each element of any chosen row (column) by its minor and prefixing a proper sign. The proper sign associated with each product is $(-1)^{i+j}$ where i is the number of the row and j is the number of the column in which the element stands. For example,

$$\Delta_3 = -a_2 \begin{vmatrix} b_1 & c_1 \\ b_3 & c_3 \end{vmatrix} + b_2 \begin{vmatrix} a_1 & c_1 \\ a_3 & c_3 \end{vmatrix} - c_2 \begin{vmatrix} a_1 & b_1 \\ a_3 & b_3 \end{vmatrix}$$

is the expansion of Δ_3 along the second row. The sign given to the first product is $-$ since a_2 stands in the second row and first column, and $(-1)^{2+1} = -1$. In all, there are six expansions of Δ_3 along its rows and columns yielding identical results when the minors are evaluated.

There are eight expansions of Δ_4 along its rows and columns, of which

$$\Delta_4 = +a_1\begin{vmatrix} b_2 & c_2 & d_2 \\ b_3 & c_3 & d_3 \\ b_4 & c_4 & d_4 \end{vmatrix} - b_1\begin{vmatrix} a_2 & c_2 & d_2 \\ a_3 & c_3 & d_3 \\ a_4 & c_4 & d_4 \end{vmatrix} + c_1\begin{vmatrix} a_2 & b_2 & d_2 \\ a_3 & b_3 & d_3 \\ a_4 & b_4 & d_4 \end{vmatrix} - d_1\begin{vmatrix} a_2 & b_2 & c_2 \\ a_3 & b_3 & c_3 \\ a_4 & b_4 & c_4 \end{vmatrix}$$

(along the first row)

$$\Delta_4 = +a_1\begin{vmatrix} b_2 & c_2 & d_2 \\ b_3 & c_3 & d_3 \\ b_4 & c_4 & d_4 \end{vmatrix} - a_2\begin{vmatrix} b_1 & c_1 & d_1 \\ b_3 & c_3 & d_3 \\ b_4 & c_4 & d_4 \end{vmatrix} + a_3\begin{vmatrix} b_1 & c_1 & d_1 \\ b_2 & c_2 & d_2 \\ b_4 & c_4 & d_4 \end{vmatrix} - a_4\begin{vmatrix} b_1 & c_1 & d_1 \\ b_2 & c_2 & d_2 \\ b_3 & c_3 & d_3 \end{vmatrix}$$

(along the first column)

$$\Delta_4 = -a_4\begin{vmatrix} b_1 & c_1 & d_1 \\ b_2 & c_2 & d_2 \\ b_3 & c_3 & d_3 \end{vmatrix} + b_4\begin{vmatrix} a_1 & c_1 & d_1 \\ a_2 & c_2 & d_2 \\ a_3 & c_3 & d_3 \end{vmatrix} - c_4\begin{vmatrix} a_1 & b_1 & d_1 \\ a_2 & b_2 & d_2 \\ a_3 & b_3 & d_3 \end{vmatrix} + d_4\begin{vmatrix} a_1 & b_1 & c_1 \\ a_2 & b_2 & c_2 \\ a_3 & b_3 & c_3 \end{vmatrix}$$

(along the fourth row)

$$\Delta_4 = -b_1\begin{vmatrix} a_2 & c_2 & d_2 \\ a_3 & c_3 & d_3 \\ a_4 & c_4 & d_4 \end{vmatrix} + b_2\begin{vmatrix} a_1 & c_1 & d_1 \\ a_3 & c_3 & d_3 \\ a_4 & c_4 & d_4 \end{vmatrix} - b_3\begin{vmatrix} a_1 & c_1 & d_1 \\ a_2 & c_2 & d_2 \\ a_4 & c_4 & d_4 \end{vmatrix} + b_4\begin{vmatrix} a_1 & c_1 & d_1 \\ a_2 & c_2 & d_2 \\ a_3 & c_3 & d_3 \end{vmatrix}$$

(along the second column)

are examples. (See Problem 24.2.)

THE COFACTOR OF AN ELEMENT of a determinant is the minor of that element together with the sign associated with the product of that element and its minor in the expansion of the determinant. The cofactors of the elements $a_1, a_2, b_1, b_3, c_1, \ldots$ will be denoted by $A_1, A_2, B_1, B_3, C_1, \ldots$. Thus the cofactor of c_1 in Δ_3 is $C_1 = +\begin{vmatrix} a_2 & b_2 \\ a_3 & b_3 \end{vmatrix}$ and the cofactor of b_3 is $B_3 = -\begin{vmatrix} a_1 & c_1 \\ a_2 & c_2 \end{vmatrix}$.

When cofactors are used, the expansions of Δ_4 given above take the more compact form

$$\begin{aligned} \Delta_4 &= a_1A_1 + b_1B_1 + c_1C_1 + d_1D_1 && \text{(along the first row)} \\ &= a_1A_1 + a_2A_2 + a_3A_3 + a_4A_4 && \text{(along the first column)} \\ &= a_4A_4 + b_4B_4 + c_4C_4 + d_4D_4 && \text{(along the fourth row)} \\ &= b_1B_1 + b_2B_2 + b_3B_3 + b_4B_4 && \text{(along the second column)} \end{aligned}$$

(See Problems 24.3–24.4.)

PROPERTIES OF DETERMINANTS. Subject always to our assumption of equivalent expansions of a determinant along any of its rows or columns, the following theorems may be proved by mathematical induction.

THEOREM I.   If two rows (two columns) of a determinant are identical, the value of the determinant is zero. For example,

$$\begin{vmatrix} 2 & 3 & 2 \\ 3 & 1 & 3 \\ 1 & 4 & 1 \end{vmatrix} = 0$$

COROLLARY I.   If each of the elements of a row (a column) of a determinant is multiplied by the cofactor of the corresponding element of another row (column), the sum of the products is zero.

THEOREM II.   If the elements of a row (a column) of a determinant are multiplied by any number m, the determinant is multiplied by m. For example,

$$5\begin{vmatrix} 2 & 3 & 4 \\ 3 & -1 & 2 \\ 1 & 4 & -3 \end{vmatrix} = \begin{vmatrix} 10 & 3 & 4 \\ 15 & -1 & 2 \\ 5 & 4 & -3 \end{vmatrix} = \begin{vmatrix} 2 & 3 & 4 \\ 15 & -5 & 10 \\ 1 & 4 & -3 \end{vmatrix}$$

THEOREM III.   If each of the elements of a row (a column) of a determinant is expressed as the sum of two or more numbers, the determinant may be written as the sum of two or more determinants. For example,

$$\begin{vmatrix} 2 & 5 & 4 \\ 4 & -2 & 3 \\ 1 & -4 & 3 \end{vmatrix} = \begin{vmatrix} -2+4 & 5 & 4 \\ 3+1 & -2 & 3 \\ 1+0 & -4 & 3 \end{vmatrix} = \begin{vmatrix} -2 & 5 & 4 \\ 3 & -2 & 3 \\ 1 & -4 & 3 \end{vmatrix} + \begin{vmatrix} 4 & 5 & 4 \\ 1 & -2 & 3 \\ 0 & -4 & 3 \end{vmatrix}$$

THEOREM IV.   If to the elements of any row (any column) of a determinant there is added m times the corresponding elements of another row (another column), the value of the determinant is unchanged. For example,

$$\begin{vmatrix} -2 & 5 & 4 \\ 3 & -2 & 2 \\ 1 & -4 & 3 \end{vmatrix} = \begin{vmatrix} -2 & 5+4(-2) & 4 \\ 3 & -2+4(3) & 2 \\ 1 & -4+4(1) & 3 \end{vmatrix} = \begin{vmatrix} -2 & -3 & 4 \\ 3 & 10 & 2 \\ 1 & 0 & 3 \end{vmatrix}$$

(See Problem 24.5.)

EVALUATION OF DETERMINANTS.   A determinant of any order may be evaluated by expanding it and all subsequent determinants (minors) thus obtained along a row or column. This procedure may be greatly simplified by the use of Theorem IV. In Problem 24.6, (a) and (b), a row (column) containing an element $+1$ or -1 is used to obtain an equivalent determinant having an element 0 in another row (column). In (c) and (d), the same theorem has been used to obtain an element $+1$ or -1; this procedure is to be followed when the given determinant is lacking in these elements.

The revised procedure consists in first obtaining an equivalent determinant in which all the elements, save one, in some row (column) are zeros and then expanding along that row (column).

EXAMPLE 1.   Evaluate $\begin{vmatrix} 1 & 4 & 3 & 1 \\ 2 & 8 & 2 & 5 \\ 4 & -4 & -1 & -3 \\ 2 & 5 & 3 & 3 \end{vmatrix}$

Using the first column since it contains the element 1 in the first row, we obtain an equivalent determinant all of whose elements, save the first, in the first row are zeros. We have

$$\begin{vmatrix} 1 & 4 & 3 & 1 \\ 2 & 8 & 2 & 5 \\ 4 & -4 & -1 & -3 \\ 2 & 5 & 3 & 3 \end{vmatrix} = \begin{vmatrix} 1 & 4+(-4)1 & 3+(-3)1 & 1+(-1)1 \\ 2 & 8+(-4)2 & 2+(-3)2 & 5+(-1)2 \\ 4 & -4+(-4)4 & -1+(-3)4 & -3+(-1)4 \\ 2 & 5+(-4)2 & 3+(-3)2 & 3+(-1)2 \end{vmatrix} = \begin{vmatrix} 1 & 0 & 0 & 0 \\ 2 & 0 & -4 & 3 \\ 4 & -20 & -13 & -7 \\ 2 & -3 & -3 & 1 \end{vmatrix}$$

$$= \begin{vmatrix} 0 & -4 & 3 \\ -20 & -13 & -7 \\ -3 & -3 & 1 \end{vmatrix} \qquad \text{(by expanding along the first row)}$$

Expanding the resulting determinant along the first row to take full advantage of the element 0, we have

$$\begin{vmatrix} 0 & -4 & 3 \\ -20 & -13 & -7 \\ -3 & -3 & 1 \end{vmatrix} = 4(-20-21) + 3(60-39) = -101$$

Solved Problems

24.1   Write the minors of the elements a_1, b_3, c_2 of Δ_3.

The minor of a_1 is $\begin{vmatrix} b_2 & c_2 \\ b_3 & c_3 \end{vmatrix}$, the minor of b_3 is $\begin{vmatrix} a_1 & c_1 \\ a_2 & c_2 \end{vmatrix}$, the minor of c_1 is $\begin{vmatrix} a_2 & b_2 \\ a_3 & b_3 \end{vmatrix}$.

24.2   Evaluate: (a) $\begin{vmatrix} -1 & 3 & -4 \\ 0 & 2 & 0 \\ 2 & -3 & 5 \end{vmatrix}$ by expanding along the second row

(b) $\begin{vmatrix} 8 & 0 & 2 & 0 \\ 5 & 1 & -3 & 0 \\ -4 & 3 & 7 & -3 \\ 4 & 0 & 6 & 0 \end{vmatrix}$ by expanding along the fourth column

(a) $\begin{vmatrix} -1 & 3 & -4 \\ 0 & 2 & 0 \\ 2 & -3 & 5 \end{vmatrix} = -0 \begin{vmatrix} 3 & -4 \\ -3 & 5 \end{vmatrix} + 2 \begin{vmatrix} -1 & -4 \\ 2 & 5 \end{vmatrix} - 0 \begin{vmatrix} -1 & 3 \\ 2 & -3 \end{vmatrix} = 2 \begin{vmatrix} -1 & -4 \\ 2 & 5 \end{vmatrix} = 2 \cdot 3 = 6$

(b) $\begin{vmatrix} 8 & 0 & 2 & 0 \\ 5 & 1 & -3 & 0 \\ -4 & 3 & 7 & -3 \\ 4 & 0 & 6 & 0 \end{vmatrix} = -(-3) \begin{vmatrix} 8 & 0 & 2 \\ 5 & 1 & -3 \\ 4 & 0 & 6 \end{vmatrix} = 3 \left\{ +1 \begin{vmatrix} 8 & 2 \\ 4 & 6 \end{vmatrix} \right\} = 3(48 - 8) = 120$

24.3 (a) Write the cofactors of the elements a_1, b_3, c_2, d_4 of Δ_4.

The cofactor of a_1 is $A_1 = + \begin{vmatrix} b_2 & c_2 & d_2 \\ b_3 & c_3 & d_3 \\ b_4 & c_4 & d_4 \end{vmatrix}$,   of b_3 is $B_3 = - \begin{vmatrix} a_1 & c_1 & d_1 \\ a_2 & c_2 & d_2 \\ a_4 & c_4 & d_4 \end{vmatrix}$

of c_2 is $C_2 = - \begin{vmatrix} a_1 & b_1 & d_1 \\ a_3 & b_3 & d_3 \\ a_4 & b_4 & d_4 \end{vmatrix}$,   of d_4 is $D_4 = + \begin{vmatrix} a_1 & b_1 & c_1 \\ a_2 & b_2 & c_2 \\ a_3 & b_3 & c_3 \end{vmatrix}$

(b) Write the expansion of Δ_4 along (1) the second row, (2) the third column, using cofactors.
(1) $\Delta_4 = a_2 A_2 + b_2 B_2 + c_2 C_2 + d_2 D_2$
(2) $\Delta_4 = c_1 C_1 + c_2 C_2 + c_3 C_3 + c_4 C_4$

24.4   Express $g_1 C_1 + g_2 C_2 + g_3 C_3 + g_4 C_4$, where the C_i are cofactors of the elements c_i of Δ_4, as a determinant.

Since the cofactors C_i contain no elements with basal letter c, we replace c_1, c_2, c_3, c_4 by g_1, g_2, g_3, g_4, respectively, in Problem 24.3(b) and obtain

$$g_1 C_1 + g_2 C_2 + g_3 C_3 + g_4 C_4 = \begin{vmatrix} a_1 & b_1 & g_1 & d_1 \\ a_2 & b_2 & g_2 & d_2 \\ a_3 & b_3 & g_3 & d_3 \\ a_4 & b_4 & g_4 & d_4 \end{vmatrix}$$

24.5   Prove by induction: If two rows (two columns) of a determinant are identical, the value of the determinant is zero.

The theorem is true for determinants of order two since $\begin{vmatrix} a_1 & a_1 \\ a_2 & a_2 \end{vmatrix} = a_1 a_2 - a_1 a_2 = 0$.

Let us assume the theorem true for determinants of order k and consider a determinant Δ of order $(k + 1)$ in which two columns are identical. When Δ is expanded along any column, other than the two with identical elements, each cofactor involved is a determinant of order k with two columns identical and, by assumption, is equal to zero. Thus, Δ is equal to zero and the theorem is proved by induction.

24.6 From the determinant $\begin{vmatrix} 1 & 4 & 3 & 1 \\ 2 & 8 & 2 & 5 \\ 4 & -4 & -1 & -3 \\ 2 & 5 & 3 & 3 \end{vmatrix}$ obtain an equivalent determinant

(a)   By adding -4 times the elements of the first column to the corresponding elements of the second column

(b)   By adding 3 times the elements of the third row to the corresponding elements of the fourth row

(c)   By adding -1 times the elements of the third column to the corresponding elements of the second column

(d)   By adding -2 times the elements of the fourth row to the corresponding elements of the second row

(a) $\begin{vmatrix} 1 & 4 & 3 & 1 \\ 2 & 8 & 2 & 5 \\ 4 & -4 & -1 & -3 \\ 2 & 5 & 3 & 3 \end{vmatrix} = \begin{vmatrix} 1 & 4+(-4)1 & 3 & 1 \\ 2 & 8+(-4)2 & 2 & 5 \\ 4 & -4+(-4)4 & -1 & -3 \\ 2 & 5+(-4)2 & 3 & 3 \end{vmatrix} = \begin{vmatrix} 1 & 0 & 3 & 1 \\ 2 & 0 & 2 & 5 \\ 4 & -20 & -1 & -3 \\ 2 & -3 & 3 & 3 \end{vmatrix}$

(b) $\begin{vmatrix} 1 & 4 & 3 & 1 \\ 2 & 8 & 2 & 5 \\ 4 & -4 & -1 & -3 \\ 2 & 5 & 3 & 3 \end{vmatrix} = \begin{vmatrix} 1 & 4 & 3 & 1 \\ 2 & 8 & 2 & 5 \\ 4 & -4 & -1 & -3 \\ 2+(3)4 & 5+(3)(-4) & 3+(3)(-1) & 3+(3)(-3) \end{vmatrix} = \begin{vmatrix} 1 & 4 & 3 & 1 \\ 2 & 8 & 2 & 5 \\ 4 & -4 & -1 & -3 \\ 14 & -7 & 0 & -6 \end{vmatrix}$

(c) $\begin{vmatrix} 1 & 4 & 3 & 1 \\ 2 & 8 & 2 & 5 \\ 4 & -4 & -1 & -3 \\ 2 & 5 & 3 & 3 \end{vmatrix} = \begin{vmatrix} 1 & 4+(-1)3 & 3 & 1 \\ 2 & 8+(-1)2 & 2 & 5 \\ 4 & -4+(-1)(-1) & -1 & -3 \\ 2 & 5+(-1)3 & 3 & 3 \end{vmatrix} = \begin{vmatrix} 1 & 1 & 3 & 1 \\ 2 & 6 & 2 & 5 \\ 4 & -3 & -1 & -3 \\ 2 & 2 & 3 & 3 \end{vmatrix}$

(d) $\begin{vmatrix} 1 & 4 & 3 & 1 \\ 2 & 8 & 2 & 5 \\ 4 & -4 & -1 & -3 \\ 2 & 5 & 3 & 3 \end{vmatrix} = \begin{vmatrix} 1 & 4 & 3 & 1 \\ 2+(-2)2 & 8+(-2)5 & 2+(-2)3 & 5+(-2)3 \\ 4 & -4 & -1 & -3 \\ 2 & 5 & 3 & 3 \end{vmatrix} = \begin{vmatrix} 1 & 4 & 3 & 1 \\ -2 & -2 & -4 & -1 \\ 4 & -4 & -1 & -3 \\ 2 & 5 & 3 & 3 \end{vmatrix}$

24.7 Evaluate:

(a) $\begin{vmatrix} 50 & 2 & -9 \\ 250 & -10 & 45 \\ -150 & 6 & 27 \end{vmatrix} = 50 \cdot 2 \cdot 9 \begin{vmatrix} 1 & 1 & -1 \\ 5 & -5 & 5 \\ -3 & 3 & 3 \end{vmatrix} = 900 \cdot 5 \cdot 3 \begin{vmatrix} 1 & 1 & -1 \\ 1 & -1 & 1 \\ -1 & 1 & 1 \end{vmatrix} = 13\,500 \begin{vmatrix} 2 & 0 & 0 \\ 1 & -1 & 1 \\ -1 & 1 & 1 \end{vmatrix}$

$$= 27\,000 \begin{vmatrix} -1 & 1 \\ 1 & 1 \end{vmatrix} = -54\,000$$

(b) $\begin{vmatrix} 3 & 2 & -3 & 1 \\ -1 & -3 & 5 & 2 \\ 2 & 1 & 6 & -3 \\ 5 & 4 & -3 & 4 \end{vmatrix} = \begin{vmatrix} 3+(-3)1 & 2+(-2)1 & -3+(3)1 & 1 \\ -1+(-3)2 & -3+(-2)2 & 5+(3)2 & 2 \\ 2+(-3)(-3) & 1+(-2)(-3) & 6+(3)(-3) & -3 \\ 5+(-3)4 & 4+(-2)4 & -3+(3)4 & 4 \end{vmatrix} = \begin{vmatrix} 0 & 0 & 0 & 1 \\ -7 & -7 & 11 & 2 \\ 11 & 7 & -3 & -3 \\ -7 & -4 & 9 & 4 \end{vmatrix}$

$$= -\begin{vmatrix} -7 & -7 & 11 \\ 11 & 7 & -3 \\ -7 & -4 & 9 \end{vmatrix} = -\begin{vmatrix} 0 & -7 & 11 \\ 4 & 7 & -3 \\ -3 & -4 & 9 \end{vmatrix} = -[7(36-9)+11(-16+21)] = -244$$

$(c)\quad \begin{vmatrix} 2 & -1 & -2 & 3 \\ 3 & 2 & 4 & -1 \\ 2 & 4 & 1 & -5 \\ 4 & -3 & 2 & 1 \end{vmatrix} = \begin{vmatrix} 0 & -1 & 0 & 0 \\ 7 & 2 & 0 & 5 \\ 10 & 4 & -7 & 7 \\ -2 & -3 & 8 & -8 \end{vmatrix} = \begin{vmatrix} 7 & 0 & 5 \\ 10 & -7 & 7 \\ -2 & 8 & -8 \end{vmatrix} = 5 \cdot 66 = 330$

Supplementary Problems

24.8　Verify: The value of a determinant is unchanged if the rows are written as columns.

Hint:　Show that $\begin{vmatrix} a_1 & b_1 & c_1 \\ a_2 & b_2 & c_2 \\ a_3 & b_3 & c_3 \end{vmatrix} = \begin{vmatrix} a_1 & a_2 & a_3 \\ b_1 & b_2 & b_3 \\ c_1 & c_2 & c_3 \end{vmatrix}$.

24.9　Prove by induction:

(a)　The expansion of a determinant of order n contains $n!$ terms.

(b)　If two rows (columns) of a determinant are interchanged, the sign of the determinant is changed.

Hint:　Show that $\begin{vmatrix} a_1 & b_1 \\ a_2 & b_2 \end{vmatrix} = - \begin{vmatrix} b_1 & a_1 \\ b_2 & a_2 \end{vmatrix}$　and proceed as in Problem 24.5.

24.10　Show, without expanding the determinants, that

(a)　$\begin{vmatrix} 3 & 2 & 1 & 4 \\ -1 & 5 & 2 & 6 \\ 2 & -4 & 7 & -5 \\ -2 & 1 & 3 & 5 \end{vmatrix} = - \begin{vmatrix} 3 & 2 & 1 & 4 \\ 2 & -4 & 7 & -5 \\ -1 & 5 & 2 & 6 \\ -2 & 1 & 3 & 5 \end{vmatrix} = - \begin{vmatrix} 3 & 2 & 1 & 4 \\ -2 & 1 & 3 & 5 \\ 2 & -4 & 7 & -5 \\ -1 & 5 & 2 & 6 \end{vmatrix} = - \begin{vmatrix} 3 & -2 & 2 & -1 \\ 2 & 1 & -4 & 5 \\ 1 & 3 & 7 & 2 \\ 4 & 5 & -5 & 6 \end{vmatrix}$

(b)　$\begin{vmatrix} 10 & 0 & -2 \\ -10 & 3 & -4 \\ -5 & -2 & 3 \end{vmatrix} = -5 \begin{vmatrix} -2 & 0 & -2 \\ 2 & 3 & -4 \\ 1 & -2 & 3 \end{vmatrix} = 10 \begin{vmatrix} 1 & 0 & 1 \\ 2 & 3 & -4 \\ 1 & -2 & 3 \end{vmatrix}$

(c)　$\begin{vmatrix} -3 & -1 & 2 \\ 1 & -3 & 3 \\ 4 & -2 & 1 \end{vmatrix} = 0$　**Hint:**　Subtract the first row from the second.

(d)　$\begin{vmatrix} a-3b & a+b & a+5b & e \\ a-2b & a+2b & a+6b & f \\ a-b & a+3b & a+7b & g \\ a & a+4b & a+8b & h \end{vmatrix} = 0$　(NOTE.　The elements of the first three columns form an A.P.)

(e)　$\begin{vmatrix} 0 & a_1 & a_2 \\ -a_1 & 0 & a_3 \\ -a_2 & -a_3 & 0 \end{vmatrix} = 0$　**Hint:**　Write the rows as columns and factor -1 from each row.

24.11　Verify: If the corresponding elements of two rows (columns) of a determinant are proportional, the value of the determinant is zero.

24.12 Evaluate each of the following determinants:

(a) $\begin{vmatrix} -3 & 6 & -1 & 1 \\ -4 & -3 & -2 & 4 \\ 5 & -4 & 1 & 3 \\ -1 & -5 & 0 & -1 \end{vmatrix}$ (b) $\begin{vmatrix} 1 & -1 & 1 & 2 \\ 3 & -2 & 4 & -3 \\ 5 & 4 & 1 & 2 \\ -3 & 0 & 3 & 1 \end{vmatrix}$ (c) $\begin{vmatrix} 3 & -1 & -1 & -4 \\ 2 & 3 & -6 & 1 \\ 4 & -1 & 3 & 1 \\ 3 & -1 & 5 & 2 \end{vmatrix}$

(d) $\begin{vmatrix} 2 & 4 & 4 & -3 & 4 \\ 1 & 3 & 1 & 0 & -1 \\ -1 & 3 & 1 & 2 & -1 \\ 4 & 8 & 11 & -10 & 9 \\ 2 & 6 & 9 & -12 & 5 \end{vmatrix}$

$Ans.$ (a) -50 (b) -397 (c) -78 (d) -316

24.13 Show that $\begin{vmatrix} 1 & 1 & 1 & 1 \\ a & b & c & d \\ a^2 & b^2 & c^2 & d^2 \\ a^3 & b^3 & c^3 & d^3 \end{vmatrix} = (b-a)(c-a)(d-a)(c-b)(d-b)(d-c)$

Chapter 25

Systems of Linear Equations

SYSTEMS OF n LINEAR EQUATIONS IN n UNKNOWNS.   Consider, for the sake of brevity, the system of four linear equations in four unknowns

$$\begin{cases} a_1x + b_1y + c_1z + d_1w = k_1 \\ a_2x + b_2y + c_2z + d_2w = k_2 \\ a_3x + b_3y + c_3z + d_3w = k_3 \\ a_4x + b_4y + c_4z + d_4w = k_4 \end{cases} \qquad (25.1)$$

in which each equation is written with the unknowns x, y, z, w in that order on the left side and the constant term on the right side. Form

$$D = \begin{vmatrix} a_1 & b_1 & c_1 & d_1 \\ a_2 & b_2 & c_2 & d_2 \\ a_3 & b_3 & c_3 & d_3 \\ a_4 & b_4 & c_4 & d_4 \end{vmatrix}, \qquad \text{the determinant of the coefficients of the unknowns,}$$

and from it the determinants

$$N_x = \begin{vmatrix} k_1 & b_1 & c_1 & d_1 \\ k_2 & b_2 & c_2 & d_2 \\ k_3 & b_3 & c_3 & d_3 \\ k_4 & b_4 & c_4 & d_4 \end{vmatrix}, \qquad N_y = \begin{vmatrix} a_1 & k_1 & c_1 & d_1 \\ a_2 & k_2 & c_2 & d_2 \\ a_3 & k_3 & c_3 & d_3 \\ a_4 & k_4 & c_4 & d_4 \end{vmatrix}, \qquad N_z = \begin{vmatrix} a_1 & b_1 & k_1 & d_1 \\ a_2 & b_2 & k_2 & d_2 \\ a_3 & b_3 & k_3 & d_3 \\ a_4 & b_4 & k_4 & d_4 \end{vmatrix},$$

$$N_w = \begin{vmatrix} a_1 & b_1 & c_1 & k_1 \\ a_2 & b_2 & c_2 & k_2 \\ a_3 & b_3 & c_3 & k_3 \\ a_4 & b_4 & c_4 & c_k \end{vmatrix}$$

by replacing the column of coefficients of the indicated unknown by the column of constants. Cramer's rule states that:

(A) If $D \neq 0$, the system (25.1) has the unique solution

$$x = N_x/D, \qquad y = N_y/D, \qquad z = N_z/D, \qquad w = N_w/D$$

(See Problems 25.1–25.2.)

(B) If $D = 0$ and at least one of $N_x, N_y, N_z, N_w \neq 0$, the system has no solution. For, if $D = 0$ and $N_x \neq 0$, then $x \cdot D = N_x$ leads to a contradiction. Such systems are called *inconsistent*. (See Problem 25.3.)

(C) If $D = 0$ and $N_x = N_y = N_z = N_w = 0$, the system may or may not have a solution. A system having an infinite number of solutions is called *dependent*.

For systems of three or four equations, the simplest procedure is to evaluate D. If $D \neq 0$, proceed as in (A); if $D = 0$, proceed as in Chapter 6. (See Problem 25.4.)

SYSTEMS OF m LINEAR EQUATIONS IN $n > m$ UNKNOWNS.   Ordinarily if there are fewer equations than unknowns, the system will have an infinite number of solutions.

To solve a consistent system of m equations, solve for m of the unknowns (in certain cases for $p < m$ of the unknowns) in terms of the others. (See Problem 25.5.)

SYSTEMS OF n EQUATIONS IN $m < n$ UNKNOWNS.   Ordinarily if there are more equations than unknowns the system is inconsistent. However, if $p \le m$ of the equations have a solution and if this solution satisfies each of the remaining equations, the system is consistent. (See Problem 25.6.)

A HOMOGENEOUS EQUATION is one in which all terms are of the same degree; otherwise, the equation is called *nonhomogeneous*. For example, the linear equation $2x + 3y - 4z = 5$ is non-homogeneous, while $2x + 3y - 4z = 0$ is homogeneous.

Every system of homogeneous linear equations

$$a_1 x + b_1 y + c_1 z + \cdots = 0$$

$$a_2 x + b_2 y + c_2 z + \cdots = 0$$

$$\vdots$$

$$a_n x + b_n y + c_n z + \cdots = 0$$

always has the *trivial solution* $x = 0$, $y = 0$, $z = 0, \ldots$.

A system of n homogeneous linear equations in n unknowns has *only* the trivial solution if D, the determinant of the coefficients, is not equal to zero. If $D = 0$, the system has nontrivial solutions as well. (See Problem 25.7.)

Solved Problems

25.1   Solve the system $\begin{cases} 3x - 2y - z - 4w = 7 & (1) \\ x + 3z + 2w = -10 & (2) \\ x + 4y + 2z + w = 0 & (3) \\ 2x + 3y + 3w = 1 & (4) \end{cases}$

We find

$$D = \begin{vmatrix} 3 & -2 & -1 & -4 \\ 1 & 0 & 3 & 2 \\ 1 & 4 & 2 & 1 \\ 2 & 3 & 0 & 3 \end{vmatrix} = \begin{vmatrix} 3 & -2 & -10 & -10 \\ 1 & 0 & 0 & 0 \\ 1 & 4 & -1 & -1 \\ 2 & 3 & -6 & -1 \end{vmatrix} = - \begin{vmatrix} -2 & -10 & -10 \\ 4 & -1 & -1 \\ 3 & -6 & -1 \end{vmatrix} = 2 \begin{vmatrix} 1 & 5 & 5 \\ 4 & -1 & -1 \\ 3 & -6 & -1 \end{vmatrix} = -210$$

$$N_x = \begin{vmatrix} 7 & -2 & -1 & -4 \\ -10 & 0 & 3 & 2 \\ 0 & 4 & 2 & 1 \\ 1 & 3 & 0 & 3 \end{vmatrix} = \begin{vmatrix} 0 & -23 & -1 & -25 \\ 0 & 30 & 3 & 32 \\ 0 & 4 & 2 & 1 \\ 1 & 3 & 0 & 3 \end{vmatrix} = - \begin{vmatrix} -23 & -1 & -25 \\ 30 & 3 & 32 \\ 4 & 2 & 1 \end{vmatrix} = - \begin{vmatrix} -23 & -1 & -2 \\ 30 & 3 & 2 \\ 4 & 2 & -3 \end{vmatrix} = -105$$

$$N_y = \begin{vmatrix} 3 & 7 & -1 & -4 \\ 1 & -10 & 3 & 2 \\ 1 & 0 & 2 & 1 \\ 2 & 1 & 0 & 3 \end{vmatrix} = \begin{vmatrix} 3 & 7 & -7 & -7 \\ 1 & -10 & 1 & 1 \\ 1 & 0 & 0 & 0 \\ 2 & 1 & -4 & 1 \end{vmatrix} = \begin{vmatrix} 7 & -7 & -7 \\ -10 & 1 & 1 \\ 1 & -4 & 1 \end{vmatrix} = 7 \begin{vmatrix} 1 & -1 & -1 \\ -10 & 1 & 1 \\ 1 & -4 & 1 \end{vmatrix} = -315$$

$$N_z = \begin{vmatrix} 3 & -2 & 7 & -4 \\ 1 & 0 & -10 & 2 \\ 1 & 4 & 0 & 1 \\ 2 & 3 & 1 & 3 \end{vmatrix} = \begin{vmatrix} 3 & -14 & 7 & -7 \\ 1 & -4 & -10 & 1 \\ 1 & 0 & 0 & 0 \\ 2 & -5 & 1 & 1 \end{vmatrix} = \begin{vmatrix} -14 & 7 & -7 \\ -4 & -10 & 1 \\ -5 & 1 & 1 \end{vmatrix} = -7 \begin{vmatrix} 2 & -1 & 1 \\ -4 & -10 & 1 \\ -5 & 1 & 1 \end{vmatrix} = 525$$

$$N_w = \begin{vmatrix} 3 & -2 & -1 & 7 \\ 1 & 0 & 3 & -10 \\ 1 & 4 & 2 & 0 \\ 2 & 3 & 0 & 1 \end{vmatrix} = \begin{vmatrix} 3 & -14 & -7 & 7 \\ 1 & -4 & 1 & -10 \\ 1 & 0 & 0 & 0 \\ 2 & -5 & -4 & 1 \end{vmatrix} = \begin{vmatrix} -14 & -7 & 7 \\ -4 & 1 & -10 \\ -5 & -4 & 1 \end{vmatrix} = -7 \begin{vmatrix} 2 & 1 & -1 \\ -4 & 1 & -10 \\ -5 & -4 & 1 \end{vmatrix} = 315$$

Then

$$x = \frac{N_x}{D} = \frac{-105}{-210} = \frac{1}{2}, \ y = \frac{N_y}{D} = \frac{-315}{-210} = \frac{3}{2}, \ z = \frac{N_z}{D} = \frac{525}{-210.} = -\frac{5}{2}, \ w = \frac{N_w}{D} = \frac{315}{-210} = -\frac{3}{2}.$$

Check. Using (1), $3(\frac{1}{2}) - 2(\frac{3}{2}) - (-\frac{5}{2}) - 4(-\frac{3}{2}) = \frac{3 - 6 + 5 + 12}{2} = 7.$

[NOTE.   The above system permits some variation in procedure. For example, having found $x = \frac{1}{2}$ and $y = \frac{3}{2}$ using determinants, the value of w may be obtained by substituting in (4)

$$2(\tfrac{1}{2}) + 3(\tfrac{3}{2}) + 3w = 1, \qquad 3w = -\tfrac{9}{2}, \qquad w = -\tfrac{3}{2}$$

and the value of z may then be obtained by substituting in (2)

$$(\tfrac{1}{2}) + 3z + 2(-\tfrac{3}{2}) = -10, \qquad 3z = -\tfrac{15}{2}, \qquad z = -\tfrac{5}{2}.$$

The solution may be checked by substituting in (1) or (3).]

25.2   Solve the system $\begin{cases} 2x + y + 5z + w = 5 & (1) \\ x + y - 3z - 4w = -1 & (2) \\ 3x + 6y - 2z + w = 8 & (3) \\ 2x + 2y + 2z - 3w = 2 & (4) \end{cases}$

We have $D = \begin{vmatrix} 2 & 1 & 5 & 1 \\ 1 & 1 & -3 & -4 \\ 3 & 6 & -2 & 1 \\ 2 & 2 & 2 & -3 \end{vmatrix} = -120, \quad N_x = \begin{vmatrix} 5 & 1 & 5 & 1 \\ -1 & 1 & -3 & -4 \\ 8 & 6 & -2 & 1 \\ 2 & 2 & 2 & -3 \end{vmatrix} = -240,$

$$N_y = \begin{vmatrix} 2 & 5 & 5 & 1 \\ 1 & -1 & -3 & -4 \\ 3 & 8 & -2 & 1 \\ 2 & 2 & 2 & -3 \end{vmatrix} = -24, \quad N_z = \begin{vmatrix} 2 & 1 & 5 & 1 \\ 1 & 1 & -1 & -4 \\ 3 & 6 & 8 & 1 \\ 2 & 2 & 2 & -3 \end{vmatrix} = 0.$$

Then $x = \dfrac{N_x}{D} = \dfrac{-240}{-120} = 2, \qquad y = \dfrac{N_y}{D} = \dfrac{-24}{-120} = \dfrac{1}{5}, \qquad$ and $\qquad z = \dfrac{N_z}{D} = \dfrac{0}{-120} = 0.$

Substituting in (1), $2(2) + (\frac{1}{5}) + 5(0) + w = 5$ and $w = \frac{4}{5}$.
Check. Using (2), $(2) + (\frac{1}{5}) - 3(0) - 4(\frac{4}{5}) = -1.$

25.3   Show that the system $\begin{cases} 2x + y + 5z + w = 2 \\ x + y - z - 4w = 1 \\ 3x + 6y + 8z + w = 3 \\ 2x + 2y + 2z - 3w = 1 \end{cases}$ is inconsistent.

Since $D = \begin{vmatrix} 2 & 1 & 5 & 1 \\ 1 & 1 & -1 & -4 \\ 3 & 6 & 8 & 1 \\ 2 & 2 & 2 & -3 \end{vmatrix} = 0$ while $N_x = \begin{vmatrix} 2 & 1 & 5 & 1 \\ 1 & 1 & -1 & -4 \\ 3 & 6 & 8 & 1 \\ 1 & 2 & 2 & -3 \end{vmatrix} = -80 \neq 0$, the system is inconsistent.

25.4   Solve when possible: (a) $\begin{cases} 2x - 3y + z = 0 & (1) \\ x + 5y - 3z = 3 & (2) \\ 5x + 12y - 8z = 9 & (3) \end{cases}$ (c) $\begin{cases} 6x - 2y + z = 1 & (1) \\ x - 4y + 2z = 0 & (2) \\ 4x + 6y - 3z = 0 & (3) \end{cases}$

(b) $\begin{cases} x + 2y + 3z = 2 & (1) \\ 2x + 4y + z = -1 & (2) \\ 3x + 6y + 5z = 2 & (3) \end{cases}$ (d) $\begin{cases} x + 2y - 3z + 5w = 11 & (1) \\ 4x - y + z - 2w = 0 & (2) \\ 2x + 4y - 6z + 10w = 22 & (3) \\ 5x + y - 2z + 3w = 11 & (4) \end{cases}$

(a)   Here $D = 0$; we shall eliminate the variable x.

$$(1) - 2(2): \quad -13y + 7z = -6$$

$$(3) - 5(2): \quad -13y + 7z = -6$$

Then $y = \dfrac{7z + 6}{13}$ and, from (2), $x = 3 - 5y + 3z = \dfrac{4z + 9}{13}$.

The solutions may be written as $x = \dfrac{4a + 9}{13}$, $y = \dfrac{7a + 6}{13}$, $z = a$, where a is arbitrary.

(b)   Here $D = 0$; we shall eliminate x.

$$(2) - 2(1): \quad -5z = -5$$

$$(3) - 3(1): \quad -4z = -4$$

Then $z = 1$ and each of the given equations reduces to $x + 2y = -1$. Note that the same situation arises when y is eliminated.

   The solution may be written $x = -1 - 2y$, $z = 1$ or as $x = -1 - 2a$, $y = a$, $z = 1$, where a is arbitrary.

(c)   Here $D = 0$; we shall eliminate z.

$$\begin{aligned} (2) - 2(1): \quad &-11x = -2 \\ (3) + 3(1): \quad &\;\;\;22x = 3 \end{aligned}$$ The system is inconsistent.

(d)   Here $D = 0$; we shall eliminate x.

$$(2) - 4(1): \quad -9y + 13z - 22w = -44$$

$$(3) - 2(1): \qquad\qquad\qquad 0 = 0$$

$$(4) - 5(1): \quad -9y + 13z - 22w = -44$$

Then $y = \dfrac{44 + 13z - 22w}{9}$ and, from (1), $x = \dfrac{11 + z - w}{9}$. The solutions are $x = \dfrac{11 + a - b}{9}$, $y = \dfrac{44 + 13a - 22b}{9}$, $z = a$, $w = b$, where a and b are arbitrary.

25.5   (a)   The system of two equations in four unknowns

$$\begin{cases} x + 2y + 3z - 4w = 5 \\ 3x - y - 5z - 5w = 1 \end{cases}$$

may be solved for any two of the unknowns in terms of the others; for example, $x = 1 + z + 2w$, $y = 2 - 2z + w$.

(b)   The system of three equations in four unknowns

$$\begin{cases} x + 2y + 3z - 4w = 5 \\ 3x - y - 5z - 5w = 1 \\ 2x + 3y + z - w = 8 \end{cases}$$

may be solved for any three of the unknowns in terms of the fourth; for example, $x = 1 + 4w$, $y = 2 - 3w$, $z = 2w$.

(c)   The system of three equations in four unknowns

$$\begin{cases} x + 2y + 3z - 4w = 5 \\ 3x - y - 5z - 5w = 1 \\ 2x - 3y - 8z - w = -4 \end{cases}$$

may be solved for any two of the unknowns in terms of the others. Note that the third equation is the same as the second minus the first. We solve any two of these equations, say the first and second, and obtain the solution given in (a) above.

25.6   Solve when possible.

(a) $\begin{cases} 3x - 2y = 1 \\ 4x + 3y = 41 \\ 6x + 2y = 23 \end{cases}$ (b) $\begin{cases} 3x + y = 1 \\ 5x - 2y = 20 \\ 4x + 5y = -17 \end{cases}$ (c) $\begin{cases} x + y + z = 2 \\ 4x + 5y - 3z = -15 \\ 5x - 3y + 4z = 23 \\ 7x - y + 6z = 27 \end{cases}$ (d) $\begin{cases} x + y = 5 \\ y + z = 8 \\ x + z = 7 \\ 5x - 5y + z = 1 \end{cases}$

(a)   The system $\begin{cases} 3x - 2y = 1 \\ 4x + 3y = 41 \end{cases}$ has solution $x = 5$, $y = 7$.

Since $6x + 2y = 6(5) + 2(7) \neq 23$, the given system is inconsistent.

(b)   The system $\begin{cases} 3x + y = 1 \\ 5x - 2y = 20 \end{cases}$ has solution $x = 2$, $y = -5$.

Since $4x + 5y = 4(2) + 5(-5) = -17$, the given system is consistent with solution $x = 2$, $y = -5$.

(c)   The system $\begin{cases} x + y + z = 2 \\ 4x + 5y - 3z = -15 \\ 5x - 3y + 4z = 23 \end{cases}$ has solution $x = 1$, $y = -2$, $z = 3$.

Since $7x - y + 6z = 7(1) - (-2) + 6(3) = 27$, the given system is consistent with solution $x = 1$, $y = -2$, $z = 3$.

(d)   The system $\begin{cases} x + y = 5 \\ y + z = 8 \\ x + z = 7 \end{cases}$ has solution $x = 2$, $y = 3$, $z = 5$.

Since $5x - 5y + z = 5(2) - 5(3) + (5) \neq 1$, the given system is inconsistent.

(NOTE.   If the constant of the fourth equation of the system were changed from 1 to 0, the resulting system would be consistent.)

25.7   Examine the following systems for nontrivial solutions:

(a) $\begin{cases} 2x - 3y + 3z = 0 \\ 3x - 4y + 5z = 0 \\ 5x + y + 2z = 0 \end{cases}$ (b) $\begin{cases} 4x + y - 2z = 0 \\ x - 2y + z = 0 \\ 11x - 4y - z = 0 \end{cases}$ (c) $\begin{cases} x + 2y + z = 0 \\ 3x + 6y + 3z = 0 \\ 5x + 10y + 5z = 0 \end{cases}$ (d) $\begin{cases} 2x + y = 0 \\ 3y - 2z = 0 \\ 2y + w = 0 \\ 4x - w = 0 \end{cases}$

(a)   Since $D = \begin{vmatrix} 2 & -3 & 3 \\ 3 & -4 & 5 \\ 5 & 1 & 2 \end{vmatrix} \neq 0$, the system has only the trivial solution.

(b)   Since $D = \begin{vmatrix} 4 & 1 & -2 \\ 1 & -2 & 1 \\ 11 & -4 & -1 \end{vmatrix} = 0$, there are nontrivial solutions.

The system $\begin{cases} 4x + y = 2z \\ x - 2y = -z \end{cases}$, for which $D = \begin{vmatrix} 4 & 1 \\ 1 & -2 \end{vmatrix} \neq 0$, has the solution $x = \frac{z}{3}$, $y = \frac{2z}{3}$.

This solution may be written as $x = \frac{a}{3}$, $y = \frac{2a}{3}$, $z = a$ or $x = a$, $y = 2a$, $z = 3a$, where a is arbitrary, or as $x : y : z = 1 : 2 : 3$.

(c)   Here $D = \begin{vmatrix} 1 & 2 & 1 \\ 3 & 6 & 3 \\ 5 & 10 & 5 \end{vmatrix} = 0$ and there are nontrivial solutions.

Since the minor of every element of D is zero, we cannot proceed as in (b). We solve the first equation for $x = -2y - z$ and write the solution as $x = -2a - b$, $y = a$, $z = b$, where a and b are arbitrary.

(d)   Here $D = \begin{vmatrix} 2 & 1 & 0 & 0 \\ 0 & 3 & -2 & 0 \\ 0 & 2 & 0 & 1 \\ 4 & 0 & 0 & -1 \end{vmatrix} = 0$ and there are nontrivial solutions.

Take $x = a$, where a is arbitrary. From the first equation, $y = -2a$; from the second, $2z = 3y = -6a$ and $z = -3a$; and from the fourth equation, $w = 4a$.

Thus the solution is $x = a$, $y = -2a$, $z = -3a$, $w = 4a$, or $x:y:z:w = 1:-2:-3:4$.

Supplementary Problems

25.8   Solve, using determinants.

(a) $\begin{cases} x + y + z = 6 \\ y + z + w = 9 \\ z + w + x = 8 \\ w + x + y = 7 \end{cases}$ (b) $\begin{cases} 3x - 2y + 2z + w = 5 \\ 2x + 4y - z - 2w = 3 \\ 3x + 7y - z + 3w = 23 \\ x - 3y + 2z - 3w = -12 \end{cases}$ (c) $\begin{cases} x + y + z + w = 2 \\ 2x + 3y - 2z - w = 5 \\ 3x - 2y + z + 3w = 4 \\ 5x + 2y + 3z - 2w = -4 \end{cases}$

Ans.   (a) $x = 1$, $y = 2$, $z = 3$, $w = 4$ (b) $x = 2$, $y = 1$, $z = -1$, $w = 3$

       (c) $x = \frac{1}{2}$, $y = 1$, $z = -\frac{3}{2}$, $w = 2$

25.9   Test for consistency, and solve when possible.

(a) $\begin{cases} 2x + 3y - 4z = 1 \\ 3x - y + 2z = -2 \\ 5x - 9y + 14z = 3 \end{cases}$ (b) $\begin{cases} x + 7y + 5z = -22 \\ x - 9y - 11z = 26 \\ x - y - 3z = 2 \end{cases}$ (c) $\begin{cases} x + y + z = 4 \\ 2x - 4y + 11z = -7 \\ 4x + 6y + z = 21 \end{cases}$

Ans.   (a) Inconsistent   (b) $x = 2z - 1$, $y = -z - 3$ (c) $x = \frac{1}{2}(3 - 5z)$, $y = \frac{1}{2}(5 + 3z)$

25.10   Solve, when possible.

(a) $\begin{cases} x - 3y + 11 = 0 \\ 3x + 2y - 33 = 0 \\ 2x - 3y + 4 = 0 \end{cases}$ (b) $\begin{cases} x - 2y - 8 = 0 \\ 3x + y - 3 = 0 \\ x - 10y + 32 = 0 \end{cases}$ (c) $\begin{cases} 2x - 3y - 7 = 0 \\ 5x + 4y + 17 = 0 \\ 4x - y + 1 = 0 \end{cases}$

Ans.   (a) $x = 7$, $y = 6$ (b)   No solution   (c) $x = -1$, $y = -3$

25.11   Solve the systems.

(a) $\begin{cases} 2x - y + 3z = 8 \\ x + 3y - 2z = -3 \end{cases}$ (b) $\begin{cases} 4x + 2y + z = 13 \\ 2x + y - 2z = -6 \end{cases}$ (c) $\begin{cases} 4x + 2y - 6z + w = 10 \\ 3x - y - 9z - w = 7 \\ 7x + y - 11z - w = 13 \end{cases}$

Ans.   (a) $x = 3 - z$, $y = -2 + z$ (b) $y = 4 - 2x$, $z = 5$

       (c) $x = 7w/10$, $y = 2 - 23w/20$, $z = -1 + w/4$

25.12 Examine for nontrivial solutions.

(a) $\begin{cases} 3x + y - 9z = 0 \\ 4x - 3y + z = 0 \\ 6x - 11y + 21z = 0 \end{cases}$ (b) $\begin{cases} 2x - 3y - 5z = 0 \\ x + 2y - 13z = 0 \\ 9x - 10y - 30z = 0 \end{cases}$ (c) $\begin{cases} 2x - 3y + 2z - 9w = 0 \\ x + 4y - z + 3w = 0 \\ 3x - 2y - 2z - 6w = 0 \\ 7x + 11y + 3z - 6w = 0 \end{cases}$

Ans. (a) $x = 2z, y = 3z$ (b) $x = y = z = 0$ (c) $x = 2w, y = -w, z = w$

Chapter 26

Partial Fractions

A RATIONAL ALGEBRAIC FRACTION is the quotient of two polynomials. A rational fraction is called *proper* if the degree of the numerator is less than that of the denominator; otherwise, the fraction is called *improper*.

An improper fraction may be written as the sum of a polynomial and a proper fraction.

For example, $\dfrac{3}{x+2}$ is a proper fraction while $\dfrac{x^2+3x+5}{x+2}$ is an improper fraction.

Note that $\dfrac{x^2+3x+5}{x+2} = x+1+\dfrac{3}{x+2}$ is the sum of a polynomial and a proper fraction.

Two or more proper fractions may be summed to yield a single fraction whose denominator is the lowest common denominator of the several fractions. For example,

(A) $\dfrac{1}{x+2}+\dfrac{2}{3x-2}=\dfrac{5x+2}{(x+2)(3x-2)}$

(B) $\dfrac{3}{x+2}+\dfrac{4}{(x+2)^2}-\dfrac{1}{x}=\dfrac{2x^2+6x-4}{x(x+2)^2}$

(C) $\dfrac{1}{x-1}+\dfrac{3x-1}{x^2+2}-\dfrac{1}{(x^2+2)^2}=\dfrac{4x^4-4x^3+11x^2-9x+7}{(x-1)(x^2+2)^2}$

The problem of this chapter is to reverse the above process, that is, to resolve a given rational fraction into a sum of simpler proper fractions, called *partial fractions*. The cases considered here will be explained by means of examples. Proofs of the validity of the methods will be found in more advanced texts.

CASE 1. FACTORS OF THE DENOMINATOR LINEAR, NONE REPEATED.
Corresponding to each factor of the denominator form a partial fraction having an unknown constant as numerator and the factor as denominator.

EXAMPLE 1.   Resolve $\dfrac{5x+2}{(x+2)(3x-2)}$ into partial fractions.

Set $\dfrac{5x+2}{(x+2)(3x-2)}=\dfrac{A}{x+2}+\dfrac{B}{3x-2}$. Then

(a) $\dfrac{5x+2}{(x+2)(3x-2)}=\dfrac{A(3x-2)+B(x+2)}{(x+2)(3x-2)}$ and

(b) $\dfrac{5x+2}{(x+2)(3x-2)}=\dfrac{(3A+B)x-2A+2B}{(x+2)(3x-2)}$

are identities which hold for all values of x except possibly for $x=-2$ and $x=\frac{2}{3}$.

FIRST SOLUTION. Equating coefficients of like term in the two members of the identity $5x+2=(3A+B)x-2A+2B$, we have $3A+B=5$, $-2A+2B=2$. Then $A=1$, $B=2$, and

$$\frac{5x+2}{(x+2)(3x-2)}=\frac{1}{x+2}+\frac{2}{3x-2}$$

SECOND SOLUTION. Consider the identity $5x + 2 = A(3x - 2) + B(x + 2)$. Since it is an identity between polynomials, it holds for *all* values of x. Now when

$$x = -2, \quad \text{coefficient of } B \text{ is } 0, \ 5(-2) + 2 = A[3(-2) - 2], \text{ and } A = 1;$$

$$x = \tfrac{2}{3}, \quad \text{coefficient of } A \text{ is } 0, \ 5(\tfrac{2}{3}) + 2 = B(\tfrac{2}{3} + 2), \text{ and } B = 2.$$

Thus, as before, $\dfrac{5x + 2}{(x + 2)(3x - 2)} = \dfrac{1}{x + 2} + \dfrac{2}{3x - 2}$. (See Problem 26.1.)

CASE 2. FACTORS OF THE DENOMINATOR LINEAR, SOME REPEATED.

Here (B) suggests that for each repeated factor $(ax + b)^k$, we set up a series of partial fractions

$$\frac{A}{(ax + b)^k} + \frac{B}{(ax + b)^{k-1}} + \cdots + \frac{K}{ax + b}$$

(See Problem 26.2.)

CASE 3. DENOMINATOR CONTAINS IRREDUCIBLE QUADRATIC FACTORS, NONE REPEATED.

For each irreducible quadratic factor $ax^2 + bx + c$ of the denominator set up a partial fraction of the form $\dfrac{Ax + B}{ax^2 + bx + c}$.

EXAMPLE 2. Resolve $\dfrac{2}{(x - 1)(x^2 + x - 4)}$ into partial fractions.

Set $\dfrac{2}{(x - 1)(x^2 + x - 4)} = \dfrac{Ax + B}{x^2 + x - 4} + \dfrac{C}{x - 1} = \dfrac{(Ax + B)(x - 1) + C(x^2 + x - 4)}{(x - 1)(x^2 + x - 4)}$

$$= \frac{(A + C)x^2 + (B - A + C)x - B - 4C}{(x - 1)(x^2 + x - 4)}$$

From the identity $2 = (A + C)x^2 + (B - A + C)x - B - 4C$ we have

(a)
$$A + C = 0 \qquad B - A + C = 0 \qquad -B - 4C = 2$$

with solution $A = 1$, $B = 2$, $C = -1$. Then $\dfrac{2}{(x - 1)(x^2 + x - 4)} = \dfrac{x + 2}{x^2 + x - 4} - \dfrac{1}{x - 1}$.

[NOTE. The value $C = -1$ might have been obtained from the above identity $2 = (Ax + B)(x - 1) + C(x^2 + x - 4)$ with $x = 1$, and then the values of A and B found by means of (a).] (See Problem 26.3.)

CASE 4. DENOMINATOR CONTAINS IRREDUCIBLE QUADRATIC FACTORS, SOME REPEATED.

Here (C) suggests that for each repeated irreducible factor $(ax^2 + bx + c)^k$ of the denominator, we set up a series of partial fractions

$$\frac{Ax + B}{(ax^2 + bx + c)^k} + \frac{Cx + D}{(ax^2 + bx + c)^{k-1}} + \cdots + \frac{Hx + K}{ax^2 + bx + c}$$

(See Problem 26.4.)

Solved Problems

26.1   Resolve into partial fractions: (a) $\dfrac{54}{x^3 - 21x + 20}$, (b) $\dfrac{x^4 - 2x^3 - 7x^2 + 5x - 24}{x^2 - 2x - 8}$.

(a) Using synthetic division, we find $x^3 - 21x + 20 = (x - 1)(x - 4)(x + 5)$. We set

$$\frac{54}{x^3 - 21x + 20} = \frac{A}{x - 1} + \frac{B}{x - 4} + \frac{C}{x + 5} = \frac{A(x - 4)(x + 5) + B(x - 1)(x + 5) + C(x - 1)(x - 4)}{x^3 - 21x + 20}$$

and consider the identity

$$54 = A(x - 4)(x + 5) + B(x - 1)(x + 5) + C(x - 1)(x - 4)$$

When $x = 1$, $54 = A(-3)(6)$ and $A = -3$; when $x = 4$, $54 = B(3 \cdot 9)$ and $B = 2$; when $x = -5$, $54 = C(-6)(-9)$ and $C = 1$. Thus

$$\frac{54}{x^3 - 21x + 20} = -\frac{3}{x - 1} + \frac{2}{x - 4} + \frac{1}{x + 5}$$

(b) The given fraction, being improper, will first be written as $x^2 + 1 + \dfrac{7x - 16}{(x - 4)(x + 2)}$. Using only the fractional term, we set

$$\frac{7x - 16}{(x - 4)(x + 2)} = \frac{A}{x - 4} + \frac{B}{x + 2} = \frac{A(x + 2) + B(x - 4)}{(x - 4)(x + 2)}$$

and consider the identity $7x - 16 = A(x + 2) + B(x - 4)$. Using $x = 4$, we find $A = 2$; using $x = -2$, we find $B = 5$. Thus

$$\frac{x^4 - 2x^3 - 7x^2 + 5x - 24}{x^2 - 2x - 8} = x^2 + 1 + \frac{2}{x - 4} + \frac{5}{x + 2}$$

26.2   Resolve into partial fractions: (a) $\dfrac{2x^2 + 6x - 4}{x(x + 2)^2}$, (b) $\dfrac{x^3 + x^2 + x - 1}{(x^2 - 1)^2}$.

(a) Set $\dfrac{2x^2 + 6x - 4}{x(x + 2)^2} = \dfrac{A}{(x + 2)^2} + \dfrac{B}{x + 2} + \dfrac{C}{x} = \dfrac{Ax + Bx(x + 2) + C(x + 2)^2}{x(x + 2)^2}$

$$= \frac{(B + C)x^2 + (A + 2B + 4C)x + 4C}{x(x + 2)^2}$$

Solution 1.   From the identity $2x^2 + 6x - 4 = (B + C)x^2 + (A + 2B + 4C)x + 4C$; by equating coefficients of like powers of x, we obtain $B + C = 2$, $A + 2B + 4C = 6$, $4C = -4$. Then $C = -1$, $B = 2 - C = 3$, $A = 6 - 2B - 4C = 4$, and

$$\frac{2x^2 + 6x - 4}{x(x + 2)^2} = \frac{4}{(x + 2)^2} + \frac{3}{x + 2} - \frac{1}{x}$$

Solution 2.   Consider the identity $2x^2 + 6x - 4 = Ax + Bx(x + 2) + C(x + 2)^2$. Using $x = -2$, we find $A = 4$; using $x = 0$, we find $C = -1$; using $x = 1$, we have $2 + 6 - 4 = A + 3B + 9C = 4 + 3B - 9$ and $B = 3$. These are the values of A, B, C found above.

Note that only two values $x = -2$ and $x = 0$ are suggested by the identity. Since three constants are to be determined, one additional value of x is needed. It may be taken at random.

(b) Set

$$\frac{x^3 + x^2 + x - 1}{(x^2 - 1)^2} = \frac{A}{(x-1)^2} + \frac{B}{x-1} + \frac{C}{(x+1)^2} + \frac{D}{x+1}$$

$$= \frac{A(x+1)^2 + B(x-1)(x+1)^2 + C(x-1)^2 + D(x+1)(x-1)^2}{(x^2-1)^2}$$

$$= \frac{(B+D)x^3 + (A+B+C-D)x^2 + (2A-B-2C-D)x + A-B+C+D}{(x^2-1)^2}$$

From the identity $x^3 + x^2 + x - 1 = (B+D)x^3 + (A+B+C-D)x^2 + (2A-B-2C-D)x + A - B + C + D$, we have

$$B + D = 1, \quad A + B + C - D = 1, \quad 2A - B - 2C - D = 1, \quad A - B + C + D = -1 \quad (1)$$

From the identity

$$x^3 + x^2 + x - 1 = A(x+1)^2 + B(x-1)(x+1)^2 + C(x-1)^2 + D(x+1)(x-1)^2 \qquad (2)$$

using $x = 1$ and $x = -1$, we find $A = \frac{1}{2}$ and $C = -\frac{1}{2}$. Then from (1),

$$B + D = 1, \quad B - D = 1, \quad \text{and} \quad D = 0, B = 1.$$

Thus

$$\frac{x^3 + x^2 + x - 1}{(x^2 - 1)^2} = \frac{\frac{1}{2}}{(x-1)^2} + \frac{1}{x-1} - \frac{\frac{1}{2}}{(x+1)^2} = \frac{1}{2(x-1)^2} + \frac{1}{x-1} - \frac{1}{2(x+1)^2}$$

[NOTE.   The constants could have been determined by solving the four equations of (1) simultaneously or by using the values $x = 1, -1, 0, 2$ in (2).]

26.3 Resolve $\dfrac{x^2 + 2x + 3}{x^4 + x^3 + 2x^2}$ into partial fractions.

Set $\dfrac{x^2 + 2x + 3}{x^4 + x^3 + 2x^2} = \dfrac{A}{x^2} + \dfrac{B}{x} + \dfrac{Cx + D}{x^2 + x + 2} = \dfrac{A(x^2 + x + 2) + Bx(x^2 + x + 2) + (Cx + D)x^2}{x^4 + x^3 + 2x^2}$

$$= \frac{(B+C)x^3 + (A+B+D)x^2 + (A+2B)x + 2A}{x^4 + x^3 + 2x^2}$$

Equating coefficients of like terms in the identity

$$x^2 + 2x + 3 = (B+C)x^3 + (A+B+D)x^2 + (A+2B)x + 2A$$

we have $B + C = 0$, $A + B + D = 1$, $A + 2B = 2$, $2A = 3$. Then $A = \frac{3}{2}$, $B = \frac{1}{4}$, $D = -\frac{3}{4}$, and

$$\frac{x^2 + 2x + 3}{x^4 + x^3 + 2x^2} = \frac{3}{2x^2} + \frac{1}{4x} - \frac{x+3}{4(x^2 + x + 2)}$$

26.4 Resolve $\dfrac{x^4 - x^3 + 8x^2 - 6x + 7}{(x-1)(x^2+2)^2}$ into partial fractions.

Set

$$\frac{x^4 - x^3 + 8x^2 - 6x + 7}{(x-1)(x^2+2)^2}$$

$$= \frac{A}{x-1} + \frac{Bx + C}{(x^2+2)^2} + \frac{Dx + E}{x^2+2}$$

$$= \frac{A(x^2+2)^2 + (Bx+C)(x-1) + (Dx+E)(x^2+2)(x-1)}{(x-1)(x^2+2)^2}$$

$$= \frac{(A+D)x^4 + (-D+E)x^3 + (4A+B+2D-E)x^2 + (-B+C-2D+2E)x + 4A-C-2E}{(x-1)(x^2+2)^2}.$$

Equating coefficients of like terms in the identity

$$x^4 - x^3 + 8x^2 - 6x + 7 = (A + D)x^4 + (-D + E)x^3 + (4A + B + 2D - E)x^2$$
$$+ (-B + C - 2D + 2E)x + 4A - C - 2E,$$

we have $A + D = 1$, $-D + E = -1$, $4A + B + 2D - E = 8$, $-B + C - 2D + 2E = -6$, $4A - C - 2E = 7$.
From the identity

$$x^4 - x^3 + 8x^2 - 6x + 7 = A(x^2 + 2)^2 + (Bx + C)(x - 1) + (Dx + E)(x^2 + 2)(x - 1)$$

Using $x = 1$, we find $A = 1$. Then $D = 1 - A = 0$, $E = -1 + D = -1$, $B = 3$, $C = -1$, and

$$\frac{x^4 - x^3 + 8x^2 - 6x + 7}{(x - 1)(x^2 + 2)^2} = \frac{1}{x - 1} + \frac{3x - 1}{(x^2 + 2)^2} - \frac{1}{x^2 + 2}$$

Supplementary Problems

Resolve into partial fractions.

26.5 $\dfrac{4x + 23}{(x - 3)(x + 2)(x + 4)} = \dfrac{1}{x - 3} - \dfrac{\frac{3}{2}}{x + 2} + \dfrac{\frac{1}{2}}{x + 4}$

26.6 $\dfrac{17x - 45}{x^3 - 2x^2 - 15x} = \dfrac{3}{x} + \dfrac{1}{x - 5} - \dfrac{4}{x + 3}$

26.7 $\dfrac{2x^2 + x + 3}{x^2 - 9} = 2 - \dfrac{3}{x + 3} + \dfrac{4}{x - 3}$

26.8 $\dfrac{x^2 + 8}{x^2(x + 2)} = \dfrac{4}{x^2} - \dfrac{2}{x} + \dfrac{3}{x + 2}$

26.9 $\dfrac{x^2 - 1}{(x - 2)^3} = \dfrac{3}{(x - 2)^3} + \dfrac{4}{(x - 2)^2} + \dfrac{1}{x - 2}$

26.10 $\dfrac{3x^3 - 2x^2 - 10x + 6}{(x + 2)^3(2x + 1)} = \dfrac{2}{(x + 2)^3} - \dfrac{10}{(x + 2)^2} + \dfrac{3}{2x + 1}$

26.11 $\dfrac{x^2 + 5x + 3}{x(x + 1)^2} = \dfrac{1}{(x + 1)^2} - \dfrac{2}{x + 1} + \dfrac{3}{x}$

26.12 $\dfrac{x^2 - 3x}{(x - 4)(x - 2)^2} = \dfrac{1}{(x - 2)^2} + \dfrac{1}{x - 4}$

26.13 $\dfrac{5x^2 - 8}{x(x^2 + 2x - 4)} = \dfrac{2}{x} + \dfrac{3x - 4}{x^2 + 2x - 4}$

26.14 $\dfrac{x^2 - 6x - 3}{x^4 + 3x^2 - 18} = \dfrac{2x + 3}{3(x^2 + 6)} - \dfrac{2x}{3(x^2 - 3)}$

26.15 $\dfrac{2x + 7}{(x^3 - 1)(x^2 + x + 1)} = \dfrac{1}{x - 1} - \dfrac{x + 2}{x^2 + x + 1} - \dfrac{3x + 4}{(x^2 + x + 1)^2}$

26.16 $\dfrac{x^3 + 4x^2 + 4}{x^3(x^2 + 2)^2} = \dfrac{1}{x^3} - \dfrac{x - 1}{(x^2 + 2)^2}$

26.17 $\dfrac{3x^2 + 8}{(x^2 + 2)^3 x^2} = \dfrac{1}{x^2} - \dfrac{1}{(x^2 + 2)^3} - \dfrac{2}{(x^2 + 2)^2} - \dfrac{1}{x^2 + 2}$

26.18 $\dfrac{x^6 + 5x^4 + x^3 + 3x^2 - 9}{(x^2 + 3)^2} = x^2 - 1 + \dfrac{x}{x^2 + 3} - \dfrac{3x}{(x^2 + 3)^2}$

Chapter 27

Infinite Sequences

GENERAL TERM OF A SEQUENCE. Frequently the law of formation of a given sequence may be stated by giving a representative or *general term* of the sequence. This general term is a function of n, where n is the number of the term in the sequence. For this reason, it is also called the nth term of the sequence.

When the general term is given, it is a simple matter to write as many terms of the sequence as desired.

EXAMPLE 1

(*a*) Write the first four terms and the tenth term of the sequence whose general term is $1/n$.

The first term ($n = 1$) is $\frac{1}{1} = 1$, the second term ($n = 2$) is $\frac{1}{2}$, and so on. The first four terms are $1, \frac{1}{2}, \frac{1}{3}, \frac{1}{4}$ and the tenth term is $\frac{1}{10}$.

(*b*) Write the first four terms and the ninth term of the sequence whose general term is $(-1)^{n-1}\dfrac{2n}{n^2+1}$.

The first term ($n = 1$) is $(-1)^{1-1}\dfrac{2\cdot 1}{1^2+1} = 1$, the second term ($n = 2$) is $(-1)^1\dfrac{2\cdot 2}{2^2+1} = -\dfrac{4}{5}$, and so on.

The first four terms are $1, -\frac{4}{5}, \frac{3}{5}, -\frac{8}{17}$ and the ninth term is $(-1)^8\dfrac{2\cdot 9}{9^2+1} = \dfrac{9}{41}$.

Note that the effect of the factor $(-1)^{n-1}$ is to produce a sequence whose terms have alternate signs, the sign of the first term being positive. The same pattern of signs is also produced by the factor $(-1)^{n+1}$. In order to produce a sequence whose terms alternate in sign, the first term being negative, the factor $(-1)^n$ is used.

When the first few terms of a sequence are given, the general term is obtained by inspection.

EXAMPLE 2. Obtain the general term for each of the sequences:

(*a*) $1, 4, 9, 16, 25, \ldots$.

The terms of the sequence are the squares of the positive integers; the general term is n^2.

(*b*) $3, 7, 11, 15, 19, 23, \ldots$.

This is an arithmetic progression having $a = 3$ and $d = 4$. The general term is $a + (n-1)d = 4n - 1$. Note, however, that the general term can be obtained about as easily by inspection.

(See Problems 27.1–27.3.)

LIMIT OF AN INFINITE SEQUENCE. From Example 4 of Chapter 9, the line $y = 2$ is a horizontal asymptote of $xy - 2x - 1 = 0$. To show this, let $P(x, y)$ move along the curve so that its abscissa takes on the values $10, 10^2, 10^3, \ldots, 10^n, \ldots$. Then the corresponding values of y are

$$2.1, 2.01, 2.001, \ldots, 2 + \frac{1}{10^n}, \ldots \tag{27.1}$$

and we infer that, by proceeding far enough along in this sequence, the difference between the terms of the sequence and 2 may be made as small as we please. This is equivalent to the following: Let ε denote a positive number, as small as we please; then there is a term of the sequence such that the difference between it and 2 is less than ε, and the same is true for all subsequent terms of the sequence. For example, let $\varepsilon = 1/10^{25}$; then the difference between the term $2 + 1/10^{26}$ and 2, $2 + 1/10^{26} - 2 = 1/10^{26}$, is less than $\varepsilon = 1/10^{25}$ and the same is true for the terms $2 + 1/10^{27}, 2 + 1/10^{28}$, and so on.

The behavior of the terms of the sequence (27.1) discussed above is indicated by the statement: *The limit of the sequence (27.1) is* 2. In general,

If, for an infinite sequence

$$s_1, s_2, s_3, \ldots, s_n, \ldots \tag{27.2}$$

and a positive number ε, however small, there exists a number s and a positive integer m such that for all $n > m$

$$|s - s_n| < \varepsilon,$$

then the limit of the sequence is s.

EXAMPLE 3. Show, using the above definition, that the limit of sequence (27.1) is 2.

Take $\varepsilon = 1/10^p$, where p is a positive integer as large as we please; thus, ε is a positive number as small as we please. We must produce a positive integer m (in other words, a term s_m) such that for $n > m$ (that is, for all subsequent terms) $|s - s_n| < \varepsilon$. Now

$$\left| 2 - \left(2 + \frac{1}{10^n} \right) \right| < \frac{1}{10^p} \quad \text{or} \quad \frac{1}{10^n} < \frac{1}{10^p}$$

requires $n > p$. Thus $m = p$ is the required value of m.

The statement that the limit of the sequence (27.2) is s describes the behavior of s_n as n increases without bound over the positive integers. Since we shall repeatedly be using the phrase "as n increases without bound" or the phrase "as n becomes infinite," which we shall take to be equivalent to the former phrase, we shall introduce the notation $n \to \infty$ for it. Thus the behavior of s_n may be described briefly by

$$\lim_{n \to \infty} s_n = s$$

(read: the limit of s_n, as n becomes infinite, is s).

We state, without proof, the following theorem:

If $\lim\limits_{n \to \infty} s_n = s$ and $\lim\limits_{n \to \infty} t_n = t$, then

(A) $\lim\limits_{n \to \infty} (s_n \pm t_n) = \lim\limits_{n \to \infty} s_n \pm \lim\limits_{n \to \infty} t_n = s \pm t$

(B) $\lim\limits_{n \to \infty} (s_n \cdot t_n) = \lim\limits_{n \to \infty} s_n \cdot \lim\limits_{n \to \infty} t_n = s \cdot t$

(C) $\lim\limits_{n \to \infty} \dfrac{s_n}{t_n} = \dfrac{\lim\limits_{n \to \infty} s_n}{\lim\limits_{n \to \infty} t_n} = \dfrac{s}{t}$,   provided $t \neq 0$

or, in words, if each of two variables approaches a limit, then the limits of the sum, difference, product, and quotient of the two variables are equal, respectively, to the sum, difference, product, and quotient of their limits provided only that, in the case of the quotient, the limit of the denominator is not zero.

This theorem makes it possible to find the limit of a sequence directly from its general term. In this connection, we shall need

$$\lim_{n \to \infty} a = a, \qquad \text{where } a \text{ is any constant} \tag{27.3}$$

$$\lim_{n \to \infty} \frac{1}{n^k} = 0, \qquad k > 0 \text{ (See Problem 27.4.)} \tag{27.4}$$

$$\lim_{n \to \infty} \frac{1}{b^n} = 0, \qquad \text{where } b \text{ is a constant} > 1 \text{ (See Problem 27.5.)} \tag{27.5}$$

(See Problem 27.6.)

THE FOLLOWING THEOREMS are useful in establishing whether or not certain sequences have a limit.

I.   Suppose M is a fixed number, such that for all values of n,

$$s_n \leq s_{n+1} \qquad \text{and} \qquad s_n \leq M;$$

then $\lim_{n \to \infty} s_n$ exists and is $\leq M$.

   If, however, s_n eventually exceeds M, no matter how large M may be, $\lim_{n \to \infty} s_n$ does not exist.

II. Suppose M is a fixed number such that, for all values of n,

$$s_n \geq s_{n+1} \qquad \text{and} \qquad s_n \geq M;$$

then $\lim_{n \to \infty} s_n$ exists and is $\geq M$.

   If, however, s_n is eventually smaller than M, no matter how small M may be, $\lim_{n \to \infty} s_n$ does not exist.

EXAMPLE 4

(a)   For the sequence $\frac{5}{2}, 3, \frac{19}{6}, \frac{13}{4}, \ldots, \left(\frac{7}{2} - \frac{1}{n}\right), \ldots,$ $s_n < s_{n+1}$ and $s_n < 4$, for all values of n; the sequence has a limit ≤ 4. In fact, $\lim_{n \to \infty} s_n = \frac{7}{2}$.

(b)   For the sequence $3, 5, 7, 9, \ldots, 2n+1, \ldots,$ $s_n < s_{n+1}$ but s_n will eventually exceed any chosen M, however large (if $M = 2^{1000} + 1$, then $2n + 1 > M$ for $n > 2^{999}$), and the sequence does not have a limit.

(See Problems 27.7–27.9.)

RECURSIVELY DEFINED SEQUENCES. Sequences can be defined recursively. For example, suppose that $a_1 = 1$ and $a_{n+1} = 2a_n$ for every natural number n. Then,

$$a_1 = 1, \qquad a_2 = a_{1+1} = 2a_1 = 2, \qquad a_3 = a_{2+1} = 2a_2 = 4, \qquad \text{etc.}$$

Thus, the sequence is $1, 2, 4, 8, \ldots$.

   One famous such sequence is the Fibonacci sequence:

$$a_1 = 1, \qquad a_2 = 1, \qquad a_{n+2} = a_{n+1} + a_n.$$

The sequence is $1, 1, 2, 3, 5, 8, 13, \ldots$.

Solved Problems

27.1   Write the first five terms and the tenth term of the sequence whose general term is

(a) $4n - 1$.

   The first term is $4 \cdot 1 - 1 = 3$, the second term is $4 \cdot 2 - 1 = 7$, the third term is $4 \cdot 3 - 1 = 11$, the fourth term is $4 \cdot 4 - 1 = 15$, the fifth term is $4 \cdot 5 - 1 = 19$; the tenth term is $4 \cdot 10 - 1 = 39$.

(b) 2^{n-1}.

   The first term is $2^{1-1} = 2^0 = 1$, the second term is $2^{2-1} = 2$, the third is $2^{3-1} = 2^2 = 4$, the fourth is $2^3 = 8$, the fifth is $2^4 = 16$; the tenth is $2^9 = 512$.

(c) $\dfrac{(-1)^{n-1}}{n+1}$.

The first term is $\dfrac{(-1)^{1-1}}{1+1} = \dfrac{(-1)^0}{2} = \dfrac{1}{2}$, the second is $\dfrac{(-1)^{2-1}}{2+1} = -\dfrac{1}{3}$, the third is $\dfrac{(-1)^2}{3+1} = \dfrac{1}{4}$, the fourth is $-\frac{1}{5}$, the fifth is $\frac{1}{6}$; the tenth is $-\frac{1}{11}$.

27.2 Write the first four terms of the sequence whose general term is

(a) $\dfrac{n+1}{n!}$.

The terms are $\dfrac{1+1}{1!}, \dfrac{2+1}{2!}, \dfrac{3+1}{3!}, \dfrac{4+1}{4!}$ or $2, \dfrac{3}{2}, \dfrac{2}{3}, \dfrac{5}{24}$.

(b) $\dfrac{x^{2n-1}}{(2n+1)!}$.

The required terms are $\dfrac{x}{3!}, \dfrac{x^3}{5!}, \dfrac{x^5}{7!}, \dfrac{x^7}{9!}$.

(c) $(-1)^{n-1} \dfrac{x^{2n-2}}{(n-1)!}$.

The terms are $\dfrac{x^0}{0!}, -\dfrac{x^2}{1!}, \dfrac{x^4}{2!}, -\dfrac{x^6}{3!}$ or $1, -x^2, \dfrac{x^4}{2}, \dfrac{x^6}{6}$.

27.3 Write the general term for each of the following sequences:

(a) $2, 4, 6, 8, 10, 12, \dots$.

The first term is $2 \cdot 1$, the second is $2 \cdot 2$, the third is $2 \cdot 3$, etc.; the general term is $2n$.

(b) $1, 3, 5, 7, 9, 11, \dots$.

Each term of the given sequence is 1 less than the corresponding term of the sequence in (a); the general term is $2n - 1$.

(c) $2, 5, 8, 11, 14, \dots$.

The first term is $3 \cdot 1 - 1$, the second term is $3 \cdot 2 - 1$, the third term is $3 \cdot 3 - 1$, and so on; the general term is $3n - 1$.

(d) $2, -5, 8, -11, 14, \dots$.

This sequence may be obtained from that in (c) by changing the signs of alternate terms beginning with the second; the general term is $(-1)^{n-1}(3n - 1)$.

(e) $2.1, 2.01, 2.001, 2.0001, \dots$.

The first term is $2 + 1/10$, the second term is $2 + 1/10^2$, the third is $2 + 1/10^3$, and so on; the general term is $2 + 1/10^n$.

(f) $\frac{1}{8}, -\frac{1}{27}, \frac{1}{64}, -\frac{1}{125}, \dots$.

The successive denominators are the cubes of $2, 3, 4, 5, \dots$ or of $1+1, 2+1, 3+1, 4+1, \dots$; the general term is $\dfrac{(-1)^{n-1}}{(n+1)^3}$.

(g) $\dfrac{3}{1}, \dfrac{4}{1\cdot 2}, \dfrac{5}{1\cdot 2\cdot 3}, \dots$.

Rewriting the sequence as $\dfrac{1+2}{1!}, \dfrac{2+2}{2!}, \dfrac{3+2}{3!}, \dots$, the general term is $\dfrac{n+2}{n!}$.

(h) $\dfrac{x}{2}, \dfrac{x^2}{6}, \dfrac{x^3}{24}, \dfrac{x^4}{120}, \dots$.

The denominators are $2!, 3!, \dots, (n+1)!, \dots$; the general term is $\dfrac{x^n}{(n+1)!}$.

(i) $x, \dfrac{-x^3}{3!}, \dfrac{x^5}{5!}, \dfrac{-x^7}{7!}, \dots$.

The exponents of x are $2\cdot 1 - 1, 2\cdot 2 - 1, 2\cdot 3 - 1, \dots$; the general term is $(-1)^{n-1}\dfrac{x^{2n-1}}{(2n-1)!}$.

(j) $3, 4, \frac{5}{2}, 1, \frac{7}{24}, \dots$.

Rewrite the sequence as $\dfrac{3}{0!}, \dfrac{4}{1!}, \dfrac{5}{2!}, \dfrac{6}{3!}, \dots$; the general term is $\dfrac{n+2}{(n-1)!}$.

27.4 Show that $\lim\limits_{n\to\infty} \dfrac{1}{n^k} = 0$ when $k > 0$.

Take $\varepsilon = 1/p^k$, where p is a positive integer as large as we please. We seek a positive number m such that for $n > m$, $|0 - 1/n^k| = 1/n^k < 1/p^k$. Since this inequality is satisfied when $n > p$, it is sufficient to take for m any number equal to or greater than p.

27.5 Show that $\lim\limits_{n\to\infty} \dfrac{1}{b^n} = 0$ when $b > 1$.

Take $\varepsilon = 1/b^p$, where p is a positive integer. Since $b > 1$, $b^p > 1$ and $\varepsilon = 1/b^p < 1$. Thus, ε may be made as small as we please by taking p sufficiently large. We seek a positive number m such that for $n > m$, $|0 - 1/b^n| = 1/b^n < 1/b^p$. Since $n > p$ satisfies the inequality, it is sufficient to take for m any number equal to or greater than p.

27.6 Evaluate each of the following:

(a) $\lim\limits_{n\to\infty} \left(\dfrac{3}{n} + \dfrac{5}{n^2} \right) = \lim\limits_{n\to\infty} \dfrac{3}{n} + \lim\limits_{n\to\infty} \dfrac{5}{n^2} = 0 + 0 = 0$

(b) $\lim\limits_{n\to\infty} \dfrac{n}{n+1} = \lim\limits_{n\to\infty} \dfrac{n/n}{\dfrac{n+1}{n}} = \lim\limits_{n\to\infty} \dfrac{1}{1 + 1/n} = \dfrac{\lim\limits_{n\to\infty} 1}{\lim\limits_{n\to\infty} (1 + 1/n)} = \dfrac{1}{1+0} = 1$

(c) $\lim\limits_{n\to\infty} \dfrac{n^2 + 2}{2n^2 - 3n} = \lim\limits_{n\to\infty} \dfrac{1 + 2/n^2}{2 - 3/n} = \dfrac{1+0}{2-0} = \dfrac{1}{2}$

(d) $\lim\limits_{n\to\infty} \left(4 - \dfrac{2^n - 1}{2^{n+1}} \right) = 4 - \lim\limits_{n\to\infty} \dfrac{2^n - 1}{2^{n+1}} = 4 - \lim\limits_{n\to\infty} \left(\dfrac{1}{2} - \dfrac{1}{2^{n+1}} \right)$

$\qquad\qquad\qquad\qquad\qquad = 4 - \dfrac{1}{2} - \lim\limits_{n\to\infty} \left(-\dfrac{1}{2^{n+1}} \right) = 4 - \dfrac{1}{2} - 0 = 3.5$

27.7 Show that every infinite arithmetic sequence fails to have a limit except when $d = 0$.

(a) If $d > 0$, then $s_n = a + (n-1)d < s_{n+1} = a + nd$; but s_n eventually exceeds any previously selected M, however large. Thus, the sequence has no limit.

(b) If $d < 0$, then $s_n > s_{n+1}$; but s_n eventually becomes smaller than any previously selected M, however small. Thus, the sequence has no limit.

(c) If $d = 0$, the sequence is $a, a, a, \ldots, a, \ldots$ with limit a.

27.8 Show that the infinite geometric sequence $3, 6, 12, \ldots, 3 \cdot 2^{n-1}, \ldots$ does not have a limit.

Here $s_n < s_{n+1}$; but $3 \cdot 2^{n-1}$ may be made to exceed any previously selected M, however large. The sequence has no limit.

27.9 Show that the following sequence does not have a limit:

$$1, 1 + \frac{1}{2}, 1 + \frac{1}{2} + \frac{1}{3}, 1 + \frac{1}{2} + \frac{1}{3} + \frac{1}{4} + \cdots, \qquad 1 + \frac{1}{2} + \frac{1}{3} + \frac{1}{4} + \cdots + \frac{1}{n}, \qquad \ldots$$

Here $s_n < s_{n+1}$. Let M, as large as we please, be chosen. Now

$$1 + \tfrac{1}{2} + \tfrac{1}{3} + \tfrac{1}{4} + \cdots = 1 + \tfrac{1}{2} + (\tfrac{1}{3} + \tfrac{1}{4}) + (\tfrac{1}{5} + \tfrac{1}{6} + \tfrac{1}{7} + \tfrac{1}{8}) + (\tfrac{1}{9} + \cdots + \tfrac{1}{16}) + (\tfrac{1}{17} + \cdots + \tfrac{1}{32}) + \cdots$$

and $\qquad\qquad \tfrac{1}{3} + \tfrac{1}{4} > \tfrac{1}{2}, \qquad \tfrac{1}{5} + \tfrac{1}{6} + \tfrac{1}{7} + \tfrac{1}{8} > \tfrac{1}{2}, \qquad \tfrac{1}{9} + \cdots + \tfrac{1}{16} > \tfrac{1}{2}, \qquad$ and so on.

Since the sum of each group exceeds $\tfrac{1}{2}$ and we may add as many groups as we please, we can eventually obtain a sum of groups which exceeds M. Thus, the sequence has no limit.

Supplementary Problems

27.10 Write the first four terms of the sequence whose general term is

(a) $\dfrac{1}{1+n}$ (c) $\dfrac{1}{3^n}$ (e) $\dfrac{n^2}{3n-2}$ (g) $(-1)^{n+1}\dfrac{1}{n!}$

(b) $\dfrac{1}{1+n\sqrt{n}}$ (d) $\dfrac{2n-1}{2n+3}$ (f) $(-1)^{n+1}\dfrac{2n+1}{2^{n+1}}$ (h) $\dfrac{n^2}{(2n)!}$

    Ans.   (a) $\frac{1}{2},\frac{1}{3},\frac{1}{4},\frac{1}{5}$ (c) $\frac{1}{3},\frac{1}{9},\frac{1}{27},\frac{1}{81}$ (f) $\frac{3}{4},-\frac{5}{8},\frac{7}{16},-\frac{9}{32}$

            (b) $\dfrac{1}{2},\dfrac{1}{1+2\sqrt{2}},\dfrac{1}{1+3\sqrt{3}},\dfrac{1}{9}$ (d) $\frac{1}{5},\frac{3}{7},\frac{5}{9},\frac{7}{11}$ (g) $1,-\frac{1}{2},\frac{1}{6},-\frac{1}{24}$

                                       (e) $1,1,\frac{9}{7},\frac{8}{5}$ (h) $\frac{1}{2},\frac{1}{6},\frac{1}{80},\frac{1}{2520}$

27.11 Write the general term of each sequence.

(a) $1,\frac{1}{3},\frac{1}{5},\frac{1}{7},\ldots$ (c) $1,\frac{3}{5},\frac{2}{5},\frac{5}{17},\frac{3}{13},\ldots$ (f) $\frac{1}{3},-\frac{1}{15},\frac{1}{35},-\frac{1}{63},\ldots$

(b) $\dfrac{4}{1\cdot3},\dfrac{5}{2\cdot4},\dfrac{6}{3\cdot5},\dfrac{7}{4\cdot6},\ldots$ (d) $\frac{1}{2},\frac{3}{8},\frac{7}{24},\frac{15}{64},\ldots$ (g) $\frac{1}{2},\frac{3}{8},\frac{5}{16},\frac{9}{32},\ldots$

                                    (e) $2,1,\frac{8}{9},1,\frac{32}{25},\frac{16}{9},\ldots$ (h) $\frac{1}{2},-x^2/4,x^4/6,-x^6/8,\ldots$

    Ans.   (a) $\dfrac{1}{2n-1}$ (c) $\dfrac{n+1}{n^2+1}$ (e) $\dfrac{2^n}{n^2}$ (g) $\dfrac{1+2^{n-1}}{2^{n+1}}$

            (b) $\dfrac{n+3}{n(n+2)}$ (d) $\dfrac{2^n-1}{n\cdot2^n}$ (f) $\dfrac{(-1)^{n+1}}{(2n-1)(2n+1)}$ (h) $(-1)^{n+1}\dfrac{x^{2n-2}}{2n}$

27.12 Evaluate.   (a) $\displaystyle\lim_{n\to\infty}\left(2-\frac{1}{n}\right)$ (d) $\displaystyle\lim_{n\to\infty}\frac{2n^2+5n-6}{n^2+n-1}$ (g) $\displaystyle\lim_{n\to\infty}\frac{1}{2^n+1}$

                (b) $\displaystyle\lim_{n\to\infty}\frac{3n+1}{3n-2}$ (e) $\displaystyle\lim_{n\to\infty}\frac{n}{n+1}$ (h) $\displaystyle\lim_{n\to\infty}\frac{2^n+1}{2^{n+1}+1}$

                (c) $\displaystyle\lim_{n\to\infty}\frac{2n}{(n+1)(n+2)}$ (f) $\displaystyle\lim_{n\to\infty}\frac{(3n)!n}{(3n+1)!}$

    Ans.   (a) 2   (b) 1   (c) 0   (d) 2   (e) 1   (f) $\frac{1}{3}$ (g) 0   (h) $\frac{1}{2}$

27.13 Explain why each of the following has no limit:

(a) $1,3,5,7,9,\ldots$ (c) $1,-2,4,-8,16,-32,\ldots$

(b) $1,0,1,0,1,0,\ldots$ (d) $\frac{1}{25},\frac{4}{25},\frac{9}{25},\frac{16}{25},\ldots$

27.14 Write out the next four terms given the recursive formula:

(a) $a_1=-1,\ a_{n+1}=\frac{1}{3}a_n$

(b) $a_1=2,\ a_2=3,\ a_{n+2}=a_{n+1}+\dfrac{a_n}{2}$

    Ans.   (a) $-\frac{1}{3},-\frac{1}{9},-\frac{1}{27},-\frac{1}{81}$

            (b) $2+\frac{3}{2},3+\frac{7}{4},\frac{7}{2}+\frac{19}{8},\frac{19}{4}+\frac{47}{16}$

Chapter 28

Infinite Series

THE INDICATED SUM of the terms of an infinite sequence is called an *infinite series*. Let

$$s_1 + s_2 + s_3 + \cdots + s_n + \cdots \qquad (28.1)$$

be such a series and define the sequence of *partial sums*

$$S_1 = s_1, \qquad S_2 = s_1 + s_2, \qquad \ldots, \qquad S_n = s_1 + s_2 + \cdots + s_n, \qquad \ldots$$

If $\lim_{n \to \infty} S_n$ exists, the series (28.1) is called *convergent*; if $\lim_{n \to \infty} S_n = S$, the series is said to converge to S. If $\lim_{n \to \infty} S_n$ does not exist, the series is called *divergent*.

EXAMPLE 1.

(a)   Every infinite geometric series

$$a + ar + ar^2 + \cdots + ar^{n-1} + \cdots$$

is convergent if $|r| < 1$ and is divergent if $|r| \geq 1$. (See Problem 28.1.)

(b)   The *harmonic series* $1 + \frac{1}{2} + \frac{1}{3} + \cdots + 1/n + \cdots$ is divergent. See Problem 27.9. (See Problem 28.2.)

A NECESSARY CONDITION THAT (28.1) BE CONVERGENT is $\lim_{n \to \infty} s_n = 0$; that is, if (28.1) is convergent then $\lim_{n \to \infty} s_n = 0$. However, this condition is *not sufficient* since the harmonic series is divergent although $\lim_{n \to \infty} s_n = \lim_{n \to \infty} (1/n) = 0$.

A SUFFICIENT CONDITION THAT (28.1) BE DIVERGENT is $\lim_{n \to \infty} s_n \neq 0$; that is, if $\lim_{n \to \infty} s_n$ exists and is different from 0, or if $\lim_{n \to \infty} s_n$ does not exist, the series is divergent. This, in turn, is not a necessary condition since the harmonic series is divergent although $\lim_{n \to \infty} s_n = 0$. (See Problem 28.3.)

SERIES OF POSITIVE TERMS

COMPARISON TEST FOR CONVERGENCE of a series of positive terms.

I.   If every term of a given series of positive terms is less than or equal to the corresponding term of a known convergent series from some point on in the series, the given series is convergent.

II.   If every term of a given series of positive terms is equal to or greater than the corresponding term of a known divergent series from some point on in the series, the given series is divergent.

The following series will be found useful in making comparison tests:

(a)   The geometric series $a + ar + ar^2 + \cdots + ar^n + \cdots$ which converges when $|r| < 1$ and diverges when $|r| \geq 1$

(b)   The p series $1 + \dfrac{1}{2^p} + \dfrac{1}{3^p} + \cdots + \dfrac{1}{n^p} + \cdots$ which converges for $p > 1$ and diverges for $p \leq 1$

(c)   Each new series tested

In comparing two series it is not sufficient to examine the first few terms of each series. *The general terms must be compared*, since the comparison must be shown from some point on. (See Problems 28.4–28.6.)

THE RATIO TEST FOR CONVERGENCE.

If, in a series of positive terms, the *test ratio*

$$r_n = \frac{s_{n+1}}{s_n}$$

approaches a limit R as $n \to \infty$, the series is convergent if $R < 1$ and is divergent if $R > 1$. If $R = 1$, the test fails to indicate convergency or divergency. (See Problem 28.7.)

SERIES WITH NEGATIVE TERMS

A SERIES WITH ALL ITS TERMS NEGATIVE may be treated as the negative of a series with all of its terms positive.

ALTERNATING SERIES.   A series whose terms are alternately positive and negative, as

$$s_1 - s_2 + s_3 - \cdots + (-1)^{n-1}s_n + \cdots \tag{28.2}$$

where each s is positive, is called an *alternating series*.

An alternating series (28.2) is convergent provided $s_n \geq s_{n+1}$, for every value of n, and $\lim\limits_{n \to \infty} s_n = 0$.

(See Problem 28.8.)

ABSOLUTELY CONVERGENT SERIES.   A series (28.1) $s_1 + s_2 + s_3 + \cdots + s_n + \cdots$ in which some of the terms are positive and some are negative is called *absolutely convergent* if the series of absolute values of the terms

$$|s_1| + |s_2| + |s_3| + \cdots + |s_n| + \cdots \tag{28.3}$$

is convergent.

CONDITIONALLY CONVERGENT SERIES.   A series (28.1), where some of the terms are positive and some are negative, is called *conditionally convergent* if it is convergent but the series of absolute values of its terms is divergent.

EXAMPLE 2.   The series $1 - \frac{1}{2} + \frac{1}{3} - \frac{1}{4} + \cdots$ is convergent, but the series of absolute values of its terms $1 + \frac{1}{2} + \frac{1}{3} + \frac{1}{4} + \cdots$ is divergent. Thus, the given series is conditionally convergent.

THE GENERALIZED RATIO TEST.   Let (28.1) $s_1 + s_2 + s_3 + \cdots + s_n + \cdots$ be a series some of whose terms are positive and some are negative. Let

$$\lim_{n \to \infty} \frac{|s_{n+1}|}{|s_n|} = R$$

The series (28.1) is absolutely convergent if $R < 1$ and is divergent if $R > 1$. If $R = 1$, the test fails. (See Problem 28.9.)

Solved Problems

28.1 Examine the infinite geometric series $a + ar + ar^2 + \cdots + ar^n + \cdots$ for convergence and divergence.

The sum of the first n terms is $S_n = \dfrac{a(1 - r^n)}{1 - r} = \dfrac{a}{1 - r}(1 - r^n)$.

If $|r| < 1$, $\lim\limits_{n \to \infty} r^n = 0$; then $\lim\limits_{n \to \infty} S_n = \dfrac{a}{1 - r}$ and the series is convergent.

If $|r| > 1$, $\lim\limits_{n \to \infty} r^n$ does not exist and $\lim\limits_{n \to \infty} S_n$ does not exist; the series is divergent.

If $r = 1$, the series is $a + a + a + \cdots + a + \cdots$; then $\lim\limits_{n \to \infty} S_n = \lim\limits_{n \to \infty} na$ does not exist.

If $r = -1$, the series is $a - a + a - a + \cdots$; then $S_n = a$ or 0 according as n is odd or even, and $\lim\limits_{n \to \infty} S_n$ does not exist.

Thus, the infinite geometric series converges to $\dfrac{a}{1 - r}$ when $|r| < 1$, and diverges when $|r| \geq 1$.

28.2 Show that the following series are convergent:

(a) $1 + \dfrac{2}{3} + \dfrac{4}{9} + \dfrac{8}{27} + \cdots + \dfrac{2^{n-1}}{3^{n-1}} + \cdots$.

This is a geometric series with ratio $r = \frac{2}{3}$; then $|r| < 1$ and the series is convergent.

(b) $2 - \frac{3}{2} + \frac{9}{8} - \frac{27}{32} + \cdots + (-1)^{n-1}2(\frac{3}{4})^{n-1} + \cdots$.

This is a geometric series with ratio $r = -\frac{3}{4}$; then $|r| < 1$ and the series is convergent.

(c) $1 + \dfrac{1}{2^p} + \dfrac{1}{2^p} + \dfrac{1}{4^p} + \dfrac{1}{4^p} + \dfrac{1}{4^p} + \dfrac{1}{8^p} + \cdots, p > 1$.

This series may be rewritten as $1 + \dfrac{2}{2^p} + \dfrac{4}{4^p} + \dfrac{8}{8^p} + \cdots$, a geometric series with ratio $r = 2/2^p$.

Since $|r| < 1$, when $p > 1$, the series is convergent.

28.3 Show that the following series are divergent:

(a) $2 + \dfrac{3}{2} + \dfrac{4}{3} + \cdots + \dfrac{n+1}{n} + \cdots$.

Since $\lim\limits_{n \to \infty} s_n = \lim\limits_{n \to \infty} \dfrac{n+1}{n} = \lim\limits_{n \to \infty}\left(1 + \dfrac{1}{n}\right) = 1 \neq 0$, the series is divergent.

(b) $\dfrac{1}{2} + \dfrac{3}{8} + \dfrac{5}{16} + \dfrac{9}{32} + \cdots + \dfrac{2^{n-1}+1}{4 \cdot 2^{n-1}} + \cdots$.

Since $\lim\limits_{n \to \infty} s_n = \lim\limits_{n \to \infty} \dfrac{2^{n-1}+1}{4 \cdot 2^{n-1}} = \lim\limits_{n \to \infty} \dfrac{1 + 1/2^{n-1}}{4} = \dfrac{1}{4} \neq 0$, the series is divergent.

28.4 Show that $1 + \dfrac{1}{2^p} + \dfrac{1}{3^p} + \dfrac{1}{4^p} + \cdots + \dfrac{1}{n^p} + \cdots$ is divergent for $p \leq 1$ and convergent for $p > 1$.

For $p = 1$, the series is the harmonic series and is divergent.

For $p < 1$, including negative values, $\dfrac{1}{n^p} \geq \dfrac{1}{n}$, for every n. Since every term of the given series is equal to or greater than the corresponding term of the harmonic series, the given series is divergent.

For $p > 1$, compare the series with the convergent series

$$1 + \dfrac{1}{2^p} + \dfrac{1}{2^p} + \dfrac{1}{4^p} + \dfrac{1}{4^p} + \dfrac{1}{4^p} + \dfrac{1}{4^p} + \dfrac{1}{8^p} + \cdots \tag{1}$$

of Solved Problem 28.2(c). Since each term of the given series is less than or equal to the corresponding term of series (1), the given series is convergent.

28.5   Use Problem 28.4 to determine whether the following series are convergent or divergent:

(a) $1 + \dfrac{1}{2\sqrt{2}} + \dfrac{1}{3\sqrt{3}} + \dfrac{1}{4\sqrt{4}} + \cdots$. The general term is $\dfrac{1}{n\sqrt{n}} = \dfrac{1}{n^{3/2}}$.

This is a p series with $p = \frac{3}{2} > 1$; the series is convergent.

(b) $1 + 4 + 9 + 16 + \cdots$. The general term is $n^2 = \dfrac{1}{n^{-2}}$.

This is a p series with $p = -2 < 1$; the series is divergent.

(c) $1 + \dfrac{\sqrt[4]{2}}{4} + \dfrac{\sqrt[4]{3}}{9} + \dfrac{\sqrt[4]{4}}{16} + \cdots$. The general term is $\dfrac{\sqrt[4]{n}}{n^2} = \dfrac{1}{n^{7/4}}$.

The series is convergent since $p = \frac{7}{4} > 1$.

28.6   Use the comparison test to determine whether each of the following is convergent or divergent:

(a) $1 + \dfrac{1}{2!} + \dfrac{1}{3!} + \dfrac{1}{4!} + \cdots$.

The general term $\dfrac{1}{n!} \le \dfrac{1}{n^2}$. Thus, the terms of the given series are less than or equal to the corresponding terms of the p series with $p = 2$. The series is convergent.

(b) $\frac{1}{2} + \frac{1}{3} + \frac{1}{5} + \frac{1}{9} + \cdots$.

The general term $\dfrac{1}{1 + 2^{n-1}} \le \dfrac{1}{2^{n-1}}$. Thus, the terms of the given series are less than or equal to the corresponding terms of the geometric series with $a = 1$ and $r = \frac{1}{2}$. The series is convergent.

(c) $\frac{2}{1} + \frac{3}{4} + \frac{4}{9} + \frac{5}{16} + \cdots$.

The general term $\dfrac{n+1}{n^2} = \dfrac{1}{n} + \dfrac{1}{n^2} > \dfrac{1}{n}$. Thus, the terms of the given series are equal to or greater than the corresponding terms of the harmonic series. The series is divergent.

(d) $\frac{1}{3} + \frac{1}{12} + \frac{1}{27} + \frac{1}{48} + \cdots$.

The general term $\dfrac{1}{3 \cdot n^2} \le \dfrac{1}{n^2}$. Thus, the terms of the given series are less than or equal to the corresponding terms of the p series with $p = 2$. The series is convergent.

(e) $1 + \dfrac{1}{2} + \dfrac{1}{3^2} + \dfrac{1}{4^3} + \dfrac{1}{5^4} + \cdots$.

The general term $\dfrac{1}{n^{n-1}} \le \dfrac{1}{n^2}$ for $n \ge 3$. Thus, neglecting the first two terms, the given series is term by term less than or equal to the corresponding terms of the p series with $p = 2$. The given series is convergent.

28.7   Apply the ratio test to each of the following. If it fails, use some other method to determine convergency or divergency.

(a) $\dfrac{1}{2} + \dfrac{1}{2} + \dfrac{3}{8} + \dfrac{1}{4} + \cdots$ or $\dfrac{1}{2} + \dfrac{2}{2^2} + \dfrac{3}{2^3} + \dfrac{4}{2^4} + \cdots$.

For this series $s_n = \dfrac{n}{2^n}$, $s_{n+1} = \dfrac{n+1}{2^{n+1}}$, and $r_n = \dfrac{s_{n+1}}{s_n} = \dfrac{n+1}{2^{n+1}} \cdot \dfrac{2^n}{n} = \dfrac{n+1}{2n}$. Then $R = \lim\limits_{n \to \infty} r_n = \lim\limits_{n \to \infty} \dfrac{n+1}{2n} = \lim\limits_{n \to \infty} \dfrac{1 + 1/n}{2} = \dfrac{1}{2} < 1$ and the series is convergent.

(b) $3 + \dfrac{9}{2} + \dfrac{9}{2} + \dfrac{27}{8} + \dfrac{81}{40} + \cdots$ or $\dfrac{3}{1!} + \dfrac{3^2}{2!} + \dfrac{3^3}{3!} + \cdots$.

Here $s_n = \dfrac{3^n}{n!}$, $s_{n+1} = \dfrac{3^{n+1}}{(n+1)!}$, and $r_n = \dfrac{3^{n+1}}{(n+1)!} \cdot \dfrac{n!}{3^n} = \dfrac{3}{n+1}$. Then $R = \lim\limits_{n \to \infty} \dfrac{3}{n+1} = 0$ and the series is convergent.

(c) $\dfrac{1}{1\cdot 1}+\dfrac{1}{2\cdot 3}+\dfrac{1}{3\cdot 5}+\dfrac{1}{4\cdot 7}+\cdots$.

Here $s_n=\dfrac{1}{n(2n-1)}$, $s_{n+1}=\dfrac{1}{(n+1)(2n+1)}$, and $r_n=\dfrac{n(2n-1)}{(n+1)(2n+1)}$.

Then $R=\lim\limits_{n\to\infty}\dfrac{n(2n-1)}{(n+1)(2n+1)}=\lim\limits_{n\to\infty}\dfrac{2-1/n}{(1+1/n)(2+1/n)}=1$ and the test fails.

Since $\dfrac{1}{n(2n-1)}\le\dfrac{1}{n^2}$, the given series is term by term less than or equal to the convergent p series, with $p=2$. The given series is convergent.

(d) $\dfrac{2}{1^2+1}+\dfrac{2^3}{2^2+2}+\dfrac{2^5}{3^2+3}+\cdots$.

Here $s_n=\dfrac{2^{2n-1}}{n^2+n}$, $s_{n+1}=\dfrac{2^{2n+1}}{(n+1)^2+(n+1)}$, and $r_n=\dfrac{2^{2n+1}}{(n+1)(n+2)}\cdot\dfrac{n(n+1)}{2^{2n-1}}=\dfrac{4n}{n+2}$. Then

$R=\lim\limits_{n\to\infty}\dfrac{4}{1+2/n}=4$ and the series is divergent.

(e) $\frac{1}{5}+\frac{2}{25}+\frac{6}{125}+\frac{24}{625}+\cdots$.

In this series $s_n=\dfrac{n!}{5^n}$, $s_{n+1}=\dfrac{(n+1)!}{5^{n+1}}$, and $r_n=\dfrac{(n+1)!}{5^{n+1}}\cdot\dfrac{5^n}{n!}=\dfrac{n+1}{5}$. Now $\lim\limits_{n\to\infty}r_n$ does not exist.

However, since $s_n\to\infty$ as $n\to\infty$, the series is divergent.

28.8 Test the following alternating series for convergence:

(a) $1-\frac{1}{3}+\frac{1}{5}-\frac{1}{7}+\cdots$.

$s_n>s_{n+1}$, for all values of n, and $\lim\limits_{n\to\infty}s_n=\lim\limits_{n\to\infty}\dfrac{1}{2n-1}=0$. The series is convergent.

(b) $\dfrac{1}{2^3}-\dfrac{2}{3^3}+\dfrac{3}{4^3}-\dfrac{4}{5^3}+\cdots$.

$s_n>s_{n+1}$, for all values of n, and $\lim\limits_{n\to\infty}s_n=\lim\limits_{n\to\infty}\dfrac{n}{(n+1)^3}=0$. The series is convergent.

28.9 Investigate the following for absolute convergence, conditional convergence, or divergence:

(a) $1-\frac{1}{2}+\frac{1}{4}-\frac{1}{8}+\cdots$.

Here $|s_n|=\dfrac{1}{2^{n-1}}$, $|s_{n+1}|=\dfrac{1}{2^n}$, and $R=\lim\limits_{n\to\infty}\dfrac{|s_{n+1}|}{|s_n|}=\lim\limits_{n\to\infty}\dfrac{2^{n-1}}{2^n}=\dfrac{1}{2}<1$. The series is absolutely convergent.

(b) $1-\dfrac{4}{1!}+\dfrac{4^2}{2!}-\dfrac{4^3}{3!}+\cdots$.

Here $|s_n|=\dfrac{4^{n-1}}{(n-1)!}$, $|s_{n+1}|=\dfrac{4^n}{n!}$, and $R=\lim\limits_{n\to\infty}\dfrac{4}{n}=0$. The series is absolutely convergent.

(c) $\dfrac{1}{2-\sqrt{2}}-\dfrac{1}{3-\sqrt{3}}+\dfrac{1}{4-\sqrt{4}}-\dfrac{1}{5-\sqrt{5}}+\cdots$.

The ratio test fails here.

Since $\dfrac{1}{n+1-\sqrt{n+1}}>\dfrac{1}{n+2-\sqrt{n+2}}$ and $\lim\limits_{n\to\infty}\dfrac{1}{n+1-\sqrt{n+1}}=0$, the series is convergent.

Since $\dfrac{1}{n+1-\sqrt{n+1}}>\dfrac{1}{n+1}$ for all values of n, the series of absolute values is term by term greater than the harmonic series, and thus is divergent. The given series is conditionally convergent.

Supplementary Problems

28.10 Investigate each of the following series for convergence or divergence:

(a) $\dfrac{1}{3}+\dfrac{1}{6}+\dfrac{1}{11}+\cdots+\dfrac{1}{2^n+n}+\cdots$

(f) $1+\dfrac{1}{\sqrt[3]{2}}+\dfrac{1}{\sqrt[3]{3}}+\cdots+\dfrac{1}{\sqrt[3]{n}}+\cdots$

(b) $\dfrac{1}{2}+\dfrac{1}{4}+\dfrac{1}{6}+\cdots+\dfrac{1}{2n}+\cdots$

(g) $\dfrac{1}{2}+\dfrac{1}{2\cdot2^2}+\dfrac{1}{3\cdot2^3}+\cdots+\dfrac{1}{n\cdot2^n}+\cdots$

(c) $1+\dfrac{1}{3}+\dfrac{1}{5}+\cdots+\dfrac{1}{2n-1}+\cdots$

(h) $\dfrac{1}{2}+\dfrac{2}{3}+\dfrac{3}{4}+\cdots+\dfrac{n}{n+1}+\cdots$

(d) $\dfrac{2}{1!}+\dfrac{2^2}{2!}+\dfrac{2^3}{3!}+\cdots+\dfrac{2^n}{n!}+\cdots$

(i) $\dfrac{1}{1\cdot3}+\dfrac{1}{3\cdot5}+\dfrac{1}{5\cdot7}+\cdots+\dfrac{1}{(2n-1)(2n+1)}+\cdots$

(e) $2+\dfrac{1}{2}+\dfrac{8}{27}+\dfrac{1}{4}+\cdots+\dfrac{2^n}{n^3}+\cdots$

(j) $1+\dfrac{2^2+1}{2^3+1}+\dfrac{3^2+1}{3^3+1}+\cdots+\dfrac{n^2+1}{n^3+1}+\cdots$

Ans.   (a) Convergent   (c) Divergent   (e) Divergent   (g) Convergent   (i) Convergent

        (b) Divergent   (d) Convergent   (f) Divergent   (h) Divergent   (j) Divergent

28.11 Investigate the following alternating series for convergence or divergence:

(a) $\dfrac{1}{4}-\dfrac{1}{10}+\dfrac{1}{28}-\dfrac{1}{82}+\cdots$

(f) $\dfrac{2}{3}-\dfrac{3}{4}\cdot\dfrac{1}{2}+\dfrac{4}{5}\cdot\dfrac{1}{3}-\dfrac{5}{6}\cdot\dfrac{1}{4}+\cdots$

(b) $2-\dfrac{3}{2}+\dfrac{4}{3}-\dfrac{5}{4}+\cdots$

(g) $\dfrac{2}{2\cdot3}-\dfrac{2^2}{3\cdot4}+\dfrac{2^3}{4\cdot5}-\dfrac{2^4}{5\cdot6}+\cdots$

(c) $1-\dfrac{1}{2}+\dfrac{1}{3}-\dfrac{1}{4}+\cdots$

(d) $\dfrac{1}{2}-\dfrac{2}{3}+\dfrac{3}{4}-\dfrac{4}{5}+\cdots$

(h) $2-\dfrac{2^3}{3!}+\dfrac{2^5}{5!}-\dfrac{2^7}{7!}+\cdots$

(e) $\dfrac{1}{4}-\dfrac{3}{6}+\dfrac{5}{8}-\dfrac{7}{10}+\cdots$

Ans.   (a) Abs. Conv.   (c) Cond. Conv.   (e) Divergent   (g) Divergent

        (b) Divergent   (d) Divergent   (f) Cond. Conv.   (h) Abs. Conv.

Chapter 29

Power Series

INFINITE SERIES OF THE FORM

$$c_0 + c_1 x + c_2 x^2 + \cdots + c_{n-1} x^{n-1} + \cdots \tag{29.1}$$

and

$$c_0 + c_1(x - a) + c_2(x - a)^2 + \cdots + c_{n-1}(x - a)^{n-1} + \cdots, \tag{29.2}$$

where $a, c_0, c_1, c_2, \ldots$ are constants, are called *power series*. The first is called a power series in x and the second a power series in $(x - a)$.

The power series (29.1) converges for $x = 0$ and (29.2) converges for $x = a$. Both series may converge for other values of x but not necessarily for every finite value of x. Our problem is to find for a given power series all values of x for which the series converges. In finding this set of values, called the *interval of convergence* of the series, the generalized ratio test of Chapter 28 will be used.

EXAMPLE 1. Find the interval of convergence of the series

$$x + x^2/2 + x^3/3 + \cdots$$

Since

$$|s_n| = \left| \frac{x^n}{n} \right| \quad \text{and} \quad |s_{n+1}| = \left| \frac{x^{n+1}}{n + 1} \right|,$$

$$R = \lim_{n \to \infty} \frac{|s_{n+1}|}{|s_n|} = \lim_{n \to \infty} \left| \frac{x^{n+1}}{n + 1} \cdot \frac{n}{x^n} \right| = \lim_{n \to \infty} \left| \frac{n}{n + 1} x \right| = |x|.$$

Then, by the ratio test, the given series is convergent for all values of x such that $|x| < 1$, that is, for $-1 < x < 1$; the series is divergent for all values of x such that $|x| > 1$, that is, for $x < -1$ and $x > 1$; and the test fails for $x = \pm 1$.

But, when $x = 1$ the series is $1 + \frac{1}{2} + \frac{1}{3} + \frac{1}{4} + \cdots$ and is divergent, and when $x = -1$ the series is $-1 + \frac{1}{2} - \frac{1}{3} + \frac{1}{4} - \cdots$ and is convergent.

Thus, the series converges on the interval $-1 \le x < 1$. This interval may be represented graphically as in Fig. 29-1. The solid line represents the interval on which the series converges, the thin lines the intervals on which the series diverges. The solid circle represents the end point for which the series converges, the open circle represents the end point at which the series diverges. (See Problems 29.1–29.8.)

Fig. 29-1

Solved Problems

In Problems 29.1–29.6, find the interval of convergence *including* the end points.

29.1 $1 + \dfrac{x}{1!} + \dfrac{x^2}{2!} + \dfrac{x^3}{3!} + \cdots$.

For this series

$$|s_n| = \left| \frac{x^{n-1}}{(n - 1)!} \right|, \quad |s_{n+1}| = \left| \frac{x^n}{n!} \right|, \quad \text{and} \quad R = \lim_{n \to \infty} \frac{|s_{n+1}|}{|s_n|} = \lim_{n \to \infty} \left| \frac{x^n}{n!} \cdot \frac{(n - 1)!}{x^{n-1}} \right| = \lim_{n \to \infty} \left| \frac{x}{n} \right| = 0.$$

The series is *everywhere convergent*, that is, it is convergent for all finite values of x.

29.2 $1 + x + 2x^2 + 3x^3 + \cdots$.

Here $|s_n| = |(n-1)x^{n-1}|$, $|s_{n+1}| = |nx^n|$, and $R = \lim\limits_{n \to \infty} \left| \dfrac{nx^n}{(n-1)x^{n-1}} \right| = \lim\limits_{n \to \infty} \left| \dfrac{n}{n-1} x \right| = |x|$.

The series converges on the interval $-1 < x < 1$ and diverges on the intervals $x < -1$ and $x > 1$.

When $x = 1$, the series is $1 + 1 + 2 + 3 + \cdots$ and is divergent.
When $x = -1$, the series is $1 - 1 + 2 - 3 + \cdots$ and is divergent.

The interval of convergence $-1 < x < 1$ is indicated in Fig. 29-2.

Fig. 29-2

29.3 $1 + \dfrac{x}{2} + \dfrac{x^2}{4} + \dfrac{x^3}{8} + \cdots$.

Here $|s_n| = \left| \dfrac{x^{n-1}}{2^{n-1}} \right|$, $|s_{n+1}| = \left| \dfrac{x^n}{2^n} \right|$, and $R = \lim\limits_{n \to \infty} \left| \dfrac{x^n}{2^n} \cdot \dfrac{2^{n-1}}{x^{n-1}} \right| = \lim\limits_{n \to \infty} \left| \dfrac{x}{2} \right| = \dfrac{1}{2}|x|$.

The series converges for all values of x such that $\frac{1}{2}|x| < 1$, that is, for $-2 < x < 2$; and diverges for $x < -2$ and $x > 2$.

For $x = 2$, the series is $1 + 1 + 1 + 1 + \cdots$ and is divergent.
For $x = -2$, the series is $1 - 1 + 1 - 1 + \cdots$ and is divergent.

The interval of convergence $-2 < x < 2$ is indicated in Fig. 29-3.

Fig. 29-3

29.4 $\dfrac{1!}{x+1} - \dfrac{2!}{(x+1)^2} + \dfrac{3!}{(x+1)^3} + \cdots$.

For this series $|s_n| = \left| \dfrac{n!}{(x+1)^n} \right|$, $|s_{n+1}| = \left| \dfrac{(n+1)!}{(x+1)^{n+1}} \right|$, and $R = \lim\limits_{n \to \infty} \left| \dfrac{(n+1)!}{(x+1)^{n+1}} \cdot \dfrac{(x+1)^n}{n!} \right| = \lim\limits_{n \to \infty} \left| \dfrac{n+1}{x+1} \right|$ does not exist.

Thus, the series diverges for every value of x.

29.5 $\dfrac{x+3}{1 \cdot 4} + \dfrac{(x+3)^2}{2 \cdot 4^2} + \dfrac{(x+3)^3}{3 \cdot 4^3} + \cdots$.

For this series $|s_n| = \left| \dfrac{(x+3)^n}{n \cdot 4^n} \right|$, $|s_{n+1}| + \left| \dfrac{(x+3)^{n+1}}{(n+1) \cdot 4^{n+1}} \right|$, and

$$R = \lim\limits_{n \to \infty} \left| \dfrac{(x+3)^{n+1}}{(n+1)4^{n+1}} \cdot \dfrac{n4^n}{(x+3)^n} \right| = \lim\limits_{n \to \infty} \left| \dfrac{n}{n+1} \cdot \dfrac{x+3}{4} \right| = \dfrac{1}{4}|x+3|.$$

The series converges for all values of x such that $\frac{1}{4}|x+3| < 1$, that is, for $-4 < x+3 < 4$ or $-7 < x < 1$, and diverges for $x < -7$ and $x > 1$.

For $x = -7$, the series is $-1 + \frac{1}{2} - \frac{1}{3} + \cdots$ and is convergent.
For $x = 1$, the series is $1 + \frac{1}{2} + \frac{1}{3} + \cdots$ and is divergent.

The interval of convergence $-7 \le x < 1$ is indicated in Fig. 29-4.

Fig. 29-4

29.6 $\dfrac{1}{1 \cdot 2 \cdot 3} - \dfrac{(x+2)^2}{2 \cdot 3 \cdot 4} + \dfrac{(x+2)^4}{3 \cdot 4 \cdot 5} - \cdots .$

Here $|s_n| = \left| \dfrac{(x+2)^{2n-2}}{n(n+1)(n+2)} \right|$, $|s_{n+1}| = \left| \dfrac{(x+2)^{2n}}{(n+1)(n+2)(n+3)} \right|$, and

$$R = \lim_{n \to \infty} \left| \frac{(x+2)^{2n}}{(n+1)(n+2)(n+3)} \cdot \frac{n(n+1)(n+2)}{(x+2)^{2n-2}} \right| = \lim_{n \to \infty} \left| \frac{n}{n+3}(x+2)^2 \right| = (x+2)^2.$$

The series converges for all values of x such that $(x+2)^2 < 1$, that is, for $-3 < x < -1$, and diverges for $x < -3$ and $x > -1$.

For $x = -3$ and $x = -1$ the series is $\dfrac{1}{1 \cdot 2 \cdot 3} - \dfrac{1}{2 \cdot 3 \cdot 4} + \dfrac{1}{3 \cdot 4 \cdot 5} - \cdots$ and is convergent.

The interval of convergence $-3 \le x \le -1$ is indicated in Fig. 29-5.

Fig. 29-5

29.7 Expand $(1 + x)^{-1}$ as a power series in x and examine for convergence.

By division, $\dfrac{1}{x+1} = 1 - x + x^2 - x^3 + x^4 - \cdots$ Then $R = \lim\limits_{n \to \infty} \left| \dfrac{x^{n+1}}{x^n} \right| = |x|.$

The series converges for $-1 < x < 1$, and diverges for $x < -1$ and $x > 1$.

For $x = 1$, the series is $1 - 1 + 1 - 1 + \cdots$ and is divergent.

For $x = -1$, the series is $1 + 1 + 1 + 1 + \cdots$ and is divergent.

The interval of convergence $-1 < x < 1$ is indicated in Fig. 29-6.

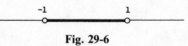

Fig. 29-6

Thus, the series $1 - x + x^2 - x^3 + x^4 - \cdots$ *represents* the function $f(x) = (1+x)^{-1}$ for all x such that $|x| < 1$. It does not represent the function for, say, $x = -4$. Note that $f(-4) = -\frac{1}{3}$, while for $x = -4$ the series is $1 + 4 + 16 + 64 + \cdots .$

29.8 Expand $(1 + x)^{1/2}$ in a power series in x and examine for convergence.

By the binomial theorem

$$(1+x)^{1/2} = 1 + \frac{1}{2}x + \frac{(\frac{1}{2})(-\frac{1}{2})}{1 \cdot 2}x^2 + \frac{(\frac{1}{2})(-\frac{1}{2})(-\frac{3}{2})}{1 \cdot 2 \cdot 3}x^3 + \frac{(\frac{1}{2})(-\frac{1}{2})(-\frac{3}{2})(-\frac{5}{2})}{1 \cdot 2 \cdot 3 \cdot 4}x^4 + \cdots$$

Except for $n = 1$,

$$|s_n| = \left| \frac{(\frac{1}{2})(-\frac{1}{2})(-\frac{3}{2}) \cdots (-2n+5)/2}{(n-1)!}x^{n-1} \right|,$$

$$|s_{n+1}| = \left| \frac{(\frac{1}{2})(-\frac{1}{2})(-\frac{3}{2}) \cdots (-2n+3)/2}{n!}x^n \right|, \quad \text{and} \quad R = \lim_{n \to \infty} \left| \frac{-2n+3}{2n}x \right| = |x|.$$

The series is convergent for $-1 < x < 1$, and divergent for $x < -1$ and $x > 1$. An investigation at the end points is beyond the scope of this book.

Supplementary Problems

In Problems 29.9–29.19 find the interval of convergence including the end points.

29.9 $1 + x^2 + x^4 + \cdots + x^{2n-2} + \cdots$

    Ans. $-1 < x < 1$

29.10 $\dfrac{x}{1 \cdot 2} + \dfrac{x^2}{2 \cdot 3} + \dfrac{x^3}{3 \cdot 4} + \cdots + \dfrac{x^n}{n(n+1)} + \cdots$

    Ans. $-1 \le x \le 1$

29.11 $\dfrac{x}{1 \cdot 3} + \dfrac{x^2}{2 \cdot 3^2} + \dfrac{x^3}{3 \cdot 3^3} + \cdots + \dfrac{x^n}{n \cdot 3^n} + \cdots$

    Ans. $-3 \le x < 3$

29.12 $\dfrac{x}{1^2 + 1} + \dfrac{x^2}{2^2 + 1} + \dfrac{x^3}{3^2 + 1} + \cdots + \dfrac{x^n}{n^2 + 1} + \cdots$

    Ans. $-1 \le x \le 1$

29.13 $\dfrac{x^2}{4} - \dfrac{x^4}{8} + \dfrac{x^6}{16} - \cdots + (-1)^{n-1}\dfrac{x^{2n}}{2^{n+1}} + \cdots$

    Ans. $-\sqrt{2} < x < \sqrt{2}$

29.14 $\dfrac{1 \cdot x}{2 \cdot 1} - \dfrac{1 \cdot 3 \cdot x^3}{2 \cdot 4 \cdot 3} + \dfrac{1 \cdot 3 \cdot 5 \cdot x^5}{2 \cdot 4 \cdot 6 \cdot 5} - \dfrac{1 \cdot 3 \cdot 5 \cdot 7 \cdot x^7}{2 \cdot 4 \cdot 6 \cdot 8 \cdot 7} + \cdots$

    Ans. $-1 \le x \le 1$

29.15 $\dfrac{1 \cdot 1}{3 \cdot 2}\,x + \dfrac{3 \cdot 2}{5 \cdot 3}\,x^2 + \dfrac{5 \cdot 3}{7 \cdot 4}\,x^3 + \dfrac{7 \cdot 4}{9 \cdot 5}\,x^4 + \cdots$

    Ans. $-1 < x < 1$

29.16 $(x - 2) + \tfrac{1}{2}(x - 2)^2 + \tfrac{1}{3}(x - 2)^3 + \tfrac{1}{4}(x - 2)^4 + \cdots$

    Ans. $1 \le x < 3$

29.17 $\dfrac{x + 1}{1 \cdot 2} - \dfrac{(x + 1)^2}{3 \cdot 2^2} + \dfrac{(x + 1)^3}{5 \cdot 2^3} - \dfrac{(x + 1)^4}{7 \cdot 2^4} + \cdots$

    Ans. $-3 < x \le 1$

29.18 $\dfrac{x - a}{b} + \dfrac{(x - a)^2}{b^2} + \dfrac{(x - a)^3}{b^3} + \dfrac{(x - a)^4}{b^4} + \cdots$

    Ans. $a - b < x < a + b$

29.19 $\dfrac{x - 2}{x} + \dfrac{1}{2}\left(\dfrac{x - 2}{x}\right)^2 + \dfrac{1}{3}\left(\dfrac{x - 2}{x}\right)^3 + \dfrac{1}{4}\left(\dfrac{x - 2}{x}\right)^4 + \cdots$

    Ans. $x \ge 1$

29.20 Show that the binomial series for $(1 + x)^k$, k any real numbers, converges for $|x| < 1$.

Given $\lim_{n \to \infty} (1 + 1/n)^n = e$, where e is the base of the natural system of logarithms, find the interval of convergence, exclusive of end points, of each of the following series:

29.21 $x + x^2/2^2 + x^3/3^3 + x^4/4^4 + \cdots + x^n/n^n + \cdots$

   Ans. All values of x.

29.22 $x - 2^2 x^2 + 3^3 x^3 - 4^4 x^4 + \cdots$

   Ans. $x = 0$

29.23 $x + \dfrac{2!x^2}{2^2} + \dfrac{3!x^3}{3^3} + \dfrac{4!x^4}{4^4} + \cdots$

   Ans. $-e < x < e$

Part III

TRIGONOMETRY

Chapter 30

Angles and Arc Length

TRIGONOMETRY, as the word implies, is concerned with the measurement of the parts of a triangle. Plane trigonometry, considered in the next several chapters, is restricted to triangles lying in planes. Spherical trigonometry deals with certain triangles which lie on spheres.

The science of trigonometry is based on certain ratios, called trigonometric functions, to be defined in the next chapter. The early applications of the trigonometric functions were to surveying, navigation, and engineering. These functions also play an important role in the study of all sorts of vibratory phenomena—sound, light, electricity, etc. As a consequence, a considerable portion of the subject matter is concerned properly with a study of the properties of and relations among the trigonometric functions.

THE PLANE ANGLE *XOP* is formed by the two intersecting half lines *OX* and *OP*. The point *O* is called the *vertex* and the half lines are called the sides of the angle. See Fig. 30-1.

Fig. 30-1

More often, a plane angle is to be thought of as generated by revolving (in a plane) a half line from the initial position *OX* to a terminal position *OP*. Then *O* is again the vertex, *OX* is called the *initial side*, and *OP* is called the *terminal* side of the angle.

An angle, so generated, is called *positive* if the direction of rotation (indicated by a curved arrow) is counterclockwise and *negative* if the direction of rotation is clockwise. The angle is positive in Figs. 30-2(*a*) and (*c*), and negative in Fig. 30-2(*b*).

(*a*)   (*b*)   (*c*)

Fig. 30-2

MEASURES OF ANGLES

A. A *degree* (°) is defined as the measure of the central angle subtended by an arc of a circle equal to $\frac{1}{360}$ of the circumference of the circle.

A *minute* (′) is $\frac{1}{60}$ of a degree; a *second* (″) is $\frac{1}{60}$ of a minute.

EXAMPLE 1

(*a*) $\frac{1}{4}(36°\,24') = 9°\,6'$ (*b*) $\frac{1}{2}(127°\,24') = \frac{1}{2}(126°\,84') = 63°\,42'$

(*c*) $\frac{1}{2}(81°\,15') = \frac{1}{2}(80°\,75') = 40°\,37.5'$ or $40°\,37'\,30''$

(*d*) $\frac{1}{4}(74°\,29'\,20'') = \frac{1}{4}(72°\,149'\,20'') = \frac{1}{4}(72°\,128'\,80'') = 18°\,37'\,20''$

We write $m\angle A$ or $m\measuredangle A$ to denote the "measure of angle A."

B. A *radian* (rad) is defined as the measure of the central angle subtended by an arc of a circle equal to the radius of the circle. See Fig. 30-3.

Fig. 30-3

The circumference of a circle $= 2\pi(\text{radius})$ and subtends an angle of $360°$. Then 2π radians $= 360°$, from which we obtain

$$1 \text{ radian} = \frac{180°}{\pi} = 57.296° = 57°\,17'\,45''$$

and

$$1 \text{ degree} = \frac{\pi}{180}\text{ radian} = 0.017453 \text{ rad,}\qquad \text{approx.}$$

where $\pi = 3.14159$.

EXAMPLE 2 (*a*) $\dfrac{7}{12}\pi \text{ rad} = \dfrac{7\pi}{12}\cdot\dfrac{180°}{\pi} = 105°$, (*b*) $50° = 50\cdot\dfrac{\pi}{180}\text{ rad} = \dfrac{5\pi}{18}\text{ rad.}$ (See Problems 30.1–30.3.)

ARC LENGTH. On a circle of radius r, a central angle of θ radians intercepts an arc of length $s = r\theta$, that is, arc length = radius × the measure of the central angle in radians. See Fig. 30-4.

Fig. 30-4

(NOTE. s and r may be measured in any convenient unit of length but they must be expressed in the same unit.)

EXAMPLE 3

(*a*) On a circle of radius 30 in., the length of arc intercepted by a central angle of $\frac{1}{3}$ radian is

$$s = r\theta = 30(\tfrac{1}{3}) = 10 \text{ in.}$$

(*b*) On the same circle a central angle of $50°$ intercepts an arc length

$$s = r\theta = 30\left(\frac{5\pi}{18}\right) = \frac{25\pi}{3}\text{ in.}$$

(c)   On the same circle an arc length $1\frac{1}{2}$ ft subtends a central angle

$$m \measuredangle\, \theta = \frac{s}{r} = \frac{18}{30} = \frac{3}{5}\,\text{rad} \qquad \text{when } s \text{ and } r \text{ are expressed in inches}$$

or $\qquad\qquad\qquad m \measuredangle\, \theta = \frac{s}{r} = \frac{\frac{3}{2}}{\frac{5}{2}} = \frac{3}{5}\,\text{rad} \qquad \text{when } s \text{ and } r \text{ are expressed in feet}$

(See Problems 30.4–30.11.)

Solved Problems

30.1   Express each angle in radian measure: (a)   30°,   (b)   135°,   (c)   25° 30′,   (d)   42° 24′ 35″.

Since $1° = \dfrac{\pi}{180}$ radian $= 0.017453$ rad,

(a) $30° = 30 \times \dfrac{\pi}{180}\,\text{rad} = \dfrac{\pi}{6}\,\text{rad} \qquad \text{or} \qquad 0.5236\,\text{rad}$

(b) $135° = 135 \times \dfrac{\pi}{180}\,\text{rad} = \dfrac{3\pi}{4}\,\text{rad} \qquad \text{or} \qquad 2.3562\,\text{rad}$

(c) $25° 30′ = 25.5° = 25.5 \times \dfrac{\pi}{180}\,\text{rad} = 0.4451\,\text{rad}$

(d) $42° 24′ 35″ = 42° + \left(\dfrac{24 \times 60 + 35}{3600}\right)° = 42.41° = 42.41 + \dfrac{\pi}{180}\,\text{rad} = 0.7402\,\text{rad}$

30.2   Express each angle in degree measure: (a) $\dfrac{\pi}{3}\,\text{rad}$,   (b) $\dfrac{5\pi}{9}\,\text{rad}$,   (c) $\dfrac{2}{5}\,\text{rad}$,   (d) $\dfrac{4}{3}\,\text{rad}$.

Since $1\,\text{rad} = \dfrac{180°}{\pi} = 57° 17′ 45″$,

(a) $\dfrac{\pi}{3}\,\text{rad} = \dfrac{\pi}{3} \times \dfrac{180°}{\pi} = 60°$ \qquad\qquad (b) $\dfrac{5\pi}{9}\,\text{rad} = \dfrac{5\pi}{9} \times \dfrac{180°}{\pi} = 100°$

(c) $\dfrac{2}{5}\,\text{rad} = \dfrac{2}{5} \times \dfrac{180°}{\pi} = \dfrac{72°}{\pi} \qquad \text{or} \qquad \dfrac{2}{5}(57° 17′ 45″) = 22° 55′ 6″$

(d) $\dfrac{4}{3}\,\text{rad} = \dfrac{4}{3} \times \dfrac{180°}{\pi} = \dfrac{240°}{\pi} \qquad \text{or} \qquad \dfrac{4}{3}(57° 17′ 45″) = 76° 23′ 40″$

30.3   A wheel is turning at the rate of 48 rpm (revolutions per minute or rev/min). Express this angular speed in (a)   rev/s   (b)   rad/min   (c)   rad/s.

(a) $48\,\text{rev/min} = \frac{48}{60}\,\text{rev/s} = \frac{4}{5}\,\text{rev/s}$

(b)   Since 1 rev $= 2\pi$ rad, $48\,\text{rev/min} = 48(2\pi)\,\text{rad/min} = 301.6\,\text{rad/min}$

(c) $48\,\text{rev/min} = \frac{4}{5}\,\text{rev/s} = \frac{4}{5}(2\pi)\,\text{rad/s} = 5.03\,\text{rad/s} \qquad \text{or}$

$\qquad 48\,\text{rev/min} = 96\pi\,\text{rad/min} = \dfrac{96\pi}{60}\,\text{rad/s} = 5.03\,\text{rad/s}$

30.4   The minute hand of a clock is 12 in long. How far does the tip of the hand move during 20 min?

During 20 min the hand moves through an angle $\theta = 120° = 2\pi/3$ rad and the tip of the hand moves over a distance $s = r\theta = 12(2\pi/3) = 8\pi$ in. $= 25.1$ in.

30.5 A central angle of a circle of radius 30 in. intercepts an arc of 6 in. Express the measure of the central angle θ in radians and in degrees.

$$\text{m} \measuredangle \, \theta = \frac{s}{r} = \frac{6}{30} = \frac{1}{5} \text{ rad} = 11° \, 27' \, 33''$$

30.6 A railroad curve is to be laid out on a circle. What radius should be used if the track is to change direction by 25° in a distance of 120 ft?

We are required to find the radius of a circle on which a central angle θ, with measure 25° or $5\pi/36$ rad, intercepts an arc of 120 ft. Then

$$r = \frac{s}{\theta} = \frac{120}{5\pi/36} = \frac{864}{\pi} \text{ ft} = 275 \text{ ft}$$

30.7 Assuming the earth to be a sphere of radius 3960 miles, find the distance of a point in latitude 36°N from the equator.

Since $36° = \dfrac{\pi}{5}$ radian, $s = r\theta = 3960\left(\dfrac{\pi}{5}\right) = 2488$ miles.

30.8 Two cities 270 miles apart lie on the same meridian. Find their difference in latitude.

$$\text{m} \measuredangle \, \theta = \frac{s}{r} = \frac{270}{3960} = \frac{3}{44} \text{ rad} \qquad \text{or} \qquad 3° \, 54.4'$$

30.9 A wheel 4 ft in diameter is rotating at 80 rpm. Find the distance (in ft) traveled by a point on the rim in 1 s, that is, the linear speed of the point (in ft/s).

$$80 \text{ rpm} = 80\left(\frac{2\pi}{60}\right) \text{ rad/s} = \frac{8\pi}{3} \text{ rad/s}$$

Then in 1 s the wheel turns through an angle θ measuring $8\pi/3$ rad and a point on the wheel will travel a distance $s = r\theta = 2(8\pi/3)$ ft $= 16.8$ ft. The linear velocity is 16.8 ft/s.

30.10 Find the diameter of a pulley which is driven at 360 rpm by a belt moving at 40 ft/s.

$$360 \text{ rev/min} = 360\left(\frac{2\pi}{60}\right) \text{ rad/s} = 12\pi \text{ rad/s}$$

Then in 1 s the pulley turns through an angle θ measuring 12π rad and a point on the rim travels a distance $s = 40$ ft

$$d = 2r = 2\left(\frac{s}{\theta}\right) = 2\left(\frac{40}{12\pi}\right) \text{ ft} = \frac{20}{3\pi} \text{ ft} = 2.12 \text{ ft}$$

30.11 A point on the rim of a turbine wheel of diameter 10 ft moves with a linear speed 45 ft/s. Find the rate at which the wheel turns (angular speed) in rad/s and in rev/s.

In 1 s a point on the rim travels a distance $s = 45$ ft. Then in 1 s the wheel turns through an angle $\theta = s/r = \frac{45}{5} = 9$ radians and its angular speed is 9 rad/s.

Since 1 rev $= 2\pi$ rad or 1 rad $= \dfrac{1}{2\pi}$ rev, 9 rad/s $= 9\left(\dfrac{1}{2\pi}\right)$ rev/s $= 1.43$ rev/s.

Supplementary Problems

30.12 Express in radian measure: (a) 25°, (b) 160°, (c) 75° 30′, (d) 112° 40′, (e) 12° 12′ 20″.

   Ans. (a) $5\pi/36$ rad or 0.4363 rad   (c) $151\pi/360$ rad or 1.3177 rad   (e) 0.2130 rad
            (b) $8\pi/9$ rad or 2.7925 rad   (d) $169\pi/270$ rad or 1.9664 rad

30.13 Express in degree measure: (a) $\pi/4$ rad, (b) $7\pi/10$ rad, (c) $5\pi/6$ rad, (d) $\frac{1}{4}$ rad, (e) $\frac{7}{5}$ rad.

   Ans. (a) 45°   (b) 126°   (c) 150°   (d) 14° 19′ 26″   (e) 80° 12′ 51″

30.14 On a circle of radius 24 in., find the length of arc subtended by a central angle (a) of $\frac{2}{3}$ rad, (b) of $3\pi/5$ rad, (c) of 75°, (d) of 130°.

   Ans. (a) 16 in.   (b) 14.4π or 45.2 in.   (c) 10π or 31.4 in.   (d) 52π/3 or 54.5 in.

30.15 A circle has a radius of 30 in. How many radians are there in an angle at the center subtended by an arc (a) of 30 in., (b) of 20 in., (c) of 50 in.?

   Ans. (a) 1 rad   (b) $\frac{2}{3}$ rad   (c) $\frac{5}{3}$ rad

30.16 Find the radius of the circle for which an arc 15 in. long subtends an angle (a) of 1 rad, (b) of $\frac{2}{3}$ rad, (c) of 3 rad, (d) of 20°, (e) of 50°.

   Ans. (a) 15 in.   (b) 22.5 in.   (c) 5 in.   (d) 43.0 in.   (e) 17.2 in.

30.17 The end of a 40-in. pendulum describes an arc of 5 in. Through what angle does the pendulum swing?

   Ans. $\frac{1}{8}$ rad or 7° 9′ 43″

30.18 A train is traveling at the rate of 12 mi/hr on a curve of radius 3000 ft. Through what angle has it turned in 1 min?

   Ans. 0.352 rad or 20° 10′

30.19 A reversed curve on a railroad track consists of two circular arcs. The central angle of one measures 20° with radius 2500 ft and the central angle of the other measures 25° with radius 3000 ft. Find the total length of the two arcs.

   Ans. 6250π/9 ft or 2182 ft

30.20 A flywheel of radius 10 in. is turning at the rate of 900 rpm. How fast does a point on the rim travel in ft/s?

   Ans. 78.5 ft/s

30.21 An automobile tire is 30 in. in diameter. How fast (rpm) does the wheel turn on the axle when the automobile maintains a speed of 45 mph?

   Ans. 504 rpm

30.22 In grinding certain tools the linear velocity of the grinding surface should not exceed 6000 ft/s. Find the maximum number of revolutions per second (a) of a 12-in. (diameter) emery wheel, (b) of an 8-in. wheel.

   Ans. (a) 6000/π rev/s or 1910 rev/s   (b) 2865 rev/s

30.23 If an automobile wheel 32 in. in diameter rotates at 800 rpm, what is the speed of the car in mph?

   Ans. 76.2 mph

Chapter 31

Trigonometric Functions of a General Angle

ANGLES IN STANDARD POSITION. With respect to a rectangular coordinate system, an angle is said to be *in standard position* when its vertex is at the origin and its initial side coincides with the positive *x* axis.

An angle is said to be a *first quadrant angle* or to be *in the first quadrant* if, when in standard position, its terminal side falls in that quadrant. Similar definitions hold for the other quadrants. For example, the angles 30°, 59°, and −330° are first quadrant angles; 119° is a second quadrant angle; −119° is a third quadrant angle; −10° and 710° are fourth quadrant angles. See Figs. 31-1 and 31-2.

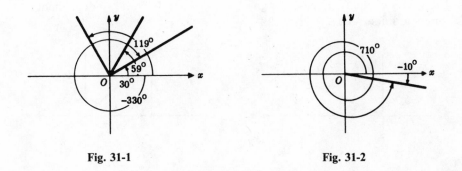

Fig. 31-1	Fig. 31-2

Two angles which, when placed in standard position, have coincident terminal sides are called *coterminal angles*. For example, 30° and −330°, −10° and 710° are pairs of coterminal angles. There are an unlimited number of angles coterminal with a given angle. (See Problem 31.1.)

The angles 0°, 90°, 180°, 270°, and all angles coterminal with them are called *quadrantal angles*.

TRIGONOMETRIC FUNCTIONS OF A GENERAL ANGLE. Let θ be an angle (not quadrantal) in standard position and let $P(x, y)$ be any point, distinct from the origin, on the terminal side of the angle. The six trigonometric functions of θ are defined, in terms of the abscissa, ordinate and distance of P, as follows:

$$\text{sine } \theta = \sin \theta = \frac{\text{ordinate}}{\text{distance}} = \frac{y}{r} \qquad \text{cotangent } \theta = \cot \theta = \frac{\text{abscissa}}{\text{ordinate}} = \frac{x}{y}$$

$$\text{cosine } \theta = \cos \theta = \frac{\text{abscissa}}{\text{distance}} = \frac{x}{r} \qquad \text{secant } \theta = \sec \theta = \frac{\text{distance}}{\text{abscissa}} = \frac{r}{x}$$

$$\text{tangent } \theta = \tan \theta = \frac{\text{ordinate}}{\text{abscissa}} = \frac{y}{x} \qquad \text{cosecant } \theta = \csc \theta = \frac{\text{distance}}{\text{ordinate}} = \frac{r}{y}$$

Note that $r = \sqrt{x^2 + y^2}$ (see Fig. 31-3).

As an immediate consequence of these definitions, we have the so-called *reciprocal relations*:

$$\sin \theta = \frac{1}{\csc \theta} \qquad \tan \theta = \frac{1}{\cot \theta} \qquad \sec \theta = \frac{1}{\cos \theta}$$

$$\cos \theta = \frac{1}{\sec \theta} \qquad \cot \theta = \frac{1}{\tan \theta} \qquad \csc \theta = \frac{1}{\sin \phi}$$

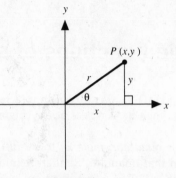

Fig. 31-3

It is evident from Figs. 31-4(a)–(d) that the values of the trigonometric functions of θ change as θ changes. The values of the functions of a given angle θ are, however, independent of the choice of the point P on its terminal side.

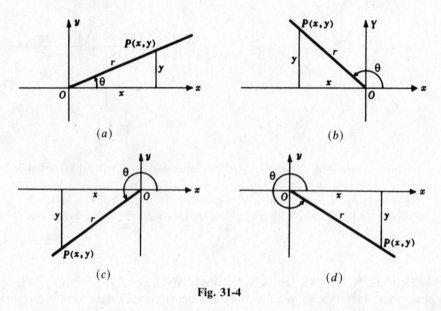

Fig. 31-4

ALGEBRAIC SIGNS OF THE FUNCTIONS. Since r is always positive, the signs of the functions in the various quadrants depend upon the signs of x and y. To determine these signs one may visualize the angle in standard position or use some device as shown in Fig. 31-5 in which only the functions having positive signs are listed.

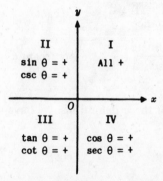

Fig. 31-5

When an angle is given, its trigonometric functions are uniquely determined. When, however, the value of one function of an angle is given, the angle is not uniquely determined. For example, if $\sin \theta = \frac{1}{2}$, then $\theta = 30°, 150°, 390°, 510° \ldots$. In general, two possible positions of the terminal side are found—for example, the terminal sides of 30° and 150° in Fig. 31-4(a), (b). The exceptions to this rule occur when the angle is quadrantal. (See Problems 31.2–31.10.)

TRIGONOMETRIC FUNCTIONS OF QUADRANTAL ANGLES.   For a quadrantal angle, the terminal side coincides with one of the axes. A point P, distinct from the origin, on the terminal side has either $x = 0$, $y \neq 0$, or $x \neq 0$, $y = 0$. In either case, two of the six functions will not be defined. For example, the terminal side of the angle 0° coincides with the positive x axis and the ordinate of P is 0. Since the ordinate occurs in the denominator of the ratio defining the cotangent and cosecant, these functions are not defined. Certain authors indicate this by writing $\cot 0° = \infty$ and others write $\cot 0° = \pm\infty$. The trigonometric functions of the quadrantal angles are given in Table 31.1.

Table 31.1

angle θ	$\sin \theta$	$\cos \theta$	$\tan \theta$	$\cot \theta$	$\sec \theta$	$\csc \theta$
0°	0	1	0		1	undefined
90°	1	0	undefined	0	undefined	1
180°	0	−1	0		−1	undefined
270°	−1	0	undefined	0	undefined	−1

Solved Problems

31.1   (a)   Construct the following angles in standard position and determine those which are coterminal: 125°, 210°, −150°, 385°, 930°, −370°, −955°, −870°.

      (b)   Give five other angles coterminal with 125°.

      (a)   The angles 125° and −955° = 125° − 3 · 360° are coterminal. The angles 210°, −150° = 210° − 360°, 930° = 210° + 2 · 360°, and −870° = 210° − 3 · 360° are coterminal. See Fig. 31-6.

Fig. 31-6

(b) $485° = 125° + 360°$, $1205° = 125° + 3 \cdot 360°$, $1925° = 125° + 5 \cdot 360°$, $-235° = 125° - 360°$, $-1315° = 125° - 4 \cdot 360°$ are coterminal with 125°.

31.2   Determine the values of the trigonometric functions of angle θ (smallest positive angle in standard position) if P is a point on the terminal side of θ and the coordinates of P are   (a) $P(3, 4)$, (b) $P(-3, 4)$,   (c) $P(-1, -3)$. See Fig. 31-7.

(a)   (b)   (c)

Fig. 31-7

(a) $r = \sqrt{3^2 + 4^2} = 5$ (b) $r = \sqrt{(-3)^2 + 4^2} = 5$ (c) $r = \sqrt{(-1)^2 + (-3)^2} = \sqrt{10}$

$\sin \theta = y/r = \frac{4}{5}$	$\sin \theta = \frac{4}{5}$	$\sin \theta = -3/\sqrt{10} = -3\sqrt{10}/10$
$\cos \theta = x/r = \frac{3}{5}$	$\cos \theta = -\frac{3}{5}$	$\cos \theta = -1/\sqrt{10} = -\sqrt{10}/10$
$\tan \theta = y/x = \frac{4}{3}$	$\tan \theta = \frac{4}{-3} = -\frac{4}{3}$	$\tan \theta = -\frac{3}{-1} = 3$
$\cot \theta = x/y = \frac{3}{4}$	$\cot \theta = -\frac{3}{4}$	$\cot \theta = -\frac{1}{3} = \frac{1}{3}$
$\sec \theta = r/x = \frac{5}{3}$	$\sec \theta = \frac{5}{-3} = -\frac{5}{3}$	$\sec \theta = \sqrt{10}/(-1) = -\sqrt{10}$
$\csc \theta = r/y = \frac{5}{4}$	$\csc \theta = \frac{5}{4}$	$\csc \theta = \sqrt{10}(-3) = -\sqrt{10}/3$

Note the reciprocal relationships. For example, in (b), $\sin \theta = 1/\csc \theta = \frac{4}{5}$, $\cos \theta = 1/\sec \theta = -\frac{3}{5}$, $\tan \theta = 1/\cot \theta = -\frac{4}{3}$, etc.

31.3   In what quadrant will θ terminate, if
   (a) $\sin \theta$ and $\cos \theta$ are both negative?   (c) $\sin \theta$ is positive and secant θ is negative?
   (b) $\sin \theta$ and $\tan \theta$ are both positive?   (d) $\sec \theta$ is negative and $\tan \theta$ is negative?

   (a)   Since $\sin \theta = y/r$ and $\cos \theta = x/r$, both x and y are negative. (Recall that r is always positive.) Thus, θ is a third quadrant angle.

   (b)   Since $\sin \theta$ is positive, y is positive; since $\tan \theta = y/x$ is positive, x is also positive. Thus, θ is a first quadrant angle.

   (c)   Since $\sin \theta$ is positive, y is positive; since $\sec \theta$ is negative, x is negative. Thus, θ is a second quadrant angle.

   (d)   Since $\sec \theta$ is negative, x is negative; since $\tan \theta$ is negative, y is then positive. Thus, θ is a second quadrant angle.

31.4   In what quadrants may θ terminate, if (a) $\sin \theta$ is positive?   (b) $\cos \theta$ is negative? (c) $\tan \theta$ is negative?   (d) $\sec \theta$ is positive?

   (a)   since $\sin \theta$ is positive, y is positive. Then x may be positive or negative and θ is a first or second quadrant angle.

   (b)   Since $\cos \theta$ is negative, x is negative. Then y may be positive or negative and θ is a second or third quadrant angle.

   (c)   Since $\tan \theta$ is negative, either y is positive and x is negative or y is negative and x is positive. Thus, θ may be a second or fourth quadrant angle.

   (d)   Since $\sec \theta$ is positive, x is positive. Thus, θ may be a first or fourth quadrant angle.

31.5   Find the values of $\cos \theta$ and $\tan \theta$, given $\sin \theta = \frac{8}{17}$ and θ in quadrant I.

Let P be a point on the terminal line of θ. Since $\sin \theta = y/r = \frac{8}{17}$, we take $y = 8$ and $r = 17$. Since θ is in quadrant I, x is positive; thus $x = \sqrt{r^2 - y^2} = \sqrt{(17)^2 - (8)^2} = 15$.

To draw Fig. 31-8, locate the point $P(15, 8)$, join it to the origin, and indicate the angle θ. Then $\cos \theta = x/r = \frac{15}{17}$ and $\tan \theta = y/x = \frac{8}{15}$.

The choice of $y = 8$, $r = 17$ is one of convenience. Note that $\frac{8}{17} = \frac{16}{34}$ and we might have taken $y = 16$, $r = 34$. Then $x = 30$, $\cos \theta = \frac{30}{34} = \frac{15}{17}$ and $\tan \theta = \frac{16}{30} = \frac{8}{15}$.

Fig. 31-8

31.6   Find the values of $\sin \theta$ and $\tan \theta$, given $\cos \theta = \frac{5}{6}$.

Since $\cos \theta$ is positive, θ is in quadrant I or IV. Since $\cos \theta = x/r = \frac{5}{6}$, we take $x = 5$, $r = 6$; $y = \pm\sqrt{(6)^2 - (5)^2} = \pm\sqrt{11}$.

(*a*)   For θ in quadrant I [Fig. 31-9(*a*)] we have $x = 5$, $y = \sqrt{11}$, $r = 6$; then $\sin \theta = y/r = \sqrt{11}/6$ and $\tan \theta = y/x = \sqrt{11}/5$.

(*b*)   For θ in quadrant IV [Fig. 31-9(*b*)] we have $x = 5$, $y = -\sqrt{11}$, $r = 6$; then $\sin \theta = y/r = -\sqrt{11}/6$ and $\tan \theta = y/x = -\sqrt{11}/5$.

| (*a*) | (*b*) |

Fig. 31-9

31.7   Find the values of $\sin \theta$ and $\cos \theta$, given $\tan \theta = -\frac{3}{4}$.

Since $\tan \theta = y/x$ is negative, θ is in quadrant II (take $x = -4$, $y = 3$) or in quadrant IV (take $x = 4$, $y = -3$). In either case $r = \sqrt{16 + 9} = 5$.

(*a*)   For θ in quadrant II [Fig. 31-10(*a*)], $\sin \theta = y/r = \frac{3}{5}$ and $\cos \theta = x/r = -\frac{4}{5}$.

(*b*)   For θ in quadrant IV [Fig. 31-10(*b*)], $\sin \theta = y/r = -\frac{3}{5}$ and $\cos \theta = x/r = \frac{4}{5}$.

| (*a*) | (*b*) |

Fig. 31-10

31.8   Find the values of the remaining functions of θ, given $\sin \theta = \sqrt{3}/2$ and $\cos \theta = -\frac{1}{2}$.

Since $\sin \theta = y/r$ is positive, y is positive. Since $\cos \theta = x/r$ is negative, x is negative. Thus, θ is in quadrant II.

Taking $x = -1$, $y = \sqrt{3}$, $r = \sqrt{(-1)^2 + (\sqrt{3})^2} = 2$ (Fig. 31-11), we have

$$\tan \theta = y/x = \sqrt{3}/-1 = -\sqrt{3} \qquad \cot \theta = 1/\tan \theta = -1/\sqrt{3} = -\sqrt{3}/3$$

$$\sec \theta = 1/\cos \theta = -2 \qquad \csc \theta = 1/\sin \theta = 2/\sqrt{3} = 2\sqrt{3}/3.$$

Fig. 31-11

31.9   Determine the values of $\cos \theta$ and $\tan \theta$ if $\sin \theta = m/n$, a negative fraction.

Since $\sin \theta$ is negative, θ is in quadrant III or IV.

(a)   In quadrant III: Take $y = m$, $r = n$, $x = -\sqrt{n^2 - m^2}$; then $\cos \theta = x/r = -\sqrt{n^2 - m^2}/n$ and $\tan \theta = y/x = -m/\sqrt{n^2 - m^2}$.

(b)   In quadrant IV: Take $y = m$, $r = n$, $x = +\sqrt{n^2 - m^2}$; then $\cos \theta = x/r = \sqrt{n^2 - m^2}/n$ and $\tan \theta = y/x = m/\sqrt{n^2 - m^2}$.

31.10   Evaluate:

(a) $\sin 0° + 2 \cos 0° + 3 \sin 90° + 4 \cos 90° + 5 \sec 0° + 6 \csc 90°$

(b) $\sin 180° + 2 \cos 180° + 3 \sin 270° + 4 \cos 270° - 5 \sec 180° - 6 \csc 270°$

(a) $0 + 2(1) + 3(1) + 4(0) + 5(1) + 6(1) = 16$

(b) $0 + 2(-1) + 3(-1) + 4(0) - 5(-1) - 6(-1) = 6$

Supplementary Problems

31.11   State the quadrant in which each angle terminates and the signs of the sine, cosine, and tangent of each angle.

(a)   125°   (b)   75°   (c)   320°   (d)   212°   (e)   460°   (f)   750°   (g)   −250°   (h)   −1000°

Ans.   (a)   II; +, −, −   (b)   I; +, +, +   (c)   IV; −, +, −   (d)   III; −, −, +   (e)   II
(f)   I   (g)   II   (h)   I

31.12   In what quadrant will θ terminate if

(a) $\sin \theta$ and $\cos \theta$ are both positive?   (e) $\tan \theta$ is positive and $\sec \theta$ is negative?

(b) $\cos \theta$ and $\tan \theta$ are both positive?   (f) $\tan \theta$ is negative and $\sec \theta$ is positive?

(c) $\sin \theta$ and $\sec \theta$ are both negative?   (g) $\sin \theta$ is positive and $\cos \theta$ is negative?

(d) $\cos \theta$ and $\cot \theta$ are both negative?   (h) $\sec \theta$ is positive and $\csc \theta$ is negative?

Ans.   (a)   I   (b)   I   (c)   III   (d)   II   (e)   III   (f)   IV   (g)   II   (h)   IV

31.13 Denote by θ the smallest positive angle whose terminal side passes through the given point and find the trigonometric functions of θ:

(a) $P(-5, 12)$ (b) $P(7, -24)$ (c) $P(2, 3)$ (d) $P(-3, -5)$

> *Ans.* (a) $\frac{12}{13}, -\frac{5}{13}, -\frac{12}{5}, -\frac{5}{12}, -\frac{13}{5}, \frac{13}{12}$
>
> (b) $-\frac{24}{25}, \frac{7}{25}, -\frac{24}{7}, -\frac{7}{24}, \frac{25}{7}, -\frac{25}{24}$
>
> (c) $3/\sqrt{13}, 2/\sqrt{13}, \frac{3}{2}, \frac{2}{3}, \sqrt{13}/2, \sqrt{13}/3$
>
> (d) $-5/\sqrt{34}, -3/\sqrt{34}, \frac{5}{3}, \frac{3}{5}, -\sqrt{34}/3, -\sqrt{34}/5$

31.14 Find the values of the trigonometric functions of θ, given

(a) $\sin \theta = \frac{7}{25}$ (d) $\cot \theta = \frac{24}{7}$ (g) $\tan \theta = \frac{3}{5}$ (j) $\csc \theta = -2/\sqrt{3}$

(b) $\cos \theta = -\frac{4}{5}$ (e) $\sin \theta = -\frac{2}{3}$ (h) $\cot \theta = \sqrt{6}/2$

(c) $\tan \theta = -\frac{5}{12}$ (f) $\cos \theta = \frac{5}{6}$ (i) $\sec \theta = -\sqrt{5}$

> *Ans.* (a) I: $\frac{7}{25}, \frac{24}{25}, \frac{7}{24}, \frac{24}{7}, \frac{25}{24}, \frac{25}{7}$; II: $\frac{7}{25}, -\frac{24}{25}, -\frac{7}{24}, -\frac{24}{7}, -\frac{25}{24}, \frac{25}{7}$
>
> (b) II: $\frac{3}{5}, -\frac{4}{5}, -\frac{3}{4}, -\frac{4}{3}, -\frac{5}{4}, \frac{5}{3}$; III: $-\frac{3}{5}, -\frac{4}{5}, \frac{3}{4}, \frac{4}{3}, -\frac{5}{4}, -\frac{5}{3}$
>
> (c) II: $\frac{5}{13}, -\frac{12}{13}, -\frac{5}{12}, -\frac{12}{5}, -\frac{13}{12}, \frac{13}{5}$; IV: $-\frac{5}{13}, \frac{12}{13}, -\frac{5}{12}, -\frac{12}{5}, \frac{13}{12}, -\frac{13}{5}$
>
> (d) I: $\frac{7}{25}, \frac{24}{25}, \frac{7}{24}, \frac{24}{7}, \frac{25}{24}, \frac{25}{7}$; III: $-\frac{7}{25}, -\frac{24}{25}, \frac{7}{24}, \frac{24}{7}, -\frac{25}{24}, -\frac{25}{7}$
>
> (e) III: $-\frac{2}{3}, -\sqrt{5}/3, 2/\sqrt{5}, \sqrt{5}/2, -3/\sqrt{5}, -\frac{3}{2}$; IV: $-\frac{2}{3}, \sqrt{5}/3, -2/\sqrt{5}, -\sqrt{5}/2, 3/\sqrt{5}, -\frac{3}{2}$
>
> (f) I: $\sqrt{11}/6, \frac{5}{6}, \sqrt{11}/5, 5/\sqrt{11}, \frac{6}{5}, 6/\sqrt{11}$; IV: $-\sqrt{11}/6, \frac{5}{6}, -\sqrt{11}/5, -5/\sqrt{11}, \frac{6}{5}, -6/\sqrt{11}$
>
> (g) I: $3/\sqrt{34}, 5/\sqrt{34}, \frac{3}{5}, \frac{5}{3}, \sqrt{34}/5, \sqrt{34}/3$; III: $-3/\sqrt{34}, -5/\sqrt{34}, \frac{3}{5}, \frac{5}{3}, -\sqrt{34}/5, -\sqrt{34}/3$
>
> (h) I: $2/\sqrt{10}, \sqrt{3}/\sqrt{5}, 2/\sqrt{6}, \sqrt{6}/2, \sqrt{5}/\sqrt{3}, \sqrt{10}/2$; III: $-2/\sqrt{10}, -\sqrt{3}/\sqrt{5}, 2/\sqrt{6}, \sqrt{6}/2, -\sqrt{5}/\sqrt{3}, -\sqrt{10}/2$
>
> (i) II: $2/\sqrt{5}, -1/\sqrt{5}, -2, -\frac{1}{2}, -\sqrt{5}, \sqrt{5}/2$; III: $-2/\sqrt{5}, -1/\sqrt{5}, 2, \frac{1}{2}, -\sqrt{5}, -\sqrt{5}/2$
>
> (j) III: $-\sqrt{3}/2, -\frac{1}{2}, \sqrt{3}, 1/\sqrt{3}, -2, -2/\sqrt{3}$; IV: $-\sqrt{3}/2, \frac{1}{2}, -\sqrt{3}, -1/\sqrt{3}, 2, -2/\sqrt{3}$

31.15 Evaluate each of the following:

(a) $\tan 180° - 2 \cos 180° + 3 \csc 270° + \sin 90° = 0$.

(b) $\sin 0° + 3 \cot 90° + 5 \sec 180° - 4 \cos 270° = -5$.

Chapter 32

Trigonometric Functions of an Acute Angle

TRIGONOMETRIC FUNCTIONS OF AN ACUTE ANGLE. In dealing with any right triangle, it will be convenient (see Fig. 32-1) to denote the vertices as A, B, C such that C is the vertex of the right triangle; to denote the angles of the triangle as A, B, C, such that $m \angle C = 90°$; and the sides opposite the angles as a, b, c, respectively. With respect to angle A, a will be called the *opposite side* and b will be called the *adjacent side*; with respect to angle B, a will be called the *adjacent side* and b the *opposite side*. Side c will always be called the *hypotenuse*.

Fig. 32-1   **Fig. 32-2**

If now the right triangle is placed in a coordinate system (Fig. 32-2) so that angle A is in standard position, point B on the terminal side of angle A has coordinates (b, a) and distance $c = \sqrt{a^2 + b^2}$. Then the trigonometric functions of angle A may be defined in terms of the sides of the right triangle, as follows:

$$\sin A = \frac{a}{c} = \frac{\text{length of opposite side}}{\text{length of hypotenuse}} \qquad \cot A = \frac{b}{a} = \frac{\text{length of adjacent side}}{\text{length of opposite side}}$$

$$\cos A = \frac{b}{c} = \frac{\text{length of adjacent side}}{\text{length of hypotenuse}} \qquad \sec A = \frac{c}{b} = \frac{\text{length of hypotenuse}}{\text{length of adjacent side}}$$

$$\tan A = \frac{a}{b} = \frac{\text{length of opposite side}}{\text{length of adjacent side}} \qquad \csc A = \frac{c}{a} = \frac{\text{length of hypotenuse}}{\text{length of opposite side}}$$

TRIGONOMETRIC FUNCTIONS OF COMPLEMENTARY ANGLES. The acute angles A and B of the right triangle ABC are complementary, that is, $m \angle A + m \angle B = 90°$. From Fig. 32-1, we have

$$\sin B = \frac{b}{c} = \cos A \qquad \cot B = \frac{a}{b} = \tan A$$

$$\cos B = \frac{a}{c} = \sin A \qquad \sec B = \frac{c}{a} = \csc A$$

$$\tan B = \frac{b}{a} = \cot A \qquad \csc B = \frac{c}{b} = \sec A$$

These relations associate the functions in pairs—sine and cosine, tangent and cotangent, secant and cosecant—each function of a pair being called the *cofunction* of the other. Thus, any function of an acute angle is equal to the corresponding cofunction of the complementary angle.

TRIGONOMETRIC FUNCTIONS OF 30°, 45°, 60°.   The results in Table 32.1 are obtained in Problems 32.8–32.9.

Table 32.1

Angle θ	$\sin\theta$	$\cos\theta$	$\tan\theta$	$\cot\theta$	$\sec\theta$	$\csc\theta$
30°	$\frac{1}{2}$	$\frac{1}{2}\sqrt{3}$	$\frac{1}{3}\sqrt{3}$	$\sqrt{3}$	$\frac{2}{3}\sqrt{3}$	2
45°	$\frac{1}{2}\sqrt{2}$	$\frac{1}{2}\sqrt{2}$	1	1	$\sqrt{2}$	$\sqrt{2}$
60°	$\frac{1}{2}\sqrt{3}$	$\frac{1}{2}$	$\sqrt{3}$	$\frac{1}{3}\sqrt{3}$	2	$\frac{2}{3}\sqrt{3}$

Problems 32.10–32.16 illustrate a number of simple applications of the trigonometric functions. For this purpose Table 32.2 will be used.

Table 32.2

Angle θ	$\sin\theta$	$\cos\theta$	$\tan\theta$	$\cot\theta$	$\sec\theta$	$\csc\theta$
15°	0.26	0.97	0.27	3.7	1.0	3.9
20°	0.34	0.94	0.36	2.7	1.1	2.9
30°	0.50	0.87	0.58	1.7	1.2	2.0
40°	0.64	0.77	0.84	1.2	1.3	1.6
45°	0.71	0.71	1.0	1.0	1.4	1.4
50°	0.77	0.64	1.2	0.84	1.6	1.3
60°	0.87	0.50	1.7	0.58	2.0	1.2
70°	0.94	0.34	2.7	0.36	2.9	1.1
75°	0.97	0.26	3.7	0.27	3.9	1.0

Solved Problems

[NOTATION: We will write AB (or c) to denote the length of AB, and $\overline{AB}$ to denote "the segment AB." $\overleftrightarrow{AB}$ denotes "the line AB."]

32.1   Find the values of the trigonometric functions of the acute angles of the right triangle ABC given $b = 24$ and $c = 25$.

Since $a^2 = c^2 - b^2 = (25)^2 - (24)^2 = 49$, $a = 7$. See Fig. 32-3. Then

$$\sin A = \frac{\text{opposite side}}{\text{hypotenuse}} = \frac{7}{25} \qquad \cot A = \frac{\text{adjacent side}}{\text{opposite side}} = \frac{24}{7}$$

$$\cos A = \frac{\text{adjacent side}}{\text{hypotenuse}} = \frac{24}{25} \qquad \sec A = \frac{\text{hypotenuse}}{\text{adjacent side}} = \frac{25}{24}$$

$$\tan A = \frac{\text{opposite side}}{\text{adjacent side}} = \frac{7}{24} \qquad \csc A = \frac{\text{hypotenuse}}{\text{opposite side}} = \frac{25}{7}$$

Fig. 32-3

and $\sin B = \frac{24}{25}$ $\cot B = \frac{7}{24}$

$$\cos B = \frac{7}{25} \qquad \sec B = \frac{25}{7}$$

$$\tan B = \frac{24}{7} \qquad \csc B = \frac{25}{24}$$

32.2   Find the values of the trigonometric functions of the acute angles of the right triangle ABC, given $a = 2$, $c = 2\sqrt{5}$.

Since $b^2 = c^2 - a^2 = (2\sqrt{5})^2 - (2)^2 = 20 - 4 = 16$, $b = 4$. See Fig. 32-4. Then

$$\sin A = \frac{2}{2\sqrt{5}} = \frac{\sqrt{5}}{5} = \cos B \qquad\qquad \cot A = \frac{4}{2} = 2 = \tan B$$

$$\cos A = \frac{4}{2\sqrt{5}} = \frac{2\sqrt{5}}{5} = \sin B \qquad\qquad \sec A = \frac{2\sqrt{5}}{4} = \frac{\sqrt{5}}{2} = \csc B$$

$$\tan A = \frac{2}{4} = \frac{1}{2} = \cot B \qquad\qquad \csc A = \frac{2\sqrt{5}}{2} = \sqrt{5} = \sec B$$

Fig. 32-4

32.3   Find the values of the trigonometric functions of the acute angle A, given $\sin A = \frac{3}{7}$.

Construct the right triangle ABC having $a = 3$, $c = 7$ and $b = \sqrt{7^2 - 3^2} = 2\sqrt{10}$ units. See Fig. 32-5. Then

$$\sin A = \frac{3}{7} \qquad\qquad \cot A = \frac{2\sqrt{10}}{3}$$

$$\cos A = \frac{2\sqrt{10}}{7} \qquad\qquad \sec A = \frac{7}{2\sqrt{10}} = \frac{7\sqrt{10}}{20}$$

$$\tan A = \frac{3}{2\sqrt{10}} = \frac{3\sqrt{10}}{20} \qquad\qquad \csc A = \frac{7}{3}$$

B

c = 7   a = 3

A   C

b = 2√10

Fig. 32-5

32.4   Find the values of the trigonometric functions of the acute angle B, given $\tan B = 1.5$.

Refer to Fig. 32-6. Construct the right triangle ABC having $b = 15$ and $a = 10$ units. (Note that $1.5 = \frac{3}{2}$ and a right triangle with $b = 3$, $a = 2$ will serve equally well.)

Fig. 32-6

Then $c = \sqrt{a^2 + b^2} = \sqrt{10^2 + 15^2} = 5\sqrt{13}$ and

$$\sin B = \frac{15}{5\sqrt{13}} = \frac{3\sqrt{13}}{13} \qquad \cot B = \tfrac{2}{3}$$

$$\cos B = \frac{10}{5\sqrt{13}} = \frac{2\sqrt{13}}{13} \qquad \sec B = \frac{5\sqrt{13}}{10} = \frac{\sqrt{13}}{2}$$

$$\tan B = \tfrac{15}{10} = \tfrac{3}{2} \qquad \csc B = \frac{5\sqrt{13}}{15} = \frac{\sqrt{13}}{3}$$

32.5 If A is acute and $\sin A = 2x/3$, determine the values of the remaining functions.

Construct the right triangle ABC having $a = 2x < 3$ and $c = 3$, as in Fig. 32-7.
Then $b = \sqrt{c^2 - a^2} = \sqrt{9 - 4x^2}$ and

$$\sin A = \frac{2x}{3}, \qquad \cos A = \frac{\sqrt{9 - 4x^2}}{3}, \qquad \tan A = \frac{2x}{\sqrt{9 - 4x^2}}, \qquad \cot A = \frac{\sqrt{9 - 4x^2}}{2x},$$

$$\sec A = \frac{3}{\sqrt{9 - 4x^2}}, \qquad \csc A = \frac{3}{2x}.$$

Fig. 32-7   Fig. 32-8

32.6 If A is acute and $\tan A = x = x/1$, determine the values of the remaining functions.

Construct the right triangle ABC having $a = x$ and $b = 1$, as in Fig. 32-8. Then $c = \sqrt{x^2 + 1}$ and

$$\sin A = \frac{x}{\sqrt{x^2 + 1}}, \qquad \cos A = \frac{1}{\sqrt{x^2 + 1}}, \qquad \tan A = x, \qquad \cot A = \frac{1}{x}, \qquad \sec A = \sqrt{x^2 + 1}, \qquad \csc A = \frac{\sqrt{x^2 + 1}}{x}.$$

32.7 If A is an acute angle:

(a) Why is $\sin A < 1$?   (d) Why is $\sin A < \tan A$?

(b) When is $\sin A = \cos A$?   (e) When is $\sin A < \cos A$?

(c) Why is $\sin A < \csc A$?   (f) When is $\tan A > 1$?

In any right triangle ABC:

(a) Side $a <$ side c; therefore $\sin A = a/c < 1$.

(b) $\sin A = \cos A$ when $a/c = b/c$; then $a = b$, $A = B$, and $A = 45°$.

(c) $\sin A < 1$ (above) and $\csc A = 1/\sin A > 1$.

(d) $\sin A = a/c$, $\tan A = a/b$, and $b < c$; therefore $a/c < a/b$ or $\sin A < \tan A$.

(e) $\sin A < \cos A$ when $a < b$; then $A < B$ or $A < 90° - A$, and $A < 45°$.

(f) $\tan A = a/b > 1$ when $a > b$; then $A > B$ and $A > 45°$.

32.8   Find the values of the trigonometric functions of 45°.

In any isosceles right triangle ABC, $A = B = 45°$ and $a = b$. See Fig. 32-9. Let $a = b = 1$; then $c = \sqrt{1 + 1} = \sqrt{2}$ and

$$\sin 45° = \frac{1}{\sqrt{2}} = \frac{1}{2}\sqrt{2} \qquad \cot 45° = 1$$

$$\cos 45° = \frac{1}{\sqrt{2}} = \frac{1}{2}\sqrt{2} \qquad \sec 45° = \sqrt{2}$$

$$\tan 45° = \tfrac{1}{1} = 1 \qquad \csc 45° = \sqrt{2}$$

Fig. 32-9

32.9   Find the values of the trigonometric functions of 30° and 60°.

In any equilateral triangle ABD (see Fig. 32-10), each angle is 60°. The bisector of any angle, as B, is the perpendicular bisector of the opposite side. Let the sides of the equilateral triangle be of length 2 units. Then in the right triangle ABC, $AB = 2$, $AC = 1$, and $BC = \sqrt{2^2 - 1^2} = \sqrt{3}$.

$$\sin 30° = \tfrac{1}{2} = \cos 60° \qquad\qquad \cot 30° = \sqrt{3} = \tan 60°$$

$$\cos 30° = \frac{\sqrt{3}}{2} = \sin 60° \qquad\qquad \sec 30° = \frac{2}{\sqrt{3}} - \frac{2\sqrt{3}}{3} = \csc 60°$$

$$\tan 30° = \frac{1}{\sqrt{3}} = \frac{\sqrt{3}}{3} = \cot 60° \qquad\qquad \csc 30° = 2 = \sec 60°$$

Fig. 32-10

32.10   When the sun is 20° above the horizon, how long is the shadow cast by a building 150 ft high?

In Fig. 32-11, $A = 20°$ and $CB = 150$. Then $\cot A = AC/CB$ and $AC = CB \cot A = 150 \cot 20° = 150(2.7) = 405$ ft.

32.11   A tree 100 ft tall casts a shadow 120 ft long. Find the measure of the angle of elevation of the sun.

In Fig. 32-12, $CB = 100$ and $AC = 120$. Then $\tan A = CB/AC = \frac{100}{120} = 0.83$ and $m \angle A = 40°$.

| Fig. 32-11 | Fig. 32-12 | Fig. 32-13 |

32.12   A ladder leans against the side of a building with its foot 12 m from the building. How far from the ground is the top of the ladder and how long is the ladder if it makes an angle of 70° with the ground?

From Fig. 32-13, $\tan A = CB/AC$; then $CB = AC \tan A = 12 \tan 70° = 12(2.7) = 32.4$. The top of the ladder is 32 m above the ground.

Sec $A = AB/AC$; then $AB = AC \sec A = 12 \sec 70° = 12(2.9) = 34.8$. The ladder is 35 m long.

32.13   From the top of a lighthouse, 120 ft above the sea, the angle of depression of a boat is 15°. How far is the boat from the lighthouse?

In Fig. 32-14, the right triangle ABC has A measuring 15° and $CB = 120$; then $\cot A = AC/CB$ and $AC = CB \cot A = 120 \cot 15° = 120(3.7) = 444$ ft.

| Fig. 32-14 | Fig. 32-15 |

32.14   Find the length of the chord of a circle of radius 20 cm subtended by a central angle of 150°.

In Fig. 32-15, OC bisects $\angle AOB$. Then $BC = AC$ and OAC is a right triangle. In $\triangle OAC$,

$$\sin \angle COA = \frac{AC}{OA} \quad \text{and} \quad AC = OA \sin \angle COA = 20 \sin 75° = 20(0.97) = 19.4$$

Then $BA = 38.8$ and the length of the chord is 39 cm.

32.15   Find the height of a tree if the angle of elevation of its top changes from 20° to 40° as the observer advances 75 ft toward its base. See Fig. 32-16.

In the right triangle ABC, $\cot A = AC/CB$; then $AC = CB \cot A$ or $DC + 75 = CB \cot 20°$.
In the right triangle DBC, $\cot D = DC/CB$; then $DC = CB \cot 40°$.

Then $DC = CB \cot 20° - 75 = CB \cot 40°$, $CB(\cot 20° - \cot 40°) = 75$,

$$CB(2.7 - 1.2) = 75, \quad \text{and} \quad CB = \frac{75}{1.5} = 50 \text{ ft.}$$

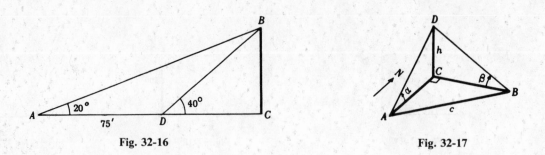

Fig. 32-16   Fig. 32-17

32.16 A tower standing on level ground is due north of point A and due west of point B, a distance c ft from A. If the angles of elevation of the top of the tower as measured from A and B are α and β, respectively, find the height h of the tower.

In the right triangle ACD of Fig. 32-17, $\cot \alpha = AC/h$; and in the right triangle BCD, $\cot \beta = BC/h$. Then $AC = h \cot \alpha$ and $BC = h \cot \beta$.

Since ABC is a right triangle, $(AC)^2 + (BC)^2 = c^2 = h^2(\cot \alpha)^2 + h^2(\cot \beta)^2$ and

$$h = \frac{c}{\sqrt{(\cot \alpha)^2 + (\cot \beta)^2}}$$

32.17 If holes are to be spaced regularly on a circle, show that the distance d between the centers of two successive holes is given by $d = 2r \sin (180°/n)$, where $r = $ radius of the circle and $n = $ number of holes. Find d when $r = 20$ in. and $n = 4$.

Let A and B be the centers of two consecutive holes on the circle of radius r and center O. See Fig. 32-18. Let the bisector of the angle O of the triangle AOB meet AB at C. In right triangle AOC, $\sin \angle AOC = AC/r = \frac{1}{2}d/r = d/2r$. Then

$$d = 2r \sin \angle AOC = 2r \sin \tfrac{1}{2} \angle AOB = 2r \sin \frac{1}{2}\left(\frac{360°}{n}\right) = 2r \sin \frac{180°}{n}.$$

When $r = 20$ and $n = 4$, $d = 2 \cdot 20 \sin 45° = 2 \cdot 20 \cdot \frac{\sqrt{2}}{2} = 20\sqrt{2}$ in.

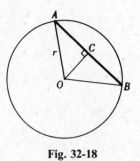

Fig. 32-18

Supplementary Problems

32.18 Find the values of the trigonometric functions of the acute angles of the right triangle ABC, given

    (a) $a = 3, b = 1$ (b) $a = 2, c = 5$ (c) $b = \sqrt{7}, c = 4$

    Ans. (a) A: $3/\sqrt{10}, 1/\sqrt{10}, 3, \frac{1}{3}, \sqrt{10}, \sqrt{10}/3$: B: $1/\sqrt{10}, 3/\sqrt{10}, \frac{1}{3}, 3, \sqrt{10}/3, \sqrt{10}$

        (b) A: $\frac{2}{5}, \sqrt{21}/5, 2/\sqrt{21}, \sqrt{21}/2, 5/\sqrt{21}, \frac{5}{2}$; B: $\sqrt{21}/5, \frac{2}{5}, \sqrt{21}/2, 2/\sqrt{21}, \frac{5}{2}, 5/\sqrt{21}$

        (c) A: $\frac{3}{4}, \sqrt{7}/4, 3/\sqrt{7}, \sqrt{7}/3, 4/\sqrt{7}, \frac{4}{3}$; B: $\sqrt{7}/4, \frac{3}{4}, \sqrt{7}/3, 3/\sqrt{7}, \frac{4}{3}, 4/\sqrt{7}$

32.19   Which is the greater and why:

(a)   sin 55° or cos 55°?   (c)   tan 15° or cot 15°?

(b)   sin 40° or cos 40°?   (d)   sec 55° or csc 55°?

Hint: Consider a right triangle having as acute angle the given angle.

Ans.   (a)   sin 55°   (b)   cos 40°   (c)   cot 15°   (d)   sec 55°

32.20   Find the value of each of the following:

(a)   sin 30° + tan 45°

(b)   cot 45° + cos 60°

(c)   sin 30° cos 60° + cos 30° sin 60°

(d)   cos 30° cos 60° − sin 30° sin 60°

(e) $\dfrac{\tan 60° - \tan 30°}{1 + \tan 60° \tan 30°}$

(f) $\dfrac{\csc 30° + \csc 60° + \csc 90°}{\sec 0° + \sec 30° + \sec 60°}$

Ans.   (a) $\frac{3}{2}$ (b) $\frac{3}{2}$ (c) 1   (d) 0   (e) $1/\sqrt{3}$ (f) 1

32.21   A man drives 500 ft along a road which is inclined 20° to the horizontal. How high above his starting point is he?

Ans.   170 ft

32.22   A tree broken over by the wind forms a right triangle with the ground. If the broken part makes an angle of 50° with the ground and if the top of the tree is now 20 ft from its base, how tall was the tree?

Ans.   56 ft

32.23   Two straight roads intersect to form an angle of 75°. Find the shortest distance from one road to a gas station on the other road 1000 m from the junction.

Ans.   970 m

32.24   Two buildings with flat roofs are 60 ft apart. From the roof of the shorter building, 40 ft in height, the angle of elevation to the edge of the roof of the taller building is 40°. How high is the taller building?

Ans.   90 ft

32.25   A ladder, with its foot in the street, makes an angle of 30° with the street when its top rests on a building on one side of the street and makes an angle of 40° with the street when its top rests on a building on the other side of the street. If the ladder is 50 ft long, how wide is the street?

Ans.   82 ft

32.26   Find the perimeter of an isosceles triangle whose base is 40 cm and whose base angle is 70°.

Ans.   156 cm

Chapter 33

Tables of Trigonometric Functions

APPROXIMATE VALUES OF THE FUNCTIONS of acute angles are given in Tables of Natural Trigonometric Functions. These tables, as published in texts, differ in several respects. Some give the values of the six functions; others are restricted to the functions sine, cosine, tangent, and cotangent; some give the values to four digits while others give values to four decimal places. We shall use the latter table here. (In case the values of secant and cosecant are not included in your table, reference to these functions is to be deleted.)

FOUR-PLACE TABLE OF NATURAL TRIGONOMETRIC FUNCTIONS

WHEN THE MEASURE OF THE ANGLE IS LESS THAN 45°, the angle is found in the left-hand column of the table and the function is read at the top of the page. When the measure of the angle is greater than 45°, the angle is found in the right-hand column and the function is read at the bottom of the page.

TO FIND THE VALUE OF A TRIGONOMETRIC FUNCTION OF A GIVEN ACUTE ANGLE. If the angle contains a number of degrees only or a number of degrees and a multiple of 10′, the value of the function is read directly from the table.

EXAMPLE 1. Find sin 24° 40′.
Opposite 24° 40′ (<45°) in the left-hand column read the entry 0.4173 in the column labeled Sin at the top of the page.

EXAMPLE 2. Find cos 72°.
Opposite 72° (>45°) in the right-hand column read the entry 0.3090 in the column labeled Cos at the bottom of the page.

EXAMPLE 3

(a)   tan 55° 20′ = 1.4460. Read *up* the page since 55° 20′ > 45°.

(b)   cot 41° 50′ = 1.1171. Read *down* the page since 41° 50′ < 45°.

If the number of minutes in the given angle is not a multiple of 10, as in 24° 43′, interpolate between the values of the functions of the two nearest angles (24° 40′ and 24° 50′) using the method of proportional parts.

EXAMPLE 4.   Find sin 24° 43′.
We find
$$\sin 24°\,40' = 0.4173$$
$$\sin 24°\,50' = \underline{0.4200}$$
$$\text{Difference for } 10' = 0.0027 = \text{tabular difference}$$

Correction = difference for 3′ = 0.3(0.0027) = 0.00081 or 0.0008 when rounded off to four decimal places.
As the angle increases, the sine of the angle increases; thus,

$$\sin 24°\,43' = 0.4173 + 0.0008 = 0.4181$$

If a five-place table is available, the value 0.41813 can be read directly from the table and then rounded off to 0.4181.

212

EXAMPLE 5.   Find cos 64° 26′.

We find

$$\cos 64° 20′ = 0.4331$$
$$\cos 64° 30′ = \underline{0.4305}$$
$$\text{Tabular difference} = 0.0026$$

Correction = 0.6(0.0026) = 0.00156 or 0.0016 to four decimal places.

As the angle increases, the cosine of the angle decreases. Thus

$$\cos 64° 26′ = 0.4331 - 0.0016 = 0.4315$$

To save time, we should proceed as follows in Example 4:

(a)   Locate sin 24° 40′ = 0.4173. For the moment, disregard the decimal point and use only the sequence 4173.
(b)   Find (mentally) the tabular difference 27, that is, the difference between the sequence 4173 corresponding to 24° 40′ and the sequence 4200 corresponding to 24° 50′.
(c)   Find 0.3(27) = 8.1 and round off to the nearest integer. This is the correction.
(d)   Add (since sine) the correction to 4173, obtaining 4181. Then sin 24° 43′ = 0.4181.

When, as in the above example, we interpolate from the smaller angle to the larger: (1) The correction is added in finding sine, tangent, and secant. (2) The correction is subtracted in finding cosine, cotangent, and cosecant. (See also Problem 33.1.)

TO FIND THE MEASURE OF AN ANGLE WHOSE FUNCTION IS GIVEN.   The process is a reversal of that given above.

EXAMPLE 6.   Reading directly from the table, we find 0.2924 = sin 17°, 2.7725 = tan 70° 10′.

EXAMPLE 7.   Find m∡A, given sin A = 0.4234.

The given value is not an entry in the table. We find, however,

$$0.4226 = \sin 25° \ 0′ \qquad 0.4226 = \sin 25° 0′$$
$$\underline{0.4253} = \sin 25° 10′ \qquad \underline{0.4234} = \sin A$$
$$\text{Tabular diff.} = 0.0027 \qquad 0.0008 = \text{partial difference}$$

$$\text{Correction} = \frac{0.0008}{0.0027}(10′) = \frac{8}{27}(10′) = 3′ \quad \text{to the nearest minute}$$

Adding (since sine) the correction, we have 25° 0′ + 3′ = 25° 3′ = m∡A.

EXAMPLE 8.   Find m∡A, given cot A = 0.6345.

We find

$$0.6330 = \cot 57° 40′ \qquad 0.6330 = \cot 57° 40′$$
$$\underline{0.6371} = \cot 57° 30′ \qquad \underline{0.6345} = \cot A$$
$$\text{Tabular diff.} = 0.0041 \qquad 0.0015 = \text{partial difference}$$

$$\text{Correction} = \frac{0.0015}{0.0041}(10′) = \frac{15}{41}(10′) = 4′ \quad \text{to the nearest minute}$$

Subtracting (since cot) the correction, we have 57° 40′ − 4′ = 57° 36′ = m∡A.

To save time, we should proceed as follows in Example 7:

(a)   Locate the next smaller entry, 0.4226 = sin 25° 0′. For the moment use only the sequence 4226.
(b)   Find the tabular difference, 27.
(c)   Find the partial difference, 8, between 4226 and the given sequence 4234.
(d)   Find $\frac{8}{27}(10′) = 3′$ and add to 25° 0′.

(See Problem 33.3.)

ERRORS IN COMPUTED RESULTS arise from

 (*a*) *Errors in the given data.* These errors are always present in data resulting from measurements.
 (*b*) *The use of prepared tables.* The entries in such tables are usually approximations of never ending decimals.

A measurement recorded as 35 m means that the result is correct to the nearest meter, that is, the true length is between 34.5 and 35.5 m. Similarly, a recorded length of 35.0 m means that the true length is between 34.95 and 35.05 m; a recorded length of 35.8 m means that the true length is between 35.75 and 35.85 m, a recorded length of 35.80 m means that the true length is between 35.795 and 35.805 m; and so on.

In the number 35 there are two significant digits, 3 and 5. They are also the significant digits in 3.5, 0.35, 0.035, 0.0035 but not in 35.0, 3.50, 0.350, 0.0350. In the numbers 35.0, 3.50, 0.350, 0.0350 there are three significant digits, 3, 5, and 0. This is another way of saying that 35 and 35.0 are not the same measurement.

It is impossible to determine the significant figures in a measurement recorded as 350, 3500, 35 000, ... For example, 350 may mean that the true result is between 345 and 355 or between 349.5 and 350.5.

ACCURACY IN COMPUTED RESULTS.   A computed result should not show more decimal places than that shown in the least accurate of the measured data. Of importance here are the following relations giving comparable degrees of accuracy in lengths and angles:

 (*a*)   Distances expressed to 2 significant digits and angles expressed to the nearest degree
 (*b*)   Distances expressed to 3 significant digits and angles expressed to the nearest 10'
 (*c*)   Distances expressed to 4 significant digits and angles expressed to the nearest 1'
 (*d*)   Distances expressed to 5 significant digits and angles expressed to the nearest 0.1'

Solved Problems

33.1   (*a*) $\sin 56° 34' = 0.8345;$ $8339 + 0.4(16) = 8339 + 6$

    (*b*) $\cos 19° 45' = 0.9412;$ $9417 - 0.5(10) = 9417 - 5$

    (*c*) $\tan 77° 12' = 4.4016;$ $43897 + 0.2(597) = 43897 + 119$

    (*d*) $\cot 40° 36' = 1.1667;$ $11708 - 0.6(68) = 11708 - 41$

    (*e*) $\sec 23° 47' = 1.0928;$ $10918 + 0.7(14) = 10918 + 10$

    (*f*) $\csc 60° 4' = 1.1539;$ $11547 - 0.4(19) = 11547 - 8$

33.2   If the correction is 6.5, 13.5, 10.5, etc., we shall round off so that the *final* result is even.

    (*a*) $\sin 28° 37' = 0.4790;$ $4772 + 0.7(25) = 4772 + 17.5$

    (*b*) $\cot 65° 53' = 0.4476;$ $4487 - 0.3(35) = 4487 - 10.5$

    (*c*) $\cos 35° 25' = 0.8150;$ $8158 - 0.5(17) = 8158 - 8.5$

    (*d*) $\sec 39° 35' = 1.2976;$ $12960 + 0.5(31) = 12960 + 15.5$

33.3 (a) $\sin A = 0.6826$, m∡$A = 43° 3'$; $43° 0' + \frac{6}{21}(10') = 43° 0' + 3'$

   (b) $\cos A = 0.5957$, m∡$A = 53° 26'$; $53° 30' - \frac{9}{24}(10') = 53° 30' - 4'$

   (c) $\tan A = 0.9470$, m∡$A = 43° 26'$; $43° 20' + \frac{35}{55}(10') = 43° 20' + 6'$

   (d) $\cot A = 1.7580$, m∡$A = 29° 38'$; $29° 40' - \frac{24}{119}(10') = 29° 40' - 2'$

   (e) $\sec A = 2.3198$, m∡$A = 64° 28'$; $64° 20' + \frac{110}{140}(10') = 64° 20' + 8'$

   (f) $\csc A = 1.5651$, m∡$A = 39° 43'$; $39° 50' - \frac{40}{55}(10') = 39° 50' - 7'$

33.4 Solve the right triangle in which $A = 35° 10'$ and $c = 72.5$. See Fig. 33-1.

Fig. 33-1

Solution: m∡$B = 90° - 35° 10' = 54° 50'$.

$$\frac{a}{c} = \sin A, \qquad a = c \sin A = 72.5(0.5760) = 41.8$$

$$\frac{b}{c} = \cos A, \qquad b = c \cos A = 72.5(0.8175) = 59.3$$

Check: $\dfrac{a}{b} = \tan A, \qquad a = b \tan A = 59.3(0.7046) = 41.8$

33.5 Solve the right triangle in which $a = 24.36$, $A = 58° 53'$. See Fig. 33-2.

Fig. 33-2

Solution: m∡$B = 90° - 58° 53' = 31° 7'$.

$$\frac{b}{a} = \cot A, \qquad b = a \cot A = 24.36(0.6036) = 14.70.$$

$$\frac{c}{a} = \csc A, \qquad c = a \csc A = 24.36(1.1681) = 28.45, \quad \text{or}$$

$$\frac{a}{c} = \sin A, \qquad c = \frac{a}{\sin A} = \frac{24.36}{0.8562} = 28.45.$$

Check: $\dfrac{b}{c} = \cos A, \qquad b = c \cos A = 28.45(0.5168) = 14.70.$

33.6   Solve the right triangle ABC in which $a = 43.9$, $b = 24.3$. See Fig. 33-3.

Fig. 33-3

Solution: $\tan A = \dfrac{43.9}{24.3} = 1.8066$; $m\measuredangle A = 61°\,2'$, $m\measuredangle B = 90° - A = 28°\,58'$.

$$\frac{c}{a} = \csc A, \qquad c = a \csc A = 43.9(1.1430) = 50.2, \quad \text{or}$$

$$\frac{a}{c} = \sin A, \qquad c = \frac{a}{\sin A} = \frac{43.9}{0.8749} = 50.2.$$

Check: $\dfrac{c}{b} = \sec A, \qquad c = b \sec A = 24.3(2.0649) = 50.2, \quad \text{or}$

$$\frac{b}{c} = \cos A, \qquad c = \frac{b}{\cos A} = \frac{24.3}{0.4843} = 50.2.$$

33.7   Solve the right triangle ABC in which $b = 15.25$, $c = 32.68$. See Fig. 33-4.

Fig. 33-4

Solution: $\sin B = \dfrac{15.25}{32.68} = 0.4666$; $m\measuredangle B = 27°\,49'$, $m\measuredangle A = 90° - m\measuredangle B = 62°\,11'$.

$$\frac{a}{b} = \cot B, \qquad a = b \cot B = 15.25(1.8953) = 28.90$$

Check: $\dfrac{a}{c} = \cos B, \qquad a = c \cos B = 32.68(0.8844) = 28.90$

33.8   The base of an isosceles triangle is 20.4 and the base angles are 48° 40′. Find the congruent sides and the altitude of the triangle. See Fig. 33-5.

Fig. 33-5

In Fig. 33-5, $\overline{BD}$ is perpendicular to $\overline{AC}$ and bisects it.
In the right triangle ABD,

$$\frac{AB}{AD} = \sec A, \qquad AB = 10.2(1.5141) = 15.4, \quad \text{or}$$

$$\frac{AD}{AB} = \cos A, \qquad AB = \frac{10.2}{0.6604} = 15.4.$$

$$\frac{DB}{AD} = \tan A, \qquad DB = 10.2(1.1369) = 11.6.$$

33.9 Considering the earth as a sphere of radius 3960 miles, find the radius r of the 40th parallel of latitude. Refer to Fig. 33-6.

In the right triangle OCB, m$\angle OBC = 40°$ and $OB = 3960$.
Then $\cos \angle OBC = CB/OB$ and $r = CB = 3960 \cos 40° = 3960(0.7660) = 3030$ miles

| Fig. 33-6 | Fig. 33-7 | Fig. 33-8 |

33.10 Find the perimeter of a regular octagon inscribed in a circle of radius 150 m.

In Fig. 33-7, two consecutive vertices A and B of the octagon are joined to the center O of the circle. The triangle OAB is isosceles with congruent sides 150 and m$\angle AOB = \dfrac{360°}{8} = 45°$. As in Problem 33.8, we bisect $\angle AOB$ to form the right triangle MOB.

Then $MB = OB \sin \angle MOB = 150 \sin 22° 30' = 150(0.3827) = 57.4$, and the perimeter of the octagon is $16MB = 16(57.4) = 918$ m.

33.11 To find the width of a river, a surveyor set up his transit at C on one bank and sighted across to a point B on the opposite bank; then turning through an angle of 90°, he laid off a distance $CA = 225$ ft. Finally, setting the transit at A, he measured $\angle CAB$ as 48° 20'. Find the width of the river.

See Fig. 33-8. In the right triangle ACB, $CB = AC \tan \angle CAB = 225 \tan 48° 20' = 225(1.1237) = 253$ ft.

33.12 In Fig. 33-9, the line AD crosses a swamp. In order to locate a point on this line, a surveyor turned through an angle 51° 16' at A and measured 1585 ft to a point C. He then turned through an angle of measure 90° at C and ran a line $\overline{CB}$. If B is on $\overline{AD}$, how far must he measure from C to reach B?

$$CB = AC \tan 51° 16' = 1585(1.2467) = 1976 \text{ ft}$$

Fig. 33-9

33.13 From a point A on level ground, the angles of elevation of the top D and bottom B of a flagpole situated on the top of a hill are measured as $47°\,54'$ and $39°\,45'$. Find the height of the hill if the height of the flagpole is 115.5 m. See Fig. 33-10.

Fig. 33-10

Let the line of the pole meet the horizontal through A in C.

In the right triangle ACD, $AC = DC \cot 47°\,54' = (115.5 + BC)(0.9036)$

In the right triangle ACB, $AC = BC \cot 39°\,45' = BC(1.2024)$

Then $(115.5 + BC)(0.9036) = BC(1.2024)$

and $BC = \dfrac{115.5(0.9036)}{1.2024 - 0.9036} = 349.3$ m

33.14 From the top of a lighthouse, 175 ft above the water, the measure of the angle of depression of a boat due south is $18°\,50'$. Calculate the speed of the boat if, after it moves due west for 2 min, the angle of depression measures $14°\,20'$.

In Fig. 33-11, $\overline{AD}$ is the lighthouse, C is the position of the boat when due south of the lighthouse, and B is the position 2 min later.

Fig. 33-11

In the right triangle CAD,
$$AC = AD \cot \angle ACD = 175 \cot 18°\,50' = 175(2.9319) = 513$$

In the right triangle BAD,
$$AB = AD \cot \angle ABD = 175 \cot 14°\,20' = 175(3.9136) = 685$$

In the right triangle ABC,
$$BC = \sqrt{(AB)^2 - (AC)^2} = \sqrt{(685)^2 - (513)^2} = 454$$

The boat travels 454 ft in 2 min; its speed is 227 ft/min.

Supplementary Problems

33.15 Find the natural trigonometric functions of each of the following angles:

   (a) 18° 47'   (b) 32° 13'   (c) 58° 24'   (d) 79° 45'

Ans.		sine	cosine	tangent	cotangent	secant	cosecant
(a)	18° 47'	0.3220	0.9468	0.3401	2.9403	1.0563	3.1057
(b)	32° 13'	0.5331	0.8460	0.6301	1.5869	1.1820	1.8757
(c)	58° 24'	0.8517	0.5240	1.6255	0.6152	1.9084	1.1741
(d)	79° 45'	0.9840	0.1780	5.5304	0.1808	5.6201	1.0162

33.16 Find the measure of the (acute) angle A, given

(a)	$\sin A = 0.5741$	Ans. $m\angle A = 35°\,2'$	(e)	$\cos A = 0.9382$	Ans. $m\angle A = 20°\,15'$
(b)	$\sin A = 0.9468$	$m\angle A = 71°\,13'$	(f)	$\cos A = 0.6200$	$m\angle A = 51°\,41'$
(c)	$\sin A = 0.3510$	$m\angle A = 20°\,33'$	(g)	$\cos A = 0.7120$	$m\angle A = 44°\,36'$
(d)	$\sin A = 0.8900$	$m\angle A = 62°\,52'$	(h)	$\cos A = 0.4651$	$m\angle A = 62°\,17'$
(i)	$\tan A = 0.2725$	$m\angle A = 15°\,15'$	(m)	$\cot A = 0.2315$	$m\angle A = 76°\,58'$
(j)	$\tan A = 1.1652$	$m\angle A = 49°\,22'$	(n)	$\cot A = 2.9715$	$m\angle A = 18°\,36'$
(k)	$\tan A = 0.5200$	$m\angle A = 27°\,28'$	(o)	$\cot A = 0.7148$	$m\angle A = 54°\,27'$
(l)	$\tan A = 2.7775$	$m\angle A = 70°\,12'$	(p)	$\cot A = 1.7040$	$m\angle A = 30°\,24'$
(q)	$\sec A = 1.1161$	$m\angle A = 26°\,22'$	(u)	$\csc A = 3.6882$	$m\angle A = 15°\,44'$
(r)	$\sec A = 1.4382$	$m\angle A = 45°\,57'$	(v)	$\csc A = 1.0547$	$m\angle A = 71°\,28'$
(s)	$\sec A = 1.2618$	$m\angle A = 37°\,35'$	(w)	$\csc A = 1.7631$	$m\angle A = 34°\,33'$
(t)	$\sec A = 2.1584$	$m\angle A = 62°\,24'$	(x)	$\csc A = 1.3436$	$m\angle A = 48°\,6'$

33.17 Solve each of the right triangles ABC, given

(a)	$m\angle A = 35°\,20'$, $c = 112$	Ans. $m\angle B = 54°\,40'$, $a = 64.8$, $b = 91.4$
(b)	$m\angle B = 48°\,40'$, $c = 225$	$m\angle A = 41°\,20'$, $a = 149$, $b = 169$
(c)	$m\angle A = 23°\,18'$, $c = 346.4$	$m\angle B = 66°\,42'$, $a = 137.0$, $b = 318.1$
(d)	$m\angle B = 54°\,12'$, $c = 182.5$	$m\angle A = 35°\,48'$, $a = 106.7$, $b = 148.0$
(e)	$m\angle A = 32°\,10'$, $a = 75.4$	$m\angle B = 57°\,50'$, $b = 120$, $c = 142$
(f)	$m\angle A = 58°\,40'$, $b = 38.6$	$m\angle B = 31°\,20'$, $a = 63.4$, $c = 74.2$
(g)	$m\angle B = 49°\,14'$, $b = 222.2$	$m\angle A = 40°\,46'$, $a = 191.6$, $c = 293.4$
(h)	$m\angle A = 66°\,36'$, $a = 112.6$	$m\angle B = 23°\,24'$, $b = 48.73$, $c = 122.7$
(i)	$m\angle A = 29°\,48'$, $b = 458.2$	$m\angle B = 60°\,12'$, $a = 262.4$, $c = 528.0$
(j)	$a = 25.4$, $b = 38.2$	$m\angle A = 33°\,37'$, $m\angle B = 56°\,23'$, $c = 45.9$
(k)	$a = 45.6$, $b = 84.8$	$m\angle A = 28°\,16'$, $m\angle B = 61°\,44'$, $c = 96.3$
(l)	$a = 38.64$, $b = 48.74$	$m\angle A = 38°\,24'$, $m\angle B = 51°\,36'$, $c = 62.21$
(m)	$a = 506.2$, $c = 984.8$	$m\angle A = 30°\,56'$, $m\angle B = 59°\,4'$, $b = 844.7$
(n)	$b = 672.9$, $c = 888.1$	$m\angle A = 40°\,44'$, $m\angle B = 49°\,16'$, $a = 579.4$

33.18 Find the base and altitude of an isosceles triangle whose vertical angle measures 65° and whose equal sides are 415 m.

   Ans.   Base = 446 m, altitude = 350 m

33.19 The base of an isosceles triangle is 15.90 in. and the base angles measure 54° 28′. Find the congruent sides and the altitude.

   Ans.   Side = 13.68 in., altitude = 11.13 in.

33.20 The radius of a circle is 21.4 ft. Find (*a*) the length of the chord subtended by a central angle of 110° 40′ and (*b*) the distance between two parallel chords on the same side of the center subtended by central angles 118° 40′ and 52° 20′.

   Ans.   (*a*)   35.2 ft   (*b*)   8.29 ft

33.21 Show that the base b of an isosceles triangle whose congruent sides are a and whose vertical angle measures θ rad is given by $b = 2a \sin \frac{1}{2}\theta$.

33.22 Show that the perimeter P of a regular polygon of n sides inscribed in a circle of radius r is given by $P = 2nr \sin(180°/n)$.

33.23 A wheel, 5 ft in diameter, rolls up an incline of 18° 20′. What is the height of the center of the wheel above the base of the incline when the wheel has rolled 5 ft up the incline?

   Ans.   3.95 ft

33.24 A wall is 15 m high and 10 m from a house. Find the length of the shortest ladder which will just touch the top of the wall and reach a window 20.5 m above the ground.

   Ans.   42.5 m

Chapter 34

Practical Applications

THE BEARING OF A POINT _B_ FROM A POINT _A_, in a horizontal plane, is usually defined as the angle (always acute) made by the ray drawn from _A_ through _B_ with the north-south line through _A_. The bearing is then read from the north or south line toward the east or west. See Fig. 34-1(_a_) through (_d_).

Bearing: N 35° E   S 35° E   S 35° W   N 35° W
    (_a_)   (_b_)   (_c_)   (_d_)

Fig. 34-1

In aeronautics the bearing of _B_ from _A_ is more often given as the angle made by the ray _AB_ with the north line through _A_, measured clockwise from the north (i.e., from the north around through the east). See Fig. 34-2(_a_) through (_d_).

Bearing: 35°   145°   215°   325°
    (_a_)   (_b_)   (_c_)   (_d_)

Fig. 34-2

VECTORS. Any physical quantity, as force or velocity, which has both magnitude and direction is called a _vector quantity_. A vector quantity may be represented by a directed line segment (arrow) called a _vector_. The _direction_ of the vector is that of the given quantity and the _length_ of the vector is proportional to the magnitude of the quantity.

EXAMPLE 1. An airplane is traveling N 40° E at 200 mph. Its velocity is represented by the vector _AB_ in Fig. 34-3.

Fig. 34-3

221

EXAMPLE 2.   A motor boat having the speed 12 mph in still water is headed directly across a river whose current is 4 mph. In Fig. 34-4, the vector *CD* represents the velocity of the current and the vector *AB* represents, to the same scale, the velocity of the boat in still water. Thus, vector *AB* is three times as long as vector *CD*.

EXAMPLE 3.   In Fig. 34-5, vector *AB* represents a force of 20 lb making an angle of 35° with the positive direction on the *x* axis and vector *CD* represents a force of 30 lb at 150° with the positive direction on the *x* axis. Both vectors are drawn to the same scale.

Two vectors are said to be equal if they have the same magnitude and direction. A vector has no fixed position in a plane and may be moved about in the plane provided only that its magnitude and direction are not changed.

Fig. 34-4   Fig. 34-5

VECTOR ADDITION.   The *resultant* or *vector sum* of a number of vectors, all in the same plane, is that vector in the plane which would produce the same effect as that produced by all of the original vectors acting together.

If two vectors α and β have the same direction, their resultant is a vector R whose magnitude is equal to the sum of the magnitudes of the two vectors and whose direction is that of the two vectors. See Fig. 34-6(a).

If two vectors have opposite directions, their resultant is a vector R whose magnitude is the difference (greater magnitude − smaller magnitude) of the magnitudes of the two vectors and whose direction is that of the vector of greater magnitude. See Fig. 34-6(b).

(a)   (b)

Fig. 34-6

In all other cases, the magnitude and direction of the resultant of two vectors is obtained by either of the following two methods.

   (*1*)   **PARALLELOGRAM METHOD.**   Place the tail ends of both vectors at any point *O* in their plane and complete the parallelogram having these vectors as adjacent sides. The directed diagonal issuing from *O* is the resultant or vector sum of the two given vectors. Thus, in Fig. 34-7(b), the vector R is the resultant of the vectors α and β of Fig. 34-7(a).

   (*2*)   **TRIANGLE METHOD.**   Choose one of the vectors and label its tail end as *O*. Place the tail end of the other vector at the arrow end of the first. The resultant is then the line segment closing the triangle and directed from *O*. Thus, in Fig. 34-7(c) and 34-7(d), R is the resultant of the vectors α and β.

PARALLELOGRAM METHOD   TRIANGLE METHOD

Fig. 34-7

EXAMPLE 4.   The resultant R of the two vectors of Example 2 represents the speed and direction in which the boat travels. Figure 34-8(a) illustrates the parallelogram method; Fig. 34-8(b) and 34-8(c) illustrate the triangle method.

Fig. 34-8

The magnitude of $R = \sqrt{(12)^2 + 4^2} = 12.6$ mph. From Fig. 34-8(a) or 34-8(b), $\tan \theta = \frac{4}{12} = 0.3333$ and m$\not\!\angle\, \theta =$ 18° 30′.

   Thus, the boat moves downstream in a line making an angle $\theta = 18° 30′$ with the direction in which it is headed or making an angle $90° - \theta = 71° 30′$ with the bank of the river.

THE COMPONENT OF A VECTOR α along a line L is the perpendicular projection of the vector α on L. It is often very useful to resolve a vector into two components along a pair of perpendicular lines.

EXAMPLE 5.   In each of Fig. 34-8(a), (b), (c) the components of R are (1) 4 mph in the direction of the current and, (2) 12 mph in the direction perpendicular to the current.

EXAMPLE 6.   In Fig. 34-9, the force F has horizontal component $F_h = F \cos 30°$ and vertical component $F_v = F \sin 30°$. Note that F is the vector sum or resultant of F_h and F_v.

Fig. 34-9

Solved Problems

34.1   A motorboat moves in the direction N 40° E for 3 hr at 20 mph. How far north and how far east does it travel?

Suppose the boat leaves A. See Fig. 34-10. Using the north-south line through A, draw the ray $\overrightarrow{AD}$ so that the bearing of D from A is N 40° E. On $\overrightarrow{AD}$ locate B such that $AB = 3(20) = 60$ miles. Through B pass a line perpendicular to the line $\overleftrightarrow{NAS}$, meeting it in C. In the right triangle ABC,

$$AC = AB \cos A = 60 \cos 40° = 60(0.7660) = 45.96$$

and
$$CB = AB \sin A = 60 \sin 40° = 60(0.6428) = 38.57.$$

Fig. 34-10

The boat travels 46 mi north and 39 mi east.

34.2   Three ships are situated as follows: A is 225 km due north of C, and B is 375 km due east of C. What is the bearing (a) of B from A, (b) of A from B?

In the right triangle ABC (see Fig. 34-11),

$$\tan \angle CAB = \tfrac{375}{225} = 1.6667 \quad \text{and} \quad \text{m} \angle CAB = 59° \, 0'.$$

Fig. 34-11

(a)   The bearing of B from A (angle SAB) is S 59° 0' E.

(b)   The bearing of A from B (angle $N'BA$) is N 59° 0' W.

34.3   Three ships are situated as follows: A is 225 km west of C while B, due south of C, bears S 25° 10' E from A. (a) How far is B from A? (b) How far is B from C? (c) What is the bearing of A from B?

From Fig. 34-12, m$\angle SAB = 25° \, 10'$ and m$\angle BAC = 64° \, 50'$. Then

$$AB = AC \sec \angle BAC = 225 \sec 64° \, 50' = 225(2.3515) = 529.1 \quad \text{or}$$

$$AB = \frac{AC}{\cos \angle BAC} = \frac{225}{\cos 64° \, 50'} = \frac{225}{0.4253} = 529.0 \quad \text{and}$$

$$CB = AC \tan \angle BAC = 225 \tan 64° \, 50' = 225(2.1283) = 478.9.$$

Fig. 34-12

(*a*) *B* is 529 km from *A*.   (*b*) *B* is 479 km from *C*.

(*c*) Since m∠*CBA* = 25° 10′, the bearing of *A* from *B* is N 25° 10′ W.

34.4 From a boat sailing due north at 16.5 mph, a wrecked ship *K* and an observation tower *T* are observed in a line due east. One hour later the wrecked ship and the tower have bearings S 34° 40′ E and S 65° 10′ E. Find the distance between the wrecked ship and the tower.

In Fig. 34-13, *C*, *K*, and *T* represent, respectively, the boat, the wrecked ship, and the tower when in a line. One hour later the boat is at *A*, 16.5 miles due north of *C*. In the right triangle *ACK*,

$$CK = 16.5 \tan 34° 40′ = 16.5(0.6916)$$

Fig. 34-13

In the right triangle *ACT*,

$$CT = 16.5 \tan 65° 10′ = 16.5(2.1609)$$

Then $KT = CT - CK = 16.5(2.1609 - 0.6916) = 24.2$ mi.

34.5 A ship is sailing due east when a light is observed bearing N 62° 10′ E. After the ship has traveled 2250 m, the light bears N 48° 25′ E. If the course is continued, how close will the ship approach the light?

In Fig. 34.14, *L* is the position of the light, *A* is the first position of the ship, *B* is the second position, and *C* is the position when nearest *L*.

In the right triangle *ACL*,

$$AC = CL \cot ∠CAL = CL \cot 27° 50′ = 1.8940\, CL.$$

Fig. 34-14

In the right triangle *BCL*,

$$BC = CL \cot \angle CBL = CL \cot 41° 35' = 1.1270 \, CL.$$

Since *AC = BC + 2250*,

$$1.8940 \, CL = 1.1270 \, CL + 2250, \quad \text{and} \quad CL = \frac{2250}{1.8940 - 1.1270} = 2934 \text{ m.}$$

34.6   Refer to Fig. 34-15. A body at *O* is being acted upon by two forces, one of 150 lb due north and the other of 200 lb due east. Find the magnitude and direction of the resultant.

Fig. 34-15

In the right triangle *OBC*,

$$OC = \sqrt{(OB)^2 + (BC)^2} = \sqrt{(200)^2 + (150)^2} = 250 \text{ lb}$$

$$\tan \angle BOC = \tfrac{150}{200} = 0.7500 \quad \text{and} \quad \text{m}\angle BOC = 36° 50'$$

The magnitude of the resultant force is 250 lb and its direction is N 53° 10' E.

34.7   An airplane is moving horizontally at 240 mph when a bullet is shot with speed 2750 ft/s at right angles to the path of the airplane. Find the resultant speed and direction of the bullet.

The speed of the airplane is $240 \text{ mi/hr} = \dfrac{240(5280)}{60(60)} \text{ ft/s} = 352 \text{ ft/s.}$

In Fig. 34-16, the vector $\overrightarrow{AB}$ represents the velocity of the airplane, the vector $\overrightarrow{AC}$ represents the initial velocity of the bullet, and the vector $\overrightarrow{AD}$ represents the resultant velocity of the bullet.

Fig. 34-16

In the right triangle ACD,

$$AD = \sqrt{(352)^2 + (2750)^2} = 2770 \text{ ft/s}$$

$$\tan \angle CAD = \tfrac{352}{2750} = 0.1280 \qquad \text{and} \qquad \text{m} \angle CAD = 7° \, 20'$$

Thus, the bullet travels at 2770 ft/s along a path making an angle of measure 82° 40′ with the path of the airplane.

34.8 A river flows due south at 125 ft/min. A motorboat, moving at 475 ft/min in still water, is headed due east across the river. (*a*) Find the direction in which the boat moves and its speed. (*b*) In what direction must the boat be headed in order that it move due east and what is its speed in that direction?

(*a*)   Refer to Fig. 34-17(*a*). In right triangle OAB,

$$OB = \sqrt{(475)^2 + (125)^2} = 491, \qquad \tan \theta = \tfrac{125}{475} = 0.2632, \qquad \text{and} \qquad \text{m} \angle \theta = 14° \, 40'.$$

(*a*)	(*b*)

Fig. 34-17

Thus the boat moves at 491 ft/min in the direction S 75° 20′ E.

(*b*)   Refer to Fig. 34-17(*b*). In right triangle OAB, $\sin \theta = \tfrac{125}{475} = 0.2632$ and $\text{m} \angle \theta = 15° \, 20'$. Thus the boat must be headed N 74° 40′ E and its speed in that direction is $OB = \sqrt{(475)^2 - (125)^2} = 485$ ft/min.

34.9 A telegraph pole is kept vertical by a guy wire which makes an angle of 25° with the pole and which exerts a pull of $F = 300$ lb on the top. Find the horizontal and vertical components F_h and F_v of the pull $\overline{F}$. See Figs. 34-18 and 34-19.

$$F_h = 300 \sin 25° = 300(0.4226) = 127 \text{ lb}$$

$$F_v = 300 \cos 25° = 300(0.9063) = 272 \text{ lb}$$

Fig. 34-18   **Fig. 34-19**

34.10 A man pulls a rope attached to a sled with a force of 100 lb. The rope makes an angle of 27° with the ground. (*a*) Find the effective pull tending to move the sled along the ground and the effective pull tending to lift the sled vertically. (*b*) Find the force which the man must exert in order that the effective force tending to move the sled along the ground is 100 lb.

(*a*) In Fig. 34-20(*a*) and 34-20(*b*), the 100-lb pull in the rope is resolved into horizontal and vertical components, F_h and F_v respectively. Then F_h is the force tending to move the sled along the ground and F_v is the force tending to lift the sled.

$$F_h = 100 \cos 27° = 100(0.8910) = 89 \text{ lb}, \qquad F_v = 100 \sin 27° = 100(0.4540) = 45 \text{ lb}.$$

(*b*) In Fig. 34-20(*c*), the horizontal component of the required force F is $F_h = 100$ lb. Then

$$\overrightarrow{F} = \frac{100}{\cos 27°} = \frac{100}{0.8910} = 112 \text{ lb}$$

(*a*)   (*b*)   (*c*)

Fig. 34-20

34.11 A block weighing $W = 500$ lb rests upon a ramp inclined 29° with the horizontal. (*a*) Find the force tending to move the block down the ramp and the force of the block on the ramp. (*b*) What minimum force must be applied to keep the block from sliding down the ramp? Neglect friction.

(*a*) Refer to Fig. 34-21. Resolve the weight W of the block into components F_1 and F_2, respectively parallel and perpendicular to the ramp. F_1 is the force tending to move the block down the ramp and F_2 is the force of the block on the ramp.

$$F_1 = W \sin 29° = 500(0.4848) = 242 \text{ lb} \qquad F_2 = W \cos 29° = 500(0.8746) = 437 \text{ lb}$$

Fig. 34-21

(*b*) 242 lb up the ramp.

Supplementary Problems

34.12 An airplane flies 100 km in the direction S 38° 10′ E. How far south and how far east of the starting point is it?

    Ans. 78.6 km south, 61.8 km east

34.13   A plane is headed due east with airspeed 240 mph. If a wind at 40 mph from the north is blowing, find the groundspeed and track.

   Ans. Groundspeed, 243 mph; track, 99° 30′ or S 80° 30′ E

34.14   A body is acted upon by a force of 75 lb, due west, and a force of 125 lb, due north. Find the magnitude and direction of the resultant force.

   Ans. 146 lb, N 31° 0′ W

34.15   Find the rectangular components of a force of 525.0 lb in a direction 38° 25′ with the horizontal.

   Ans. 411.3 lb, 326.2 lb

34.16   A barge is being towed north at the rate of 18 mph. A man walks across the deck from west to east at the rate of 6 ft/s. Find the magnitude and direction of the actual velocity.

   Ans. 27 ft/s, N 12° 50′ E

34.17   A ship at *A* is to sail to *C*, 56 km north and 258 km east of *A*. After sailing N 25° 10′ E for 120 km to *P*, the ship is headed toward *C*. Find the distance of *P* from *C* and the required course to reach *C*.

   Ans. 214 km, S 75° 40′E

34.18   A guy wire 78 m long runs from the top of a telephone pole 56 m high to the ground and pulls on the pole with a force of 290 lb. What is the horizontal pull on the top of the pole?

   Ans. 202 lb

34.19   A weight of 200 lb is placed on a smooth plane inclined at an angle measuring 38° with the horizontal and held in place by a rope parallel to the surface and fastened to a peg in the plane. Find the pull on the string.

   Ans. 123 lb

34.20   A man wishes to raise a 300-lb weight to the top of a wall 20 ft high by dragging it up an incline. What is the length of the shortest inclined plane he can use if his pulling strength is 140 lb?

   Ans. 43 ft

34.21   A 150-lb shell is dragged up a runway inclined 40° to the horizontal. Find (*a*) the force of the shell against the runway and (*b*) the force required to drag the shell.

   Ans. (*a*) 115 lb (*b*) 96 lb

Chapter 35

Reduction to Functions of Positive Acute Angles

COTERMINAL ANGLES. Let θ be any angle; then

$$\sin(\theta + n360°) = \sin\theta \qquad \cot(\theta + n360°) = \cot\theta$$

$$\cos(\theta + n360°) = \cos\theta \qquad \sec(\theta + n360°) = \sec\theta$$

$$\tan(\theta + n360°) = \tan\theta \qquad \csc(\theta + n360°) = \csc\theta$$

where n is any positive or negative integer or zero.

EXAMPLES. $\sin 400° = \sin(40° + 360°) = \sin 40°$

$$\cos 850° = \cos(130° + 2 \cdot 360°) = \cos 130°$$

$$\tan(-1000°) = \tan(80° - 3 \cdot 360°) = \tan 80°$$

FUNCTIONS OF A NEGATIVE ANGLE. Let θ be an angle; then

$$\sin(-\theta) = -\sin\theta \qquad \cot(-\theta) = -\cot\theta$$

$$\cos(-\theta) = \cos\theta \qquad \sec(-\theta) = \sec\theta$$

$$\tan(-\theta) = -\tan\theta \qquad \csc(-\theta) = -\csc\theta$$

EXAMPLES. $\sin(-50°) = -\sin 50°, \cos(-30°) = \cos 30°, \tan(-200°) = -\tan 200°$.

REDUCTION FORMULAS. Let θ be an angle; then

$$\sin(90° - \theta) = \cos\theta \qquad \sin(90° + \theta) = \cos\theta$$

$$\cos(90° - \theta) = \sin\theta \qquad \cos(90° + \theta) = -\sin\theta$$

$$\tan(90° - \theta) = \cot\theta \qquad \tan(90° + \theta) = -\cot\theta$$

$$\cot(90° - \theta) = \tan\theta \qquad \cot(90° + \theta) = -\tan\theta$$

$$\sec(90° - \theta) = \csc\theta \qquad \sec(90° + \theta) = -\csc\theta$$

$$\csc(90° - \theta) = \sec\theta \qquad \csc(90° + \theta) = \sec\theta$$

$$\sin(180° - \theta) = \sin\theta \qquad \sin(180° + \theta) = -\sin\theta$$

$$\cos(180° - \theta) = -\cos\theta \qquad \cos(180° + \theta) = -\cos\theta$$

$$\tan(180° - \theta) = -\tan\theta \qquad \tan(180° + \theta) = \tan\theta$$

$$\cot(180° - \theta) = -\cot\theta \qquad \cot(180° + \theta) = \cot\theta$$

$$\sec(180° - \theta) = -\sec\theta \qquad \sec(180° + \theta) = -\sec\theta$$

$$\csc(180° - \theta) = \csc\theta \qquad \csc(180° + \theta) = -\csc\theta$$

GENERAL REDUCTION FORMULA.   Any trigonometric function of $(n \cdot 90° \pm \theta)$, where θ is any angle, is *numerically* equal

(a)   To the same function of θ if n is an even integer

(b)   To the corresponding cofunction of θ if n is an odd integer

The algebraic sign in each case is the same as the sign of the given function for that quadrant in which $n \cdot 90° \pm \theta$ lies when θ is a positive acute angle.

EXAMPLES

(1) $\sin(180° - \theta) = \sin(2 \cdot 90° - \theta) = \sin \theta$ since $180°$ is an even multiple of $90°$ and, when θ is positive acute, the terminal side of $180° - \theta$ lies in quadrant II.

(2) $\cos(180° + \theta) = \cos(2 \cdot 90° + \theta) = -\cos \theta$ since $180°$ is an even multiple of $90°$ and, when θ is positive acute, the terminal side of $180° + \theta$ lies in quadrant III.

(3) $\tan(270° - \theta) = \tan(3 \cdot 90° - \theta) = \cot \theta$ since $270°$ is an odd multiple of $90°$ and, when θ is positive acute, the terminal side of $270° - \theta$ lies in quadrant III.

(4) $\cos(270° + \theta) = \cos(3 \cdot 90° + \theta) = \sin \theta$ since $270°$ is an odd multiple of $90°$ and, when θ is positive acute, the terminal side of $270° + \theta$ lies in quadrant IV.

Solved Problems

35.1   Express each of the following in terms of a function of θ:

(a) $\sin(\theta - 90°)$ (d) $\cos(-180° + \theta)$ (g) $\sin(540° + \theta)$ (j) $\cos(-450° - \theta)$

(b) $\cos(\theta - 90°)$ (e) $\sin(-270° - \theta)$ (h) $\tan(720° - \theta)$ (k) $\csc(-900° + \theta)$

(c) $\sec(-\theta - 90°)$ (f) $\tan(\theta - 360°)$ (i) $\tan(720° + \theta)$ (l) $\sin(-540° - \theta)$

(a) $\sin(\theta - 90°) = \sin(-90° + \theta) = \sin(-1 \cdot 90° + \theta) = -\cos \theta$, the sign being negative since, when θ is positive acute, the terminal side of $\theta - 90°$ lies in quadrant IV.

(b) $\cos(\theta - 90°) = \cos(-90° + \theta) = \cos(-1 \cdot 90° + \theta) = \sin \theta$.

(c) $\sec(-\theta - 90°) = \sec(-90° - \theta) = \sec(-1 \cdot 90° - \theta) = -\csc \theta$, the sign being negative since, when θ is positive acute, the terminal side of $-\theta - 90°$ lies in quadrant III.

(d) $\cos(-180° + \theta) = \cos(-2 \cdot 90° + \theta) = -\cos \theta$. (quadrant III)

(e) $\sin(-270° - \theta) = \sin(-3 \cdot 90° - \theta) = \cos \theta$. (quadrant I)

(f) $\tan(\theta - 360°) = \tan(-4 \cdot 90° + \theta) = \tan \theta$. (quadrant I)

(g) $\sin(540° + \theta) = \sin(6 \cdot 90° + \theta) = -\sin \theta$. (quadrant III)

(h) $\tan(720° - \theta) = \tan(8 \cdot 90° - \theta) = -\tan \theta = \tan(2 \cdot 360° - \theta) = \tan(-\theta) = -\tan \theta$.

(i) $\tan(720° + \theta) = \tan(8 \cdot 90° + \theta) = \tan \theta = \tan(2 \cdot 360° + \theta) = \tan \theta$.

(j) $\cos(-450° - \theta) = \cos(-5 \cdot 90° - \theta) = -\sin \theta$.

(k) $\csc(-900° + \theta) = \csc(-10 \cdot 90° + \theta) = -\csc \theta$.

(l) $\sin(-540° - \theta) = \sin(-6 \cdot 90° - \theta) = \sin \theta$.

35.2   Express each of the following in terms of functions of a positive acute angle in two ways:

(a) $\sin 130°$ (c) $\sin 200°$ (e) $\tan 165°$ (g) $\sin 670°$ (i) $\csc 865°$ (k) $\cos(-680°)$

(b) $\tan 325°$ (d) $\cos 310°$ (f) $\sec 250°$ (h) $\cot 930°$ (j) $\sin(-100°)$ (l) $\tan(-290°)$

(a) $\sin 130° = \sin(2 \cdot 90° - 50°) = \sin 50° = \sin(1 \cdot 90° + 40°) = \cos 40°$

(b) $\tan 325° = \tan(4 \cdot 90° - 35°) = -\tan 35° = \tan(3 \cdot 90° + 55°) = -\cot 55°$

(c) $\sin 200° = \sin(2 \cdot 90° + 20°) = -\sin 20° = \sin(3 \cdot 90° - 70°) = -\cos 70°$

(d) $\cos 310° = \cos(4 \cdot 90° - 50°) = \cos 50° = \cos(3 \cdot 90° + 40°) = \sin 40°$

(e) $\tan 165° = \tan(2 \cdot 90° - 15°) = -\tan 15° = \tan(1 \cdot 90° + 75°) = -\cot 75°$

(f) $\sec 250° = \sec(2 \cdot 90° + 70°) = -\sec 70° = \sec(3 \cdot 90° - 20°) = -\csc 20°$

(g) $\sin 670° = \sin(8 \cdot 90° - 50°) = -\sin 50° = \sin(7 \cdot 90° + 40°) = -\cos 40°$
or $\sin 670° = \sin(310° + 360°) = \sin 310° = \sin(4 \cdot 90° - 50°) = -\sin 50°.$

(h) $\cot 930° = \cot(10 \cdot 90° + 30°) = \cot 30° = \cot(11 \cdot 90° - 60°) = \tan 60°$
or $\cot 930° = \cot(210° + 2 \cdot 360°) = \cot 210° = \cot(2 \cdot 90° + 30°) = \cot 30°.$

(i) $\csc 865° = \csc(10 \cdot 90° - 35°) = \csc 35° = \csc(9 \cdot 90° + 55°) = \sec 55°$
or $\csc 865° = \csc(145° + 2 \cdot 360°) = \csc 145° = \csc(2 \cdot 90° - 35°) = \csc 35°.$

(j) $\sin(-100°) = \sin(-2 \cdot 90° + 80°) = -\sin 80° = \sin(-1 \cdot 90° - 10°) = -\cos 10°$
or $\sin(-100°) = -\sin 100° = -\sin(2 \cdot 90° - 80°) = -\sin 80°$ or $\sin(-100°) = \sin(-100° + 360°) = \sin 260° = \sin(2 \cdot 90° + 80°) = -\sin 80°.$

(k) $\cos(-680°) = \cos(-8 \cdot 90° + 40°) = \cos 40° = \cos(-7 \cdot 90° - 50°) = \sin 50°$
or $\cos(-680°) = \cos(-680° + 2 \cdot 360°) = \cos 40°.$

(l) $\tan(-290°) = \tan(-4 \cdot 90° + 70°) = \tan 70° = \tan(-3 \cdot 90° - 20°) = \cot 20°$
or $\tan(-290°) = \tan(-290° + 360°) = \tan 70°.$

35.3 Find the exact values of the sine, cosine, and tangent of

(a) 120°,   (b) 210°,   (c) 315°,   (d) −135°,   (e) −240°,   (f) −330°.

Call θ, always positive acute, the *related angle* of ϕ when $\phi = 180° - \theta$, $180° + \theta$, or $360° - \theta$. Then any function of ϕ is numerically equal to the same function of θ. The algebraic sign in each case is that of the function in the quadrant in which the terminal side of ϕ lies.

(a) $120° = 180° - 60°$. The related angle is 60°; 120° is in quadrant II. $\sin 120° = \sin 60° = \sqrt{3}/2$, $\cos 120° = -\cos 60° = -\frac{1}{2}$, $\tan 120° = -\tan 60° = -\sqrt{3}$.

(b) $210° = 180° + 30°$. The related angle is 30°; 210° is in quadrant III. $\sin 210° = -\sin 30° = -\frac{1}{2}$, $\cos 210° = -\cos 30° = -\sqrt{3}/2$, $\tan 210° = \tan 30° = \sqrt{3}/3$.

(c) $315° = 360° - 45°$. The related angle is 45°; 315° is in quadrant IV. $\sin 315° = -\sin 45° = -\sqrt{2}/2$, $\cos 315° = \cos 45° = \sqrt{2}/2$, $\tan 315° = -\tan 45° = -1$.

(d) Any function of −135° is the same function of $-135° + 360° = 225° = \phi$. $225° = 180° + 45°$. The related angle is 45°; 225° is in quadrant III. $\sin(-135°) = -\sin 45° = -\sqrt{2}/2$, $\cos(-135°) = -\cos 45° = -\sqrt{2}/2$, $\tan(-135°) = 1$.

(e) Any function of −240° is the same function of $-240° + 360° = 120°$. $120° = 180° - 60°$. The related angle is 60°; 120° is in quadrant II. $\sin(-240°) = \sin 60° = \sqrt{3}/2$, $\cos(-240°) = -\cos 60° = -\frac{1}{2}$, $\tan(-240°) = -\tan 60° = -\sqrt{3}$.

(f) Any function of −330° is the same function of $-330° + 360° = 30°$. $\sin(-330°) = \sin 30° = \frac{1}{2}$, $\cos(-330°) = \cos 30° = \sqrt{3}/2$, $\tan(-330°) = \tan 30° = \sqrt{3}/3$.

35.4 Using the table of natural functions, find

(a) $\sin 125° 14' = \sin(180° - 54° 46') = \sin 54° 46' = 0.8168$

(b) $\cos 169° 40' = \cos(180° - 10° 20') = -\cos 10° 20' = -0.9838$

(c) $\tan 200° 23' = \tan(180° + 20° 23') = \tan 20° 23' = 0.3716$

(d) $\cot 250° 44' = \cot(180° + 70° 44') = \cot 70° 44' = 0.3495$

(e) $\cos 313° 18' = \cos(360° - 46° 42') = \cos 46° 42' = 0.6858$

(f) $\sin 341° 52' = \sin(360° - 18° 8') = -\sin 18° 8' = -0.3112$

35.5 If $\tan 25° = a$, find

(a) $\dfrac{\tan 155° - \tan 115°}{1 + \tan 155° \tan 115°} = \dfrac{-\tan 25° - (-\cot 25°)}{1 + (-\tan 25°)(-\cot 25°)} = \dfrac{-a + 1/a}{1 + a(1/a)} = \dfrac{-a^2 + 1}{a + a} = \dfrac{1 - a^2}{2a}.$

(b) $\dfrac{\tan 205° - \tan 115°}{\tan 245° - \tan 335°} = \dfrac{\tan 25° - (-\cot 25°)}{\cot 25° + (-\tan 25°)} = \dfrac{a + 1/a}{1/a - a} = \dfrac{a^2 + 1}{1 - a^2}.$

35.6 If $m\angle A + m\angle B + m\angle C = 180°$, then

(a) $\sin(B + C) = \sin(180° - A) = \sin A.$

(b) $\sin \frac{1}{2}(B + C) = \sin \frac{1}{2}(180° - A) = \sin(90° - \frac{1}{2}A) = \cos \frac{1}{2}A.$

35.7 Show that $\sin \theta$ and $\tan \frac{1}{2}\theta$ have the same sign.

(a) Suppose $m\angle \theta = n \cdot 180°$. If n is even (including zero), say $2m$, then $\sin(2m \cdot 180°) = \tan(m \cdot 180°) = 0$. The case when n is odd is excluded since then $\tan \frac{1}{2}\theta$ is not defined.

(b) Suppose $m\angle \theta = n \cdot 180° + \phi$, where $0 < m\angle \phi < 180°$. If n is even, including zero, θ is in quadrant I or quadrant II and $\sin \theta$ is positive while $\frac{1}{2}\theta$ is in quadrant I or quadrant III and $\tan \frac{1}{2}\theta$ is positive. If n is odd, θ is in quadrant III or IV and $\sin \theta$ is negative while $\frac{1}{2}\theta$ is in quadrant II or IV and $\tan \frac{1}{2}\theta$ is negative.

35.8 Find all positive values of θ less than 360° for which $\sin \theta = -\frac{1}{2}$.

There will be two angles (see Chapter 31), one in the third quadrant and one in the fourth quadrant. The related angle of each has its sine equal to $+\frac{1}{2}$ and is 30°. Thus the required angles are θ with measure $180° + 30° = 210°$ and θ with measure $360° - 30° = 330°$.

(NOTE.   To obtain *all* values of θ for which $\sin \theta = -\frac{1}{2}$, add $n \cdot 360°$ to each of the above solutions; thus $\theta = 210° + n \cdot 360°$ and $\theta = 330° + n \cdot 360°$, where n is any integer.)

35.9 Find all positive values of θ less than 360° for which $\cos \theta = 0.9063$.

There are two solutions, $m\angle \theta = 25°$ in quadrant I and $m\angle \theta = 360° - 25° = 335°$ in quadrant IV.

35.10 Find all positive values of $\frac{1}{4}\theta$ less than 360°, given $\sin \theta = 0.6428$.

The two positive angles less than 360° for which $\sin \theta = 0.6428$ are $m\angle \theta = 40°$ and $m\angle \theta = 180° - 40° = 140°$. But if $\frac{1}{4}\theta$ is to include all values less than 360°, θ must include all values less than $4 \cdot 360° = 1440°$. Hence, for θ we take the two angles above and all coterminal angles with measures less than 1440°, that is,

$$\theta = 40°, 400°, 760°, 1120°; 140°, 500°, 860°, 1220° \quad \text{and}$$

$$\tfrac{1}{4}\theta = 10°, 100°, 190°, 280°; 35°, 125°, 215°, 305°.$$

35.11 Find all positive values of θ less than 360° which satisfy $\sin 2\theta = \cos \frac{1}{2}\theta$.

Since $\cos \frac{1}{2}\theta = \sin(90° - \frac{1}{2}\theta) = \sin 2\theta$, $2\theta = 90° - \frac{1}{2}\theta$, $450° - \frac{1}{2}\theta$, $810° - \frac{1}{2}\theta$, $1170° - \frac{1}{2}\theta$, ... Then $\frac{5}{2}\theta = 90°, 450°, 810°, 1170°, ...$ and $m\angle \theta = 36°, 180°, 324°, 468°, ...$.

Since $\cos \frac{1}{2}\theta = \sin(90° + \frac{1}{2}\theta) = \sin 2\theta$, $2\theta = 90° + \frac{1}{2}\theta$, $450° + \frac{1}{2}\theta$, $810° + \frac{1}{2}\theta$, ... Then $\frac{3}{2}\theta = 90°, 450°, 810°, ...$ and $\theta = 60°, 300°, 540°, ...$.

The required solutions have measures 36°, 180°, 324°; 60°, 300°.

Supplementary Problems

35.12 Express each of the following in terms of functions of a positive acute angle:

(a) $\sin 145°$ (d) $\cot 155°$ (g) $\sin(-200°)$ (j) $\cot 610°$

(b) $\cos 215°$ (e) $\sec 325°$ (h) $\cos(-760°)$ (k) $\sec 455°$

(c) $\tan 440°$ (f) $\csc 190°$ (i) $\tan(-1385°)$ (l) $\csc 825°$

Ans. (a) $\sin 35°$ or $\cos 55°$ (g) $\sin 20°$ or $\cos 70°$

(b) $-\cos 35°$ or $-\sin 55°$ (h) $\cos 40°$ or $\sin 50°$

(c) $\tan 80°$ or $\cot 10°$ (i) $\tan 55°$ or $\cot 35°$

(d) $-\cot 25°$ or $-\tan 65°$ (j) $\cot 70°$ or $\tan 20°$

(e) $\sec 35°$ or $\csc 55°$ (k) $-\sec 85°$ or $-\csc 5°$

(f) $-\csc 10°$ or $-\sec 80°$ (l) $\csc 75°$ or $\sec 15°$

35.13 Find the exact values of the sine, cosine, and tangent of

(a) $150°$ (b) $225°$ (c) $300°$ (d) $-120°$ (e) $-210°$ (f) $-315°$

Ans. (a) $\frac{1}{2}, -\sqrt{3}/2, -1/\sqrt{3}$ (d) $-\sqrt{3}/2, -\frac{1}{2}, \sqrt{3}$

(b) $-\sqrt{2}/2, -\sqrt{2}/2, 1$ (e) $\frac{1}{2}, -\sqrt{3}/2, -1/\sqrt{3}$

(c) $-\sqrt{3}/2, \frac{1}{2}, -\sqrt{3}$ (f) $\sqrt{2}/2, \sqrt{2}/2, 1$

35.14 Using appropriate tables, find

(a) $\sin 155° 13' = 0.4192$ (f) $\log \sin 129° 44.8' = 9.88586\text{-}10$

(b) $\cos 104° 38' = -0.2526$ (g) $\log \sin 110° 32.7' = 9.97146\text{-}10$

(c) $\tan 305° 24' = -1.4071$ (h) $\log \sin 162° 35.6' = 9.47589\text{-}10$

(d) $\sin 114° 18' = 0.9114$ (i) $\log \sin 138° 30.5' = 9.82119\text{-}10$

(e) $\cos 166° 51' = -0.9738$ (j) $\log \sin 174° 22.7' = 8.99104\text{-}10$

35.15 Find all angles, $0 \le \theta < 360°$, for which

(a) $\sin \theta = \sqrt{2}/2$ (b) $\cos \theta = -1$ (c) $\sin \theta = -0.6180$ (d) $\cos \theta = 0.5125$

(e) $\tan \theta = -1.5301$

Ans. (a) $45°, 135°$ (c) $218° 10', 321° 50'$ (e) $123° 10', 303° 10'$

(b) $180°$ (d) $59° 10', 300° 50'$

35.16 When θ is a second quadrant angle for which $\tan \theta = -\frac{2}{3}$, show that

(a) $\dfrac{\sin(90° - \theta) - \cos(180° - \theta)}{\tan(270° + \theta) + \cot(360° - \theta)} = -\dfrac{2}{\sqrt{13}}$ (b) $\dfrac{\tan(90° + \theta) + \cos(180° + \theta)}{\sin(270° - \theta) - \cot(-\theta)} = \dfrac{2 + \sqrt{13}}{2 - \sqrt{13}}$

Chapter 36

Variations and Graphs of the Trigonometric Functions

LINE REPRESENTATIONS OF THE TRIGONOMETRIC FUNCTIONS. Let θ be any given angle in standard position. (See the Figs. 36-1 through 36-4 for θ in each of the quadrants.) With the vertex O as center describe a circle of radius one unit cutting the initial side $\overrightarrow{OX}$ of θ at A, the positive y axis at B, and the terminal side of θ at P. Draw $\overline{MP}$ perpendicular to $\overrightarrow{OX}$, draw also the tangents to the circle at A and B meeting the terminal side of θ or its extension through O in the points Q and R, respectively.

Fig. 36-1

Fig. 36-2

Fig. 36-3

Fig. 36-4

In each of the figures, the right triangles OMP, OAQ, and OBR are similar, and

$$\sin \theta = \frac{MP}{OP} = MP \qquad \cot \theta = \frac{OM}{MP} = \frac{BR}{OB} = BR$$

$$\cos \theta = \frac{OM}{OP} = OM \qquad \sec \theta = \frac{OP}{OM} = \frac{OQ}{OA} = OQ$$

$$\tan \theta = \frac{MP}{OM} = \frac{AQ}{OA} = AQ \qquad \csc \theta = \frac{OP}{MP} = \frac{OR}{OB} = OR$$

Then, $\overline{MP}$, $\overline{OM}$, $\overline{AQ}$, etc., are directed line segments, the magnitude of a function being given by the length of the corresponding segment and the sign being given by the indicated direction. The directed segments $\overline{OQ}$ and $\overline{OR}$ are to be considered positive when measured on the terminal side of the angle and negative when measured on the terminal side extended.

VARIATIONS OF THE TRIGONOMETRIC FUNCTIONS.   Let P move counterclockwise about the unit circle, starting at A, so that m $\angle$ θ = m $\angle$ XOP varies continuously from 0° to 360°. Using Figs. 36-1 through 36-4 Table 36.1 is derived.

Table 36.1

As θ increases from	0° to 90°	90° to 180°	180° to 270°	270° to 360°
sin θ	I from 0 to 1	D from 1 to 0	D from 0 to −1	I from −1 to 0
cos θ	D from 1 to 0	D from 0 to −1	I from −1 to 0	I from 0 to 1
tan θ	I from 0 without limit (0 to +∞)	I from large negative values to 0 (−∞ to 0)	I from 0 without limit (0 to +∞)	I from large negative values to 0 (−∞ to 0)
cot θ	D from large positive values to 0 (+∞ to 0)	D from 0 without limit (0 to −∞)	D from large positive values to 0 (+∞ to 0)	D from 0 without limit (0 to −∞)
sec θ	I from 1 without limit (1 to +∞)	I from large negative values to −1 (−∞ to −1)	D from −1 without limit (−1 to −∞)	D from large positive values to 1 (+∞ to 1)
csc θ	D from large positive values to 1 (+∞ to 1)	I from 1 without limit (1 to +∞)	I from large negative values to −1 (−∞ to −1)	D from −1 without limit (−1 to −∞)

I = increases; D = decreases.

GRAPHS OF THE TRIGONOMETRIC FUNCTIONS.   In Table 36.2, values of the angle x are given in radians.

Table 36.2

x	$y = \sin x$	$y = \cos x$	$y = \tan x$	$y = \cot x$	$y = \sec x$	$y = \csc x$
0	0	1.00	0	±∞	1.00	±∞
$\pi/6$	0.50	0.87	0.58	1.73	1.15	2.00
$\pi/4$	0.71	0.71	1.00	1.00	1.41	1.41
$\pi/3$	0.87	0.50	1.73	0.58	2.00	1.15
$\pi/2$	1.00	0	±∞	0	±∞	1.00
$2\pi/3$	0.87	−0.50	−1.73	−0.58	−2.00	1.15
$3\pi/4$	0.71	−0.71	−1.00	−1.00	−1.41	1.41
$5\pi/6$	0.50	−0.87	−0.58	−1.73	−1.15	2.00
π	0	−1.00	0	+∞	−1.00	+∞
$7\pi/6$	−0.50	−0.87	0.58	1.73	−1.15	−2.00
$5\pi/4$	−0.71	−0.71	1.00	1.00	−1.41	−1.41
$4\pi/3$	−0.87	−0.50	1.73	0.58	−2.00	−1.15
$3\pi/2$	−1.00	0	±∞	0	±∞	−1.00
$5\pi/3$	−0.87	0.50	−1.73	−0.58	2.00	−1.15
$7\pi/4$	−0.71	0.71	−1.00	−1.00	1.41	−1.41
$11\pi/6$	−0.50	0.87	−0.58	−1.73	1.15	−2.00
2π	0	1.00	0	±∞	1.00	±∞

Note 1.   Since $\sin\left(\frac{1}{2}\pi + x\right) = \cos x$, the graph of $y = \cos x$ may be obtained most easily by shifting the graph of $y = \sin x$ a distance $\frac{1}{2}\pi$ to the left. See Fig. 36-5.

Fig. 36-5

Note 2.   Since $\csc\left(\frac{1}{2}\pi + x\right) = \sec x$, the graph of $y = \csc x$ may be obtained by shifting the graph of $y = \sec x$ a distance $\frac{1}{2}\pi$ to the right. Notice, too, the relationship between the graphs for tan x and cot x. See Figs. 36-6 through 36-9.

$y = \tan x$

Fig. 36-6

$y = \cot x$

Fig. 36-7

$y = \sec x$

Fig. 36-8

$y = \csc x$

Fig. 36-9

PERIODIC FUNCTIONS.   Any function of a variable x, $f(x)$, which repeats its values in definite cycles, is called *periodic*. The smallest range of values of x which corresponds to a complete cycle of values of the function is called the period of the function. It is evident from the graphs of the trigonometric functions that the sine, cosine, secant, and cosecant are of period 2π while the tangent and cotangent are of period π.

THE GENERAL SINE CURVE. The *amplitude (maximum ordinate) and period (wavelength) of* $y = \sin x$ are, respectively, 1 and 2π. For a given value of x, the value of $y = a \sin x, a > 0$, is a times the value of $y = \sin x$. Thus, the amplitude of $y = a \sin x$ is a and the period is 2π. Since when $bx = 2\pi$, $x = 2\pi/b$, the amplitude of $y = \sin bx$, $b > 0$, is 1 and the period is $2\pi/b$.

The general sine curve (sinusoid) of equation

$$y = a \sin bx, \qquad a > 0, \quad b > 0,$$

has amplitude a and period $2\pi/b$. Thus the graph of $y = 3 \sin 2x$ has amplitude 3 and period $2\pi/2 = \pi$. Figure 36-10 exhibits the graphs of $y = \sin x$ and $y = 3 \sin 2x$ on the same axes.

Fig. 36-10

COMPOSITION OF SINE CURVES. More complicated forms of wave motions are obtained by combining two or more sine curves. The method of adding corresponding ordinates is illustrated in the following example.

EXAMPLE. Construct the graph of $y = \sin x + 3 \sin 2x$. See Fig. 36-10.

First the graphs of $y_1 = \sin x$ and $y_2 = 3 \sin 2x$ are constructed on the same axes. Then, corresponding to a given value $x = OA_1$, the ordinate A_1B of $y = \sin x + 3 \sin 2x$ is the *algebraic* sum of the ordinates A_1B_1 of $y_1 = \sin x$ and A_1C_1 of $y_2 = 3 \sin 2x$.

Also, $A_2B = A_2B_2 + A_2C_2$, $A_3B = A_3B_3 + A_3C_3$, etc.

Solved Problems

36.1   Sketch the graphs of the following for one wavelength:

   (*a*) $y = 4 \sin x$ (*c*) $y = 3 \sin \frac{1}{2}x$ (*e*) $y = 3 \cos \frac{1}{2}x = 3 \sin \left(\frac{1}{2}x + \frac{1}{2}\pi\right)$

   (*b*) $y = \sin 3x$ (*d*) $y = 2 \cos x = 2 \sin \left(x + \frac{1}{2}\pi\right)$

   In each case we use the same curve and then put in the y axis and choose the units on each axis to satisfy the requirements of amplitude and period of each curve.

   (*a*) $y = 4 \sin x$ has amplitude = 4 and period = 2π. See Fig. 36-11(*a*).

   (*b*) $y = \sin 3x$ has amplitude = 1 and period = $2\pi/3$. See Fig. 36-11(*b*).

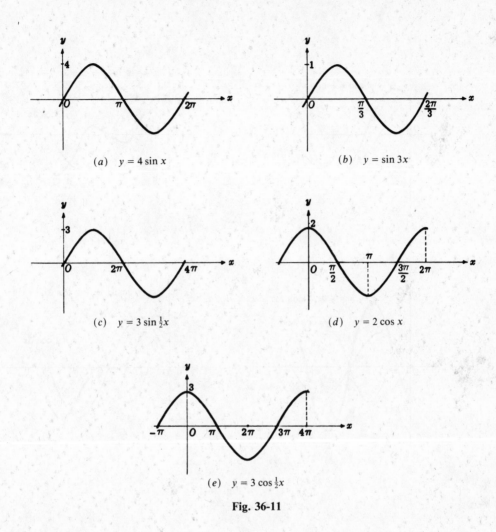

(a) $y = 4 \sin x$

(b) $y = \sin 3x$

(c) $y = 3 \sin \tfrac{1}{2}x$

(d) $y = 2 \cos x$

(e) $y = 3 \cos \tfrac{1}{2}x$

Fig. 36-11

(c) $y = 3 \sin \tfrac{1}{2}x$ has amplitude = 3 and period = $2\pi/\tfrac{1}{2} = 4\pi$. See Fig. 36-11(c).

(d) $y = 2 \cos x$ has amplitude = 2 and period = 2π. Note the position of the y axis. See Fig. 36-11(d).

(e) $y = 3 \cos \tfrac{1}{2}x$ has amplitude = 3 and period = 4π. See Fig. 36-11(e).

36.2   Construct the graph of each of the following:

(a) $y = \sin x + \cos x$ (c) $y = \sin 2x - \cos 3x$

(b) $y = \sin 2x + \cos 3x$ (d) $y = 3 \sin 2x + 2 \cos 3x$

See Fig. 36-12(a)–(d).

Supplementary Problems

36.3   Sketch the graph of each of the following for one wavelength:

(a) $y = 3 \sin x$ (b) $y = \sin 2x$ (c) $y = 4 \sin (x/2)$ (d) $y = 4 \cos x$

(e) $y = 2 \cos (x/3)$

Fig. 36-12

36.4 Construct the graph of each of the following for one wavelength:

(a) $y = \sin x + 2\cos x$ (d) $y = \sin 2x + \sin 3x$

(b) $y = \sin 3x + \cos 2x$ (e) $y = \sin 3x - \cos 2x$

(c) $y = \sin x + \sin 2x$ (f) $y = 2\sin 3x + 3\cos 2x$

Chapter 37

Fundamental Relations and Identities

FUNDAMENTAL RELATIONS

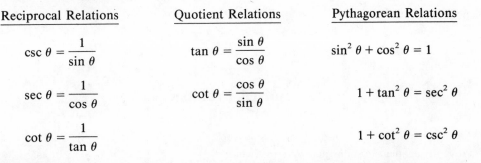

Reciprocal Relations	Quotient Relations	Pythagorean Relations
$\csc \theta = \dfrac{1}{\sin \theta}$	$\tan \theta = \dfrac{\sin \theta}{\cos \theta}$	$\sin^2 \theta + \cos^2 \theta = 1$
$\sec \theta = \dfrac{1}{\cos \theta}$	$\cot \theta = \dfrac{\cos \theta}{\sin \theta}$	$1 + \tan^2 \theta = \sec^2 \theta$
$\cot \theta = \dfrac{1}{\tan \theta}$		$1 + \cot^2 \theta = \csc^2 \theta$

The above relations hold for every value of θ for which the functions involved are defined.

Thus, $\sin^2 \theta + \cos^2 \theta = 1$ holds for every value of θ while $\tan \theta = \sin \theta / \cos \theta$ holds for all values of θ for which $\tan \theta$ is defined, i.e., for all $\theta \neq n \cdot 90°$ where n is odd. Note that for the excluded values of θ, $\cos \theta = 0$ and $\sin \theta \neq 0$.

For proofs of the quotient and Pythagorean relations, see Problems 37.1–37.2. The reciprocal relations were treated in Chapter 31. (See also Problems 37.3–37.6.)

SIMPLIFICATION OF TRIGONOMETRIC EXPRESSIONS.   It is frequently desirable to transform or reduce a given expression involving trigonometric functions to a simpler form.

EXAMPLE 1

   (a)   Using $\csc \theta = \dfrac{1}{\sin \theta}$, $\cos \theta \csc \theta = \cos \theta \dfrac{1}{\sin \theta} = \dfrac{\cos \theta}{\sin \theta} = \cot \theta$.

   (b)   Using $\tan \theta = \dfrac{\sin \theta}{\cos \theta}$, $\cos \theta \tan \theta = \cos \theta \dfrac{\sin \theta}{\cos \theta} = \sin \theta$.

EXAMPLE 2.   Using the relation $\sin^2 \theta + \cos^2 \theta = 1$,

   (a) $\sin^3 \theta + \sin \theta \cos^2 \theta = (\sin^2 \theta + \cos^2 \theta) \sin \theta = (1) \sin \theta = \sin \theta$.

   (b) $\dfrac{\cos^2 \theta}{1 - \sin \theta} = \dfrac{1 - \sin^2 \theta}{1 - \sin \theta} = \dfrac{(1 - \sin \theta)(1 + \sin \theta)}{1 - \sin \theta} = 1 + \sin \theta$.

(NOTE.   The relation $\sin^2 \theta + \cos^2 \theta = 1$ may be written as $\sin^2 \theta = 1 - \cos^2 \theta$ and as $\cos^2 \theta = 1 - \sin^2 \theta$. Each form is equally useful.) (See Problems 37.7–37.9.)

TRIGONOMETRIC IDENTITIES.   A relation involving the trigonometric functions which is valid for all values of the angle for which the functions are defined is called a trigonometric identity. The eight fundamental relations above are trigonometric identities; so also are

$$\cos \theta \csc \theta = \cot \theta \qquad \text{and} \qquad \cos \theta \tan \theta = \sin \theta$$

of Example 1 above.

A trigonometric identity is verified by transforming one member (your choice) into the other. In general, one begins with the more complicated side.

Success in verifying identities requires

(*a*)   Complete familiarity with the fundamental relations
(*b*)   Complete familiarity with the processes of factoring, adding fractions, etc.
(*c*)   Practice

(See Problems 37.10–37.17.)

Solved Problems

37.1   Prove the quotient relations: $\tan \theta = \dfrac{\sin \theta}{\cos \theta}$, $\cot \theta = \dfrac{\cos \theta}{\sin \theta}$.

For any angle θ, $\sin \theta = y/r$, $\cos \theta = x/r$, $\tan \theta = y/x$, and $\cot \theta = x/y$, where $P(x, y)$ is any point on the terminal side of θ at a distance r from the origin.

Then $\tan \theta = \dfrac{y}{x} = \dfrac{y/r}{x/r} = \dfrac{\sin \theta}{\cos \theta}$ and $\cot \theta = \dfrac{x}{y} = \dfrac{x/r}{y/r} = \dfrac{\cos \theta}{\sin \theta}$.

$\left(\text{Also, } \cot \theta = \dfrac{1}{\tan \theta} = \dfrac{\cos \theta}{\sin \theta}.\right)$

37.2   Prove the Pythagorean relations: (*a*) $\sin^2 \theta + \cos^2 \theta = 1$ (*b*) $1 + \tan^2 \theta - \sec^2 \theta$
(*c*) $1 + \cot^2 \theta = \csc^2 \theta$.

For $P(x, y)$ defined as in Problem 37.1, we have A) $x^2 + y^2 = r^2$.

(*a*)   Dividing A) by r^2, $(x/r)^2 + (y/r)^2 = 1$ and $\sin^2 \theta + \cos^2 \theta = 1$.

(*b*)   Dividing A) by x^2, $1 + (y/x)^2 = (r/x)^2$ and $1 + \tan^2 \theta = \sec^2 \theta$. Also, dividing $\sin^2 \theta + \cos^2 \theta = 1$ by $\cos^2 \theta$,

$$\left(\frac{\sin \theta}{\cos \theta}\right)^2 + 1 = \left(\frac{1}{\cos \theta}\right)^2 \quad \text{or} \quad \tan^2 \theta + 1 = \sec^2 \theta.$$

(*c*)   Dividing A) by y^2, $(x/y)^2 + 1 = (r/y)^2$ and $\cot^2 \theta + 1 = \csc^2 \theta$. Also, dividing $\sin^2 \theta + \cos^2 \theta = 1$ by $\sin^2 \theta$,

$$1 + \left(\frac{\cos \theta}{\sin \theta}\right)^2 = \left(\frac{1}{\sin \theta}\right)^2 \quad \text{or} \quad 1 + \cot^2 \theta = \csc^2 \theta.$$

37.3   Express each of the other functions of θ in terms of $\sin \theta$.

$$\cos^2 \theta = 1 - \sin^2 \theta \quad \text{and} \quad \cos \theta = \pm\sqrt{1 - \sin^2 \theta}$$

$$\tan \theta = \frac{\sin \theta}{\cos \theta} = \frac{\sin \theta}{\pm\sqrt{1 - \sin^2 \theta}} \qquad \cot \theta = \frac{1}{\tan \theta} = \frac{\pm\sqrt{1 - \sin^2 \theta}}{\sin \theta}$$

$$\sec \theta = \frac{1}{\cos \theta} = \frac{1}{\pm\sqrt{1 - \sin^2 \theta}} \qquad \csc \theta = \frac{1}{\sin \theta}$$

Note that $\cos \theta = \pm\sqrt{1 - \sin^2 \theta}$. Writing $\cos \theta = \sqrt{1 - \sin^2 \theta}$ limits angle θ to those quadrants (first and fourth) in which the cosine is positive.

37.4   Express each of the other functions of θ in terms of $\tan \theta$.

$$\sec^2 \theta = 1 + \tan^2 \theta \quad \text{and} \quad \sec \theta = \pm\sqrt{1 + \tan^2 \theta}, \quad \cos \theta = \frac{1}{\sec \theta} = \frac{1}{\pm\sqrt{1 + \tan^2 \theta}},$$

$$\frac{\sin \theta}{\cos \theta} = \tan \theta \quad \text{and} \quad \sin \theta = \tan \theta \cos \theta = \tan \theta \frac{1}{\pm\sqrt{1 + \tan^2 \theta}} = \frac{\tan \theta}{\pm\sqrt{1 + \tan^2 \theta}},$$

$$\csc \theta = \frac{1}{\sin \theta} = \frac{\pm\sqrt{1 + \tan^2 \theta}}{\tan \theta}, \quad \cot \theta = \frac{1}{\tan \theta}.$$

37.5   Using the fundamental relations, find the values of the functions of θ, given $\sin \theta = \frac{3}{5}$.

From $\cos^2 \theta = 1 - \sin^2 \theta$, $\cos \theta = \pm\sqrt{1 - \sin^2 \theta} = \pm\sqrt{1 - (\frac{3}{5})^2} = \pm\sqrt{\frac{16}{25}} = \pm\frac{4}{5}$.

Now $\sin \theta$ and $\cos \theta$ are both positive when θ is a first quadrant angle while $\sin \theta = +$ and $\cos \theta = -$ when θ is a second quadrant angle. Thus,

First Quadrant		Second Quadrant	
$\sin \theta = \frac{3}{5}$	$\cot \theta = \frac{4}{3}$	$\sin \theta = \frac{3}{5}$	$\cot \theta = -\frac{4}{3}$
$\cos \theta = \frac{4}{5}$	$\sec \theta = \frac{5}{4}$	$\cos \theta = -\frac{4}{5}$	$\sec \theta = -\frac{5}{4}$
$\tan \theta = \frac{\frac{3}{5}}{\frac{4}{5}} = \frac{3}{4}$	$\csc \theta = \frac{5}{3}$	$\tan \theta = -\frac{3}{4}$	$\csc \theta = \frac{5}{3}$

37.6   Using the fundamental relations, find the values of the functions of θ, given $\tan \theta = -\frac{5}{12}$.

Since $\tan \theta = -$, θ is either a second or fourth quadrant angle.

Second Quadrant	Fourth Quadrant
$\tan \theta = -\frac{5}{12}$	$\tan \theta = -\frac{5}{12}$
$\cot \theta = 1/\tan \theta = -\frac{12}{5}$	$\cot \theta = -\frac{12}{5}$
$\sec \theta = -\sqrt{1 + \tan^2 \theta} = -\frac{13}{12}$	$\sec \theta = \frac{13}{12}$
$\cos \theta = 1/\sec \theta = -\frac{12}{13}$	$\cos \theta = \frac{12}{13}$
$\csc \theta = \sqrt{1 + \cot^2 \theta} = \frac{13}{5}$	$\csc \theta = -\frac{13}{5}$
$\sin \theta = 1/\csc \theta = \frac{5}{13}$	$\sin \theta = -\frac{5}{13}$

37.7   Perform the indicated operations.

(a) $(\sin \theta - \cos \theta)(\sin \theta + \cos \theta) = \sin^2 \theta - \cos^2 \theta$

(b) $(\sin A + \cos A)^2 = \sin^2 A + 2 \sin A \cos A + \cos^2 A$

(c) $(\sin x + \cos y)(\sin y - \cos x) = \sin x \sin y - \sin x \cos x + \sin y \cos y - \cos x \cos y$

(d) $(\tan^2 A - \cot A)^2 = \tan^4 A - 2 \tan^2 A \cot A + \cot^2 A$

(e) $1 + \dfrac{\cos \theta}{\sin \theta} = \dfrac{\sin \theta + \cos \theta}{\sin \theta}$

(f) $1 - \dfrac{\sin \theta}{\cos \theta} + \dfrac{2}{\cos^2 \theta} = \dfrac{\cos^2 \theta - \sin \theta \cos \theta + 2}{\cos^2 \theta}$

37.8   Factor.

(a) $\sin^2 \theta - \sin \theta \cos \theta = \sin \theta(\sin \theta - \cos \theta)$

(b) $\sin^2 \theta + \sin^2 \theta \cos^2 \theta = \sin^2 \theta(1 + \cos^2 \theta)$

(c) $\sin^2 \theta + \sin \theta \sec \theta - 6 \sec^2 \theta = (\sin \theta + 3 \sec \theta)(\sin \theta - 2 \sec \theta)$

(d) $\sin^3 \theta \cos^2 \theta - \sin^2 \theta \cos^3 \theta + \sin \theta \cos^2 \theta = \sin \theta \cos^2 \theta(\sin^2 \theta - \sin \theta \cos \theta + 1)$

(e) $\sin^4 \theta - \cos^4 \theta = (\sin^2 \theta + \cos^2 \theta)(\sin^2 \theta - \cos^2 \theta) = (\sin^2 \theta + \cos^2 \theta)(\sin \theta - \cos \theta)(\sin \theta + \cos \theta)$

37.9 Simplify each of the following:

(a) $\sec\theta - \sec\theta\sin^2\theta = \sec\theta(1 - \sin^2\theta) = \sec\theta\cos^2\theta = \dfrac{1}{\cos\theta}\cos^2\theta = \cos\theta$

(b) $\sin\theta\sec\theta\cot\theta = \sin\theta\,\dfrac{1}{\cos\theta}\,\dfrac{\cos\theta}{\sin\theta} = \dfrac{\sin\theta\cos\theta}{\cos\theta\sin\theta} = 1$

(c) $\sin^2\theta(1 + \cot^2\theta) = \sin^2\theta\csc^2\theta = \sin^2\theta\,\dfrac{1}{\sin^2\theta} = 1$

(d) $\sin^2\theta\sec^2\theta - \sec^2\theta = (\sin^2\theta - 1)\sec^2\theta = -\cos^2\theta\sec^2\theta = -\cos^2\theta\,\dfrac{1}{\cos^2\theta} = -1$

(e) $(\sin\theta + \cos\theta)^2 + (\sin\theta - \cos\theta)^2 = \sin^2\theta + 2\sin\theta\cos\theta + \cos^2\theta + \sin^2\theta$
$$- 2\sin\theta\cos\theta + \cos^2\theta$$
$$= 2(\sin^2\theta + \cos^2\theta) = 2$$

(f) $\tan^2\theta\cos^2\theta + \cot^2\theta\sin^2\theta = \dfrac{\sin^2\theta}{\cos^2\theta}\cos^2\theta + \dfrac{\cos^2\theta}{\sin^2\theta}\sin^2\theta = \sin^2\theta + \cos^2\theta = 1$

(g) $\tan\theta + \dfrac{\cos\theta}{1 + \sin\theta} = \dfrac{\sin\theta}{\cos\theta} + \dfrac{\cos\theta}{1 + \sin\theta} = \dfrac{\sin\theta(1 + \sin\theta) + \cos^2\theta}{\cos\theta(1 + \sin\theta)}$

$$= \dfrac{\sin\theta + \sin^2\theta + \cos^2\theta}{\cos\theta(1 + \sin\theta)} = \dfrac{\sin\theta + 1}{\cos\theta(1 + \sin\theta)} = \dfrac{1}{\cos\theta} = \sec\theta$$

Verify the following identities:

37.10 $\sec^2\theta\csc^2\theta = \sec^2\theta + \csc^2\theta$

$$\sec^2\theta + \csc^2\theta = \dfrac{1}{\cos^2\theta} + \dfrac{1}{\sin^2\theta} = \dfrac{\sin^2\theta + \cos^2\theta}{\sin^2\theta\cos^2\theta} = \dfrac{1}{\sin^2\theta\cos^2\theta} = \dfrac{1}{\sin^2\theta}\,\dfrac{1}{\cos^2\theta} = \csc^2\theta\sec^2\theta$$

37.11 $\sec^4\theta - \sec^2\theta = \tan^4\theta + \tan^2\theta$

$$\tan^4\theta + \tan^2\theta = \tan^2\theta(\tan^2\theta + 1) = \tan^2\theta\sec^2\theta = (\sec^2\theta - 1)\sec^2\theta = \sec^4\theta - \sec^2\theta$$

or $\sec^4\theta - \sec^2\theta = \sec^2\theta(\sec^2\theta - 1) = \sec^2\theta\tan^2\theta = (1 + \tan^2\theta)\tan^2\theta = \tan^2\theta + \tan^4\theta.$

37.12 $2\csc x = \dfrac{\sin x}{1 + \cos x} + \dfrac{1 + \cos x}{\sin x}$

$$\dfrac{\sin x}{1 + \cos x} + \dfrac{1 + \cos x}{\sin x} = \dfrac{\sin^2 x + (1 + \cos x)^2}{\sin x(1 + \cos x)} = \dfrac{\sin^2 x + 1 + 2\cos x + \cos^2 x}{\sin x(1 + \cos x)}$$

$$= \dfrac{2 + 2\cos x}{\sin x(1 + \cos x)} = \dfrac{2(1 + \cos x)}{\sin x(1 + \cos x)} = \dfrac{2}{\sin x} = 2\csc x$$

37.13 $\dfrac{1 - \sin x}{\cos x} = \dfrac{\cos x}{1 + \sin x}$

$$\dfrac{\cos x}{1 + \sin x} = \dfrac{\cos^2 x}{\cos x(1 + \sin x)} = \dfrac{1 - \sin^2 x}{\cos x(1 + \sin x)} = \dfrac{(1 - \sin x)(1 + \sin x)}{\cos x(1 + \sin x)} = \dfrac{1 - \sin x}{\cos x}$$

37.14 $\dfrac{\sec A - \csc A}{\sec A + \csc A} = \dfrac{\tan A - 1}{\tan A + 1}$

$$\dfrac{\sec A - \csc A}{\sec A + \csc A} = \dfrac{\dfrac{1}{\cos A} - \dfrac{1}{\sin A}}{\dfrac{1}{\cos A} + \dfrac{1}{\sin A}} = \dfrac{\dfrac{\sin A}{\cos A} - 1}{\dfrac{\sin A}{\cos A} + 1} = \dfrac{\tan A - 1}{\tan A + 1}$$

37.15 $\dfrac{\tan x - \sin x}{\sin^3 x} = \dfrac{\sec x}{1 + \cos x}$

$$\frac{\tan x - \sin x}{\sin^3 x} = \frac{\dfrac{\sin x}{\cos x} - \sin x}{\sin^3 x} = \frac{\sin x - \sin x \cos x}{\cos x \sin^3 x} = \frac{\sin x (1 - \cos x)}{\cos x \sin^3 x}$$

$$= \frac{1 - \cos x}{\cos x \sin^2 x} = \frac{1 - \cos x}{\cos x (1 - \cos^2 x)} = \frac{1}{\cos x (1 + \cos x)} = \frac{\sec x}{1 + \cos x}$$

37.16 $\dfrac{\cos A \cot A - \sin A \tan A}{\csc A - \sec A} = 1 + \sin A \cos A$

$$\frac{\cos A \cot A - \sin A \tan A}{\csc A - \sec A} = \frac{\cos A \dfrac{\cos A}{\sin A} - \sin A \dfrac{\sin A}{\cos A}}{\dfrac{1}{\sin A} - \dfrac{1}{\cos A}} = \frac{\cos^3 A - \sin^3 A}{\cos A - \sin A}$$

$$= \frac{(\cos A - \sin A)(\cos^2 A + \cos A \sin A + \sin^2 A)}{\cos A - \sin A}$$

$$= \cos^2 A + \cos A \sin A + \sin^2 A = 1 + \cos A \sin A$$

37.17 $\dfrac{\sin \theta - \cos \theta + 1}{\sin \theta + \cos \theta - 1} = \dfrac{\sin \theta + 1}{\cos \theta}$

$$\frac{\sin \theta + 1}{\cos \theta} = \frac{(\sin \theta + 1)(\sin \theta + \cos \theta - 1)}{\cos \theta (\sin \theta + \cos \theta - 1)} = \frac{\sin^2 \theta + \sin \theta \cos \theta + \cos \theta - 1}{\cos \theta (\sin \theta + \cos \theta - 1)}$$

$$= \frac{-\cos^2 \theta + \sin \theta \cos \theta + \cos \theta}{\cos \theta (\sin \theta + \cos \theta - 1)} = \frac{\cos \theta (\sin \theta - \cos \theta + 1)}{\cos \theta (\sin \theta + \cos \theta - 1)}$$

$$= \frac{\sin \theta - \cos \theta + 1}{\sin \theta + \cos \theta - 1}$$

Supplementary Problems

37.18 Find the values of the trigonometric functions of θ, given $\sin \theta = \frac{2}{3}$.

    Ans.   Quad I: $\frac{2}{3}, \sqrt{5}/3, 2/\sqrt{5}, \sqrt{5}/2, 3/\sqrt{5}, \frac{3}{2}$
                  Quad II: $\frac{2}{3}, -\sqrt{5}/3, -2/\sqrt{5}, -\sqrt{5}/2, -3/\sqrt{5}, \frac{3}{2}$

37.19 Find the values of the trigonometric functions of θ, given $\cos \theta = -\frac{5}{6}$.

    Ans.   Quad II: $\sqrt{11}/6, -\frac{5}{6}, -\sqrt{11}/5, -5/\sqrt{11}, -\frac{6}{5}, 6/\sqrt{11}$
                  Quad III: $-\sqrt{11}/6, -\frac{5}{6}, \sqrt{11}/5, 5/\sqrt{11}, -\frac{6}{5}, -6/\sqrt{11}$

37.20 Find the values of the trigonometric functions of θ, given $\tan \theta = \frac{5}{4}$.

    Ans.   Quad I: $5/\sqrt{41}, 4/\sqrt{41}, \frac{5}{4}, \frac{4}{5}, \sqrt{41}/4, \sqrt{41}/5$
                  Quad III: $-5/\sqrt{41}, -4/\sqrt{41}, \frac{5}{4}, \frac{4}{5}, -\sqrt{41}/4, -\sqrt{41}/5$

37.21   Find the values of the trigonometric functions of θ, given $\cot \theta = -\sqrt{3}$.

> *Ans.*   Quad II: $\frac{1}{2}, -\sqrt{3}/2, -1/\sqrt{3}, -\sqrt{3}, -2/\sqrt{3}, 2$
>   Quad IV: $-\frac{1}{2}, \sqrt{3}/2, -1/\sqrt{3}, -\sqrt{3}, 2/\sqrt{3}, -2$

37.22   Find the value of $\dfrac{\sin \theta + \cos \theta - \tan \theta}{\sec \theta + \csc \theta - \cot \theta}$ when $\tan \theta = -\frac{4}{3}$.

> *Ans.*   Quad II: $\frac{23}{5}$;
>   Quad IV: $\frac{34}{35}$

Verify the following identities:

37.23 $\sin \theta \sec \theta = \tan \theta$

37.24 $(1 - \sin^2 A)(1 + \tan^2 A) = 1$

37.25 $(1 - \cos \theta)(1 + \sec \theta) \cot \theta = \sin \theta$

37.26 $\csc^2 x (1 - \cos^2 x) = 1$

37.27 $\dfrac{\sin \theta}{\csc \theta} + \dfrac{\cos \theta}{\sec \theta} = 1$

37.28 $\dfrac{1 - 2\cos^2 A}{\sin A \cos A} = \tan A - \cot A$

37.29 $\tan^2 x \csc^2 x \cot^2 x \sin^2 x = 1$

37.30 $\sin A \cos A (\tan A + \cot A) = 1$

37.31 $1 - \dfrac{\cos^2 \theta}{1 + \sin \theta} = \sin \theta$

37.32 $\dfrac{1}{\sec \theta + \tan \theta} = \sec \theta - \tan \theta$

37.33 $\dfrac{1}{1 - \sin A} + \dfrac{1}{1 + \sin A} = 2 \sec^2 A$

37.34 $\dfrac{1 - \cos x}{1 + \cos x} = \dfrac{\sec x - 1}{\sec x + 1} = (\cot x - \csc x)^2$

37.35 $\tan \theta \sin \theta + \cos \theta = \sec \theta$

37.36 $\tan \theta - \csc \theta \sec \theta (1 - 2 \cos^2 \theta) = \cot \theta$

37.37 $\dfrac{\sin \theta}{\sin \theta + \cos \theta} = \dfrac{\sec \theta}{\sec \theta + \csc \theta}$

37.38 $\dfrac{\sin x + \tan x}{\cot x + \csc x} = \sin x \tan x$

37.39 $\dfrac{\sec x + \csc x}{\tan x + \cot x} = \sin x + \cos x$

37.40 $\dfrac{\sin^3 \theta + \cos^3 \theta}{\sin \theta + \cos \theta} = 1 - \sin \theta \cos \theta$

37.41 $\cot \theta + \dfrac{\sin \theta}{1 + \cos \theta} = \csc \theta$

37.42 $\dfrac{\sin \theta \cos \theta}{\cos^2 \theta - \sin^2 \theta} = \dfrac{\tan \theta}{1 - \tan^2 \theta}$

37.43 $(\tan x + \tan y)(1 - \cot x \cot y) + (\cot x + \cot y)(1 - \tan x \tan y) = 0$

37.44 $(x \sin \theta - y \cos \theta)^2 + (x \cos \theta + y \sin \theta)^2 = x^2 + y^2$

37.45 $(2r \sin \theta \cos \theta)^2 + r^2(\cos^2 \theta - \sin^2 \theta)^2 = r^2$

37.46 $(r \sin \theta \cos \phi)^2 + (r \sin \theta \sin \phi)^2 + (r \cos \theta)^2 = r^2$

Chapter 38

Trigonometric Functions of Two Angles

ADDITION FORMULAS

$$\sin (\alpha + \beta) = \sin \alpha \cos \beta + \cos \alpha \sin \beta$$

$$\cos (\alpha + \beta) = \cos \alpha \cos \beta - \sin \alpha \sin \beta$$

$$\tan (\alpha + \beta) = \frac{\tan \alpha + \tan \beta}{1 - \tan \alpha \tan \beta}$$

(For a proof of these formulas, see Problems 38.1–38.2.)

SUBTRACTION FORMULAS

$$\sin (\alpha - \beta) = \sin \alpha \cos \beta - \cos \alpha \sin \beta$$

$$\cos (\alpha - \beta) = \cos \alpha \cos \beta + \sin \alpha \sin \beta$$

$$\tan (\alpha - \beta) = \frac{\tan \alpha - \tan \beta}{1 + \tan \alpha \tan \beta}$$

(For a proof of these formulas, see Problem 38.3.)

DOUBLE-ANGLE FORMULAS

$$\sin 2\alpha = 2 \sin \alpha \cos \alpha$$

$$\cos 2\alpha = \cos^2 \alpha - \sin^2 \alpha = 1 - 2 \sin^2 \alpha = 2 \cos^2 \alpha - 1$$

$$\tan 2\alpha = \frac{2 \tan \alpha}{1 - \tan^2 \alpha}$$

(For a proof of these formulas, see Problem 38.9.)

HALF-ANGLE FORMULAS

$$\sin \tfrac{1}{2}\theta = \pm \sqrt{\frac{1 - \cos \theta}{2}}$$

$$\cos \tfrac{1}{2}\theta = \pm \sqrt{\frac{1 + \cos \theta}{2}}$$

$$\tan \tfrac{1}{2}\theta = \pm \sqrt{\frac{1 - \cos \theta}{1 + \cos \theta}} = \frac{\sin \theta}{1 + \cos \theta} = \frac{1 - \cos \theta}{\sin \theta}$$

(For a proof of these formulas, see Problem 38.10.)

Solved Problems

38.1   Prove (1) $\sin(\alpha + \beta) = \sin\alpha\cos\beta + \cos\alpha\sin\beta$ and (2) $\cos(\alpha + \beta) = \cos\alpha\cos\beta - \sin\alpha\sin\beta$ when α and β are positive acute angles.

Let α and β be the measures of positive acute angles such that $\alpha + \beta < 90°$ [Fig. 38-1(a)] and $\alpha + \beta > 90°$ [Fig. 38-1(b)].

(a) (b)

Fig. 38-1

To construct these figures, place angle α in standard position and then place angle β with its vertex at O and with its initial side along the terminal side of angle α. Let P be any point on the terminal side of angle $(\alpha + \beta)$: Draw $\overline{PA}$ perpendicular to $\overline{OX}$, $\overline{PB}$ perpendicular to the terminal side of angle α, $\overline{BC}$ perpendicular to $\overline{OX}$, and $\overline{BC}$ perpendicular to $\overline{AP}$.

Now $m\angle APB = \alpha$ since corresponding sides (OA and AP, OB and BP) are perpendicular. Then

$$\sin(\alpha + \beta) = \frac{AP}{OP} = \frac{AD + DP}{OP} = \frac{CB + DP}{OP} = \frac{CB}{OP} + \frac{DP}{OP} = \frac{CB}{OB}\cdot\frac{OB}{OP} + \frac{DP}{BP}\cdot\frac{BP}{OP}$$

$$= \sin\alpha\cos\beta + \cos\alpha\sin\beta$$

and $$\cos(\alpha + \beta) = \frac{OA}{OP} = \frac{OC - AC}{OP} = \frac{OC - DB}{OP} = \frac{OC}{OP} - \frac{DB}{OP} = \frac{OC}{OB}\cdot\frac{OB}{OP} - \frac{DB}{BP}\cdot\frac{BP}{OP}$$

38.2   Prove $\tan(\alpha + \beta) = \dfrac{\tan\alpha + \tan\beta}{1 - \tan\alpha\tan\beta}$.

$$\tan(\alpha + \beta) = \frac{\sin(\alpha + \beta)}{\cos(\alpha + \beta)} = \frac{\sin\alpha\cos\beta + \cos\alpha\sin\beta}{\cos\alpha\cos\beta - \sin\alpha\sin\beta}$$

$$= \frac{\dfrac{\sin\alpha\cos\beta}{\cos\alpha\cos\beta} + \dfrac{\cos\alpha\sin\beta}{\cos\alpha\cos\beta}}{\dfrac{\cos\alpha\cos\beta}{\cos\alpha\cos\beta} - \dfrac{\sin\alpha\sin\beta}{\cos\alpha\cos\beta}}$$

$$= \frac{\tan\alpha + \tan\beta}{1 - \tan\alpha\tan\beta}$$

38.3   Prove the subtraction formulas.

$$\sin(\alpha - \beta) = \sin[\alpha + (-\beta)] = \sin\alpha\cos(-\beta) + \cos\alpha\sin(-\beta)$$

$$= \sin\alpha(\cos\beta) + \cos\alpha(-\sin\beta) = \sin\alpha\cos\beta - \cos\alpha\sin\beta$$

$$\cos(\alpha - \beta) = \cos[\alpha + (-\beta)] = \cos\alpha\cos(-\beta) - \sin\alpha\sin(-\beta)$$

$$= \cos\alpha(\cos\beta) - \sin\alpha(-\sin\beta) = \cos\alpha\cos\beta + \sin\alpha\sin\beta$$

$$\tan(\alpha - \beta) = \tan[\alpha + (-\beta)] = \frac{\tan\alpha + \tan(-\beta)}{1 - \tan\alpha\tan(-\beta)} = \frac{\tan\alpha + (-\tan\beta)}{1 - \tan\alpha(-\tan\beta)}$$

$$= \frac{\tan\alpha - \tan\beta}{1 + \tan\alpha\tan\beta}$$

38.4   Find the values of sine, cosine, and tangent of 15°, using (a)   15° = 45° − 30° and   (b)   15° = 60° − 45°.

(a) $\sin 15° = \sin(45° - 30°) = \sin 45°\cos 30° - \cos 45°\sin 30°$

$$= \frac{1}{\sqrt{2}}\cdot\frac{\sqrt{3}}{2} - \frac{1}{\sqrt{2}}\cdot\frac{1}{2} = \frac{\sqrt{3}-1}{2\sqrt{2}} = \frac{\sqrt{2}}{4}(\sqrt{3}-1)$$

$$\cos 15° = \cos(45° - 30°) = \cos 45°\cos 30° + \sin 45°\sin 30°$$

$$= \frac{1}{\sqrt{2}}\cdot\frac{\sqrt{3}}{2} + \frac{1}{\sqrt{2}}\cdot\frac{1}{2} = \frac{\sqrt{2}}{4}(\sqrt{3}+1)$$

$$\tan 15° = \tan(45° - 30°) = \frac{\tan 45° - \tan 30°}{1 + \tan 45°\tan 30°} = \frac{1 - 1/\sqrt{3}}{1 + 1(1/\sqrt{3})} = \frac{\sqrt{3}-1}{\sqrt{3}+1} = 2 - \sqrt{3}$$

(b) $\sin 15° = \sin(60° - 45°) = \sin 60°\cos 45° - \cos 60°\sin 45°$

$$= \frac{\sqrt{3}}{2}\cdot\frac{1}{\sqrt{2}} - \frac{1}{2}\cdot\frac{1}{\sqrt{2}} = \frac{\sqrt{2}}{4}(\sqrt{3}-1)$$

$$\cos 15° = \cos(60° - 45°) = \cos 60°\cos 45° + \sin 60°\sin 45°$$

$$= \frac{1}{2}\cdot\frac{1}{\sqrt{2}} + \frac{\sqrt{3}}{2}\cdot\frac{1}{\sqrt{2}} = \frac{\sqrt{2}}{4}(\sqrt{3}+1)$$

$$\tan 15° = \tan(60° - 45°) = \frac{\tan 60° - \tan 45°}{1 + \tan 60°\tan 45°} = \frac{\sqrt{3}-1}{\sqrt{3}+1} = 2 - \sqrt{3}$$

38.5   Prove (a) $\sin(45° + \theta) - \sin(45° - \theta) = \sqrt{2}\sin\theta$,   ($b$) $\sin(30° + \theta) + \cos(60° + \theta) = \cos\theta$.

(a) $\sin(45° + \theta) - \sin(45° - \theta) = (\sin 45°\cos\theta + \cos 45°\sin\theta) - (\sin 45°\cos\theta - \cos 45°\sin\theta)$

$$= 2\cos 45°\sin\theta = 2\frac{1}{\sqrt{2}}\sin\theta = \sqrt{2}\sin\theta$$

(b) $\sin(30° + \theta) + \cos(60° + \theta) = (\sin 30°\cos\theta + \cos 30°\sin\theta) + (\cos 60°\cos\theta - \sin 60°\sin\theta)$

$$= \left(\frac{1}{2}\cos\theta + \frac{\sqrt{3}}{2}\sin\theta\right) + \left(\frac{1}{2}\cos\theta - \frac{\sqrt{3}}{2}\sin\theta\right) = \cos\theta$$

38.6   Simplify (a) $\sin(\alpha + \beta) + \sin(\alpha - \beta)$, ($b$) $\cos(\alpha + \beta) - \cos(\alpha - \beta)$, ($c$) $\dfrac{\tan(\alpha + \beta) - \tan\alpha}{1 + \tan(\alpha + \beta)\tan\alpha}$,

(d) $(\sin\alpha\cos\beta - \cos\alpha\sin\beta)^2 + (\cos\alpha\cos\beta + \sin\alpha\sin\beta)^2$.

(a) $\sin(\alpha + \beta) + \sin(\alpha - \beta) = (\sin \alpha \cos \beta + \cos \alpha \sin \beta) + (\sin \alpha \cos \beta - \cos \alpha \sin \beta)$

$$= 2 \sin \alpha \cos \beta$$

(b) $\cos(\alpha + \beta) - \cos(\alpha - \beta) = (\cos \alpha \cos \beta - \sin \alpha \sin \beta) - (\cos \alpha \cos \beta + \sin \alpha \sin \beta)$

$$= -2 \sin \alpha \sin \beta$$

(c) $\dfrac{\tan(\alpha + \beta) - \tan \alpha}{1 + \tan(\alpha + \beta)\tan \alpha} = \tan[(\alpha + \beta) - \alpha] = \tan \beta$

(d) $(\sin \alpha \cos \beta - \cos \alpha \sin \beta)^2 + (\cos \alpha \cos \beta + \sin \alpha \sin \beta)^2 = \sin^2(\alpha - \beta) + \cos^2(\alpha - \beta) = 1$

38.7 Find $\sin(\alpha + \beta)$, $\cos(\alpha + \beta)$, $\sin(\alpha - \beta)$, $\cos(\alpha - \beta)$ and determine the quadrants in which $(\alpha + \beta)$ and $(\alpha - \beta)$ terminate, given

(a) $\sin \alpha = \frac{4}{5}$, $\cos \beta = \frac{5}{13}$; α and β in quadrant I.

(b) $\sin \alpha = \frac{2}{3}$, $\cos \beta = \frac{3}{4}$; α in quadrant II, β in quadrant IV.

(a) $\cos \alpha = \frac{3}{5}$ and $\sin \beta = \frac{12}{13}$. See Figs. 38-2 and 38-3.

Fig. 38-2   Fig. 38-3

$\sin(\alpha + \beta) = \sin \alpha \cos \beta + \cos \alpha \sin \beta = \frac{4}{5} \cdot \frac{5}{13} + \frac{3}{5} \cdot \frac{12}{13} = \frac{56}{65}$

$\cos(\alpha + \beta) = \cos \alpha \cos \beta - \sin \alpha \sin \beta = \frac{3}{5} \cdot \frac{5}{13} - \frac{4}{5} \cdot \frac{12}{13} = -\frac{33}{65}$

$(\alpha + \beta)$ in quadrant II

$\sin(\alpha - \beta) = \sin \alpha \cos \beta - \cos \alpha \sin \beta = \frac{4}{5} \cdot \frac{5}{13} - \frac{3}{5} \cdot \frac{12}{13} = -\frac{16}{65}$

$\cos(\alpha - \beta) = \cos \alpha \cos \beta + \sin \alpha \sin \beta = \frac{3}{5} \cdot \frac{5}{13} + \frac{4}{5} \cdot \frac{12}{13} = \frac{63}{65}$

$(\alpha - \beta)$ in quadrant IV

(b) $\cos \alpha = -\sqrt{5}/3$ and $\sin \beta = -\sqrt{7}/4$. See Figs. 38-4 and 38-5.

Fig. 38-4   Fig. 38-5

$$\sin(\alpha + \beta) = \sin\alpha\cos\beta + \cos\alpha\sin\beta = \frac{2}{3}\cdot\frac{3}{4} + \left(-\frac{\sqrt{5}}{3}\right)\left(-\frac{\sqrt{7}}{4}\right) = \frac{6 + \sqrt{35}}{12}$$

$$\cos(\alpha + \beta) = \cos\alpha\cos\beta - \sin\alpha\sin\beta = \left(-\frac{\sqrt{5}}{3}\right)\frac{3}{4} - \frac{2}{3}\left(-\frac{\sqrt{7}}{4}\right) = \frac{-3\sqrt{5} + 2\sqrt{7}}{12}$$

$$(\alpha + \beta) \text{ in quadrant II}$$

$$\sin(\alpha - \beta) = \sin\alpha\cos\beta - \cos\alpha\sin\beta = \frac{2}{3}\cdot\frac{3}{4} - \left(-\frac{\sqrt{5}}{3}\right)\left(-\frac{\sqrt{7}}{4}\right) = \frac{6 - \sqrt{35}}{12}$$

$$\cos(\alpha - \beta) = \cos\alpha\cos\beta + \sin\alpha\sin\beta = \left(-\frac{\sqrt{5}}{3}\right)\frac{3}{4} + \frac{2}{3}\left(-\frac{\sqrt{7}}{4}\right) = \frac{-3\sqrt{5} - 2\sqrt{7}}{12}$$

$$(\alpha - \beta) \text{ in quadrant II}$$

38.8 Prove (a) $\cot(\alpha + \beta) = \dfrac{\cot\alpha\cot\beta - 1}{\cot\beta + \cot\alpha}$, (b) $\cot(\alpha - \beta) = \dfrac{\cot\alpha\cot\beta + 1}{\cot\beta - \cot\alpha}$.

(a) $$\cot(\alpha + \beta) = \frac{1}{\tan(\alpha + \beta)} = \frac{1 - \tan\alpha\tan\beta}{\tan\alpha + \tan\beta} = \frac{1 - \dfrac{1}{\cot\alpha\cot\beta}}{\dfrac{1}{\cot\alpha} + \dfrac{1}{\cot\beta}} = \frac{\cot\alpha\cot\beta - 1}{\cot\beta + \cot\alpha}$$

(b) $$\cot(\alpha - \beta) = \cot[\alpha + (-\beta)] = \frac{\cot\alpha\cot(-\beta) - 1}{\cot(-\beta) + \cot\alpha} = \frac{-\cot\alpha\cot\beta - 1}{-\cot\beta + \cot\alpha} = \frac{\cot\alpha\cot\beta + 1}{\cot\beta - \cot\alpha}$$

38.9 Prove the double-angle formulas.

In $\sin(\alpha + \beta) = \sin\alpha\cos\beta + \cos\alpha\sin\beta$, $\cos(\alpha + \beta) = \cos\alpha\cos\beta - \sin\alpha\sin\beta$, and $\tan(\alpha + \beta) = \dfrac{\tan\alpha + \tan\beta}{1 - \tan\alpha\tan\beta}$ put $\beta = \alpha$. Then

$$\sin 2\alpha = \sin\alpha\cos\alpha + \cos\alpha\sin\alpha = 2\sin\alpha\cos\alpha,$$

$$\cos 2\alpha = \cos\alpha\cos\alpha - \sin\alpha\sin\alpha$$

$$= \cos^2\alpha - \sin^2\alpha = (1 - \sin^2\alpha) - \sin^2\alpha = 1 - 2\sin^2\alpha$$

$$= \cos^2\alpha - (1 - \cos^2\alpha) = 2\cos^2\alpha - 1,$$

$$\tan 2\alpha = \frac{\tan\alpha + \tan\alpha}{1 - \tan\alpha\tan\alpha} = \frac{2\tan\alpha}{1 - \tan^2\alpha}.$$

38.10 Prove the half-angle formulas.

In $\cos 2\alpha = 1 - 2\sin^2\alpha$, put $\alpha = \frac{1}{2}\theta$. Then

$$\cos\theta = 1 - 2\sin^2\tfrac{1}{2}\theta, \qquad \sin^2\tfrac{1}{2}\theta = \frac{1 - \cos\theta}{2}, \qquad \text{and} \qquad \sin\tfrac{1}{2}\theta = \pm\sqrt{\frac{1 - \cos\theta}{2}}.$$

In $\cos 2\alpha = 2\cos^2\alpha - 1$, put $\alpha = \frac{1}{2}\theta$. Then

$$\cos\theta = 2\cos^2\tfrac{1}{2}\theta - 1, \qquad \cos^2\tfrac{1}{2}\theta = \frac{1 + \cos\theta}{2}, \qquad \text{and} \qquad \cos\tfrac{1}{2}\theta = \pm\sqrt{\frac{1 + \cos\theta}{2}}.$$

Finally,
$$\tan \tfrac{1}{2}\theta = \frac{\sin \tfrac{1}{2}\theta}{\cos \tfrac{1}{2}\theta} = \pm \sqrt{\frac{1 - \cos \theta}{1 + \cos \theta}}$$

$$= \pm \sqrt{\frac{(1 - \cos \theta)(1 + \cos \theta)}{(1 + \cos \theta)(1 + \cos \theta)}} = \pm \sqrt{\frac{1 - \cos^2 \theta}{(1 + \cos \theta)^2}} = \frac{\sin \theta}{1 + \cos \theta}$$

$$= \pm \sqrt{\frac{(1 - \cos \theta)(1 - \cos \theta)}{(1 + \cos \theta)(1 - \cos \theta)}} = \pm \sqrt{\frac{(1 - \cos \theta)^2}{1 - \cos^2 \theta}} = \frac{1 - \cos \theta}{\sin \theta}.$$

The signs $\pm$ are not needed here since $\tan \tfrac{1}{2}\theta$ and $\sin \theta$ always have the same sign (Problem 37.7) and $1 - \cos \theta$ is always positive.

38.11 Using the half-angle formulas, find the exact values of (*a*) $\sin 15°$,   (*b*) $\sin 292\tfrac{1}{2}°$.

(*a*)
$$\sin 15° = \sqrt{\frac{1 - \cos 30°}{2}} = \sqrt{\frac{1 - \sqrt{3}/2}{2}} = \frac{1}{2}\sqrt{2 - \sqrt{3}}$$

(*b*)
$$\sin 292\tfrac{1}{2}° = -\sqrt{\frac{1 - \cos 585°}{2}} = -\sqrt{\frac{1 - \cos 225°}{2}}$$

$$= -\sqrt{\frac{1 + 1/\sqrt{2}}{2}} = -\frac{1}{2}\sqrt{2 + \sqrt{2}}$$

38.12 Find the values of sine, cosine, and tangent of $\tfrac{1}{2}\theta$, given (*a*) $\sin \theta = \tfrac{5}{13}$, θ in quadrant II and (*b*) $\cos \theta = \tfrac{3}{7}$, θ in quadrant IV.

(*a*) $\sin \theta = \tfrac{5}{13}$, $\cos \theta = -\tfrac{12}{13}$, and $\tfrac{1}{2}\theta$ in quadrant I. See Fig. 38-6.

Fig. 38-6

$$\sin \tfrac{1}{2}\theta = \sqrt{\frac{1 - \cos \theta}{2}} = \sqrt{\frac{1 + \tfrac{12}{13}}{2}} = \sqrt{\frac{25}{26}} = \frac{5\sqrt{26}}{26}$$

$$\cos \tfrac{1}{2}\theta = \sqrt{\frac{1 + \cos \theta}{2}} = \sqrt{\frac{1 - \tfrac{12}{13}}{2}} = \sqrt{\frac{1}{26}} = \frac{\sqrt{26}}{26}$$

$$\tan \tfrac{1}{2}\theta = \frac{1 - \cos \theta}{\sin \theta} = \frac{1 + \tfrac{12}{13}}{\tfrac{5}{13}} = 5$$

(*b*) $\sin \theta = -2\sqrt{10}/7$, $\cos \theta = \tfrac{3}{7}$, and $\tfrac{1}{2}\theta$ in quadrant II. See Fig. 38-7.

Fig. 38-7

$$\sin\tfrac{1}{2}\theta = \sqrt{\frac{1-\cos\theta}{2}} = \sqrt{\frac{1-\tfrac{3}{7}}{2}} = \frac{\sqrt{14}}{7}$$

$$\cos\tfrac{1}{2}\theta = -\sqrt{\frac{1+\cos\theta}{2}} = -\sqrt{\frac{1+\tfrac{3}{7}}{2}} = \frac{\sqrt{35}}{7}$$

$$\tan\tfrac{1}{2}\theta = \frac{1-\cos\theta}{\sin\theta} = \frac{1-\tfrac{3}{7}}{-2\sqrt{10}/7} = -\frac{\sqrt{10}}{5}$$

38.13 Show that (a) $\sin\theta = 2\sin\tfrac{1}{2}\theta\cos\tfrac{1}{2}\theta$,   (b) $\sin A = \pm\sqrt{\dfrac{1-\cos 2A}{2}}$,   (c) $\tan 4x = \dfrac{\sin 8x}{1+\cos 8x}$
(d) $\cos 6\theta = 1 - 2\sin^2 3\theta$ (e) $\sin^2\tfrac{1}{2}\theta = \tfrac{1}{2}(1-\cos\theta)$, $\cos^2\tfrac{1}{2}\theta = \tfrac{1}{2}(1+\cos\theta)$.

(a) This is obtained from $\sin 2\alpha = 2\sin\alpha\cos\alpha$ by putting $\alpha = \tfrac{1}{2}\theta$.

(b) This is obtained from $\sin\tfrac{1}{2}\theta = \pm\sqrt{\dfrac{1-\cos\theta}{2}}$ by putting $\theta = 2A$.

(c) This is obtained from $\tan\tfrac{1}{2}\theta = \dfrac{\sin\theta}{1+\cos\theta}$ by putting $\theta = 8x$.

(d) This is obtained from $\cos 2\alpha = 1 - 2\sin^2\alpha$ by putting $\alpha = 3\theta$.

(e) These formulas are obtained by squaring $\sin\tfrac{1}{2}\theta = \pm\sqrt{\dfrac{1-\cos\theta}{2}}$ and $\cos\tfrac{1}{2}\theta = \pm\sqrt{\dfrac{1+\cos\theta}{2}}$.

38.14 Express (a) $\sin 3\alpha$ in terms of $\sin\alpha$,   (b) $\cos 4\alpha$ in terms of $\cos\alpha$.

(a) $\sin 3\alpha = \sin(2\alpha + \alpha) = \sin 2\alpha\cos\alpha + \cos 2\alpha\sin\alpha$

$\qquad = (2\sin\alpha\cos\alpha)\cos\alpha + (1-2\sin^2\alpha)\sin\alpha = 2\sin\alpha\cos^2\alpha + (1-2\sin^2\alpha)\sin\alpha$

$\qquad = 2\sin\alpha(1-\sin^2\alpha) + (1-2\sin^2\alpha)\sin\alpha = 3\sin\alpha - 4\sin^3\alpha$

(b) $\qquad\cos 4\alpha = \cos 2(2\alpha) = 2\cos^2 2\alpha - 1 = 2(2\cos^2\alpha - 1)^2 - 1 = 8\cos^4\alpha - 8\cos^2\alpha + 1$

38.15 Prove $\cos 2x = \cos^4 x - \sin^4 x$.

$$\cos^4 x - \sin^4 x = (\cos^2 x + \sin^2 x)(\cos^2 x - \sin^2 x) = \cos^2 x - \sin^2 x = \cos 2x$$

38.16 Prove $1 - \tfrac{1}{2}\sin 2x = \dfrac{\sin^3 x + \cos^3 x}{\sin x + \cos x}$.

$$\frac{\sin^3 x + \cos^3 x}{\sin x + \cos x} = \frac{(\sin x + \cos x)(\sin^2 x - \sin x\cos x + \cos^2 x)}{\sin x + \cos x}$$

$$= 1 - \sin x\cos x = 1 - \tfrac{1}{2}(2\sin x\cos x) = 1 - \tfrac{1}{2}\sin 2x$$

38.17 Prove $\cos\theta = \sin(\theta + 30°) + \cos(\theta + 60°)$.

$$\sin(\theta + 30°) + \cos(\theta + 60°) = (\sin\theta\cos 30° + \cos\theta\sin 30°) + (\cos\theta\cos 60° - \sin\theta\sin 60°)$$

$$= \frac{\sqrt{3}}{2}\sin\theta + \frac{1}{2}\cos\theta + \frac{1}{2}\cos\theta - \frac{\sqrt{3}}{2}\sin\theta = \cos\theta$$

38.18 Prove $\cos x = \dfrac{1 - \tan^2\tfrac{1}{2}x}{1 + \tan^2\tfrac{1}{2}x}$.

$$\frac{1 - \tan^2\tfrac{1}{2}x}{1 + \tan^2\tfrac{1}{2}x} = \frac{1 - \dfrac{\sin^2\tfrac{1}{2}x}{\cos^2\tfrac{1}{2}x}}{\sec^2\tfrac{1}{2}x} = \frac{\left(1 - \dfrac{\sin^2\tfrac{1}{2}x}{\cos^2\tfrac{1}{2}x}\right)\cos^2\tfrac{1}{2}x}{\sec^2\tfrac{1}{2}x\cos^2\tfrac{1}{2}x} = \cos^2\tfrac{1}{2}x - \sin^2\tfrac{1}{2}x = \cos x$$

38.19   Prove $2 \tan 2x = \dfrac{\cos x + \sin x}{\cos x - \sin x} - \dfrac{\cos x - \sin x}{\cos x + \sin x}$.

$$\frac{\cos x + \sin x}{\cos x - \sin x} - \frac{\cos x - \sin x}{\cos x + \sin x}$$

$$= \frac{(\cos x + \sin x)^2 - (\cos x - \sin x)^2}{(\cos x - \sin x)(\cos x + \sin x)}$$

$$= \frac{(\cos^2 x + 2 \sin x \cos x + \sin^2 x) - (\cos^2 x - 2 \sin x \cos x + \sin^2 x)}{\cos^2 x - \sin^2 x}$$

$$= \frac{4 \sin x \cos x}{\cos^2 x - \sin^2 x} = \frac{2 \sin 2x}{\cos 2x} = 2 \tan 2x$$

38.20   Prove $\sin^4 A = \frac{3}{8} - \frac{1}{2} \cos 2A + \frac{1}{8} \cos 4A$.

$$\sin^4 A = (\sin^2 A)^2 = \left(\frac{1 - \cos 2A}{2} \right)^2 = \frac{1 - 2 \cos 2A + \cos^2 2A}{4}$$

$$= \frac{1}{4} \left(1 - 2 \cos 2A + \frac{1 + \cos 4A}{2} \right) = \frac{3}{8} - \frac{1}{2} \cos 2A + \frac{1}{8} \cos 4A$$

38.21   Prove $\tan^6 x = \tan^4 x \sec^2 x - \tan^2 x \sec^2 x + \sec^2 x - 1$.

$$\tan^6 x = \tan^4 x \tan^2 x = \tan^4 x (\sec^2 x - 1) = \tan^4 x \sec^2 x - \tan^2 x \tan^2 x$$

$$= \tan^4 x \sec^2 x - \tan^2 x (\sec^2 x - 1) = \tan^4 x \sec^2 x - \tan^2 x \sec^2 x + \tan^2 x$$

$$= \tan^4 x \sec^2 x - \tan^2 x \sec^2 x + \sec^2 x - 1$$

38.22   When $A + B + C = 180°$, show that $\sin 2A + \sin 2B + \sin 2C = 4 \sin A \sin B \sin C$.

Since $C = 180° - (A + B)$,

$$\sin 2A + \sin 2B + \sin 2C = \sin 2A + \sin 2B + \sin [360° - 2(A + B)]$$

$$= \sin 2A + \sin 2B - \sin 2(A + B)$$

$$= \sin 2A + \sin 2B - \sin 2A \cos 2B - \cos 2A \sin 2B$$

$$= (\sin 2A)(1 - \cos 2B) + (\sin 2B)(1 - \cos 2A)$$

$$= 2 \sin 2A \sin^2 B + 2 \sin 2B \sin^2 A$$

$$= 4 \sin A \cos A \sin^2 B + 4 \sin B \cos B \sin^2 A$$

$$= 4 \sin A \sin B (\sin A \cos B + \cos A \sin B)$$

$$= 4 \sin A \sin B \sin (A + B)$$

$$= 4 \sin A \sin B \sin [180° - (A + B)]$$

$$= 4 \sin A \sin B \sin C$$

38.23 When $A + B + C = 180°$, show that $\tan A + \tan B + \tan C = \tan A \tan B \tan C$.

Since $C = 180° - (A + B)$,

$$\tan A + \tan B + \tan C = \tan A + \tan B + \tan[180° - (A + B)] = \tan A + \tan B - \tan(A + B)$$

$$= \tan A + \tan B - \frac{\tan A + \tan B}{1 - \tan A \tan B} = (\tan A + \tan B)\left(1 - \frac{1}{1 - \tan A \tan B}\right)$$

$$= (\tan A + \tan B)\left(-\frac{\tan A \tan B}{1 - \tan A \tan B}\right) = -\tan A \tan B \frac{\tan A + \tan B}{1 - \tan A \tan B}$$

$$= -\tan A \tan B \tan(A + B) = \tan A \tan B \tan[180° - (A + B)]$$

$$= \tan A \tan B \tan C$$

Supplementary Problems

38.24 Find the values of sine, cosine, and tangent of (a) 75° (b) 255°.

Ans. $(a)\dfrac{\sqrt{2}}{4}(\sqrt{3}+1), \dfrac{\sqrt{2}}{4}(\sqrt{3}-1), 2+\sqrt{3}$ (b) $-\dfrac{\sqrt{2}}{4}(\sqrt{3}+1), -\dfrac{\sqrt{2}}{4}(\sqrt{3}-1), 2+\sqrt{3}$

38.25 Find the values of $\sin(\alpha + \beta)$, $\cos(\alpha + \beta)$, and $\tan(\alpha + \beta)$, given

(a) $\sin \alpha = \frac{3}{5}$, $\cos \beta = \frac{5}{13}$, α and β in quadrant I.

Ans. $\frac{63}{65}, -\frac{16}{65}, -\frac{63}{16}$

(b) $\sin \alpha = \frac{8}{17}$, $\tan \beta = \frac{5}{12}$, α and β in quadrant I.

Ans. $\frac{171}{221}, \frac{140}{221}, \frac{171}{140}$

(c) $\cos \alpha = -\frac{12}{13}$, $\cot \beta = \frac{24}{7}$, α in quadrant II, β in quadrant III.

Ans. $-\frac{36}{325}, \frac{323}{325}, -\frac{36}{323}$

(d) $\sin \alpha = \frac{1}{3}$, $\sin \beta = \frac{2}{5}$, α in quadrant I, β in quadrant II.

Ans. $\dfrac{4\sqrt{2}-\sqrt{21}}{15}, -\dfrac{2+2\sqrt{42}}{15}, -\dfrac{4\sqrt{2}-\sqrt{21}}{2+2\sqrt{42}}$

38.26 Find the values of $\sin(\alpha - \beta)$, $\cos(\alpha - \beta)$, and $\tan(\alpha - \beta)$, given

(a) $\sin \alpha = \frac{3}{5}$, $\sin \beta = \frac{5}{13}$, α and β in quadrant I.

Ans. $\frac{16}{65}, \frac{63}{65}, \frac{16}{63}$

(b) $\sin \alpha = \frac{8}{17}$, $\tan \beta = \frac{5}{12}$, α and β in quadrant I.

Ans. $\frac{21}{221}, \frac{220}{221}, \frac{21}{220}$

(c) $\cos \alpha = -\frac{12}{13}$, $\cot \beta = \frac{24}{7}$, α in quadrant II, β in quadrant I.

Ans. $\frac{204}{325}, -\frac{253}{325}, -\frac{204}{253}$

(d) $\sin \alpha = \frac{1}{3}$, $\sin \beta = \frac{2}{5}$, α in quadrant II, β in quadrant I.

Ans. $\dfrac{4\sqrt{2}+\sqrt{21}}{15}, -\dfrac{2\sqrt{42}-2}{15}, -\dfrac{4\sqrt{2}+\sqrt{21}}{2\sqrt{42}-2}$

38.27 Prove

(a) $\sin(\alpha + \beta) - \sin(\alpha - \beta) = 2\cos\alpha\sin\beta$

(b) $\cos(\alpha + \beta) + \cos(\alpha - \beta) = 2\cos\alpha\cos\beta$

(c) $\tan(45° - \theta) = \dfrac{1 - \tan\theta}{1 + \tan\theta}$

(d) $\dfrac{\tan(\alpha + \beta)}{\cot(\alpha - \beta)} = \dfrac{\tan^2\alpha - \tan^2\beta}{1 - \tan^2\alpha\tan^2\beta}$

(e) $\tan(\alpha + \beta + \gamma) = \tan[(\alpha + \beta) + \gamma] = \dfrac{\tan\alpha + \tan\beta + \tan\gamma - \tan\alpha\tan\beta\tan\gamma}{1 - \tan\alpha\tan\beta - \tan\beta\tan\gamma - \tan\gamma\tan\alpha}$

(f) $\dfrac{\sin(x + y)}{\cos(x - y)} = \dfrac{\tan x + \tan y}{1 + \tan x\tan y}$

(g) $\tan(45° + \theta) = \dfrac{\cos\theta + \sin\theta}{\cos\theta - \sin\theta}$

(h) $\sin(\alpha + \beta)\sin(\alpha - \beta) = \sin^2\alpha - \sin^2\beta$

38.28 If A and B are the measures of acute angles, find $A + B$ given

(a) $\tan A = \frac{1}{4}$, $\tan B = \frac{3}{5}$.   **Hint:** $\tan(A + B) = 1$.

   Ans. 45°

(b) $\tan A = \frac{5}{3}$, $\tan B = 4$.

   Ans. 135°

38.29 If $\tan(x + y) = 33$ and $\tan x = 3$, show that $\tan y = 0.3$.

38.30 Find the values of $\sin 2\theta$, $\cos 2\theta$, and $\tan 2\theta$, given

(a) $\sin\theta = \frac{3}{5}$, θ in quadrant I.

   Ans. $\frac{24}{25}, \frac{7}{25}, \frac{24}{7}$

(b) $\sin\theta = \frac{3}{5}$, θ in quadrant II.

   Ans. $-\frac{24}{25}, \frac{7}{25}, -\frac{24}{7}$

(c) $\sin\theta = -\frac{1}{2}$, θ in quadrant IV.

   Ans. $-\sqrt{3}/2, \frac{1}{2}, -\sqrt{3}$

(d) $\tan\theta = -\frac{1}{5}$, θ in quadrant II.

   Ans. $-\frac{5}{13}, \frac{12}{13}, -\frac{5}{12}$

(e) $\tan\theta = u$, θ in quadrant I.

   Ans. $\dfrac{2u}{1 + u^2}, \dfrac{1 - u^2}{1 + u^2}, \dfrac{2u}{1 - u^2}$

38.31 Prove

(a) $\tan\theta\sin 2\theta = 2\sin^2\theta$

(b) $\cot\theta\sin 2\theta = 1 + \cos 2\theta$

(c) $\dfrac{\sin^3 x - \cos^3 x}{\sin x - \cos x} = 1 + \frac{1}{2}\sin 2x$

(d) $\dfrac{1 - \sin 2A}{\cos 2A} = \dfrac{1 - \tan A}{1 + \tan A}$

(e) $\cos 2\theta = \dfrac{1 - \tan^2 \theta}{1 + \tan^2 \theta}$

(f) $\dfrac{1 + \cos 2\theta}{\sin 2\theta} = \cot \theta$

(g) $\cos 3\theta = 4 \cos^3 \theta - 3 \cos \theta$

(h) $\cos^4 x = \frac{3}{8} + \frac{1}{2} \cos 2x + \frac{1}{8} \cos 4x$

38.32 Find the values of sine, cosine, and tangent of

(a) $30°$, given $\cos 60° = \frac{1}{2}$.

   Ans. $\frac{1}{2}, \sqrt{3}/2, 1/\sqrt{3}$

(b) $105°$, given $\cos 210° = -\sqrt{3}/2$.

   Ans. $\frac{1}{2}\sqrt{2 + \sqrt{3}}, -\frac{1}{2}\sqrt{2 - \sqrt{3}}, -(2 + \sqrt{3})$

(c) $\frac{1}{2}\theta$, given $\sin \theta = \frac{3}{5}$, θ in quadrant I.

   Ans. $1/\sqrt{10}, 3/\sqrt{10}, \frac{1}{3}$

(d) θ, given $\cot 2\theta = \frac{7}{24}$, 2θ in quadrant I.

   Ans. $\frac{3}{5}, \frac{4}{5}, \frac{3}{4}$

(e) θ, given $\cot 2\theta = -\frac{5}{12}$, 2θ in quadrant II.

   Ans. $3/\sqrt{13}, 2/\sqrt{13}, \frac{3}{2}$

38.33 Prove

(a) $\cos x = 2 \cos^2 \frac{1}{2}x - 1 = 1 - 2 \sin^2 \frac{1}{2}x$

(b) $\sin x = 2 \sin \frac{1}{2}x \cos \frac{1}{2}x$

(c) $(\sin \frac{1}{2}\theta - \cos \frac{1}{2}\theta)^2 = 1 - \sin \theta$

(d) $\tan \frac{1}{2}\theta = \csc \theta - \cot \theta$

(e) $\dfrac{1 - \tan \frac{1}{2}\theta}{1 + \tan \frac{1}{2}\theta} = \dfrac{1 - \sin \theta}{\cos \theta} = \dfrac{\cos \theta}{1 + \sin \theta}$

(f) $\dfrac{2 \tan \frac{1}{2}x}{1 + \tan^2 \frac{1}{2}x} = \sin x$

38.34 In the right triangle ABC, in which C is the right angle, prove

$$\sin 2A = \frac{2ab}{c^2}, \qquad \cos 2A = \frac{b^2 - a^2}{c^2}, \qquad \sin \frac{1}{2}A = \sqrt{\frac{c - b}{2c}}, \qquad \cos \frac{1}{2}A = \sqrt{\frac{c + b}{2c}}.$$

38.35 Prove (a) $\dfrac{\sin 3x}{\sin x} - \dfrac{\cos 3x}{\cos x} = 2$,   (b) $\tan 50° - \tan 40° = 2 \tan 10°$.

38.36 If $A + B + C = 180°$, prove

(a) $\sin A + \sin B + \sin C = 4 \cos \frac{1}{2}A \cos \frac{1}{2}B \cos \frac{1}{2}C$

(b) $\cos A + \cos B + \cos C = 1 + 4 \sin \frac{1}{2}A \sin \frac{1}{2}B \sin \frac{1}{2}C$

(c) $\sin^2 A + \sin^2 B - \sin^2 C = 2 \sin A \sin B \cos C$

(d) $\tan \frac{1}{2}A \tan \frac{1}{2}B + \tan \frac{1}{2}B \tan \frac{1}{2}C + \tan \frac{1}{2}C \tan \frac{1}{2}A = 1$

Chapter 39

Sum, Difference, and Product Formulas

PRODUCTS OF SINES AND COSINES

$$\sin \alpha \cos \beta = \tfrac{1}{2}[\sin (\alpha + \beta) + \sin (\alpha - \beta)]$$
$$\cos \alpha \sin \beta = \tfrac{1}{2}[\sin (\alpha + \beta) - \sin (\alpha - \beta)]$$
$$\cos \alpha \cos \beta = \tfrac{1}{2}[\cos (\alpha + \beta) + \cos (\alpha - \beta)]$$
$$\sin \alpha \sin \beta = -\tfrac{1}{2}[\cos (\alpha + \beta) - \cos (\alpha - \beta)]$$

(For proofs of these formulas, see Problem 39.1.)

SUM AND DIFFERENCE OF SINES AND COSINES

$$\sin A + \sin B = 2 \sin \tfrac{1}{2}(A + B) \cos \tfrac{1}{2}(A - B)$$
$$\sin A - \sin B = 2 \cos \tfrac{1}{2}(A + B) \sin \tfrac{1}{2}(A - B)$$
$$\cos A + \cos B = 2 \cos \tfrac{1}{2}(A + B) \cos \tfrac{1}{2}(A - B)$$
$$\cos A - \cos B = -2 \sin \tfrac{1}{2}(A + B) \sin \tfrac{1}{2}(A - B)$$

(For proofs of these formulas, see Problem 39.2.)

Solved Problems

39.1   Derive the product formulas.

Since $\sin (\alpha + \beta) + \sin (\alpha - \beta) = (\sin \alpha \cos \beta + \cos \alpha \cos \beta) + (\sin \alpha \cos \beta - \cos \alpha \sin \beta)$
$$= 2 \sin \alpha \cos \beta,$$

$$\sin \alpha \cos \beta = \tfrac{1}{2}[\sin (\alpha + \beta) + \sin (\alpha - \beta)]$$

Since $\sin (\alpha + \beta) - \sin (\alpha - \beta) = 2 \cos \alpha \sin \beta$,

$$\cos \alpha \sin \beta = \tfrac{1}{2}[\sin (\alpha + \beta) - \sin (\alpha - \beta)]$$

Since $\cos (\alpha + \beta) + \cos (\alpha - \beta) = (\cos \alpha \cos \beta - \sin \alpha \sin \beta) + (\cos \alpha \cos \beta + \sin \alpha \sin \beta)$
$$= 2 \cos \alpha \cos \beta,$$

$$\cos \alpha \cos \beta = \tfrac{1}{2}[\cos (\alpha + \beta) + \cos (\alpha - \beta)]$$

Since $\cos (\alpha + \beta) - \cos (\alpha - \beta) = -2 \sin \alpha \sin \beta$,

$$\sin \alpha \sin \beta = -\tfrac{1}{2}[\cos (\alpha + \beta) - \cos (\alpha - \beta)]$$

39.2 Derive the sum and difference formulas.

Let $\alpha + \beta = A$ and $\alpha - \beta = B$ so that $\alpha = \frac{1}{2}(A + B)$ and $\beta = \frac{1}{2}(A - B)$. Then (see Problem 39.1)

$\sin(\alpha + \beta) + \sin(\alpha - \beta) = 2\sin\alpha\cos\beta$ becomes $\sin A + \sin B = 2\sin\frac{1}{2}(A + B)\cos\frac{1}{2}(A - B)$

$\sin(\alpha + \beta) - \sin(\alpha - \beta) = 2\cos\alpha\sin\beta$ becomes $\sin A - \sin B = 2\cos\frac{1}{2}(A + B)\sin\frac{1}{2}(A - B)$

$\cos(\alpha + \beta) + \cos(\alpha - \beta) = 2\cos\alpha\cos\beta$ becomes $\cos A + \cos B = 2\cos\frac{1}{2}(A + B)\cos\frac{1}{2}(A - B)$

$\cos(\alpha + \beta) - \cos(\alpha - \beta) = -2\sin\alpha\cos\beta$ becomes $\cos A - \cos B = -2\sin\frac{1}{2}(A + B)\sin\frac{1}{2}(A - B)$

39.3 Express each of the following as a sum or difference:

(a) $\sin 40° \cos 30°$,   (b) $\cos 110° \sin 55°$,   (c) $\cos 50° \cos 35°$,   (d) $\sin 55° \sin 40°$.

(a) $\sin 40° \cos 30° = \frac{1}{2}[\sin(40° + 30°) + \sin(40° - 30°)] = \frac{1}{2}(\sin 70° + \sin 10°)$

(b) $\cos 110° \sin 55° = \frac{1}{2}[\sin(110° + 55°) - \sin(110° - 55°)] = \frac{1}{2}(\sin 165° - \sin 55°)$

(c) $\cos 50° \cos 35° = \frac{1}{2}[\cos(50° + 35°) + \cos(50° - 35°)] = \frac{1}{2}(\cos 85° + \cos 15°)$

(d) $\sin 55° \sin 40° = -\frac{1}{2}[\cos(55° + 40°) - \cos(55° - 40°)] = -\frac{1}{2}(\cos 95° - \cos 15°)$

39.4 Express each of the following as a product:

(a) $\sin 50° + \sin 40°$,   (b) $\sin 70° - \sin 20°$,   (c) $\cos 55° + \cos 25°$,   (d) $\cos 35° - \cos 75°$.

(a) $\sin 50° + \sin 40° = 2\sin\frac{1}{2}(50° + 40°)\cos\frac{1}{2}(50° - 40°) = 2\sin 45° \cos 5°$

(b) $\sin 70° - \sin 20° = 2\cos\frac{1}{2}(70° + 20°)\sin\frac{1}{2}(70° - 20°) = 2\cos 45° \sin 25°$

(c) $\cos 55° + \cos 25° = 2\cos\frac{1}{2}(55° + 25°)\cos\frac{1}{2}(55° - 25°) = 2\cos 40° \cos 15°$

(d) $\cos 35° - \cos 75° = -2\sin\frac{1}{2}(35° + 75°)\sin\frac{1}{2}(35° - 75°) = -2\sin 55° \sin(-20°)$
$$= 2\sin 55° \sin 20°$$

39.5 Prove $\dfrac{\sin 4A + \sin 2A}{\cos 4A + \cos 2A} = \tan 3A$.

$$\frac{\sin 4A + \sin 2A}{\cos 4A + \cos 2A} = \frac{2\sin\frac{1}{2}(4A + 2A)\cos\frac{1}{2}(4A - 2A)}{2\cos\frac{1}{2}(4A + 2A)\cos\frac{1}{2}(4A - 2A)} = \frac{\sin 3A}{\cos 3A} = \tan 3A$$

39.6 Prove $\dfrac{\sin A - \sin B}{\sin A + \sin B} = \dfrac{\tan\frac{1}{2}(A - B)}{\tan\frac{1}{2}(A + B)}$.

$$\frac{\sin A - \sin B}{\sin A + \sin B} = \frac{2\cos\frac{1}{2}(A + B)\sin\frac{1}{2}(A - B)}{2\sin\frac{1}{2}(A + B)\cos\frac{1}{2}(A - B)} = \cot\frac{1}{2}(A + B)\tan\frac{1}{2}(A - B) = \frac{\tan\frac{1}{2}(A - B)}{\tan\frac{1}{2}(A + B)}$$

39.7 Prove $\cos^3 x \sin^2 x = \frac{1}{16}(2\cos x - \cos 3x - \cos 5x)$.

$$\cos^3 x \sin^2 x = (\sin x \cos x)^2 \cos x = \frac{1}{4}\sin^2 2x \cos x = \frac{1}{4}(\sin 2x)(\sin 2x \cos x)$$
$$= \frac{1}{4}(\sin 2x)[\frac{1}{2}(\sin 3x + \sin x)] = \frac{1}{8}(\sin 3x \sin 2x + \sin 2x \sin x)$$
$$= \frac{1}{8}\{-\frac{1}{2}(\cos 5x - \cos x) + [-\frac{1}{2}(\cos 3x - \cos x)]\}$$
$$= \frac{1}{16}(2\cos x - \cos 3x - \cos 5x)$$

39.8 Prove $1 + \cos 2x + \cos 4x + \cos 6x = 4\cos x \cos 2x \cos 3x$.

$$1 + (\cos 2x + \cos 4x) + \cos 6x = 1 + 2\cos 3x \cos x + \cos 6x = (1 + \cos 6x) + 2\cos 3x \cos x$$
$$= 2\cos^2 3x + 2\cos 3x \cos x = 2\cos 3x(\cos 3x + \cos x)$$
$$= 2\cos 3x(2\cos 2x \cos x) = 4\cos x \cos 2x \cos 3x$$

39.9 Transform $4 \cos x + 3 \sin x$ into the form $c \cos(x - \alpha)$.

Since $c \cos(x - \alpha) = c(\cos x \cos \alpha + \sin x \sin \alpha)$, set $c \cos \alpha = 4$ and $c \sin \alpha = 3$. Then $\cos \alpha = 4/c$ and $\sin \alpha = 3/c$. Since $\sin^2 \alpha + \cos^2 \alpha = 1$, $c = 5$ and -5.

Using $c = 5$, $\cos \alpha = \frac{4}{5}$, $\sin \alpha = \frac{3}{5}$, and $\alpha = 36° \, 52'$. Thus, $4 \cos x + 3 \sin x = 5 \cos(x - 36° \, 52')$.

Using $c = -5$, $\alpha = 216° \, 52'$ and $4 \cos x + 3 \sin x = -5 \cos(x - 216° \, 52')$.

39.10 Find the maximum and minimum values of $4 \cos x + 3 \sin x$ on the interval $0 \le x \le 2\pi$.

From Problem 39.9, $4 \cos x + 3 \sin x = 5 \cos(x - 36° \, 52')$.

Now on the prescribed interval, $\cos \theta$ attains its maximum value 1 when $\theta = 0$ and its minimum value -1 when $\theta = \pi$. Thus, the maximum value of $4 \cos x + 3 \sin x$ is 5 which occurs when $x - 36° \, 52' = 0$ or when $x = 36° \, 52'$ while the minimum value is -5 which occurs when $x - 36° \, 52' = \pi$ or when $x = 216° \, 52'$.

Supplementary Problems

39.11 Express each of the following products as a sum or difference of sines or of cosines.

(a) $\sin 35° \cos 25° = \frac{1}{2}(\sin 60° + \sin 10°)$

(b) $\sin 25° \cos 75° = \frac{1}{2}(\sin 100° - \sin 50°)$

(c) $\cos 50° \cos 70° = \frac{1}{2}(\cos 120° + \cos 20°)$

(d) $\sin 130° \sin 55° = -\frac{1}{2}(\cos 185° - \cos 75°)$

(e) $\sin 4x \cos 2x = \frac{1}{2}(\sin 6x + \sin 2x)$

(f) $\sin(x/2) \cos(3x/2) = \frac{1}{2}(\sin 2x - \sin x)$

(g) $\cos 7x \cos 4x = \frac{1}{2}(\cos 11x + \cos 3x)$

(h) $\sin 5x \sin 4x = -\frac{1}{2}(\cos 9x - \cos x)$

39.12 Show that

(a) $2 \sin 45° \cos 15° = \frac{1}{2}(\sqrt{3} + 1)$ and $\cos 15° = \frac{\sqrt{2}}{4}(\sqrt{3} + 1)$,

(b) $2 \sin 82\frac{1}{2}° \cos 37\frac{1}{2}° = \frac{1}{2}(\sqrt{3} + \sqrt{2})$ (c) $2 \sin 127\frac{1}{2}° \sin 97\frac{1}{2}° = \frac{1}{2}(\sqrt{3} + \sqrt{2})$

39.13 Express each of the following as a product:

(a) $\sin 50° + \sin 20° = 2 \sin 35° \cos 15°$ (e) $\sin 4x + \sin 2x = 2 \sin 3x \cos x$

(b) $\sin 75° - \sin 35° = 2 \cos 55° \sin 20°$ (f) $\sin 7\theta - \sin 3\theta = 2 \cos 5\theta \sin 2\theta$

(c) $\cos 65° + \cos 15° = 2 \cos 40° \cos 25°$ (g) $\cos 6\theta + \cos 2\theta = 2 \cos 4\theta \cos 2\theta$

(d) $\cos 80° - \cos 70° = -2 \sin 75° \sin 5°$ (h) $\cos(3x/2) - \cos(9x/2) = 2 \sin 3x \sin(3x/2)$

39.14 Show that

(a) $\sin 40° + \sin 20° = \cos 10°$ (c) $\cos 465° + \cos 165° = -\sqrt{6}/2$

(b) $\sin 105° + \sin 15° = \sqrt{6}/2$ (d) $\dfrac{\sin 75° - \sin 15°}{\cos 75° + \cos 15°} = \dfrac{1}{\sqrt{3}}$

39.15 Prove

(a) $\dfrac{\sin A + \sin 3A}{\cos A + \cos 3A} = \tan 2A$ (c) $\dfrac{\sin A + \sin B}{\sin A - \sin B} = \dfrac{\tan \frac{1}{2}(A + B)}{\tan \frac{1}{2}(A - B)}$

(b) $\dfrac{\sin 2A + \sin 4A}{\cos 2A + \cos 4A} = \tan 3A$ (d) $\dfrac{\cos A + \cos B}{\cos A - \cos B} = -\cot \frac{1}{2}(A - B) \cot \frac{1}{2}(A + B)$

(e) $\sin \theta + \sin 2\theta + \sin 3\theta = \sin 2\theta + (\sin \theta + \sin 3\theta) = \sin 2\theta(1 + 2\cos \theta)$

(f) $\cos \theta + \cos 2\theta + \cos 3\theta = \cos 2\theta(1 + 2\cos \theta)$

(g) $\sin 2\theta + \sin 4\theta + \sin 6\theta = (\sin 2\theta + \sin 4\theta) + 2 \sin 3\theta \cos 3\theta$
$\qquad\qquad\qquad\qquad = 4 \cos \theta \cos 2\theta \sin 3\theta$

(h) $\dfrac{\sin 3x + \sin 5x + \sin 7x + \sin 9x}{\cos 3x + \cos 5x + \cos 7x + \cos 9x} = \dfrac{(\sin 3x + \sin 9x) + (\sin 5x + \sin 7x)}{(\cos 3x + \cos 9x) + (\cos 5x + \cos 7x)} = \tan 6x$

39.16 Prove

(a) $\cos 130° + \cos 110° + \cos 10° = 0$ (b) $\cos 220° + \cos 100° + \cos 20° = 0$

39.17 Prove

(a) $\cos^2 \theta \sin^3 \theta = \frac{1}{16}(2 \sin \theta + \sin 3\theta - \sin 5\theta)$

(b) $\cos^2 \theta \sin^4 \theta = \frac{1}{32}(2 - \cos 2\theta - 2 \cos 4\theta + \cos 6\theta)$

(c) $\cos^5 \theta = \frac{1}{16}(10 \cos \theta + 5 \cos 3\theta + \cos 5\theta)$

(d) $\sin^5 \theta = \frac{1}{16}(10 \sin \theta - 5 \sin 3\theta + \sin 5\theta)$

39.18 Transform

(a) $4 \cos x + 3 \sin x$ into the form $c \sin (x + \alpha)$.   *Ans.* $5 \sin (x + 53° 8')$

(b) $4 \cos x + 3 \sin x$ into the form $c \sin (x - \alpha)$,   *Ans.* $5 \sin (x - 306° 52')$

(c) $\sin x - \cos x$ into the form $c \sin (x - \alpha)$.   *Ans.* $\sqrt{2} \sin (x - 45°)$

(d) $5 \cos 3t + 12 \sin 3t$ into the form $c \cos (3t - \alpha)$.   *Ans.* $13 \cos (3t - 67° 23')$

39.19 Find the maximum and minimum values of each sum of Problem 39.18 and a value of x or t between 0 and 2π at which each occurs.

    Ans. (a) Maximum = 5, when $x = 36° 52'$ (i.e., when $x + 53° 8' = 90°$); minimum = -5, when $x = 216° 52'$

           (b) Same as (a)

           (c) Maximum = $\sqrt{2}$, when $x = 135°$; minimum = $-\sqrt{2}$, when $x = 315°$

           (d) Maximum = 13, when $t = 22° 28'$; minimum = -13, when $t = 82° 28'$

Chapter 40

Oblique Triangles

AN OBLIQUE TRIANGLE is one which does not contain a right angle. Such a triangle contains either three acute angles or two acute angles and one obtuse angle.

The convention of denoting the measures of the angles by A, B, C and the lengths of the corresponding opposite sides by a, b, c will be used here. See Figs. 40-1 and 40-2.

Fig. 40-1	Fig. 40-2

When three parts, not all angles, are known, the triangle is uniquely determined, except in one case to be noted below. The four cases of oblique triangles are

Case I. Given one side and two angles

Case II. Given two sides and the angle opposite one of them

Case III. Given two sides and the included angle

Case IV. Given the three sides

THE LAW OF SINES. In any triangle, the sides are proportional to the sines of the opposite angles, i.e.,

$$\frac{a}{\sin A} = \frac{b}{\sin B} = \frac{c}{\sin C}$$

The following relations follow readily:

$$\frac{a}{b} = \frac{\sin A}{\sin B}, \qquad \frac{b}{c} = \frac{\sin B}{\sin C}, \qquad \frac{c}{a} = \frac{\sin C}{\sin A}.$$

(For a proof of the law of sines, see Problem 40.1.)

PROJECTION FORMULAS. In any triangle ABC,

$$a = b \cos C + c \cos B$$

$$b = c \cos A + a \cos C$$

$$c = a \cos B + b \cos A$$

(For the derivation of these formulas, see Problem 40.3.)

CASE I. Given one side and two angles

EXAMPLE. Suppose a, B, and C are given.

To find A, use $A = 180° - (B + C)$.

To find b, use $\dfrac{b}{a} = \dfrac{\sin B}{\sin A}$ whence $b = \dfrac{a \sin B}{\sin A}$.

To find c, use $\dfrac{c}{a} = \dfrac{\sin C}{\sin A}$ whence $c = \dfrac{a \sin C}{\sin A}$.

(See Problems 40.4–40.6.)

CASE II. Given two sides and the angle opposite one of them

EXAMPLE. Suppose b, c, and B are given.

From $\dfrac{\sin C}{\sin B} = \dfrac{c}{b}$, $\quad \sin C = \dfrac{c \sin B}{b}$.

If $\sin C > 1$, no angle C is determined.
If $\sin C = 1$, $C = 90°$ and a right triangle is determined.
If $\sin C < 1$, two angles are determined: an acute angle C and an obtuse angle $C' = 180° - C$. Thus, there may be one or two triangles determined. This is known as the "ambiguous case."

This case is discussed geometrically in Problem 40.7. The results obtained may be summarized as follows:

When the given angle is *acute*, there will be

(a)   *One* solution if the side opposite the given angle is equal to or greater than the other given side
(b)   *No* solution, *one* solution (right triangle), or *two* solutions if the side opposite the given angle is less than the other given side.

When the given angle is *obtuse*, there will be

(c)   *no* solution when the side opposite the given angle is less than or equal to the other given side
(d)   *one* solution if the side opposite the given angle is greater than the other given side.

EXAMPLE

(1)   When $b = 30$, $c = 20$, and $B = 40°$, there is one solution since B is acute and $b > c$.

(2)   When $b = 20$, $c = 30$, and $B = 40°$, there is either no solution, one solution, or two solutions. The particular subcase is determined after computing $\sin C = \dfrac{c \sin B}{b}$.

(3)   When $b = 30$, $c = 20$, and $B = 140°$, there is one solution.
(4)   When $b = 20$, $c = 30$, and $B = 140°$, there is no solution.

This, the so-called ambiguous case, is solved by the law of sines and may be checked by the projection formulas. (See Problems 40.8–40.10.)

THE LAW OF COSINES. In any triangle ABC, the square of any side is equal to the sum of the squares of the other two sides diminished by twice the product of these sides and the cosine of their included angle, i.e.,

$$a^2 = b^2 + c^2 - 2bc \cos A$$

$$b^2 = c^2 + a^2 - 2ca \cos B$$

$$c^2 = a^2 + b^2 - 2ab \cos C$$

(For the derivation of these formulas, see Problem 40.11.)

CASE III.   Given two sides and the included angle

EXAMPLE.   Suppose a, b, and C are given.

To find c, use $c^2 = a^2 + b^2 - 2ab \cos C$.

To find A, use $\sin A = \dfrac{a \sin C}{c}$. To find B, use $\sin B = \dfrac{b \sin C}{c}$.

To check, use $A + B + C = 180°$.
(See Problems 40.12–40.14.)

CASE IV.   Given the three sides

EXAMPLE.   With a, b, and c given, solve the law of cosines for each of the angles.

To find the angles, use $\cos A = \dfrac{b^2 + c^2 - a^2}{2bc}$, $\cos B = \dfrac{c^2 + a^2 - b^2}{2ca}$, $\cos C = \dfrac{a^2 + b^2 - c^2}{2ab}$.

To check, use $A + B + C = 180°$.
(See Problems 40.15–40.16.)

Solved Problems

40.1   Derive the law of sines.

Let ABC be any oblique triangle. In Fig. 40-3(a), angles A and B are acute while in Fig. 40-3(b), angle B is obtuse. Draw $\overline{CD}$ perpendicular to $\overline{AB}$ or $\overrightarrow{AB}$ extended and denote its length by h.

(a)   (b)

Fig. 40-3

In the right triangle ACD of either figure, $h = b \sin A$ while in the right triangle BCD, $h = a \sin B$ since in Fig. 40-3(b), $h = a \sin \angle DBC = a \sin(180° - B) = a \sin B$. Thus,

$$a \sin B = b \sin A \qquad \text{or} \qquad \frac{a}{\sin A} = \frac{b}{\sin B}.$$

In a similar manner (by drawing a perpendicular from B to $\overline{AC}$ or a perpendicular from A to $\overline{BC}$), we obtain

$$\frac{a}{\sin A} = \frac{c}{\sin C} \qquad \text{or} \qquad \frac{b}{\sin B} = \frac{c}{\sin C}.$$

Thus, finally,

$$\frac{a}{\sin A} = \frac{b}{\sin B} = \frac{c}{\sin C}.$$

40.2   Derive one of the projection formulas.

Refer to Fig. 40-3. In the right triangle ACD of either figure, $AD = b \cos A$. In the right triangle BCD of Fig. 40-3(a), $DB = a \cos B$. Thus, in Fig. 40-3(a),

$$c = AB = AD + DB = b \cos A + a \cos B = a \cos B + b \cos A$$

In the right triangle BCD of Fig. 40-3(b), $BD = a \cos \angle DBC = a \cos(180° - B) = -a \cos B$. Thus, in Fig. 40-3($b$),

$$c = AB = AD - BD = b \cos A - (-a \cos B) = a \cos B + b \cos A$$

CASE I

40.3   Solve the triangle ABC, given $c = 25$, $A = 35°$, and $B = 68°$. See Fig. 40-4.

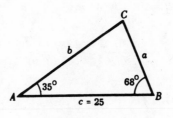

Fig. 40-4

To find C: $\qquad C = 180° - (A + B) = 180° - 103° = 77°.$

To find a: $\qquad a = \dfrac{c \sin A}{\sin C} = \dfrac{25 \sin 35°}{\sin 77°} = \dfrac{25(0.5736)}{0.9744} = 15$

To find b: $\qquad b = \dfrac{c \sin B}{\sin C} = \dfrac{25 \sin 68°}{\sin 77°} = \dfrac{25(0.9272)}{0.9744} = 24$

To check by projection formula:

$$c = a \cos B + b \cos A = 15 \cos 68° + 24 \cos 35° = 15(0.3746) + 24(0.8192) = 25.3$$

The required parts are $a = 15$, $b = 24$, and $C = 77°$.

40.4 A and B are two points on opposite banks of a river. From A a line $AC = 275$ ft is laid off and the angles $CAB = 125° 40'$ and $ACB = 48° 50'$ are measured. Find the length of $\overline{AB}$. See Fig. 40-5.

Fig. 40-5

In the triangle ABC, $B = 180° - (C + A) = 5° 30'$ and

$$AB = c = \frac{b \sin C}{\sin B} = \frac{275 \sin 48° 50'}{\sin 5° 30'} = \frac{275(0.7528)}{0.0958} = 2160 \text{ ft}$$

40.5   A tower 125 ft high is on a cliff on the bank of a river. From the top of the tower the angle of depression of a point on the opposite shore is 28° 40′ and from the base of the tower the angle of depression of the same point is 18° 20′. Find the width of the river and the height of the cliff.

In Fig. 40-6 *BC* represents the tower, $\overline{DB}$ represents the cliff, and *A* is the point on the opposite shore.

Fig. 40-6

In triangle *ABC*, $C = 90° - 28° 40′ = 61° 20′$, $B = 90° + 18° 20′ = 108° 20′$, $A = 180° - (B + C) = 10° 20′$.

$$c = \frac{a \sin C}{\sin A} = \frac{125 \sin 61° 20′}{\sin 10° 20′} = \frac{125(0.8774)}{0.1794} = 611$$

In the right triangle *ABD*, $DB = c \sin 18° 20′ = 611(0.3145) = 192$, $AD = c \cos 18° 20′ = 611(0.9492) = 580$.
The river is 580 ft wide and the cliff is 192 ft high.

40.6   Discuss the several special cases when two sides and the angle opposite one of them are given.

Let *b*, *c*, and *B* be the given parts. Construct the given angle *B* and lay off the side $BA = c$. With *A* as center and radius equal to *b* (the side opposite the given angle) describe an arc. Figures 40-7(*a*)–(*e*) illustrate the special cases which may occur when the given angle *B* is acute while Figures 40-7(*f*)–(*g*) illustrate the cases when *B* is obtuse.

Fig. 40-7

The given angle B is acute.

Fig. 40-7(*a*).   When $b < AD = c \sin B$, the arc does not meet BX and no triangle is determined.

Fig. 40-7(*b*).   When $b = AD$, the arc is tangent to BX and one triangle—a right triangle with the right angle at C—is determined.

Fig. 40-7(*c*).   When $b > AD$ and $b < c$, the arc meets BX in two points C and C' on the same side of B. Two triangles ABC, in which C is acute, and ABC' in which $C' = 180° - C$ is obtuse, are determined.

Fig. 40-7(*d*).   When $b > AD$ and $b = c$, the arc meets $\overrightarrow{BX}$ in C and B. One triangle (isosceles) is determined.

Fig. 40-7(*e*).   When $b > c$, the arc meets BX in C and $\overrightarrow{BX}$ extended in C'. Since the triangle ABC' does not contain the given angle B, only one triangle ABC is determined.

The given angle is obtuse.

Fig. 40-7(*f*).   When $b < c$ or $b = c$, no triangle is formed.

Fig. 40-7(*g*).   When $b > c$, only one triangle is formed as in Fig. 40.7(*e*).

CASE II

40.7   Solve the triangle ABC, given $c = 628$, $b = 480$, and $C = 55°\,10'$. Refer to Fig. 40-8.

Fig. 40-8

Fig. 40-9

Since C is acute and $c > b$, there is only one solution.

For B: $\sin B = \dfrac{b \sin C}{c} = \dfrac{480 \sin 55°\,10'}{628} = \dfrac{480(0.8208)}{628} = 0.6274$ and $B = 38°\,50'$

For A: $A = 180° - (B + C) = 86°\,0'$

For a: $a = \dfrac{b \sin A}{\sin B} = \dfrac{480 \sin 86°\,0'}{\sin 38°\,50'} = \dfrac{480(0.9976)}{0.6271} = 764$

The required parts are $B = 38°\,50'$, $A = 86°\,0'$, and $a = 764$.

40.8   Solve the triangle ABC, given $a = 525$, $c = 421$, and $A = 130°\,50'$. Refer to Fig. 40-9.

Since A is obtuse and $a > c$, there is one solution.

For C: $\sin C = \dfrac{c \sin A}{a} = \dfrac{421 \sin 130°\,50'}{525} = \dfrac{421(0.7566)}{525} = 0.6067$ and $C = 37°\,20'$

For B: $B = 180° - (C + A) = 11°\,50'$

For b: $b = \dfrac{a \sin B}{\sin A} = \dfrac{525 \sin 11°\,50'}{\sin 130°\,50'} = \dfrac{525(0.2051)}{0.7566} = 142$

The required parts are $C = 37°\,20'$, $B = 11°\,50'$, and $b = 142$.

40.9   Solve the triangle ABC, given $a = 31.5$, $b = 51.8$, and $A = 33°\,40'$. Refer to Fig. 40-10.

Fig. 40-10

Since A is acute and $a < b$, there is the possibility of two solutions.

For B: $\sin B = \dfrac{b \sin A}{a} = \dfrac{51.8 \sin 33°\,40'}{31.5} = \dfrac{51.8(0.5544)}{31.5} = 0.9117$

There are two solutions, $B = 65°\,40'$ and $B' = 180° - 65°\,40' = 114°\,20'$.

For C: $C = 180° - (A + B) = 80°\,40'$

For C': $C' = 180° - (A + B') = 32°\,0'$

For c: $c = \dfrac{a \sin C}{\sin A} = \dfrac{31.5 \sin 80°\,40'}{\sin 33°\,40'} = \dfrac{31.5(0.9868)}{0.5544} = 56.1$

For c': $c' = \dfrac{a \sin C'}{\sin A} = \dfrac{31.5 \sin 32°\,0'}{\sin 33°\,40'} = \dfrac{31.5(0.5299)}{0.5544} = 30.1$

The required parts are

For triangle ABC: $B = 65°\,40'$, $C = 80°\,40'$,   and $c = 56.1$.

For triangle ABC': $B' = 114°\,20'$, $C' = 32°\,0'$,   and $c' = 30.1$.

40.10   Derive the law of cosines.

In the right triangle ACD of either figure, $b^2 = h^2 + (AD)^2$.

In the right triangle BCD of Fig. 40-11(a), $h = a \sin B$ and $DB = a \cos B$. Then $AD = AB - DB = c - a \cos B$ and

$$b^2 = h^2 + (AD)^2 = a^2 \sin^2 B + c^2 - 2ca \cos B + a^2 \cos^2 B$$
$$= a^2(\sin^2 B + \cos^2 B) + c^2 - 2ca \cos B = c^2 + a^2 - 2ca \cos B$$

Fig. 40-11

In the right triangle BCD of Fig. 40-11(b), $h = a \sin \angle CBD = a \sin(180° - B) = a \sin B$ and $BD = a \cos \angle CBD = a \cos(180° - B) = -a \cos B$. Then $AD = AB + BD = c - a \cos B$ and $b^2 = c^2 + a^2 - 2ca \cos B$.

The remaining equations may be obtained by cyclic changes of the letters.

CASE III

40.11   Solve the triangle ABC, given $a = 132$, $b = 224$, and $C = 28° 40'$. See Fig. 40-12.

Fig. 40-12

For c: $c^2 = a^2 + b^2 - 2ab \cos C$

$$= (132)^2 + (224)^2 - 2(132)(224) \cos 28° 40'$$

$$= (132)^2 + (224)^2 - 2(132)(224)(0.8774)$$

$$= 15\,714 \quad \text{and} \quad c = 125$$

For A: $\sin A = \dfrac{a \sin C}{c} = \dfrac{132 \sin 28° 40'}{125} = \dfrac{132(0.4797)}{125} = 0.5066$ and $A = 30° 30'$

For B: $\sin B = \dfrac{b \sin C}{c} = \dfrac{224 \sin 28° 40'}{125} = \dfrac{224(0.4797)}{125} = 0.8596$ and $B = 120° 40'$

(Since $b > a$, A is acute; since $A + C < 90°$, $B > 90°$.)
Check: $A + B + C = 179° 50'$. The required parts are $A = 30° 30'$, $B = 120° 40'$, $c = 125$.

40.12   Two forces of 17.5 lb and 22.5 lb act on a body. If their directions make an angle of $50° 10'$ with each other, find the magnitude of their resultant and the angle which it makes with the larger force.

In the parallelogram $ABCD$, (see Fig. 40-13) $A + B = C + D = 180°$ and $B = 180° - 50° 10' = 129° 50'$. In the triangle ABC,

$$b^2 = c^2 + a^2 - 2ca \cos B \quad [\cos 129° 50' = -\cos (180° - 129° 50') = -\cos 50° 10']$$

$$= (22.5)^2 + (17.5)^2 - 2(22.5)(17.5)(-0.6406) = 1317 \quad \text{and} \quad b = 36.3$$

$$\sin A = \frac{a \sin B}{b} = \frac{17.5 \sin 129° 50'}{36.3} = \frac{17.5(0.7679)}{36.3} = 0.3702 \quad \text{and} \quad A = 21° 40'$$

The resultant is a force of 36.3 lb; the required angle is $21° 40'$.

Fig. 40-13

40.13 From A a pilot flies 125 mi in the direction N 38° 20′ W and turns back. Through an error, he then flies 125 mi in the direction S 51° 40′ E. How far and in what direction must he now fly to reach his intended destination A?

Denote the turn back point as B and his final position as C. In the triangle ABC (see Fig. 40-14),

$$b^2 = c^2 + a^2 - 2ca \cos B = (125)^2 + (125)^2 - 2(125)(125) \cos 13° 20′$$
$$= 2(125)^2(1 - 0.9730) = 843.7 \quad \text{and} \quad b = 29.0$$

$$\sin A = \frac{a \sin B}{b} = \frac{125 \sin 13° 20′}{29.0} = \frac{125(0.2306)}{29.0} = 0.9940 \quad \text{and} \quad A = 83° 40′$$

Since $\angle CAN_1 = A - \angle N_1 AB = 45° 20′$, the pilot must fly a course S 45° 20′ W for 29.0 mi in going from C to A.

Fig. 40-14

CASE IV

40.14 Solve the triangle ABC, given $a = 30.3$, $b = 40.4$, and $c = 62.6$. Refer to Fig. 40-15.

Fig. 40-15

For A: $\cos A = \dfrac{b^2 + c^2 - a^2}{2bc} = \dfrac{(40.4)^2 + (62.6)^2 - (30.3)^2}{2(40.4)(62.6)} = 0.9159 \quad$ and $\quad A = 23° 40′$

For B: $\cos B = \dfrac{c^2 + a^2 - b^2}{2ca} = \dfrac{(62.6)^2 + (30.3)^2 - (40.4)^2}{2(62.6)(30.3)} = 0.8448 \quad$ and $\quad B = 32° 20′$

For C: $\cos C = \dfrac{a^2 + b^2 - c^2}{2ab} = \dfrac{(30.3)^2 + (40.4)^2 - (62.6)^2}{2(30.3)(40.4)} = -0.5590 \quad$ and $\quad C = 124° 0′$

Check: $A + B + C = 180°$.

40.15 The distances of a point C from two points A and B, which cannot be measured directly, are required. The line CA is continued through A for a distance 175 m to D, the line $\overleftrightarrow{CB}$ is continued through B for 225 m to E, and the distances $AB = 300$ m, $DB = 326$ m, and $DE = 488$ m are measured. Find AC and BC. Refer to Fig. 40-16.

Fig. 40-16

Triangle *ABC* may be solved for the required parts after the angles ∠*BAC* and ∠*ABC* have been found. The first angle is the supplement of ∠*BAD* and the second is the supplement of the sum of ∠*ABD* and ∠*DBE*.

In the triangle *ABD* whose sides are known,

$$\cos \angle BAD = \frac{(175)^2 + (300)^2 - (326)^2}{2(175)(300)} = 0.1367 \quad \text{and} \quad \angle BAD = 82° 10'$$

$$\cos \angle ABD = \frac{(300)^2 + (326)^2 - (175)^2}{2(300)(326)} = 0.8469 \quad \text{and} \quad \angle ABD = 32° 10'$$

In the triangle *BDE* whose sides are known,

$$\cos \angle DBE = \frac{(225)^2 + (326)^2 - (488)^2}{2(225)(326)} = -0.5538 \quad \text{and} \quad \angle DBE = 123° 40'$$

In the triangle *ABC*: *AB* = 300,

$$\angle BAC = 180° - \angle BAD = 97° 50'$$

$$\angle ABC = 180° - (\angle ABD + \angle DBE) = 24° 10'$$

$$\angle ACB = 180° - (\angle BAC + \angle ABC) = 58° 0'$$

Then
$$AC = \frac{AB \sin \angle ABC}{\sin \angle ACB} = \frac{300 \sin 24° 10'}{\sin 58° 0'} = \frac{300(0.4094)}{0.8480} = 145$$

and
$$BC = \frac{AB \sin \angle BAC}{\sin \angle ACB} = \frac{300 \sin 97° 50'}{\sin 58° 0'} = \frac{300(0.9907)}{0.8480} = 350$$

The required distances are *AC* = 145 m and *BC* = 350 m.

Supplementary Problems

Solve each of the following oblique triangles *ABC*, given:

40.16   *a* = 125, *A* = 54° 40′, *B* = 65° 10′.   *Ans.*   *b* = 139, *c* = 133, *C* = 60° 10′

40.17   *b* = 321, *A* = 75° 20′, *C* = 38° 30′.   *Ans.*   *a* = 339, *c* = 218, *B* = 66° 10′

40.18   *b* = 215, *c* = 150, *B* = 42° 40′.   *Ans.*   *a* = 300, *A* = 109° 10′, *C* = 28° 10′

40.19 $a = 512$, $b = 426$, $A = 48°\,50'$.   *Ans.* $c = 680$, $B = 38°\,50'$, $C = 92°\,20'$

40.20 $b = 50.4$, $c = 33.3$, $B = 118°\,30'$.   *Ans.* $a = 25.1$, $A = 26°\,0'$, $C = 35°\,30'$

40.21 $b = 40.2$, $a = 31.5$, $B = 112°\,20'$.   *Ans.* $c = 15.7$, $A = 46°\,30'$, $C = 21°\,10'$

40.22 $b = 51.5$, $a = 62.5$, $B = 40°\,40'$.   *Ans.* $c = 78.9$, $A = 52°\,20'$, $C = 87°\,0'$, $c' = 16.0$, $A' = 127°\,40'$, $C' = 11°\,40'$

40.23 $a = 320$, $c = 475$, $A = 35°\,20'$.   *Ans.* $b = 552$, $B = 85°\,30'$, $C' = 59°\,10'$, $b' = 224$, $B' = 23°\,50'$, $C' = 120°\,50'$

40.24 $b = 120$, $c = 270$, $A = 118°\,40'$.   *Ans.* $a = 344$, $B = 17°\,50'$, $C = 43°\,30'$

40.25 $a = 24.5$, $b = 18.6$, $c = 26.4$.   *Ans.* $A = 63°\,10'$, $B = 42°\,40'$, $C = 74°\,10'$

40.26 $a = 6.34$, $b = 7.30$, $c = 9.98$.   *Ans.* $A = 39°\,20'$, $B = 46°\,50'$, $C = 93°\,50'$

40.27   Two ships have radio equipment with a range of 200 mi. One is 155 mi N 42° 40′ E and the other is 165 mi N 45° 10′ W of a shore station. Can the two ships communicate directly?

> *Ans.*   No; they are 222 mi apart.

40.28   A ship sails 15.0 mi on a course S 40° 10′ W and then 21.0 mi on a course N 28° 20′ W. Find the distance and direction of the last position from the first.

> *Ans.*   20.9 mi, N 70° 40′ W

40.29   A lighthouse is 10 mi northwest of a dock. A ship leaves the dock at 9 A.M. and steams west at 12 mi per hr. At what time will it be 8 mi from the lighthouse?

> *Ans.*   9:16 A.M. and 9:54 A.M.

40.30   Two forces of 115 lb and 215 lb acting on an object have a resultant of magnitude 275 lb. Find the angle between the directions in which the given forces act.

> *Ans.*   70° 50′

40.31   A tower 150 m high is situated at the top of a hill. At a point 650 m down the hill the angle between the surface of the hill and the line of sight to the top of the tower is 12° 30′. Find the inclination of the hill to a horizontal plane.

> *Ans.*   7° 50′

Chapter 41

Inverse Trigonometric Functions

INVERSE TRIGONOMETRIC FUNCTIONS. The equation

$$x = \sin y \qquad (41.1)$$

defines a unique value of x for each given angle y. But when x is given, the equation may have no solution or many solutions. For example: if $x = 2$, there is no solution, since the sine of an angle never exceeds 1; if $x = \frac{1}{2}$, there are many solutions $y = 30°, 150°, 390°, 510°, -210°, -330°, \ldots$.

To express y as a function of x, we will write

$$y = \arcsin x \qquad (41.2)$$

In spite of the use of the word *arc*, (41.2) is to be interpreted as stating that "y is an angle whose sine is x." Similarly we shall write $y = \arccos x$ if $x = \cos y$, $y = \arctan x$ if $x = \tan y$, etc. An alternate notation for $y = \arcsin x$ is $y = \sin^{-1} x$ (and, similarly $y = \cos^{-1} x$, etc. for the other functions).

GRAPHS OF THE INVERSE TRIGONOMETRIC FUNCTIONS. The graph of $y = \arcsin x$ is the graph of $x = \sin y$ and differs from the graph of $y = \sin x$ of Chapter 38 in that the roles of x and y are interchanged. Thus, the graph of $y = \arcsin x$ is a sine curve drawn on the y axis instead of the x axis.

Similarly the graphs of the remaining inverse trigonometric functions are those of the corresponding trigonometric functions except that the roles of x and y are interchanged.

PRINCIPAL VALUES. It is at times necessary to consider the inverse trigonometric functions as single-valued (i.e., one value of y corresponding to each admissible value of x). To do this, we agree to select one out of the many angles corresponding to the given value of x. For example, when $x = \frac{1}{2}$, we shall agree to select the value $y = 30°$ and when $x = -\frac{1}{2}$, we shall agree to select the value $y = -30°$. This selected value is called the *principal value* of arcsin x. When only the principal value is called for, we shall write Arcsin x, Arccos x, etc. The portions of the graphs on which the principal values of each of the inverse trigonometric functions lie are shown in Figs. 41-1(a) through (f) by a heavier line.

If $\qquad\qquad y = \text{Sin}^{-1} x$, then $\qquad -\dfrac{\pi}{2} \le y \le \dfrac{\pi}{2}$.

If $\qquad\qquad y = \text{Cos}^{-1} x$, then $\qquad 0 \le y \le \pi$;

and if $\qquad\qquad y = \text{Tan}^{-1} x$, then $\qquad -\dfrac{\pi}{2} < y < \dfrac{\pi}{2}$.

Similarly,

if $\qquad\qquad y = \text{Sec}^{-1} x$ then $\qquad 0 \le y \le \pi, \quad y \ne \dfrac{\pi}{2}$;

if $\qquad\qquad y = \text{Cot}^{-1} x$ then $\qquad 0 < y < \pi$;

and if $\qquad\qquad y = \text{Csc}^{-1} x$, then $\qquad -\dfrac{\pi}{2} \le y \le \dfrac{\pi}{2}, \quad y \ne 0$.

For example, $\text{Sin}^{-1} \dfrac{\sqrt{3}}{2} = \dfrac{\pi}{3}$, $\text{Arctan } 1 = \dfrac{\pi}{4}$, $\text{Sin}^{-1} \dfrac{-\sqrt{3}}{2} = \dfrac{-\pi}{3}$, $\text{Arccos}\left(\dfrac{-1}{2}\right) = \dfrac{2\pi}{3}$, $\text{Sec}^{-1}\left(\dfrac{-2}{\sqrt{3}}\right) = \dfrac{-5\pi}{6}$ and $\text{Arccsc}(-\sqrt{2}) + \dfrac{-\pi}{4}$.

273

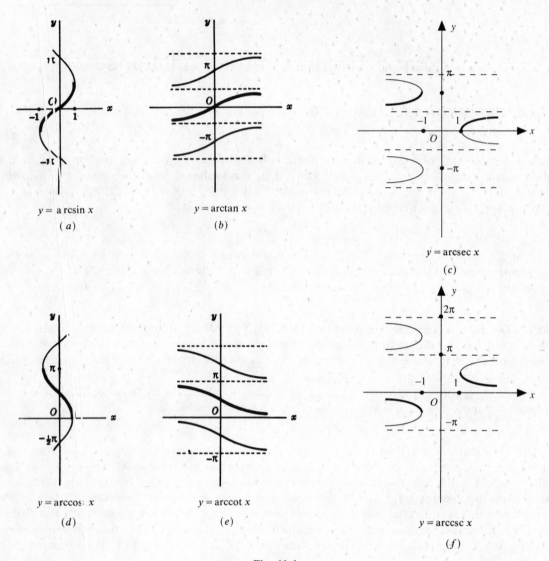

$y = \arcsin x$ $y = \arctan x$

(a)   (b)

$y = \operatorname{arcsec} x$

(c)

$y = \arccos x$ $y = \operatorname{arccot} x$

(d)   (e)

$y = \operatorname{arccsc} x$

(f)

Fig. 41-1

GENERAL VALUES OF THE INVERSE TRIGONOMETRIC FUNCTIONS. Let y be an inverse trigonometric function of x. Since the value of a trigonometric function of y is known, there are determined in general two positions for the terminal side of the angle y (see Chapter 31). Let y_1 and y_2 respectively be angles determined by the two positions of the terminal side. Then the totality of values of y consist of the angles y_1 and y_2, together with all angles coterminal with them, that is,

$$y_1 + 2n\pi \qquad \text{and} \qquad y_2 + 2n\pi$$

where n is any positive or negative integer, or is zero.

One of the values y_1 or y_2 may always be taken as the principal value of the inverse trigonometric function.

EXAMPLE. Write expressions for the general value of (a) $\arcsin \frac{1}{2}$, (b) $\arccos(-1)$, (c) $\arctan(-1)$.

(a) The principal value of $\arcsin \frac{1}{2}$ is $\pi/6$, and a second value (not coterminal with the principal value) is $5\pi/6$. The general value of $\arcsin \frac{1}{2}$ is given by $\pi/6 + 2n\pi$, $5\pi/6 + 2n\pi$, where n is any positive or negative integer, or is zero.

(b) The principal value is π and there is no other value not coterminal with it. Thus, the general value is given by $\pi + 2n\pi$, where n is a positive or negative integer, or is zero.

(c) The principal value is $-\pi/4$, and a second value (not coterminal with the principal value) is $3\pi/4$. Thus, the general value is given by $-\pi/4 + 2n\pi$, $3\pi/4 + 2n\pi$, where n is a positive or negative integer, or is zero.

Solved Problems

41.1 Find the principal value of each of the following:

(a) Arcsin $0 = 0$ (e) Arcsec $2 = \pi/3$ (i) Arctan $(-1) = -\pi/4$

(b) Arccos $(-1) = \pi$ (f) Arccsc $(-\sqrt{2}) = -3\pi/4$ (j) Arccot $0 = \pi/2$

(c) Arctan $\sqrt{3} = \pi/3$ (g) Arccos $0 = \pi/2$ (k) Arcsec $(-\sqrt{2}) = -3\pi/4$

(d) Arccot $\sqrt{3} = \pi/6$ (h) Arcsin $(-1) = -\pi/2$ (l) Arccsc $(-2) = -5\pi/6$

41.2 Express the principal value of each of the following to the nearest minute:

(a) Arcsin $0.3333 = 19°\,28'$ (g) Arcsin $(-0.6439) = -40°\,5'$

(b) Arccos $0.4000 = 66°\,25'$ (h) Arccos $(-0.4519) = 116°\,52'$

(c) Arctan $1.5000 = 56°\,19'$ (i) Arctan $(-1.4400) = -55°\,13'$

(d) Arccot $1.1875 = 40°\,6'$ (j) Arccot $(-0.7340) = 126°\,17'$

(e) Arcsec $1.0324 = 14°\,24'$ (k) Arcsec $(-1.2067) = -145°\,58'$

(f) Arccsc $1.5082 = 41°\,32'$ (l) Arccsc $(-4.1923) = -166°\,12'$

41.3 Verify each of the following:

(a) $\sin(\text{Arcsin }\tfrac{1}{2}) = \sin \pi/6 = \tfrac{1}{2}$ (e) Arccos $[\cos(-\pi/4)] =$ Arccos $\sqrt{2}/2 = \pi/4$

(b) $\cos[\text{Arccos}(-\tfrac{1}{2})] = \cos 2\pi/3 = -\tfrac{1}{2}$ (f) Arcsin $(\tan 3\pi/4) =$ Arcsin $(-1) = -\pi/2$

(c) $\cos[\text{Arcsin}(-\sqrt{2}/2)] = \cos(-\pi/4) = \sqrt{2}/2$ (g) Arccos $[\tan(-5\pi/4)] =$ Arccos $(-1) = \pi$

(d) Arcsin $(\sin \pi/3) =$ Arcsin $\sqrt{3}/2 = \pi/3$

41.4 Verify each of the following:

(a) Arcsin $\sqrt{2}/2 -$ Arcsin $\tfrac{1}{2} = \pi/4 - \pi/6 = \pi/12$

(b) Arccos $0 +$ Arctan $(-1) = \pi/2 + (-\pi/4) = \pi/4 =$ Arctan 1

41.5 Evaluate each of the following:

(a) $\cos(\text{Arcsin }\tfrac{3}{5})$,   (b) $\sin[\text{Arccos}(-\tfrac{2}{3})]$,   (c) $\tan[\text{Arcsin}(-\tfrac{3}{4})]$.

(a) Let $\theta =$ Arcsin $\tfrac{3}{5}$; then $\sin \theta = \tfrac{3}{5}$, θ being a first quadrant angle. From Fig. 41-2(a), $\cos(\text{Arcsin }\tfrac{3}{5}) = \cos \theta = \tfrac{4}{5}$.

(b) Let $\theta =$ Arccos $(-\tfrac{2}{3})$; then $\cos \theta = -\tfrac{2}{3}$, θ being a second quadrant angle. From Fig. 41-2(b), $\sin[\text{Arccos}(-\tfrac{2}{3})] = \sin \theta = \sqrt{5}/3$.

(c) Let $\theta =$ Arcsin $(-\tfrac{3}{4})$; then $\sin \theta = -\tfrac{3}{4}$, θ being a fourth quadrant angle. From Fig. 41-2(c), $\tan[\text{Arcsin}(-\tfrac{3}{4})] = \tan \theta = -3/\sqrt{7} = -3\sqrt{7}/7$.

Fig. 41-2

41.6   Evaluate $\sin(\text{Arcsin}\,\frac{12}{13} + \text{Arcsin}\,\frac{4}{5})$.

Let $\theta = \text{Arcsin}\,\frac{12}{13}$ and $\phi = \text{Arcsin}\,\frac{4}{5}$. Then $\sin\theta = \frac{12}{13}$ and $\sin\phi = \frac{4}{5}$, θ and ϕ being first quadrant angles. From Fig. 41-3 and Fig. 41-4,

$$\sin(\text{Arcsin}\,\tfrac{12}{13} + \text{Arcsin}\,\tfrac{4}{5}) = \sin(\theta + \phi) = \sin\theta\cos\phi + \cos\theta\sin\phi$$

$$= \tfrac{12}{13}\cdot\tfrac{3}{5} + \tfrac{5}{13}\cdot\tfrac{4}{5} = \tfrac{56}{65}$$

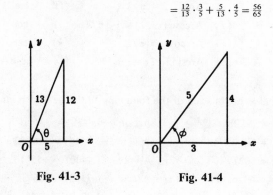

Fig. 41-3   Fig. 41-4

41.7   Evaluate $\cos(\text{Arctan}\,\frac{15}{8} - \text{Arcsin}\,\frac{7}{25})$.

Let $\theta = \text{Arctan}\,\frac{15}{8}$ and $\phi = \text{Arcsin}\,\frac{7}{25}$. Then $\tan\theta = \frac{15}{8}$ and $\sin\phi = \frac{7}{25}$, θ and ϕ being first quadrant angles. From Fig. 41-5 and Fig. 41-6,

$$\cos(\text{Arctan}\,\tfrac{15}{8} - \text{Arcsin}\,\tfrac{7}{25}) = \cos(\theta - \phi) = \cos\theta\cos\phi + \sin\theta\sin\phi$$

$$= \tfrac{8}{17}\cdot\tfrac{24}{25} + \tfrac{15}{17}\cdot\tfrac{7}{25} = \tfrac{297}{425}$$

Fig. 41-5   Fig. 41-6

41.8   Evaluate $\sin(2\,\text{Arctan}\,3)$.

Let $\theta = \text{Arctan}\,3$; then $\tan\theta = 3$, θ being a first quadrant angle. From Fig. 41-7,

$$\sin(2\,\text{Arctan}\,3) = \sin 2\theta = 2\sin\theta\cos\theta = 2\left(\frac{3}{\sqrt{10}}\right)\left(\frac{1}{\sqrt{10}}\right) = \frac{3}{5}$$

Fig. 41-7

41.9 Show that Arcsin $1/\sqrt{5}$ + Arcsin $2/\sqrt{5}$ = $\pi/2$.

Let θ = Arcsin $1/\sqrt{5}$ and ϕ = Arcsin $2/\sqrt{5}$; then sin θ = $1/\sqrt{5}$ and sin ϕ = $2/\sqrt{5}$, each angle terminating in the first quadrant. We are to show that $\theta + \phi$ = $\pi/2$ or, taking the sines of both members, that sin $(\theta + \phi)$ = sin $\pi/2$.

From Fig. 41-8 and Fig. 41-9,

$$\sin(\theta + \phi) = \sin\theta\cos\phi + \cos\theta\sin\phi = \frac{1}{\sqrt{5}}\cdot\frac{1}{\sqrt{5}} + \frac{2}{\sqrt{5}}\cdot\frac{2}{\sqrt{5}} = 1 = \sin\frac{\pi}{2}$$

Fig. 41-8   **Fig. 41-9**

41.10 Show that 2 Arctan $\frac{1}{2}$ = Arctan $\frac{4}{3}$.

Let θ = Arctan $\frac{1}{2}$ and ϕ = Arctan $\frac{4}{3}$; then tan θ = $\frac{1}{2}$ and tan ϕ = $\frac{4}{3}$. We are to show that 2θ = ϕ or, taking the tangents of both members, that tan 2θ = tan ϕ. Now

$$\tan 2\theta = \frac{2\tan\theta}{1-\tan^2\theta} = \frac{2(\frac{1}{2})}{1-(\frac{1}{2})^2} = \frac{4}{3} = \tan\phi$$

41.11 Show that Arcsin $\frac{77}{85}$ − Arcsin $\frac{3}{5}$ = Arccos $\frac{15}{17}$.

Let θ = Arcsin $\frac{77}{85}$, ϕ = Arcsin $\frac{3}{5}$, and ψ = Arccos $\frac{15}{17}$; then sin θ = $\frac{77}{85}$, sin ϕ = $\frac{3}{5}$, and cos ψ = $\frac{15}{17}$, each angle terminating in the first quadrant. Taking the sine of both members of the given relation, we are to show that sin $(\theta - \phi)$ = sin ψ. From Figs. 41-10, 41-11, and 41-12,

$$\sin(\theta - \phi) = \sin\theta\cos\phi - \cos\theta\sin\phi = \frac{77}{85}\cdot\frac{4}{5} - \frac{36}{85}\cdot\frac{3}{5} = \frac{8}{17} = \sin\psi$$

Fig. 41-10   **Fig. 41-11**   **Fig. 41-12**

41.12 Show that $\operatorname{Arccot} \frac{43}{32} - \operatorname{Arctan} \frac{1}{4} = \operatorname{Arccos} \frac{12}{13}$.

Let $\theta = \operatorname{Arccot} \frac{43}{32}$, $\phi = \operatorname{Arctan} \frac{1}{4}$, and $\psi = \operatorname{Arccos} \frac{12}{13}$ (see Fig. 41-13); then $\cot \theta = \frac{43}{32}$, $\tan \phi = \frac{1}{4}$, and $\cos \psi = \frac{12}{13}$, each angle terminating in the first quadrant. Taking the tangent of both members of the given relation, we are to show that $\tan(\theta - \phi) = \tan \psi$.

$$\tan(\theta - \phi) = \frac{\tan \theta - \tan \phi}{1 + \tan \theta \tan \phi} = \frac{\frac{32}{43} - \frac{1}{4}}{1 + (\frac{32}{43})(\frac{1}{4})} = \frac{5}{12} = \tan \psi$$

Fig. 41-13

41.13 Show that $\operatorname{Arctan} \frac{1}{2} + \operatorname{Arctan} \frac{1}{5} + \operatorname{Arctan} \frac{1}{8} = \pi/4$.

We shall show that $\operatorname{Arctan} \frac{1}{2} + \operatorname{Arctan} \frac{1}{5} = \pi/4 - \operatorname{Arctan} \frac{1}{8}$.

$$\tan(\operatorname{Arctan} \tfrac{1}{2} + \operatorname{Arctan} \tfrac{1}{5}) = \frac{\frac{1}{2} + \frac{1}{5}}{1 - (\frac{1}{2})(\frac{1}{5})} = \frac{7}{9}$$

and

$$\tan(\pi/4 - \operatorname{Arctan} \tfrac{1}{8}) = \frac{1 - \frac{1}{8}}{1 + \frac{1}{8}} = \frac{7}{9}$$

41.14 Show that $2 \operatorname{Arctan} \frac{1}{3} + \operatorname{Arctan} \frac{1}{7} = \operatorname{Arcsec} \sqrt{34}/5 + \operatorname{Arccsc} \sqrt{17}$.

Let $\theta = \operatorname{Arctan} \frac{1}{3}$, $\phi = \operatorname{Arctan} \frac{1}{7}$, $\lambda = \operatorname{Arcsec} \sqrt{34}/5$, and $\psi = \operatorname{Arccsc} \sqrt{17}$; then $\tan \theta = \frac{1}{3}$, $\tan \phi = \frac{1}{7}$, $\sec \lambda = \sqrt{34}/5$, and $\csc \psi = \sqrt{17}$, each angle terminating in the first quadrant.

Taking the tangent of both members of the given relation, we are to show that $\tan(2\theta + \phi) = \tan(\lambda + \psi)$. Now

$$\tan 2\theta = \frac{2 \tan \theta}{1 - \tan^2 \theta} = \frac{2(\frac{1}{3})}{1 - (\frac{1}{3})^2} = \frac{3}{4}$$

$$\tan(2\theta + \phi) = \frac{\tan 2\theta + \tan \phi}{1 - \tan 2\theta \tan \phi} = \frac{\frac{3}{4} + \frac{1}{7}}{1 - (\frac{3}{4})(\frac{1}{7})} = 1$$

and, using Fig. 41-14 and Fig. 41-15,

$$\tan(\lambda + \psi) = \frac{\frac{3}{5} + \frac{1}{4}}{1 - (\frac{3}{5})(\frac{1}{4})} = 1$$

Fig. 41-14

Fig. 41-15

41.15 Find the general value of each of the following:

(a) $\arcsin \sqrt{2}/2 = \pi/4 + 2n\pi, \; 3\pi/4 + 2n\pi$ (d) $\arcsin(-1) = -\pi/2 + 2n\pi$

(b) $\arccos \frac{1}{2} = \pi/3 + 2n\pi, \; 5\pi/3 + 2n\pi$ (e) $\arccos 0 = \pi/2 + 2n\pi, \; 3\pi/2 + 2n\pi$

(c) $\arctan 0 = 2n\pi, \; (2n+1)\pi$ (f) $\arctan(-\sqrt{3}) = -\pi/3 + 2n\pi, \; 2\pi/3 + 2n\pi$

where n is a positive or negative integer, or is zero.

41.16 Show that the general value of

(a) $\arcsin x = n\pi + (-1)^n \operatorname{Arcsin} x,$

(b) $\arccos x = 2n\pi \pm \operatorname{Arccos} x,$

(c) $\arctan x = n\pi + \operatorname{Arctan} x,$

where n is any positive or negative integer, or is zero.

(a) Let $\theta = \operatorname{Arcsin} x$. Then since $\sin(\pi - \theta) = \sin \theta$, all values of $\arcsin x$ are given by

$$(1) \quad \theta + 2m\pi \qquad \text{and} \qquad (2) \quad \pi - \theta + 2m\pi = (2m+1)\pi - \theta$$

   Now, when $n = 2m$, that is, n is an even integer, (1) may be written as $n\pi + \theta = n\pi + (-1)^n\theta$; and when $n = 2m + 1$, that is, n is an odd integer, (2) may be written as $n\pi - \theta = n\pi + (-1)^n\theta$. Thus, $\arcsin x = n\pi + (-1)^n \operatorname{Arcsin} x$, where n is any positive or negative integer, or is zero.

(b) Let $\theta = \operatorname{Arccos} x$. Then since $\cos(-\theta) = \cos \theta$, all values of $\arccos x$ are given by $\theta + 2n\pi$ and $-\theta + 2n\pi$ or $2n\pi \pm \theta = 2n\pi \pm \operatorname{Arccos} x$, where n is any positive or negative integer, or is zero.

(c) Let $\theta = \operatorname{Arctan} x$. Then since $\tan(\pi + \theta) = \tan \theta$, all values of $\arctan x$ are given by $\theta + 2m\pi$ and $(\pi + \theta) + 2m\pi = \theta + (2m+1)\pi$ or, as in (a), by $n\pi + \operatorname{Arctan} x$, where n is any positive or negative integer, or is zero.

41.17 Express the general value of each of the functions of Problem 41.15, using the form of Problem 41.16.

(a) $\arcsin \sqrt{2}/2 = n\pi + (-1)^n \pi/4$ (d) $\arcsin(-1) = n\pi + (-1)^n(-\pi/2)$

(b) $\arccos \frac{1}{2} = 2n\pi \pm \pi/3$ (e) $\arccos 0 = 2n\pi \pm \pi/2$

(c) $\arctan 0 = n\pi$ (f) $\arctan(-\sqrt{3}) = n\pi - \pi/3$

where n is any positive or negative integer, or is zero.

Supplementary Problems

41.18 Write the following in inverse function notation:

(a) $\sin \theta = \frac{3}{4},$ (b) $\cos \alpha = -1,$ (c) $\tan x = -2,$ (d) $\cot \beta = \frac{1}{2}.$

   Ans. (a) $\theta = \arcsin \frac{3}{4}$ (b) $\alpha = \arccos(-1)$ (c) $x = \arctan(-2)$ (d) $\beta = \operatorname{arccot} \frac{1}{2}$

41.19 Find the principal value of each of the following:

(a) $\operatorname{Arcsin} \sqrt{3}/2$ (d) $\operatorname{Arccot} 1$ (g) $\operatorname{Arctan}(-\sqrt{3})$ (j) $\operatorname{Arccsc}(-1)$

(b) $\operatorname{Arccos}(-\sqrt{2}/2)$ (e) $\operatorname{Arcsin}(-\frac{1}{2})$ (h) $\operatorname{Arccot} 0$

(c) $\operatorname{Arctan} 1/\sqrt{3}$ (f) $\operatorname{Arccos}(-\frac{1}{2})$ (i) $\operatorname{Arcsec}(-\sqrt{2})$

   Ans. (a) $\pi/3$ (b) $3\pi/4$ (c) $\pi/6$ (d) $\pi/4$ (e) $-\pi/6$ (f) $2\pi/3$
   (g) $-\pi/3$ (h) $\pi/2$ (i) $-3\pi/4$ (j) $-\pi/2$

41.20 Evaluate each of the following:

(*a*) $\sin [\text{Arcsin} (-\tfrac{1}{2})]$ (*f*) $\sin (\text{Arccos} \tfrac{4}{5})$ (*k*) $\text{Arctan} (\cot 230°)$

(*b*) $\cos (\text{Arccos} \sqrt{3}/2)$ (*g*) $\cos [\text{Arcsin} (-\tfrac{12}{13})]$ (*l*) $\text{Arccot} (\tan 100°)$

(*c*) $\tan [\text{Arctan} (-1)]$ (*h*) $\sin (\text{Arctan} 2)$ (*m*) $\sin (2 \text{Arcsin} \tfrac{2}{3})$

(*d*) $\sin [\text{Arccos} (-\sqrt{3}/2)]$ (*i*) $\text{Arccos} (\sin 220°)$ (*n*) $\cos (2 \text{Arcsin} \tfrac{3}{5})$

(*e*) $\tan (\text{Arcsin} 0)$ (*j*) $\text{Arcsin} [\cos (-105°)]$ (*o*) $\sin (\tfrac{1}{2} \text{Arccos} \tfrac{4}{5})$

Ans. (*a*) $-\tfrac{1}{2}$ (*f*) $\tfrac{3}{5}$ (*k*) 40°

(*b*) $\sqrt{3}/2$ (*g*) $\tfrac{5}{13}$ (*l*) 170°

(*c*) -1 (*h*) $2/\sqrt{5}$ (*m*) $4\sqrt{5}/9$

(*d*) $\tfrac{1}{2}$ (*i*) 130°   (*n*) $\tfrac{7}{25}$

(*e*) 0   (*j*) $-15°$ (*o*) $1/\sqrt{10}$

41.21 Show that

(*a*) $\sin (\text{Arcsin} \tfrac{5}{13} + \text{Arcsin} \tfrac{4}{5}) = \tfrac{63}{65}$

(*b*) $\cos (\text{Arccos} \tfrac{15}{17} - \text{Arccos} \tfrac{7}{25}) = \tfrac{297}{425}$

(*c*) $\sin \left(\text{Arcsin} \dfrac{1}{2} - \text{Arccos} \dfrac{1}{3} \right) = \dfrac{1 - 2\sqrt{6}}{6}$

(*d*) $\tan (\text{Arctan} \tfrac{3}{4} + \text{Arccot} \tfrac{15}{8}) = \tfrac{77}{36}$

(*e*) $\cos \left(\text{Arctan} \dfrac{-4}{3} + \text{Arcsin} \dfrac{12}{13} \right) = \dfrac{63}{65}$

(*f*) $\tan \left(\text{Arcsin} \dfrac{-3}{5} - \text{Arccos} \dfrac{5}{13} \right) = \dfrac{63}{16}$

(*g*) $\tan \left(2 \text{Arcsin} \dfrac{4}{5} + \text{Arccos} \dfrac{12}{13} \right) = -\dfrac{253}{204}$

(*h*) $\sin (2 \text{Arcsin} \tfrac{4}{5} - \text{Arccos} \tfrac{12}{13}) = \tfrac{323}{325}$

41.22 Show that

(*a*) $\text{Arctan} \dfrac{1}{2} + \text{Arctan} \dfrac{1}{3} = \dfrac{\pi}{4}$ (*e*) $\text{Arccos} \dfrac{12}{13} + \text{Arctan} \dfrac{1}{4} = \text{Arccot} \dfrac{43}{32}$

(*b*) $\text{Arcsin} \dfrac{4}{5} + \text{Arctan} \dfrac{3}{4} = \dfrac{\pi}{2}$ (*f*) $\text{Arcsin} \dfrac{3}{5} + \text{Arcsin} \dfrac{15}{17} = \text{Arccos} \dfrac{-13}{85}$

(*c*) $\text{Arctan} \dfrac{4}{3} - \text{Arctan} \dfrac{1}{7} = \dfrac{\pi}{4}$ (*g*) $\text{Arctan} \alpha + \text{Arctan} \dfrac{1}{\alpha} = \dfrac{\pi}{2}$ $(a > 0)$

(*d*) $2 \text{Arctan} \dfrac{1}{3} + \text{Arctan} \dfrac{1}{7} = \dfrac{\pi}{4}$

41.23 Prove: The area of the segment cut from a circle of radius r by a chord at a distance d from the center is given by $K = r^2 \text{Arccos} \dfrac{d}{r} - d\sqrt{r^2 - d^2}$.

Chapter 42

Trigonometric Equations

TRIGONOMETRIC EQUATIONS, i.e., equations involving trigonometric functions of unknown angles, are called

- (*a*) Identical equations or *identities*, if they are satisfied by all values of the unknown angles for which the functions are defined.
- (*b*) Conditional equations, or equations, if they are satisfied only by particular values of the unknown angles.

For example,

- (*a*) $\sin x \csc x = 1$ is an identity, being satisfied by every value of x for which $\csc x$ is defined.
- (*b*) $\sin x = 0$ is a conditional equation since it is not satisfied by $x = \frac{1}{4}\pi$ or $\frac{1}{2}\pi$.

Hereafter in this chapter we shall use the term "equation" instead of "conditional equation."

A SOLUTION OF A TRIGONOMETRIC EQUATION, as $\sin x = 0$, is a value of the angle x which satisfies the equation. Two solutions of $\sin x = 0$ are $x = 0$ and $x = \pi$.

If a given equation has one solution, it has in general an unlimited number of solutions. Thus, the complete solution of $\sin x = 0$ is given by

$$x = 0 + 2n\pi, \qquad x = \pi + 2\pi$$

where n is any positive or negative integer or is zero.

In this chapter we shall list only the particular solutions for which $0 \le x < 2\pi$.

PROCEDURES FOR SOLVING TRIGONOMETRIC EQUATIONS. There is no general method for solving trigonometric equations. Three standard procedures are illustrated below and other procedures are introduced in the solved problems.

(*A*) The equation may be factorable.

EXAMPLE 1. Solve $\sin x - 2 \sin x \cos x = 0$.

Factoring, $\sin x - 2 \sin x \cos x = \sin x(1 - 2 \cos x) = 0$, and setting each factor equal to zero, we have

$$\sin x = 0 \quad \text{and} \quad x = 0, \pi; \qquad 1 - 2\cos x = 0 \quad \text{or} \quad \cos x = \tfrac{1}{2} \quad \text{and} \quad x = \frac{\pi}{3}, \frac{5\pi}{3}.$$

Check. For $x = 0$, $\qquad \sin x - 2 \sin x \cos x = 0 - 2(0)(1) = 0$;
for $x = \pi/3$, $\qquad \sin x - 2 \sin x \cos x = \tfrac{1}{2}\sqrt{3} - 2(\tfrac{1}{2}\sqrt{3})(\tfrac{1}{2}) = 0$;
for $x = \pi$, $\qquad \sin x - 2 \sin x \cos x = 0 - 2(0)(-1) = 0$;
for $x = 5\pi/3$, $\qquad \sin x - 2 \sin x \cos x = -\tfrac{1}{2}\sqrt{3} - 2(-\tfrac{1}{2}\sqrt{3})(\tfrac{1}{2}) = 0$.

Thus, the required solutions $(0 \le x < 2\pi)$ are $x = 0, \ \pi/3, \ \pi, \ 5\pi/3$.

(*B*) The various functions occurring in the equation may be expressed in terms of a single function.

EXAMPLE 2. Solve $2 \tan^2 x + \sec^2 x = 2$.

Replacing $\sec^2 x$ by $1 + \tan^2 x$, we have $2 \tan^2 x + (1 + \tan^2 x) = 2$, $3 \tan^2 x = 1$, and $\tan x = \pm 1/\sqrt{3}$. From $\tan x = 1/\sqrt{3}$, $x = \pi/6$ and $7\pi/6$; from $\tan x = -1/\sqrt{3}$, $x = 5\pi/6$ and $11\pi/6$. After checking each of these values in the original equation, we find that the required solutions $(0 \le x < 2\pi)$ are $x = \pi/6, 5\pi/6, 7\pi/6, 11\pi/6$.

The necessity of the check is illustrated in

EXAMPLE 3.   Solve $\sec x + \tan x = 0$.

Multiplying the equation $\sec x + \tan x = \dfrac{1}{\cos x} + \dfrac{\sin x}{\cos x} = 0$ by $\cos x$, we have $1 + \sin x = 0$ or $\sin x = -1$; then $x = 3\pi/2$. However, neither $\sec x$ nor $\tan x$ is defined when $x = 3\pi/2$ and the equation has no solution.

(C)   Both members of the equation are squared.

EXAMPLE 4.   Solve $\sin x + \cos x = 1$.

If the procedure of (B) were used, we would replace $\sin x$ by $\pm\sqrt{1 - \cos^2 x}$ or $\cos x$ by $\pm\sqrt{1 - \sin^2 x}$ and thereby introduce radicals. To avoid this, we write the equation in the form $\sin x = 1 - \cos x$ and square both members. We have

$$\sin^2 x = 1 - 2\cos x + \cos^2 x \tag{42.1}$$
$$1 - \cos^2 x = 1 - 2\cos x + \cos^2 x$$
$$2\cos^2 x - 2\cos x = 2\cos x(\cos x - 1) = 0$$

From $\cos x = 0$, $x = \pi/2, 3\pi/2$; from $\cos x = 1$, $x = 0$.

Check.   For $x = 0$, $\sin x + \cos x = 0 + 1 = 1$;
        for $x = \pi/2$, $\sin x + \cos x = 1 + 0 = 1$;
        for $x = 3\pi/2$, $\sin x + \cos x = -1 + 0 \neq 1$.

Thus, the required solutions are $x = 0,\ \pi/2$.

The value $x = 3\pi/2$, called an *extraneous solution*, was introduced by squaring the two members. Note that (42.1) is also obtained when both members of $\sin x = \cos x - 1$ are squared and that $x = 3\pi/2$ satisfies this latter relation.

Solved Problems

Solve each of the trigonometric equations 42.1–42.22 for all x such that $0 \leq x < 2\pi$. (If all solutions are required, adjoin $+2n\pi$, where n is zero or any positive or negative integer, to each result given.) In a number of solutions, the details of the check have been omitted.

42.1 $2\sin x - 1 = 0$.

Here $\sin x = \frac{1}{2}$ and $x = \pi/6, 5\pi/6$.

42.2 $\sin x \cos x = 0$.

From $\sin x = 0$, $x = 0, \pi$; from $\cos x = 0$, $x = \pi/2, 3\pi/2$. The required solutions are $x = 0, \pi/2, \pi, 3\pi/2$.

42.3 $(\tan x - 1)(4\sin^2 x - 3) = 0$.

From $\tan x - 1 = 0$, $\tan x = 1$ and $x = \pi/4, 5\pi/4$; from $4\sin^2 x - 3 = 0$, $\sin x = \pm\sqrt{3}/2$ and $x = \pi/3, 2\pi/3, 4\pi/3, 5\pi/3$. The required solutions are $x = \pi/4, \pi/3, 2\pi/3, 5\pi/4, 4\pi/3, 5\pi/3$.

42.4 $\sin^2 x + \sin x - 2 = 0$.

Factoring, $(\sin x + 2)(\sin x - 1) = 0$.
From $\sin x + 2 = 0$, $\sin x = -2$ and there is no solution; from $\sin x - 1 = 0$, $\sin x = 1$ and $x = \pi/2$. The required solution is $x = \pi/2$.

42.5 $3 \cos^2 x = \sin^2 x$.

First Solution.   Replacing $\sin^2 x$ by $1 - \cos^2 x$, we have $3 \cos^2 x = 1 - \cos^2 x$ or $4 \cos^2 x = 1$. Then $\cos x = \pm\frac{1}{2}$ and the required solutions are $x = \pi/3,\ 2\pi/3,\ 4\pi/3,\ 5\pi/3$.

Second Solution.   Dividing the equation by $\cos^2 x$, we have $3 = \tan^2 x$. Then $\tan x = \pm\sqrt{3}$ and the solutions above are obtained.

42.6 $2 \sin x - \csc x = 1$.

Multiplying the equation by $\sin x$, $2 \sin^2 x - 1 = \sin x$, and rearranging, we have

$$2 \sin^2 x - \sin x - 1 = (2 \sin x + 1)(\sin x - 1) = 0$$

From $2 \sin x + 1 = 0$, $\sin x = -\frac{1}{2}$ and $x = 7\pi/6,\ 11\pi/6$; from $\sin x = 1$, $x = \pi/2$.

Check.   For $x = \pi/2$, $2 \sin x - \csc x = 2(1) - 1 = 1$;

            for $x = 7\pi/6$ and $11\pi/6$, $2 \sin x - \csc x = 2(-\frac{1}{2}) - (-2) = 1$.

The solutions are $x = \pi/2,\ 7\pi/6,\ 11\pi/6$.

42.7 $2 \sec x = \tan x + \cot x$.

Transforming to sines and cosines, and clearing fractions, we have

$$\frac{2}{\cos x} = \frac{\sin x}{\cos x} + \frac{\cos x}{\sin x} \qquad \text{or} \qquad 2 \sin x = \sin^2 x + \cos^2 x = 1$$

Then $\sin x = \frac{1}{2}$ and $x = \pi/6,\ 5\pi/6$.

42.8 $\tan x + 3 \cot x = 4$.

Multiplying by $\tan x$ and rearranging, $\tan^2 x - 4 \tan x + 3 = (\tan x - 1)(\tan x - 3) = 0$.

From $\tan x - 1 = 0$, $\tan x = 1$ and $x = \pi/4,\ 5\pi/4$; from $\tan x - 3 = 0$, $\tan x = 3$ and $x = 71°\,34'$, $251°\,34'$.

Check.   For $x = \pi/4$ and $5\pi/4$, $\tan x + 3 \cot x = 1 + 3(1) = 4$;

            for $x = 71°\,34'$ and $251°\,34'$, $\tan x + 3 \cot x = 3 + 3(\frac{1}{3}) = 4$.

The solutions are $45°,\ 71°\,34',\ 225°,\ 251°\,34'$.

42.9 $\csc x + \cot x = \sqrt{3}$.

First Solution.   Writing the equation in the form $\csc x = \sqrt{3} - \cot x$ and squaring, we have

$$\csc^2 x = 3 - 2\sqrt{3} \cot x + \cot^2 x$$

Replacing $\csc^2 x$ by $1 + \cot^2 x$ and combining, this becomes $2\sqrt{3} \cot x - 2 = 0$. Then $\cot x = 1/\sqrt{3}$ and $x = \pi/3,\ 4\pi/3$.

Check.   For $x = \pi/3$, $\csc x + \cot x = 2/\sqrt{3} + 1/\sqrt{3} = \sqrt{3}$;

            for $x = 4\pi/3$, $\csc x + \cot x = -2/\sqrt{3} + 1/\sqrt{3} \neq \sqrt{3}$. The required solution is $x = \pi/3$.

Second Solution.   Upon making the indicated replacement, the equation becomes

$$\frac{1}{\sin x} + \frac{\cos x}{\sin x} = \sqrt{3}$$

and, clearing of fractions, $1 + \cos x = \sqrt{3} \sin x$.

Squaring both members, we have $1 + 2 \cos x + \cos^2 x = 3 \sin^2 x = 3(1 - \cos^2 x)$ or

$$4 \cos^2 x + 2 \cos x - 2 = 2(2 \cos x - 1)(\cos x + 1) = 0$$

From $2 \cos x - 1 = 0$, $\cos x = \frac{1}{2}$ and $x = \pi/3,\ 5\pi/3$; from $\cos x + 1 = 0$, $\cos x = -1$ and $x = \pi$.

Now $x = \pi/3$ is the solution. The values $x = \pi$ and $5\pi/3$ are to be excluded since $\csc \pi$ is not defined while $\csc 5\pi/3$ and $\cot 5\pi/3$ are both negative.

42.10 $\cos x - \sqrt{3} \sin x = 1$.

First Solution.   Putting the equation in the form $\cos x - 1 = \sqrt{3} \sin x$ and squaring, we have

$$\cos^2 x - 2 \cos x + 1 = 3 \sin^2 x = 3(1 - \cos^2 x)$$

Then, combining and factoring,

$$4 \cos^2 x - 2 \cos x - 2 = 2(2 \cos x + 1)(\cos x - 1) = 0$$

From $2 \cos x + 1 = 0$, $\cos x = -\frac{1}{2}$ and $x = 2\pi/3, 4\pi/3$; from $\cos x - 1 = 0$, $\cos x = 1$ and $x = 0$.
Check.   For $x = 0$, $\cos x - \sqrt{3} \sin x = 1 - \sqrt{3}(0) = 1$;
for $x = 2\pi/3$, $\cos x - \sqrt{3} \sin x = -\frac{1}{2} - \sqrt{3}(\sqrt{3}/2) \neq 1$;
for $x = 4\pi/3$, $\cos x - \sqrt{3} \sin x = -\frac{1}{2} - \sqrt{3}(-\sqrt{3}/2) = 1$.
The required solutions are $x = 0, 4\pi/3$.
Second Solution.   The left member of the given equation may be put in the form

$$\sin \theta \cos x + \cos \theta \sin x = \sin (\theta + x)$$

in which θ is a known angle, by dividing the given equation by $r > 0$, $\dfrac{1}{r} \cos x + \left(\dfrac{-\sqrt{3}}{r}\right) \sin x = \dfrac{1}{r}$, and

setting $\sin \theta = \dfrac{1}{r}$ and $\cos \theta = \dfrac{-\sqrt{3}}{r}$. Since $\sin^2 \theta + \cos^2 \theta = 1$, $\left(\dfrac{1}{r}\right)^2 + \left(\dfrac{-\sqrt{3}}{r}\right)^2 = 1$ and $r = 2$. Now $\sin \theta = \frac{1}{2}$,
$\cos \theta = -\sqrt{3}/2$ so that the given equation may be written as $\sin (\theta + x) = \frac{1}{2}$ with $\theta = 5\pi/6$. Then $\theta + x = 5\pi/6 + x = \arcsin \frac{1}{2} = \pi/6, 5\pi/6, 13\pi/6, 17\pi/6, \ldots$ and $x = -2\pi/3, 0, 4\pi/3, 2\pi, \ldots$. As before, the required solutions are $x = 0, 4\pi/3$.

Note that r is the positive square root of the sum of the squares of the coefficients of $\cos x$ and $\sin x$ when the equation is written in the form $a \cos x + b \sin x = c$, that is,

$$r = \sqrt{a^2 + b^2}$$

The equation will have no solution if $\dfrac{c}{\sqrt{a^2 + b^2}}$ is greater than 1 or less than -1.

42.11 $2 \cos x = 1 - \sin x$.

First Solution.   As in Problem 42.10, we obtain

$$4 \cos^2 x = 1 - 2 \sin x + \sin^2 x$$

$$4(1 - \sin^2 x) = 1 - 2 \sin x + \sin^2 x$$

$$5 \sin^2 x - 2 \sin x - 3 = (5 \sin x + 3)(\sin x - 1) = 0$$

From $5 \sin x + 3 = 0$, $\sin x = -\frac{3}{5} = -0.6000$ and $x = 216° 52', 323° 8'$; from $\sin x - 1 = 0$, $\sin x = 1$ and $x = \pi/2$.
Check.   For $x = \pi/2$, $2(0) = 1 - 1$;
for $x = 216° 52'$, $2(-\frac{4}{5}) \neq 1 - (-\frac{3}{5})$;
for $x = 323° 8'$, $2(\frac{4}{5}) = 1 - (-\frac{3}{5})$.
The required solutions are $x = 90°, 323° 8'$.
Second Solution.   Writing the equation as $2 \cos x + \sin x = 1$ and dividing by $r = \sqrt{2^2 + 1^2} = \sqrt{5}$, we have

$$\frac{2}{\sqrt{5}} \cos x + \frac{1}{\sqrt{5}} \sin x = \frac{1}{\sqrt{5}} \tag{1}$$

Let $\sin \theta = 2/\sqrt{5}$, $\cos \theta = 1/\sqrt{5}$; then (1) becomes

$$\sin \theta \cos x + \cos \theta \sin x = \sin (\theta + x) = \frac{1}{\sqrt{5}}$$

with $\theta = 63° 26'$. Now $\theta + x = 63° 26' + x = \arcsin (1/\sqrt{5}) = \arcsin 0.4472 = 26° 34', 153° 26', 386° 34', \ldots$ and $x = 90°, 323° 8'$ as before.

EQUATIONS INVOLVING MULTIPLE ANGLES

42.12 $\sin 3x = -\frac{1}{2}\sqrt{2}$.

Since we require x such that $0 \le x < 2\pi$, $3x$ must be such that $0 \le 3x < 6\pi$. Then $3x = 5\pi/4, 7\pi/4,$ $13\pi/4, 15\pi/4, 21\pi/4, 23\pi/4$ and $x = 5\pi/12, 7\pi/12, 13\pi/12, 5\pi/4, 7\pi/4, 23\pi/12$. Each of these values is a solution.

42.13 $\cos \frac{1}{2}x = \frac{1}{2}$.

Since we require x such that $0 \le x < 2\pi$, $\frac{1}{2}x$ must be such that $0 \le \frac{1}{2}x < \pi$. Then $\frac{1}{2}x = \pi/3$ and $x = 2\pi/3$.

42.14 $\sin 2x + \cos x = 0$.

Substituting for $\sin 2x$, we have $2 \sin x \cos x + \cos x = \cos x(2 \sin x + 1) = 0$. From $\cos x = 0$, $x = \pi/2$, $3\pi/2$; from $\sin x = -\frac{1}{2}$, $x = 7\pi/6, 11\pi/6$. The required solutions are $x = \pi/2, 7\pi/6, 3\pi/2, 11\pi/6$.

42.15 $2 \cos^2 \frac{1}{2}x = \cos^2 x$.

Substituting $1 + \cos x$ for $2 \cos^2 \frac{1}{2}x$, the equation becomes $\cos^2 x - \cos x - 1 = 0$; then $\cos x = \dfrac{1 \pm \sqrt{5}}{2} =$ $1.6180, -0.6180$. Since $\cos x$ cannot exceed 1, we consider $\cos x = -0.6180$ and obtain the solutions $x = 128°\,10', 231°\,50'$.

(NOTE.   To solve $\sqrt{2} \cos \frac{1}{2}x = \cos x$ and $\sqrt{2} \cos \frac{1}{2}x = -\cos x$, we square and obtain the equation of this problem. The solution of the first of these equations is $231°\,50'$ and the solution of the second is $128°\,10'$.)

42.16 $\cos 2x + \cos x + 1 = 0$.

Substituting $2 \cos^2 x - 1$ for $\cos 2x$, we have $2 \cos^2 x + \cos x = \cos x(2 \cos x + 1) = 0$. From $\cos x = 0$, $x = \pi/2, 3\pi/2$; from $\cos x = -\frac{1}{2}$, $x = 2\pi/3, 4\pi/3$. The required solutions are $x = \pi/2, 2\pi/3, 3\pi/2, 4\pi/3$.

42.17 $\tan 2x + 2 \sin x = 0$.

Using $\tan 2x = \dfrac{\sin 2x}{\cos 2x} = \dfrac{2 \sin x \cos x}{\cos 2x}$, we have

$$\frac{2 \sin x \cos x}{\cos 2x} + 2 \sin x = 2 \sin x \left(\frac{\cos x}{\cos 2x} + 1 \right) = 2 \sin x \left(\frac{\cos x + \cos 2x}{\cos 2x} \right) = 0$$

From $\sin x = 0$, $x = 0, \pi$; from $\cos x + \cos 2x = \cos x + 2 \cos^2 x - 1 = (2 \cos x - 1)(\cos x + 1) = 0$, $x = \pi/3, 5\pi/3$, and π. The required solutions are $x = 0, \pi/3, \pi, 5\pi/3$.

42.18 $\sin 2x = \cos 2x$.

First Solution.   Let $2x = \theta$; then we are to solve $\sin \theta = \cos \theta$ for $0 \le \theta < 4\pi$. Then $\theta = \pi/4, 5\pi/4,$ $9\pi/4, 13\pi/4$ and $x = \theta/2 = \pi/8, 5\pi/8, 9\pi/8, 13\pi/8$ are the solutions.

Second Solution.   Dividing by $\cos 2x$, the equation becomes $\tan 2x = 1$ for which $2x = \pi/4, 5\pi/4,$ $9\pi/4, 13\pi/4$ as in the first solution.

42.19 $\sin 2x = \cos 4x$.

Since $\cos 4x = \cos 2(2x) = 1 - 2 \sin^2 2x$, the equation becomes

$$2 \sin^2 2x + \sin 2x - 1 = (2 \sin 2x - 1)(\sin 2x + 1) = 0$$

From $2 \sin 2x - 1 = 0$ or $\sin 2x = \frac{1}{2}$, $2x = \pi/6, 5\pi/6, 13\pi/6, 17\pi/6$ and $x = \pi/12, 5\pi/12, 13\pi/12, 17\pi/12$; from $\sin 2x + 1 = 0$ or $\sin 2x = -1$, $2x = 3\pi/2, 7\pi/2$ and $x = 3\pi/4, 7\pi/4$. All of these values are solutions.

42.20 $\sin 3x = \cos 2x.$

To avoid the substitution for $\sin 3x$, we use one of the procedures below.
First Solution.   Since $\cos 2x = \sin\left(\frac{1}{2}\pi - 2x\right)$ and also $\cos 2x = \sin\left(\frac{1}{2}\pi + 2x\right)$, we consider

(a) $\sin 3x = \sin\left(\frac{1}{2}\pi - 2x\right)$, obtaining $3x = \pi/2 - 2x, 5\pi/2 - 2x, 9\pi/2 - 2x, \ldots$
(b) $\sin 3x = \sin\left(\frac{1}{2}\pi + 2x\right)$, obtaining $3x = \pi/2 + 2x, 5\pi/2 + 2x, 9\pi/2 + 2x, \ldots$

From (a), $5x = \pi/2, 5\pi/2, 9\pi/2, 13\pi/2, 17\pi/2$ (since $5x < 10\pi$); and from (b), $x = \pi/2$. The required solutions are $x = \pi/10, \pi/2, 9\pi/10, 13\pi/10, 17\pi/10$.
Second Solution.   Since $\sin 3x = \cos\left(\frac{1}{2}\pi - 3x\right)$ and $\cos 2x = \cos(-2x)$, we consider

(c) $\cos 2x = \cos\left(\frac{1}{2}\pi - 3x\right)$, obtaining $5x = \pi/2, 5\pi/2, 9\pi/2, 13\pi/2, 17\pi/2$
(d) $\cos(-2x) = \cos\left(\frac{1}{2}\pi - 3x\right)$, obtaining $x = \pi/2$, as before

42.21 $\tan 4x = \cot 6x.$

Since $\cot 6x = \tan\left(\frac{1}{2}\pi - 6x\right)$, we consider the equation $\tan 4x = \tan\left(\frac{1}{2}\pi - 6x\right)$. Then $4x = \pi/2 - 6x,$ $3\pi/2 - 6x, 5\pi/2 - 6x, \ldots$, the function $\tan\theta$ being of period π. Thus, $10x = \pi/2, 3\pi/2, 5\pi/2, 7\pi/2,$ $9\pi/2, \ldots, 39\pi/2$ and the required solutions are $x = \pi/20, 3\pi/20, \pi/4, 7\pi/20, \ldots, 39\pi/20$.

42.22 $\sin 5x - \sin 3x - \sin x = 0.$

Replacing $\sin 5x - \sin 3x$ by $2\cos 4x \sin x$ (Chapter 41), the given equation becomes

$$2\cos 4x \sin x - \sin x = \sin x (2\cos 4x - 1) = 0$$

From $\sin x = 0$, $x = 0, \pi$; from $2\cos 4x - 1 = 0$ or $\cos 4x = \frac{1}{2}$, $4x = \pi/3, 5\pi/3, 7\pi/3, 11\pi/3, 13\pi/3,$ $17\pi/3, 19\pi/3, 23\pi/3$ and $x = \pi/12, 5\pi/12, 7\pi/12, 11\pi/12, 13\pi/12, 17\pi/12, 19\pi/12, 23\pi/12$. Each of the values obtained is a solution.

42.23   Solve the system

$$\begin{matrix} r\sin\theta = 2 \\ r\cos\theta = 3 \end{matrix} \quad \text{for } r > 0 \quad \text{and} \quad 0 \le \theta < 2\pi \qquad \begin{matrix} (1) \\ (2) \end{matrix}$$

Squaring the two equations and adding, $r^2\sin\theta + r^2\cos\theta = r^2 = 13$ and $r = \sqrt{13} = 3.606$.
When $r > 0$, $\sin\theta$ and $\cos\theta$ are both >0 and θ is acute. Dividing (1) by (2), $\tan\theta = \frac{2}{3} = 0.6667$ and $\theta = 33°41'$.

42.24   Solve the system

$$\begin{matrix} r\sin\theta = 3 \\ r = 4(1 + \sin\theta) \end{matrix} \quad \text{for } r > 0 \quad \text{and} \quad 0 \le \theta < 2\pi \qquad \begin{matrix} (1) \\ (2) \end{matrix}$$

Dividing (2) by (1),

$$\frac{1}{\sin\theta} = \frac{4(1 + \sin\theta)}{3} \quad \text{or} \quad 4\sin^2\theta + 4\sin\theta - 3 = 0$$

and

$$(2\sin\theta + 3)(2\sin\theta - 1) = 0$$

From $2\sin\theta - 1 = 0$, $\sin\theta = \frac{1}{2}$, $\theta = \pi/6$ and $5\pi/6$; using (1), $r\left(\frac{1}{2}\right) = 3$ and $r = 6$. Note that $2\sin\theta + 3 = 0$ is excluded since when $r > 0$, $\sin\theta > 0$ by (1). The required solutions are $\theta = \pi/6$, $r = 6$ and $\theta = 5\pi/6$, $r = 6$.

42.25 Solve the system

$$\begin{array}{ll} \sin x + \sin y = 1.2 \\ \cos x + \cos y = 1.5 \end{array} \quad \text{for } 0 \le x, y < 2\pi \qquad \begin{array}{r}(1)\\(2)\end{array}$$

Since each sum on the left is greater than 1, each of the four functions is positive and both x and y are acute.

Using the appropriate formulas of Chapter 41, we obtain

$$2 \sin \tfrac{1}{2}(x+y) \cos \tfrac{1}{2}(x-y) = 1.2 \qquad (1')$$

$$2 \cos \tfrac{1}{2}(x+y) \cos \tfrac{1}{2}(x-y) = 1.5. \qquad (2')$$

Dividing $(1')$ by $(2')$, $\dfrac{\sin \tfrac{1}{2}(x+y)}{\cos \tfrac{1}{2}(x+y)} = \tan \tfrac{1}{2}(x+y) = \dfrac{1.2}{1.5} = 0.8000$ and $\tfrac{1}{2}(x+y) = 38° 40'$ since $\tfrac{1}{2}(x+y)$ is

also acute. Substituting for $\sin \tfrac{1}{2}(x+y) = 0.6248$ in $(1')$, we have $\cos \tfrac{1}{2}(x-y) = \dfrac{0.6}{0.6248} = 0.9603$ and

$\tfrac{1}{2}(x-y) = 16° 12'$.

Then $x = \tfrac{1}{2}(x+y) + \tfrac{1}{2}(x-y) = 54° 52'$ and $y = \tfrac{1}{2}(x+y) - \tfrac{1}{2}(x-y) = 22° 28'$.

42.26 Solve Arccos $2x$ = Arcsin x.

If x is positive, $\alpha = $ Arccos $2x$ and $\beta = $ Arcsin x terminate in quadrant I; if x is negative, α terminates in quadrant II and β terminates in quadrant IV. Thus, x must be positive.

For x positive, $\sin \beta = x$ and $\cos \beta = \sqrt{1 - x^2}$. Taking the cosine of both members of the given equation, we have

$$\cos (\text{Arccos } 2x) = \cos (\text{Arcsin } x) = \cos \beta \qquad \text{or} \qquad 2x = \sqrt{1 - x^2}$$

Squaring, $4x^2 = 1 - x^2$, $5x^2 = 1$, and $x = \sqrt{5}/5 = 0.4472$.

Check. Arccos $2x$ = Arccos $0.8944 = 26° 30'$ = Arcsin 0.4472, approximating the angle to the nearest $10'$.

42.27 Solve Arccos $(2x^2 - 1) = 2$ Arccos $\tfrac{1}{2}$.

Let $\alpha = $ Arccos $(2x^2 - 1)$ and $\beta = $ Arccos $\tfrac{1}{2}$; then $\cos \alpha = 2x^2 - 1$ and $\cos \beta = \tfrac{1}{2}$.
Taking the cosine of both members of the given equation,

$$\cos \alpha = 2x^2 - 1 = \cos 2\beta = 2 \cos^2 \beta - 1 = 2(\tfrac{1}{2})^2 - 1 = -\tfrac{1}{2}$$

Then $2x^2 = \tfrac{1}{2}$ and $x = \pm\tfrac{1}{2}$.

Check.   For $x = \pm\tfrac{1}{2}$, Arccos $(-\tfrac{1}{2}) = 2$ Arccos $\tfrac{1}{2}$ or $120° = 2(60°)$.

42.28 Solve Arccos $2x$ - Arccos $x = \pi/3$.

If x is positive, $0 < $ Arccos $2x < $ Arccos x; if x is negative, Arccos $2x > $ Arccos $x > 0$. Thus, x must be negative.

Let $\alpha = $ Arccos $2x$ and $\beta = $ Arccos x; then $\cos \alpha = 2x, \sin \alpha = \sqrt{1 - 4x^2}, \cos \beta = x$ and $\sin \beta = \sqrt{1 - x^2}$ since both α and β terminate in quadrant II.

Taking the cosine of both members of the given equation,

$$\cos (\alpha - \beta) = \cos \alpha \cos \beta + \sin \alpha \sin \beta = 2x^2 + \sqrt{1 - 4x^2} \sqrt{1 - x^2} = \cos \frac{\pi}{3} = \frac{1}{2}$$

or

$$\sqrt{1 - 4x^2} \sqrt{1 - x^2} = \tfrac{1}{2} - 2x^2$$

Squaring, $1 - 5x^2 + 4x^4 = \tfrac{1}{4} - 2x^2 + 4x^4$, $3x^2 = \tfrac{3}{4}$, and $x = -\tfrac{1}{2}$.

Check. Arccos (-1) - Arccos $(-\tfrac{1}{2}) = \pi - 2\pi/3 = \pi/3$.

42.29   Solve Arcsin $2x = \frac{1}{4}\pi -$ Arcsin x.

Let $\alpha =$ Arcsin $2x$ and $\beta =$ arcsin x; then $\sin \alpha = 2x$ and $\sin \beta = x$. If x is negative, α and β terminate in quadrant IV; thus, x must be positive and β acute.

Taking the sine of both members of the given equation,

$$\sin \alpha = \sin \left(\tfrac{1}{4}\pi - \beta\right) = \sin \tfrac{1}{4}\pi \cos \beta - \cos \tfrac{1}{4}\pi \sin \beta$$

or $\qquad\qquad 2x = \tfrac{1}{2}\sqrt{2}\sqrt{1-x^2} - \tfrac{1}{2}\sqrt{2}x \qquad$ and $\qquad (2\sqrt{2}+1)x = \sqrt{1-x^2}$

Squaring, $(8 + 4\sqrt{2} + 1)x^2 = 1 - x^2$, $x^2 = 1/(10 + 4\sqrt{2})$, and $x = 0.2527$.

Check.   Arcsin $0.5054 = 30° 22'$; Arcsin $0.2527 = 14° 38'$, and $\frac{1}{4}\pi - 14° 38' = 30° 22'$.

42.30   Solve Arctan $x +$ Arctan $(1 - x) =$ Arctan $\frac{4}{3}$.

Let $\alpha =$ Arctan x and $\beta =$ Arctan $(1 - x)$; then $\tan \alpha = x$ and $\tan \beta = 1 - x$. Taking the tangent of both members of the given equation,

$$\tan (\alpha + \beta) = \frac{\tan \alpha + \tan \beta}{1 - \tan \alpha \tan \beta} = \frac{x + (1 - x)}{1 - x(1 - x)} = \frac{1}{1 - x + x^2} = \tan(\text{Arctan } \tfrac{4}{3}) = \tfrac{4}{3}$$

Then $3 = 4 - 4x + 4x^2$, $4x^2 - 4x + 1 = (2x - 1)^2 = 0$, and $x = \frac{1}{2}$.

Check.   Arctan $\frac{1}{2} +$ Arctan $(1 - \frac{1}{2}) = 2$ Arctan $0.5000 = 53° 8'$ and Arctan $\frac{4}{3} =$ Arctan $1.3333 = 53° 8'$.

Supplementary Problems

Solve each of the following equations for all x such that $0 \le x < 2\pi$:

42.31 $\sin x = \sqrt{3}/2$. $\qquad\qquad\qquad$ *Ans.* $\pi/3, 2\pi/3$

42.32 $\cos^2 x = \frac{1}{2}$. $\qquad\qquad\qquad$ *Ans.* $\pi/4, 3\pi/4, 5\pi/4, 7\pi/4$

42.33 $\sin x \cos x = 0$. $\qquad\qquad\qquad$ *Ans.* $0, \pi/2, \pi, 3\pi/2$

42.34 $(\tan x - 1)(2 \sin x + 1) = 0$. $\qquad$ *Ans.* $\pi/4, 7\pi/6, 5\pi/4, 11\pi/6$

42.35 $2 \sin^2 x - \sin x - 1 = 0$. $\qquad$ *Ans.* $\pi/2, 7\pi/6, 11\pi/6$

42.36 $\sin 2x + \sin x = 0$. $\qquad\qquad$ *Ans.* $0, 2\pi/3, \pi, 4\pi/3$

42.37 $\cos x + \cos 2x = 0$. $\qquad\qquad$ *Ans.* $\pi/3, \pi, 5\pi/3$

42.38 $2 \tan x \sin x - \tan x = 0$. $\qquad$ *Ans.* $0, \pi/6, 5\pi/6, \pi$

42.39 $2 \cos x + \sec x = 3$. $\qquad\qquad$ *Ans.* $0, \pi/3, 5\pi/3$

42.40 $2 \sin x + \csc x = 3$. $\qquad\qquad$ *Ans.* $\pi/6, \pi/2, 5\pi/6$

42.41 $\sin x + 1 = \cos x$.   *Ans.* $0, 3\pi/2$

42.42 $\sec x - 1 = \tan x$.   *Ans.* 0

42.43 $2 \cos x + 3 \sin x = 2$.   *Ans.* $0°, 112° 37'$

42.44 $3 \sin x + 5 \cos x + 5 = 0$.   *Ans.* $180°, 241° 56'$

42.45 $1 + \sin x = 2 \cos x$.   *Ans.* $36° 52', 270°$

42.46 $3 \sin x + 4 \cos x = 2$.   *Ans.* $103° 18', 330° 27'$

42.47 $\sin 2x = -\sqrt{3}/2$.   *Ans.* $2\pi/3, 5\pi/6, 5\pi/3, 11\pi/6$

42.48 $\tan 3x = 1$.   *Ans.* $\pi/12, 5\pi/12, 3\pi/4, 13\pi/12, 17\pi/12, 7\pi/4$

42.49 $\cos x/2 = \sqrt{3}/2$.   *Ans.* $\pi/3$

42.50 $\cot x/3 = -1/\sqrt{3}$.   *Ans.* No solution in given interval

42.51 $\sin x \cos x = \frac{1}{2}$ *Ans.* $\pi/4, 5\pi/4$

42.52 $\sin x/2 + \cos x = 1$.   *Ans.* $0, \pi/3, 5\pi/3$

42.53 $\sin 3x + \sin x = 0$.   *Ans.* $0, \pi/2, \pi, 3\pi/2$

42.54 $\cos 2x + \cos 3x = 0$.   *Ans.* $\pi/5, 3\pi/5, \pi, 7\pi/5, 9\pi/5$

42.55 $\sin 2x + \sin 4x = 2 \sin 3x$.   *Ans.* $0, \pi/3, 2\pi/3, \pi, 4\pi/3, 5\pi/3$

42.56 $\cos 5x + \cos x = 2 \cos 2x$.   *Ans.* $0, \pi/4, 2\pi/3, 3\pi/4, 5\pi/4, 4\pi/3, 7\pi/4$

42.57 $\sin x + \sin 3x = \cos x + \cos 3x$.   *Ans.* $\pi/8, \pi/2, 5\pi/8, 9\pi/8, 3\pi/2, 13\pi/8$

Solve each of the following systems for $r \geq 0$ and $0 \leq \theta < 2\pi$:

42.58 $r = a \sin \theta$ *Ans.* $\theta = \pi/6, r = a/2$
           $r = a \cos 2\theta$ $\theta = 5\pi/6, r = a/2; \theta = 3\pi/2, r = -a$

42.59 $r = a \cos \theta$ *Ans.* $\theta = \pi/2, r = 0; \theta = 3\pi/2, r = 0$
           $r = a \sin 2\theta$ $\theta = \pi/6, r = \sqrt{3}a/2$
                                                    $\theta = 5\pi/6, r = -\sqrt{3}a/2$

42.60 $r = 4(1 + \cos \theta)$ *Ans.* $\theta = \pi/3,\ r = 6$
         $r = 3 \sec \theta$ $\theta = 5\pi/3,\ r = 6$

Solve each of the following equations:

42.61   Arctan $2x$ + Arctan $x = \pi/4$.   *Ans.* $x = 0.281$

42.62   Arcsin x + Arctan $x = \pi/2$.   *Ans.* $x = 0.786$

42.63   Arccos x + Arctan $x = \pi/2$.   *Ans.* $x = 0$

Chapter 43

Complex Numbers

PURE IMAGINARY NUMBERS. The square root of a negative number (i.e., $\sqrt{-1}$, $\sqrt{-5}$, $\sqrt{-9}$) is called a *pure imaginary number*. Since by definition $\sqrt{-5} = \sqrt{5} \cdot \sqrt{-1}$ and $\sqrt{-9} = \sqrt{9} \cdot \sqrt{-1} = 3\sqrt{-1}$, it is convenient to introduce the symbol $i = \sqrt{-1}$ and to adopt $\sqrt{-5} = i\sqrt{5}$ and $\sqrt{-9} = 3i$ as the standard form for these numbers.

The symbol i has the property $i^2 = -1$; and for higher integral powers we have $i^3 = i^2 \cdot i = (-1)i = -i$, $i^4 = (i^2)^2 = (-1)^2 = 1$, $i^5 = i^4 \cdot i = i$, etc.

The use of the standard form simplifies the operations on pure imaginaries and eliminates the possibility of certain common errors. Thus, $\sqrt{-9} \cdot \sqrt{4} = \sqrt{-36} = 6i$ since $\sqrt{-9} \cdot \sqrt{4} = 3i(2) = 6i$ but $\sqrt{-9} \cdot \sqrt{-4} \neq \sqrt{36}$ since $\sqrt{-9} \cdot \sqrt{-4} = (3i)(2i) = 6i^2 = -6$.

Notice the cyclic nature of the powers of $i \cdot i^m$ equals $i, -1, -i, 1$ for every natural number, $m \cdot i^6 = -1$, $i^7 = -i$, $i^8 = 1$, etc.

COMPLEX NUMBERS. A number $a + bi$, where a and b are real numbers, is called a *complex number*. The first term a is called the *real part* of the complex number and the second term bi is called the *pure imaginary part*.

Complex numbers may be thought of as including all real numbers and all pure imaginary numbers. For example, $5 = 5 + 0i$ and $3i = 0 + 3i$.

Two complex numbers $a + bi$ and $c + di$ are said to be *equal* if and only if $a = c$ and $b = d$.

The *conjugate* of a complex number $a + bi$ is the complex number $a - bi$. Thus, $2 + 3i$ and $2 - 3i$, $-3 + 4i$ and $-3 - 4i$ are pairs of conjugate complex numbers.

ALGEBRAIC OPERATIONS

(1) **ADDITION.** To add complex numbers, add the real parts and add the pure imaginary parts.

EXAMPLE 1. $(2 + 3i) + (4 - 5i) = (2 + 4) + (3 - 5)i = 6 - 2i$.

(2) **SUBTRACTION.** To subtract two complex numbers, subtract the real parts and subtract the pure imaginary parts.

EXAMPLE 2. $(2 + 3i) - (4 - 5i) = (2 - 4) + [3 - (-5)]i = -2 + 8i$.

(3) **MULTIPLICATION.** To multiply two complex numbers, carry out the multiplication as if the numbers were ordinary binomials and replace i^2 by -1.

EXAMPLE 3. $(2 + 3i)(4 - 5i) = 8 + 2i - 15i^2 = 8 + 2i - 15(-1) = 23 + 2i$.

(4) **DIVISION.** To divide two complex numbers, multiply both numerator and denominator of the fraction by the conjugate of the denominator.

EXAMPLE 4. $\dfrac{2 + 3i}{4 - 5i} = \dfrac{(2 + 3i)(4 + 5i)}{(4 - 5i)(4 + 5i)} = \dfrac{(8 - 15) + (10 + 12)i}{16 + 25} = -\dfrac{7}{41} + \dfrac{22}{41}i$.

$$\left[\text{Note the form of the result; it is neither } \frac{-7 + 22i}{41} \text{ nor } \frac{1}{41}(-7 + 22i). \right]$$

(See Problems 43.1–43.9.)

GRAPHIC REPRESENTATION OF COMPLEX NUMBERS.   The complex number $x + yi$ may be represented graphically by the point P (see Fig. 43-1) whose rectangular coordinates are (x, y).

The point O, having coordinates $(0, 0)$ represents the complex number $0 + 0i = 0$. All points on the x axis have coordinates of the form $(x, 0)$ and correspond to real numbers $x + 0i = x$. For this reason, the x axis is called the *axis of reals*. All points on the y axis have coordinates of the form $(0, y)$ and correspond to pure imaginary numbers $0 + yi = yi$. The y axis is called the *axis of imaginaries*. The plane on which the complex numbers are represented is called the *complex plane*. See Fig. 43-1.

In addition to representing a complex number by a point P in the complex plane, the number may be represented by the directed line segment or vector OP. See Fig. 43-2.

Fig. 43-1   Fig. 43-2

GRAPHIC REPRESENTATION OF ADDITION AND SUBTRACTION.   Review the material on vectors in Chapter 34. Let $z_1 = x_1 + iy_1$ and $z_2 = x_2 + iy_2$ be two complex numbers. The vector representation of these numbers suggests the familiar parallelogram law for determining graphically the sum $z_1 + z_2 = (x_1 + iy_1) + (x_2 + iy_2)$. See Fig. 43-3.

Since $z_1 - z_2 = (x_1 + iy_1) - (x_2 + iy_2) = (x_1 + iy_1) + (-x_2 - iy_2)$, the difference $z_1 - z_2$ of the two complex numbers may be obtained graphically by applying the parallelogram law to $x_1 + iy_1$ and $-x_2 - iy_2$. (See Fig. 43-4.)

In Fig. 43-5 both the sum $OR = z_1 + z_2$ and the difference $OS = z_1 - z_2$ are shown. Note that the segments OS and P_2P_1 (the other diagonal of OP_2RP_1) are equal. (See Problem 43.11.)

Fig. 43-3

Fig. 43-4   Fig. 43-5

POLAR OR TRIGONOMETRIC FORM OF COMPLEX NUMBERS.   Let the complex number $x + yi$ be represented (Fig. 43-6) by the vector OP. This vector (and hence the complex number) may be described in terms of the length r of the vector and *any* positive angle θ which the vector makes with the positive x axis (axis of positive reals). The number $r = \sqrt{x^2 + y^2}$ is called the *modulus* or *absolute value* of the complex number. The angle θ, called the *amplitude* of the complex number, is usually chosen as the smallest, positive angle for which $\tan \theta = y/x$ but at times it will be found more convenient to choose some other angle coterminal with it.

From Fig. 43-6, $x = r \cos \theta$ and $y = r \sin \theta$; then $z = x + yi = r \cos \theta + ir \sin \theta = r (\cos \theta + i \sin \theta)$. We call $z = r (\cos \theta + i \sin \theta)$ the *polar* or *trigonometric form* and $z = x + yi$ the *rectangular form* of the complex number z.

Fig. 43-6   Fig. 43-7

EXAMPLE 5.   Express $z = 1 - i\sqrt{3}$ in polar form. (See Fig. 43-7.)

The modulus is $r = \sqrt{(1)^2 + (-\sqrt{3})^2} = 2$. Since $\tan \theta = y/x = -\sqrt{3}/1 = -\sqrt{3}$, the amplitude θ is either 120° or 300°. Now we know that P lies in quadrant IV; hence, $\theta = 300°$ and the required polar form is

$$z = r(\cos \theta + i \sin \theta) = 2(\cos 300° + i \sin 300°)$$

Note that z may also be represented in polar form by

$$z = 2[\cos (300° + n360°) + i \sin (300° + n360°)]$$

where n is any integer.

EXAMPLE 6.   Express the complex number $z = 8(\cos 210° + i \sin 210°)$ in rectangular form.

Since $\cos 210° = -\sqrt{3}/2$ and $\sin 210° = -\frac{1}{2}$,

$$z = 8(\cos 210° + i \sin 210°) = 8\left[-\frac{\sqrt{3}}{2} + i\left(-\frac{1}{2} \right) \right] = -4\sqrt{3} - 4i$$

is the required rectangular form.
(See Problems 43.12–43.13.)

MULTIPLICATION AND DIVISION IN POLAR FORM

MULTIPLICATION.   The modulus of the product of two complex numbers is the product of their moduli, and the amplitude of the product is the sum of their amplitudes.

DIVISION.   The modulus of the quotient of two complex numbers is the modulus of the dividend divided by the modulus of the divisor, and the amplitude of the quotient is the amplitude of the dividend minus the amplitude of the divisor. For a proof of these theorems, see Problem 43.14.

EXAMPLE 7.   Find (*a*) the product $z_1 z_2$, (*b*) the quotient z_1 / z_2, and (*c*) the quotient z_2 / z_1 where $z_1 = 2(\cos 300° + i \sin 300°)$ and $z_2 = 8(\cos 210° + i \sin 210°)$.

(*a*)   The modulus of the product is 2(8) = 16. The amplitude is 300° + 210° = 510° but, following the convention, we shall use the smallest positive coterminal angle 510° − 360° = 150°. Thus $z_1 z_2 = 16(\cos 150° + i \sin 150°)$.

(b) The modulus of the quotient z_1/z_2 is $\frac{2}{8} = \frac{1}{4}$ and the amplitude is $300° - 210° = 90°$. Thus $z_1/z_2 = \frac{1}{4}(\cos 90° + i \sin 90°)$.

(c) The modulus of the quotient z_2/z_1 is $\frac{8}{2} = 4$. The amplitude is $210° - 300° = -90°$ but we shall use the smallest positive coterminal angle $-90° + 360° = 270°$. Thus $z_2/z_1 = 4(\cos 270° + i \sin 270°)$.

[NOTE.   From Examples 5 and 6 the numbers are $z_1 = 1 - i\sqrt{3}$ and $z_2 = -4\sqrt{3} - 4i$ in rectangular form. Then

$$z_1z_2 = (1 - i\sqrt{3})(-4\sqrt{3} - 4i) = -8\sqrt{3} + 8i = 16(\cos 150° + i \sin 150°)$$

as in (a), and

$$\frac{z_2}{z_1} = \frac{-4\sqrt{3} - 4i}{1 - i\sqrt{3}} = \frac{(-4\sqrt{3} - 4i)(1 + i\sqrt{3})}{(1 - i\sqrt{3})(1 + i\sqrt{3})} = \frac{-16i}{4} = -4i$$

$$= 4(\cos 270° + i \sin 270°) \qquad \text{as in } (c)]$$

(See Problems 43.15–43.16.)

DE MOIVRE'S THEOREM.   If n is any rational number,

$$[r(\cos \theta + i \sin \theta)]^n = r^n(\cos n\theta + i \sin n\theta)$$

A proof of this theorem is beyond the scope of this book; a verification for $n = 2$ and $n = 3$ is given in Problem 43.17.

EXAMPLE 8.

$$(\sqrt{3} - i)^{10} = [2(\cos 330° + i \sin 330°)]^{10}$$

$$= 2^{10}(\cos 10 \cdot 330° + i \sin 10 \cdot 330°)$$

$$= 1024(\cos 60° + i \sin 60°) = 1024\left(\frac{1}{2} + \frac{i\sqrt{3}}{2}\right)$$

$$= 512 + 512i\sqrt{3}$$

(See Problem 43.18.)

ROOTS OF COMPLEX NUMBERS.   We state, without proof, the theorem:
A complex number $a + bi = r(\cos \theta + i \sin \theta)$ has exactly n distinct nth roots.

The procedure for determining these roots is given in Example 9.

EXAMPLE 9.   Find all fifth roots of $4 - 4i$.
The usual polar form of $4 - 4i$ is $4\sqrt{2}(\cos 315° + i \sin 315°)$ but we shall need the more general form $4\sqrt{2}[\cos(315° + k360°) + i \sin(315° + k360°)]$, where k is any integer, including zero.
Using De Moivre's theorem, a fifth root of $4 - 4i$ is given by

$$\{4\sqrt{2}[\cos(315° + k360°) + i \sin(315° + k360°)]\}^{1/5} = (4\sqrt{2})^{1/5}\left(\cos \frac{315° + k360°}{5} + i \sin \frac{315° + k360°}{5}\right)$$

$$= \sqrt{2}[\cos(63° + k72°) + i \sin(63° + k72°)]$$

Assigning in turn the values $k = 0, 1, \ldots,$ we find

$$k = 0: \quad \sqrt{2}(\cos 63° + i \sin 63°) = R_1$$

$$k = 1: \quad \sqrt{2}(\cos 135° + i \sin 135°) = R_2$$

$$k = 2: \quad \sqrt{2}(\cos 207° + i \sin 207°) = R_3$$

$$k = 3: \quad \sqrt{2}(\cos 279° + i \sin 279°) = R_4$$

$$k = 4: \quad \sqrt{2}(\cos 351° + i \sin 351°) = R_5$$

$$k = 5: \quad \sqrt{2}(\cos 423° + i \sin 423°) = \sqrt{2}(\cos 63° + i \sin 63°) = R_1, \qquad \text{etc.}$$

Thus, the five fifth roots are obtained by assigning the values 0, 1, 2, 3, 4 (i.e., 0, 1, 2, 3, ..., $n-1$) to k. (See also Problem 43.19.)

The modulus of each of the roots is $\sqrt{2}$; hence these roots lie on a circle of radius $\sqrt{2}$ with center at the origin. The difference in amplitude of two consecutive roots is $72°$; hence the roots are equally spaced on this circle, as shown in Fig. 43-8.

Fig. 43-8

Solved Problems

In Problems 43.1–43.6, perform the indicated operations, simplify, and write the result in the form $a + bi$.

43.1 $(3 - 4i) + (-5 + 7i) = (3 - 5) + (-4 + 7)i = -2 + 3i$

43.2 $(4 + 2i) - (-1 + 3i) = [4 - (-1)] + (2 - 3)i = 5 - i$

43.3 $(2 + i)(3 - 2i) = (6 + 2) + (-4 + 3)i = 8 - i$

43.4 $(3 + 4i)(3 - 4i) = 9 + 16 = 25$

43.5 $\dfrac{1 + 3i}{2 + i} = \dfrac{(1 + 3i)(2 - i)}{(2 + i)(2 - i)} = \dfrac{(2 + 3) + (-1 + 6)i}{4 + 1} = 1 + i$

43.6 $\dfrac{3 - 2i}{2 - 3i} = \dfrac{(3 - 2i)(2 + 3i)}{(2 - 3i)(2 + 3i)} = \dfrac{(6 + 6) + (9 - 4)i}{4 + 9} = \dfrac{12}{13} + \dfrac{5}{13}i$

43.7   Find x and y such that $2x - yi = 4 + 3i$.

       Here $2x = 4$ and $-y = 3$; then $x = 2$ and $y = -3$.

43.8   Show that the conjugate complex numbers $2 + i$ and $2 - i$ are roots of the equation $x^2 - 4x + 5 = 0$.

       For $x = 2 + i$: $(2 + i)^2 - 4(2 + i) + 5 = 4 + 4i + i^2 - 8 - 4i + 5 = 0$.
       For $x = 2 - i$: $(2 - i)^2 - 4(2 - i) + 5 = 4 - 4i + i^2 - 8 + 4i + 5 = 0$.
       Since each number satisfies the equation, it is a root of the equation.

43.9 Show that the conjugate of the sum of two complex numbers is equal to the sum of their conjugates.

Let the complex numbers be $a + bi$ and $c + di$. Their sum is $(a + c) + (b + d)i$ and the conjugate of the sum is $(a + c) - (b + d)i$.

The conjugates of the two given numbers are $a - bi$ and $c - di$, and their sum is $(a + c) + (-b - d)i = (a + c) - (b + d)i$.

43.10 Represent graphically (as a vector) the following complex numbers: (a) $3 + 2i$ (b) $2 - i$ (c) $-2 + i$ (d) $-1 - 3i$.

We locate, in turn, the points whose coordinates are $(3, 2)$, $(2, -1)$, $(-2, 1)$, $(-1, -3)$ and join each to the origin O.

43.11 Perform graphically the following operations:
(a) $(3 + 4i) + (2 + 5i)$,   (b) $(3 + 4i) + (2 - 3i)$,   (c) $(4 + 3i) - (2 + i)$,   (d) $(4 + 3i) - (2 - i)$.

For (a) and (b), draw as in Figs. 43-9(a) and 43-9(b) the two vectors and apply the parallelogram law.

For (c) draw the vectors representing $4 + 3i$ and $-2 - i$ and apply the parallelogram law as in Fig. 43-9(c).

For (d) draw the vectors representing $4 + 3i$ and $-2 + i$ and apply the parallelogram law as in Fig. 43-9(d).

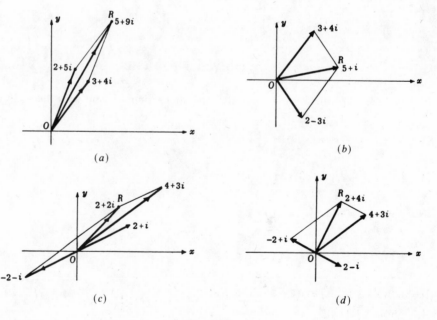

Fig. 43-9

43.12 Express each of the following complex numbers z in polar form:
(a) $-1 + i\sqrt{3}$,   (b) $6\sqrt{3} + 6i$,   (c) $2 - 2i$,   (d) $-3 = -3 + 0i$,   (e) $4i = 0 + 4i$,   (f) $-3 - 4i$.

(a) P lies in the second quadrant; $r = \sqrt{(-1)^2 + (\sqrt{3})^2} = 2$; $\tan \theta = \sqrt{3}/(-1) = -\sqrt{3}$ and $\theta = 120°$. Thus, $z = 2(\cos 120° + i \sin 120°)$.

(b) P lies in the first quadrant; $r = \sqrt{(6\sqrt{3})^2 + 6^2} = 12$; $\tan \theta = 6/6\sqrt{3} = 1/\sqrt{3}$ and $\theta = 30°$. Thus, $z = 12(\cos 30° + i \sin 30°)$.

(c) P lies in the fourth quadrant; $r = \sqrt{2^2 + (-2)^2} = 2\sqrt{2}$; $\tan \theta = -\frac{2}{2} = -1$ and $\theta = 315°$. Thus, $z = 2\sqrt{2}(\cos 315° + i \sin 315°)$.

(d) P lies on the negative x axis and $\theta = 180°$: $r = \sqrt{(-3)^2 + 0^2} = 3$. Thus, $z = 3(\cos 180° + i \sin 180°)$.

(e) P lies on the positive y axis and $\theta = 90°$; $r = \sqrt{0^2 + 4^2} = 4$. Thus, $z = 4(\cos 90° + i \sin 90°)$.

(f) P lies in the third quadrant; $r = \sqrt{(-3)^2 + (-4)^2} = 5$; $\tan \theta = -4/(-3) = 1.3333$, $\theta = 233° 8'$. Thus,
        $z = 5(\cos 233° 8' + i \sin 233° 8')$.

43.13   Express each of the following complex numbers z in rectangular form:

   (a) $4(\cos 240° + i \sin 240°)$ (c) $3(\cos 90° + i \sin 90°)$

   (b) $2(\cos 315° + i \sin 315°)$ (d) $5(\cos 128° + i \sin 128°)$

   (a) $4(\cos 240° + i \sin 240°) = 4[-\frac{1}{2} + i(-\sqrt{3}/2)] = -2 - 2i\sqrt{3}$

   (b) $2(\cos 315° + i \sin 315°) = 2[1/\sqrt{2} + i(-1/\sqrt{2})] = \sqrt{2} - i\sqrt{2}$

   (c) $3(\cos 90° + i \sin 90°) = 3[0 + i(1)] = 3i$

   (d) $5(\cos 128° + i \sin 128°) = 5[-0.6157 + i(0.7880)] = -3.0785 + 3.9400i$

43.14   Prove:

   (a)   The modulus of the product of two complex numbers is the product of their moduli, and the amplitude
           of the product is the sum of their amplitudes.

   (b)   The modulus of the quotient of two complex numbers is the modulus of the dividend divided by the
           modulus of the divisor, and the amplitude of the quotient is the amplitude of the dividend minus
           the amplitude of the divisor.

   Let $z_1 = r_1(\cos \theta_1 + i \sin \theta_1)$ and $z_2 = r_2(\cos \theta_2 + i \sin \theta_2)$.

   (a)
   $$z_1 z_2 = r_1(\cos \theta_1 + i \sin \theta_1) \cdot r_2(\cos \theta_2 + i \sin \theta_2)$$
   $$= r_1 r_2[(\cos \theta_1 \cos \theta_2 - \sin \theta_1 \sin \theta_2) + i(\sin \theta_1 \cos \theta_2 + \cos \theta_1 \sin \theta_2)]$$
   $$= r_1 r_2[\cos (\theta_1 + \theta_2) + i \sin (\theta_1 + \theta_2)]$$

   (b)
   $$\frac{r_1(\cos \theta_1 + i \sin \theta_1)}{r_2(\cos \theta_2 + i \sin \theta_2)} = \frac{r_1(\cos \theta_1 + i \sin \theta_1)(\cos \theta_2 - i \sin \theta_2)}{r_2(\cos \theta_2 + i \sin \theta_2)(\cos \theta_2 - i \sin \theta_2)}$$

   $$= \frac{r_1}{r^2} \cdot \frac{(\cos \theta_1 \cos \theta_2 + \sin \theta_1 \sin \theta_2) + i(\sin \theta_1 \cos \theta_2 - \cos \theta_1 \sin \theta_2)}{\cos^2 \theta_2 + \sin^2 \theta_2}$$

   $$= \frac{r_1}{r_2}[\cos (\theta_1 - \theta_2) + i \sin (\theta_1 - \theta_2)]$$

43.15   Perform the indicated operations, giving the result in both polar and rectangular form.

   (a) $5(\cos 170° + i \sin 170°) \cdot (\cos 55° + i \sin 55°)$

   (b) $2(\cos 50° + i \sin 50°) \cdot 3(\cos 40° + i \sin 40°)$

   (c) $6(\cos 110° + i \sin 110°) \cdot \frac{1}{2}(\cos 212° + i \sin 212°)$

   (d) $10(\cos 305° + i \sin 305°) \div 2(\cos 65° + i \sin 65°)$

   (e) $4(\cos 220° + i \sin 220°) \div 2(\cos 40° + i \sin 40°)$

   (f) $6(\cos 230° + i \sin 230°) \div 3(\cos 75° + i \sin 75°)$

   (a)   The modulus of the product is $5(1) = 5$ and the amplitude is $170° + 55° = 225°$.
           In polar form the product is $5(\cos 225° + i \sin 225°)$ and in rectangular form the product
           is $5(-\sqrt{2}/2 - i\sqrt{2}/2) = -5\sqrt{2}/2 - 5i\sqrt{2}/2$.

   (b)   The modulus of the product is $2(3) = 6$ and the amplitude is $50° + 40° = 90°$.
           In polar form the product is $6(\cos 90° + i \sin 90°)$ and in rectangular form it is $6(0 + i) = 6i$.

   (c)   The modulus of the product is $6(\frac{1}{2}) = 3$ and the amplitude is $110° + 212° = 322°$.
           In polar form the product is $3(\cos 322° + i \sin 322°)$ and in rectangular form it is $3(0.7880 -$
           $0.6157i) = 2.3640 - 1.8471i$.

(d) The modulus of the quotient is $\frac{10}{2} = 5$ and the amplitude is $305° - 65° = 240°$.

In polar form the product is $5(\cos 240° + i \sin 240°)$ and in rectangular form it is $5(-\frac{1}{2} - i\sqrt{3}/2) = -\frac{5}{2} - 5i\sqrt{3}/2$.

(e) The modulus of the quotient is $\frac{4}{2} = 2$ and the amplitude is $220° - 40° = 180°$.

In polar form the quotient is $2(\cos 180° + i \sin 180°)$ and in rectangular form it is $2(-1 + 0i) = -2$.

(f) The modulus of the quotient is $\frac{6}{3} = 2$ and the amplitude is $230° - 75° = 155°$.

In polar form the quotient is $2(\cos 155° + i \sin 155°)$ and in rectangular form it is $2(-0.9063 + 0.4226i) = -1.8126 + 0.8452i$.

43.16 Express each of the numbers in polar form, perform the indicated operation, and give the result in rectangular form.

(a) $(-1 + i\sqrt{3})(\sqrt{3} + i)$ (d) $-2 \div (-\sqrt{3} + i)$ (g) $(3 + 2i)(2 + i)$

(b) $(3 - 3i\sqrt{3})(-2 - 2i\sqrt{3})$ (e) $6i \div (-3 - 3i)$ (h) $(2 + 3i) \div (2 - 3i)$

(c) $(4 - 4i\sqrt{3}) \div (-2\sqrt{3} + 2i)$ (f) $(1 + i\sqrt{3})(1 + i\sqrt{3})$

(a)
$$(-1 + i\sqrt{3})(\sqrt{3} + i) = 2(\cos 120° + i \sin 120°) \cdot 2(\cos 30° + i \sin 30°)$$
$$= 4(\cos 150° + i \sin 150°) = 4(-\sqrt{3}/2 + \tfrac{1}{2}i) = -2\sqrt{3} + 2i$$

(b)
$$(3 - 3i\sqrt{3})(-2 - 2i\sqrt{3}) = 6(\cos 300° + i \sin 300°) \cdot 4(\cos 240° + i \sin 240°)$$
$$= 24(\cos 540° + i \sin 540°) = 24(-1 + 0i) = -24$$

(c)
$$(4 - 4i\sqrt{3}) \div (-2\sqrt{3} + 2i) = 8(\cos 300° + i \sin 300°) \div 4(\cos 150° + i \sin 150°)$$
$$= 2(\cos 150° + i \sin 150°) = 2(-\sqrt{3}/2 + \tfrac{1}{2}i) = -\sqrt{3} + i$$

(d)
$$-2 \div (-\sqrt{3} + i) = 2(\cos 180° + i \sin 180°) \div 2(\cos 150° + i \sin 150°)$$
$$= \cos 30° + i \sin 30° = \tfrac{1}{2}\sqrt{3} + \tfrac{1}{2}i$$

(e)
$$6i \div (-3 - 3i) = 6(\cos 90° + i \sin 90°) \div 3\sqrt{2}(\cos 225° + i \sin 225°)$$
$$= \sqrt{2}(\cos 225° + i \sin 225°) = -1 - i$$

(f)
$$(1 + i\sqrt{3})(1 + i\sqrt{3}) = 2(\cos 60° + i \sin 60°) \cdot 2(\cos 60° + i \sin 60°)$$
$$= 4(\cos 120° + i \sin 120°) = 4(-\tfrac{1}{2} + \tfrac{1}{2}i\sqrt{3}) = -2 + 2i\sqrt{3}$$

(g)
$$(3 + 2i)(2 + i) = \sqrt{13}(\cos 33° 41' + i \sin 33° 41') \cdot \sqrt{5}(\cos 26° 34' + i \sin 26° 34')$$
$$= \sqrt{65}(\cos 60° 15' + i \sin 60° 15')$$
$$= \sqrt{65}(0.4962 + 0.8682i) = 4.001 + 7.000i = 4 + 7i$$

(h)
$$\frac{2 + 3i}{2 - 3i} = \frac{\sqrt{13}(\cos 56° 19' + i \sin 56° 19')}{\sqrt{13}(\cos 303° 41' + i \sin 303° 41')} = \frac{\cos 416° 19' + i \sin 416° 19'}{\cos 303° 41'' + i \sin 303° 41'}$$
$$= \cos 112° 38' + i \sin 112° 38' = -0.3849 + 0.9230i$$

43.17 Verify De Moivre's theorem for $n = 2$ and $n = 3$.

Let $z = r(\cos \theta + i \sin \theta)$.

For $n = 2$: $z^2 = [r(\cos \theta + i \sin \theta)][r(\cos \theta + i \sin \theta)]$
$$= r^2[(\cos^2 \theta - \sin^2 \theta) + i(2 \sin \theta \cos \theta)] = r^2(\cos 2\theta + i \sin 2\theta)$$

For $n = 3$: $z^3 = z^2 \cdot z = [r^2(\cos 2\theta + i \sin 2\theta)][r(\cos \theta + i \sin \theta)]$
$$= r^3[(\cos 2\theta \cos \theta - \sin 2\theta \sin \theta) + i(\sin 2\theta \cos \theta + \cos 2\theta \sin \theta)]$$
$$= r^3(\cos 3\theta + i \sin 3\theta).$$

The theorem may be established for n a positive integer by mathematical induction.

43.18 Evaluate each of the following using De Moivre's theorem and express each result in rectangular form:

(a) $(1 + i\sqrt{3})^4$,   (b) $(\sqrt{3} - i)^5$,   (c) $(-1 + i)^{10}$,   (d) $(2 + 3i)^4$.

(a) $(1 + i\sqrt{3})^4 = [2(\cos 60° + i \sin 60°)]^4 = 2^4(\cos 4 \cdot 60° + i \sin 4 \cdot 60°)$
            $= 2^4(\cos 240° + i \sin 240°) = -8 - 8i\sqrt{3}$

(b) $(\sqrt{3} - i)^5 = [2(\cos 330° + i \sin 330°)]^5 = 32(\cos 1650° + i \sin 1650°)$
            $= 32(\cos 210° + i \sin 210°) = -16\sqrt{3} - 16i$

(c) $(-1 + i)^{10} = [\sqrt{2}(\cos 135° + i \sin 135°)]^{10} = 32(\cos 270° + i \sin 270°) = -32i$

(d) $(2 + 3i)^4 = [\sqrt{13}(\cos 56° 19' + i \sin 56° 19')]^4 = 13^2(\cos 225° 16' + i \sin 225° 16')$
            $= 169(-0.7038 - 7104i) = -118.9 - 120.1i$

43.19   Find the indicated roots in rectangular form, except when this would necessitate the use of tables.

(a) Square roots of $2 - 2i\sqrt{3}$ (e) Fourth roots of i

(b) Fourth roots of $-8 - 8i\sqrt{3}$ (f) Sixth roots of -1

(c) Cube roots of $-4\sqrt{2} + 4i\sqrt{2}$ (g) Fourth roots of $-16i$

(d) Cube roots of 1 (h) Fifth roots of $1 + 3i$

(a) $2 - 2i\sqrt{3} = 4[\cos(300° + k360°) + i \sin(300° + k360°)]$

and $(2 - 2i\sqrt{3})^{1/2} = 2[\cos(150° + k180°) + i \sin(150° + k\,180°)]$

Putting $k = 0$ and 1, the required roots are

$$R_1 = 2(\cos 150° + i \sin 150°) = 2(-\tfrac{1}{2}\sqrt{3} + \tfrac{1}{2}i) = -\sqrt{3} + i$$

$$R_2 = 2(\cos 330° + i \sin 330°) = 2(\tfrac{1}{2}\sqrt{3} - \tfrac{1}{2}i) = \sqrt{3} - i$$

(b) $-8 - 8i\sqrt{3} = 16[\cos(240° + k360°) + i \sin(240° + k360°)]$

and $(-8 - 8i\sqrt{3})^{1/4} = 2[\cos(60° + k90°) + i \sin(60° + k90°)]$

Putting $k = 0, 1, 2, 3$, the required roots are

$$R_1 = 2(\cos 60° + i \sin 60°) = 2(\tfrac{1}{2} + i\tfrac{1}{2}\sqrt{3}) = 1 + i\sqrt{3}$$

$$R_2 = 2(\cos 150° + i \sin 150°) = 2(-\tfrac{1}{2}\sqrt{3} + \tfrac{1}{2}i) = -\sqrt{3} + i$$

$$R_3 = 2(\cos 240° + i \sin 240°) = 2(-\tfrac{1}{2} - i\tfrac{1}{2}\sqrt{3}) = -1 - i\sqrt{3}$$

$$R_4 = 2(\cos 330° + i \sin 330°) = 2(\tfrac{1}{2}\sqrt{3} - \tfrac{1}{2}i) = \sqrt{3} - i$$

(c) $-4\sqrt{2} + 4i\sqrt{2} = 8[\cos(135° + k360°) + i \sin(135° + k360°)]$

and $(-4\sqrt{2} + 4i\sqrt{2})^{1/3} = 2[\cos(45° + k120°) + i \sin(45° + k120°)]$

Putting $k = 0, 1, 2$, the required roots are

$$R_1 = 2(\cos 45° + i \sin 45°) = 2(1/\sqrt{2} + i/\sqrt{2}) = \sqrt{2} + i\sqrt{2}$$

$$R_2 = 2(\cos 165° + i \sin 165°)$$

$$R_3 = 2(\cos 285° + i \sin 285°)$$

(d) $1 = \cos(0° + k360°) + i \sin(0° + k360°)$ and $1^{1/3} = \cos(k120°) + i \sin(k120°)$. Putting $k = 0, 1, 2$, the required roots are

$$R_1 = \cos 0° + i \sin 0° = 1$$

$$R_2 = \cos 120° + i \sin 120° = -\tfrac{1}{2} + i\tfrac{1}{2}\sqrt{3}$$

$$R_3 = \cos 240° + i \sin 240° = -\tfrac{1}{2} - i\tfrac{1}{2}\sqrt{3}$$

Note that $R_2^2 = \cos 2(120°) + i \sin 2(120°) = R_3,$

$R_3^2 = \cos 2(240°) + i \sin 2(240°) = R_2,$

and $R_2 R_3 = (\cos 120° + i \sin 120°)(\cos 240° + i \sin 240°)$
            $= \cos 0° + i \sin 0° = R_1.$

(e) $i = \cos(90° + k360°) + i\sin(90° + k360°)$ and $i^{1/4} = \cos(22\frac{1}{2}° + k90°) + i\sin(22\frac{1}{2}° + k90°)$. Thus, the required roots are

$$R_1 = \cos 22\frac{1}{2}° + i\sin 22\frac{1}{2}° \qquad R_3 = \cos 202\frac{1}{2}° + i\sin 202\frac{1}{2}°$$

$$R_2 = \cos 112\frac{1}{2}° + i\sin 112\frac{1}{2}° \qquad R_4 = \cos 292\frac{1}{2}° + i\sin 292\frac{1}{2}°$$

(f) $-1 = \cos(180° + k360°) + i\sin(180° + k360°)$ and $(-1)^{1/6} = \cos(30° + k60°) + i\sin(30° + k60°)$. Thus, the required roots are

$$R_1 = \cos 30° + i\sin 30° = \tfrac{1}{2}\sqrt{3} + \tfrac{1}{2}i$$

$$R_2 = \cos 90° + i\sin 90° = i$$

$$R_3 = \cos 150° + i\sin 150° = -\tfrac{1}{2}\sqrt{3} + \tfrac{1}{2}i$$

$$R_4 = \cos 210° + i\sin 210° = -\tfrac{1}{2}\sqrt{3} - \tfrac{1}{2}i$$

$$R_5 = \cos 270° + i\sin 270° = -i$$

$$R_6 = \cos 330° + i\sin 330° = \tfrac{1}{2}\sqrt{3} - \tfrac{1}{2}i$$

Note that $R_2^2 = R_5^2 = \cos 180° + i\sin 180°$ and thus R_2 and R_5 are the square roots of -1; that $R_1^3 = R_3^3 = R_5^3 = \cos 90° + i\sin 90° = i$ and thus R_1, R_3, R_5 are the cube roots of i; and that $R_2^3 = R_4^3 = R_6^3 = \cos 270° + i\sin 270° = -i$ and thus R_2, R_4, R_6 are the cube roots of $-i$.

(g) $-16i = 16[\cos(270° + k360°) + i\sin(270° + k360°)]$ and $(-16i)^{1/4} = 2[\cos(67\frac{1}{2}° + k90°) + i\sin(67\frac{1}{2}° + k90°)]$. Thus, the required roots are

$$R_1 = 2(\cos 67\frac{1}{2}° + i\sin 67\frac{1}{2}°) \qquad R_3 = 2(\cos 247\frac{1}{2}° + i\sin 247\frac{1}{2}°)$$

$$R_2 = 2(\cos 157\frac{1}{2}° + i\sin 157\frac{1}{2}°) \qquad R_4 = 2(\cos 337\frac{1}{2}° + i\sin 337\frac{1}{2}°)$$

(h) $1 + 3i = \sqrt{10}[\cos(71° 34' + k360°) + i\sin(71° 34' + k360°)]$ and $(1 + 3i)^{1/5} = \sqrt[10]{10}[\cos 14° 19' + k72°) + i\sin(14° 19' + k72°)]$. The required roots are

$$R_1 = \sqrt[10]{10}(\cos 14° 19' + i\sin 14° 19')$$

$$R_2 = \sqrt[10]{10}\cos 86° 19' + i\sin 86° 19')$$

$$R_3 = \sqrt[10]{10}(\cos 158° 19' + i\sin 158° 19')$$

$$R_4 = \sqrt[10]{10}(\cos 230° 19' + i\sin 230° 19')$$

$$R_5 = \sqrt[10]{10}(\cos 302° 19' + i\sin 302° 19').$$

Supplementary Problems

43.20 Perform the indicated operations, writing the results in the form $a + bi$.

(a) $(6 - 2i) + (2 + 3i) = 8 + i$

(b) $(6 - 2i) - (2 + 3i) = 4 - 5i$

(c) $(3 + 2i) + (-4 - 3i) = -1 - i$

(d) $(3 - 2i) - (4 - 3i) = -1 + i$

(e) $3(2 - i) = 6 - 3i$

(f) $2i(3 + 4i) = -8 + 6i$

(g) $(2 + 3i)(1 + 2i) = -4 + 7i$

(h) $(2 - 3i)(5 + 2i) = 16 - 11i$

(i) $(3 - 2i)(-4 + i) = -10 + 11i$

(j) $(2 + 3i)(3 + 2i) = 13i$

(k) $(2 + \sqrt{-5})(3 - 2\sqrt{-4}) = (6 + 4\sqrt{5}) + (3\sqrt{5} - 8)i$

(l) $(1 + 2\sqrt{-3})(2 - \sqrt{-3}) = 8 + 3\sqrt{3}i$

(m) $(2 - i)^2 = 3 - 4i$

(n) $(4 + 2i)^2 = 12 + 16i$

(o) $(1 + i)^2(2 + 3i) = -6 + 4i$

(p) $\dfrac{2 + 3i}{1 + i} = \dfrac{5}{2} + \dfrac{1}{2}i$

(q) $\dfrac{3 - 2i}{3 - 4i} = \dfrac{17}{25} + \dfrac{6}{25}i$

(r) $\dfrac{3 - 2i}{2 + 3i} = -i$

43.21 Show that $3 + 2i$ and $3 - 2i$ are roots of $x^2 - 6x + 13 = 0$.

43.22 Perform graphically the following operations:
   (a) $(2 + 3i) + (1 + 4i)$ (c) $(2 + 3i) - (1 + 4i)$
   (b) $(4 - 2i) + (2 + 3i)$ (d) $(4 - 2i) - (2 + 3i)$

43.23 Express each of the following complex numbers in polar form:
   (a) $3 + 3i = 3\sqrt{2}(\cos 45° + i \sin 45°)$ (e) $-8 = 8(\cos 180° + i \sin 180°)$
   (b) $1 + \sqrt{3}i = 2(\cos 60° + i \sin 60°)$ (f) $-2i = 2(\cos 270° + i \sin 270°)$
   (c) $-2\sqrt{3} - 2i = 4(\cos 210° + i \sin 210°)$ (g) $-12 + 5i = 13(\cos 157° 23' + i \sin 157° 23')$
   (d) $\sqrt{2} - i\sqrt{2} = 2(\cos 315° + i \sin 315°)$ (h) $-4 - 3i = 5(\cos 216° 52' + i \sin 216° 52')$

43.24 Perform the indicated operation and express the results in the form $a + bi$.
   (a) $3(\cos 25° + i \sin 25°)8(\cos 200° + i \sin 200°) = -12\sqrt{2} - 12\sqrt{2}i$
   (b) $4(\cos 50° + i \sin 50°)2(\cos 100° + i \sin 100°) = -4\sqrt{3} + 4i$
   (c) $\dfrac{4(\cos 190° + i \sin 190°)}{2(\cos 70° + i \sin 70°)} = -1 + i\sqrt{3}$
   (d) $\dfrac{12(\cos 200° + i \sin 200°)}{3(\cos 350° + i \sin 350°)} = -2\sqrt{3} - 2i$

43.25 Use the polar form in finding each of the following products and quotients, and express each result in the form $a + bi$:

   (a) $(1 + i)(\sqrt{2} - i\sqrt{2}) = 2\sqrt{2}$ (d) $\dfrac{4 + 4\sqrt{3}i}{\sqrt{3} + i} = 2\sqrt{3} + 2i$

   (b) $(-1 - i\sqrt{3})(-4\sqrt{3} + 4i) = 8\sqrt{3} + 8i$ (e) $\dfrac{-1 + i\sqrt{3}}{\sqrt{2} + i\sqrt{2}} = 0.2588 + 0.9659i$

   (c) $\dfrac{1 - i}{1 + i} = -i$ (f) $\dfrac{3 + i}{2 + i} = 1.4 - 0.2i$

43.26 Use De Moivre's theorem to evaluate each of the following and express each result in the form $a + bi$:
   (a) $[2(\cos 6° + i \sin 6°)]^5 = 16\sqrt{3} + 16i$ (f) $(\sqrt{3}/2 + i/2)^9 = -i$
   (b) $[\sqrt{2}(\cos 75° + i \sin 75°)]^4 = 2 - 2\sqrt{3}i$ (g) $(3 + 4i)^4 = -526.9 - 336.1i$
   (c) $(1 + i)^8 = 16$
   (d) $(1 - i)^6 = 8i$ (h) $\dfrac{(1 - i\sqrt{3})^3}{(-2 + 2i)^4} = \dfrac{1}{8}$ (i) $\dfrac{(1 + i)(\sqrt{3} + i)^3}{(1 - i\sqrt{3})^3} = 1 - i$
   (e) $(\frac{1}{2} - i\sqrt{3}/2)^{20} = -\frac{1}{2} - i\sqrt{3}/2$

43.27 Find all the indicated roots, expressing the results in the form $a + bi$ unless tables would be needed to do so.
   (a) The square roots of i.   *Ans.* $\sqrt{2}/2 + i\sqrt{2}/2, -\sqrt{2}/2 - i\sqrt{2}/2$
   (b) The square roots of $1 + i\sqrt{3}$.   *Ans.* $\sqrt{6}/2 + i\sqrt{2}/2, -\sqrt{6}/2 - i\sqrt{2}/2$
   (c) The cube roots of -8.   *Ans.* $1 + i\sqrt{3}, -2, 1 - i\sqrt{3}$
   (d) The cube roots of $27i$.   *Ans.* $3\sqrt{3}/2 + 3i/2, -3\sqrt{3}/2 + 3i/2, -3i$
   (e) The cube roots of $-4\sqrt{3} + 4i$.

   Ans. $2(\cos 50° + i \sin 50°), 2(\cos 170° + i \sin 170°), 2(\cos 290° + i \sin 290°)$

(f) The fifth roots of $1 + i$.

   Ans. $\sqrt[10]{2}(\cos 9° + i \sin 9°)$, $\sqrt[10]{2}(\cos 81° + i \sin 81°)$, etc.

(g) The sixth roots of $-\sqrt{3} + i$.

   Ans. $\sqrt[6]{2}(\cos 25° + i \sin 25°)$, $\sqrt[6]{2}(\cos 85° + i \sin 85°)$, etc.

43.28 Find the tenth roots of 1 and show that the product of any two of them is again one of the tenth roots of 1.

43.29 Show that the reciprocal of any one of the tenth roots of 1 is again a tenth root of 1.

43.30 Denote either of the complex cube roots of 1 [Problem 43.19(d)] by ω_1 and the other by ω_2. Show that $\omega_1^2 \omega_2 = \omega_1$ and $\omega_1 \omega_2^2 = \omega_2$.

43.31 Show that $(\cos \theta + i \sin \theta)^{-n} = \cos n\theta - i \sin n\theta$.

43.32 Use the fact that the segments OS and $P_2 P_1$ in Fig. 43-5 are equal to devise a second procedure for constructing the differences $OS = z_1 - z_2$ of two complex numbers z_1 and z_2.

Part IV

ANALYTIC GEOMETRY

Chapter 44

Coordinates and Loci

THE PROJECTION of a point P on a line l is the foot M of the perpendicular to l through P. The projection of the line segment PQ on a line l is the line segment $\overline{MN}$ where M is the projection of P and N is the projection of Q on l. See Fig. 44-1.

| Fig. 44-1 | Fig. 44-2 |

THE PROJECTION OF THE LINE SEGMENT joining $P_1(x_1, y_1)$ and $P_2(x_2, y_2)$ on the x axis or on any line parallel to the x axis is $(x_2 - x_1)$ and the projection of P_1P_2 on the y axis or any line parallel to the y axis is $(y_2 - y_1)$.

Thus, in Fig. 44-2, $M_1M_2 = P_1R = x_2 - x_1$ and $N_1N_2 = RP_2 = y_2 - y_1$. These are directed distances, that is, $M_1M_2 = -M_2M_1$. (See Problems 44.1–44.2.)

THE LENGTH d OF THE LINE SEGMENT $\overline{P_1P_2}$, assumed not parallel to a coordinate axis, is given by

$$d = \sqrt{(x_2 - x_1)^2 + (y_2 - y_1)^2}$$

In Fig. 44-2, $d = \sqrt{(P_1R)^2 + (RP_2)^2}$. Note that this is not a directed distance. (See Problems 44.3–44.5.)

A POINT P is said to divide a line segment $\overline{P_1P_2}$ in the ratio $r_1 : r_2$ if P is on the line P_1P_2 and if $P_1P/PP_2 = r_1/r_2$. The point $P(x, y)$ dividing the segment joining $P_1(x_1, y_1)$ and $P_2(x_2, y_2)$ in the ratio $r_1 : r_2$ is given by

$$x = \frac{r_1x_2 + r_2x_1}{r_1 + r_2}, \qquad y = \frac{r_1y_2 + r_2y_1}{r_1 + r_2}$$

The midpoint $P(x, y)$ of the segment $\overline{P_1P_2}$ is given by

$$x = \tfrac{1}{2}(x_1 + x_2), \qquad y = \tfrac{1}{2}(y_1 + y_2)$$

(See Problems 44.6–44.8.)

THE INCLINATION θ of a straight line is the smallest positive angle (measured counterclockwise) from the positive end of the x axis to the line. The range of θ is given by $0° \leq \theta \leq 180°$. This is equivalent to assuming upward direction along the line as positive and defining the inclination of the line as the angle between the positive directions on the line and x axis.

304

THE SLOPE OF A STRAIGHT LINE is the tangent of its inclination, that is, $m = \tan \theta$. If $P_1(x_1, y_1)$ and $P_2(x_2, y_2)$ are any two distinct points on a line, its slope (see Fig. 44-2) is given by

$$m = \tan \theta = \tan \angle RP_1P_2 = \frac{RP_2}{P_1R} = \frac{y_2 - y_2}{x_2 - x_1}$$

The vertical line has no slope. (See Problem 44.9.)

THE ANGLE α which a line l_1 of inclination θ_1 makes with a line l_2 of inclination θ_2 is defined as the smallest positive angle through which l_2 must be turned to be parallel to l_1. Thus, $\alpha = \theta_1 - \theta_2$ when $\theta_1 > \theta_2$ and $\alpha = \theta_1 - \theta_2 + 180°$ when $\theta_1 < \theta_2$. The range of α is given by $0° \le \alpha < 180°$. See Fig. 44-3(a).

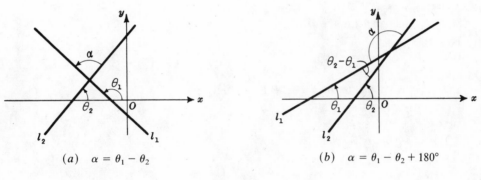

(a)　$\alpha = \theta_1 - \theta_2$　　　　　　　(b)　$\alpha = \theta_1 - \theta_2 + 180°$

Fig. 44-3

The tangent of the angle α is given by

$$\tan \alpha = \frac{\tan \theta_1 - \tan \theta_2}{1 + \tan \theta_1 \tan \theta_2} = \frac{m_1 - m_2}{1 + m_1 m_2} \tag{44.1}$$

(NOTE.　The angle α is sometimes defined as $\alpha = \theta_1 - \theta_2$ with the consequent range $-180° \le \alpha \le 180°$.) See Fig. 44-3(b). (See Problems 44.10–44.14.)

THE AREA OF A TRIANGLE whose vertices are $P_1(x_1, y_1)$, $P_2(x_2, y_2)$, and $P_3(x_3, y_3)$ is given, except possibly for sign, by

$$\tfrac{1}{2}(x_1y_2 + x_2y_3 + x_3y_1 - x_1y_3 - x_2y_1 - x_3y_2) - \frac{1}{2}\begin{vmatrix} x_1 & y_1 & 1 \\ x_2 & y_2 & 1 \\ x_3 & y_3 & 1 \end{vmatrix}$$

(See Problem 44.15.)

THE BASIC PROBLEMS of analytic geometry are

1.　To find the locus of a given equation (see Chapter 9)
2.　To find the equation of a locus defined by a given geometric condition

(See Problems 44.17–44.18.)

Solved Problems

44.1   For each pair of points, find the projection of $\overline{P_1P_2}$ on the coordinate axes.

(a) $P_1(2, 5)$, $P_2(6, 3)$
       The projection on the x axis is $x_2 - x_1 = 6 - 2 = 4$,
       The projection on the y axis is $y_2 - y_1 = 3 - 5 = -2$.

(b) $P_1(3, 4)$, $P_2(2, -1)$
       The projections on the x and y axes are, respectively, $2 - 3 = -1$ and $-1 - 4 = -5$.

44.2   For each pair of points, find the directed length of $\overline{P_1P_2}$.

(a) $P_1(-3, 4)$, $P_2(5, 4)$
       Since P_1P_2 is parallel to the x axis, $P_1P_2 = x_2 - x_1 = 5 - (-3) = 8$.

(b) $P_1(6, 3)$, $P_2(-4, 3)$
       Here $P_1P_2 = -4 - 6 = -10$.

(c) $P_1(5, 2)$, $P_2(5, 8)$
       Since P_1P_2 is parallel to the y axis, $P_1P_2 = y_2 - y_1 = 8 - 2 = 6$.

(d) $P_1(-3, -2)$, $P_2(-3, -8)$
       Here $P_1P_2 = -8 - (-2) = -6$.

44.3   Find the distance between the following pairs of points:

(a) $(2, 1)$, $(8, 7)$ (b) $(-3, -2)$, $(-5, -8)$ (c) $(0, -2)$, $(5, -6)$
(d) $(0, 0)$, $(3, 5)$ (e) $(0, 0)$, (x, y)

(a)   Identifying the first point with P_1 and the second with P_2,

$$d = \sqrt{(x_2 - x_1)^2 + (y_2 - y_1)^2} = \sqrt{(8 - 2)^2 + (7 - 1)^2} = \sqrt{36 + 36} = 6\sqrt{2}$$

(It is immaterial here which point is identified with P_1.)

(b) $d = \sqrt{[-5 - (-3)]^2 + [(-8 - (-2)]^2} = \sqrt{4 + 36} = 2\sqrt{10}$

(c) $d = \sqrt{(5 - 0)^2 + [-6 - (-2)]^2} = \sqrt{25 + 16} = \sqrt{41}$

(d) $d = \sqrt{(3 - 0)^2 + (5 - 0)^2} = \sqrt{9 + 25} = \sqrt{34}$ (e) $d = \sqrt{x^2 + y^2}$

44.4   Show that the points $A(3, 5)$, $B(1, -1)$, and $C(-4, -16)$ lie on a straight line by showing that $AB + BC = AC$.

$$AB = \sqrt{(1 - 3)^2 + (-1 - 5)^2} = \sqrt{40} = 2\sqrt{10}, \quad BC = \sqrt{(-4 - 1)^2 + [-16 - (-1)]^2} = \sqrt{250} = 5\sqrt{10},$$

and $AC = \sqrt{(-4 - 3)^2 + (-16 - 5)^2} = \sqrt{490} = 7/\sqrt{10}.$

Thus, $AB + BC = 2\sqrt{10} + 5\sqrt{10} = 7\sqrt{10} = AC.$

44.5   Show that the triangle whose vertices are $A(4, 3)$, $B(6, -2)$, $C(-11, -3)$ is a right triangle.

$$AB = \sqrt{(6 - 4)^2 + (-2 - 3)^2} = \sqrt{29}, \quad BC = \sqrt{(-11 - 6)^2 + [-3 - (-2)]^2} = \sqrt{290},$$
$$AC = \sqrt{(-11 - 4)^2 + (-3 - 3)^2} = \sqrt{261}$$

Since $(AB)^2 + (AC)^2 = 29 + 261 = 290 = (BC)^2$, the triangle is a right triangle with hypotenuse $\overline{BC}$.

44.6   Derive the point of division formula: $x = \dfrac{r_1 x_2 + r_2 x_1}{r_1 + r_2}$, $y = \dfrac{r_1 y_2 + r_2 y_1}{r_1 + r_2}$.

The point $P(x, y)$ divides the segment $P_1 P_2$, joining $P_1(x_1, y_1)$ and $P_2(x_2, y_2)$, in the ratio $P_1 P / P P_2 = r_1 / r_2$. See Fig. 44-4.

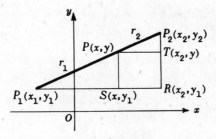

Fig. 44-4

Now $\dfrac{P_1 P}{P P_2} = \dfrac{P_1 S}{PT} = \dfrac{x - x_1}{x_2 - x} = \dfrac{r_1}{r_2}$; then

$$x(r_1 + r_2) = r_1 x_2 + r_2 x_1 \qquad \text{and} \qquad x = \frac{r_1 x_2 + r_2 x_1}{r_1 + r_2}$$

Similarly, $\dfrac{P_1 P}{P P_2} = \dfrac{SP}{TP_2} = \dfrac{r_1}{r_2}$ and $y = \dfrac{r_1 y_2 + r_2 y_1}{r_1 + r_2}$.

44.7   Find the coordinates of the point $P(x, y)$ which divides the segment $\overline{P_1 P_2}$ in the given ratio.

(a) $P_1(1, 3)$, $P_2(7, 9)$; $1:2$ $x = \dfrac{r_1 x_2 + r_2 x_1}{r_1 + r_2} = \dfrac{1 \cdot 7 + 2 \cdot 1}{1 + 2} = 3$, $y = \dfrac{r_1 y_2 + r_2 y_1}{r_1 + r_2} = \dfrac{1 \cdot 9 + 2 \cdot 3}{1 + 2} = 5$;

$$P(3, 5)$$

(b) $P_1(5, -4)$, $P_2(-3, -6)$; $-1:2$ $x = \dfrac{-1(-3) + 2 \cdot 5}{-1 + 2} = 13$, $y = \dfrac{-1(-6) + 2(-4)}{-1 + 2} = -2$; $P(13, -2)$

(c) $P_1(5, -5)$, $P_2(2, -3)$; $4:-3$ $x = \dfrac{4 \cdot 2 + (-3)5}{4 + (-3)} = -7$, $y = \dfrac{4(-3) + (-3)(-5)}{4 + (-3)} = 3$; $P(-7, 3)$

See Fig. 44-5.

(a) INTERNAL DIVISION   (b) EXTERNAL DIVISION   (c) EXTERNAL DIVISION

Fig. 44-5

44.8 Find the coordinates of the midpoint M of the segment joining the pair of points.

(a) $(1, 2), (5, 8)$ $x = \frac{1}{2}(x_1 + x_2) = \frac{1}{2}(1 + 5) = 3,\ y = \frac{1}{2}(y_1 + y_2) = \frac{1}{2}(2 + 8) = 5;\ M(3, 5)$

(b) $(2, -3), (-5, 7)$ $x = \frac{1}{2}(2 - 5) = -\frac{3}{2},\ y = \frac{1}{2}(-3 + 7) = 2;\ M(-\frac{3}{2}, 2)$

(c) $(a, 0), (0, b)$ $x = \frac{1}{2}(a + 0) = \frac{1}{2}a,\ y = \frac{1}{2}(0 + b) = \frac{1}{2}b;\ M(\frac{1}{2}a, \frac{1}{2}b)$

44.9 Find the slope m and inclination θ of the line passing through the points

(a) $(6, 1)$ and $(1, -4)$ (b) $(-3, 2)$ and $(4, -1)$

(a) Here $m = \tan \theta = \dfrac{y_2 - y_1}{x_2 - x_1} = \dfrac{-4 - 1}{1 - 6} = 1$ and $\theta = 45°$

(b) Here $m = \tan \theta = \dfrac{-1 - 2}{4 - (-3)} = -\dfrac{3}{7} = -0.4286$ and $\theta = 156° \, 48'$

44.10 Find the angle α which a line whose slope is $\frac{3}{5}$ makes with a line whose slope is 4.

Take $m_1 = \frac{3}{5}$ and $m_2 = 4$; then $\tan \alpha = \dfrac{m_1 - m_2}{1 + m_1 m_2} = \dfrac{\frac{3}{5} - 4}{1 + (\frac{3}{5})4} = -1$ and $\alpha = 135°$.

44.11 Find the slope of the line which makes an angle of 120° with a line of slope $-\frac{1}{3}$.

Denote the required slope by m_1; then $\alpha = 120°$ and $m_2 = -\frac{1}{3}$. Hence

$$\tan 120° = -\sqrt{3} = \frac{m_1 - (-\frac{1}{3})}{1 + m_1(-\frac{1}{3})} = \frac{3m_1 + 1}{3 - m_1} \quad \text{and} \quad m_1 = -\frac{6 + 5\sqrt{3}}{3}$$

44.12 Find the interior angles of the triangle whose vertices are $A(4, 3)$, $B(-2, 2)$, and $C(2, -8)$.

Locate the triangle and compute the slope of each side. At each vertex indicate the required angle by means of a curved arrow tipped to indicate positive direction about that vertex. The head of the arrow is on the side whose slope is to be taken as m_1 and the tail of the arrow is on the side whose slope is to be taken as m_2, as shown in Fig. 44-6. Then

$$\tan A = \frac{m_1 - m_2}{1 + m_1 \cdot m_2} = \frac{\frac{11}{2} - \frac{1}{6}}{1 + (\frac{11}{2})(\frac{1}{6})} = \frac{64}{23} = 2.7826 \quad \text{and} \quad A = 70° \, 14'$$

$$\tan B = \frac{\frac{1}{6} - (-\frac{5}{2})}{1 + (\frac{1}{6})(-\frac{5}{2})} = \frac{32}{7} = 4.5714 \quad \text{and} \quad B = 77° \, 40'$$

$$\tan C = \frac{-\frac{5}{2} - \frac{11}{2}}{1 + (-\frac{5}{2})(\frac{11}{2})} = \frac{32}{51} = 0.6275 \quad \text{and} \quad C = 32° \, 6'$$

Fig. 44-6

44.13   The vertices of a quadrilateral are $P_1(3, 2)$, $P_2(-4, 5)$, $P_3(-7, -1)$, $P_4(0, -3)$. Show, by finding numerical values, that the perimeter of the quadrilateral whose vertices are the midpoints of the sides is equal to the sum of the diagonals of the given quadrilateral. Refer to Fig. 44-7.

Fig. 44-7

The midpoints of the sides $\overline{P_1P_2}$, $\overline{P_2P_3}$, $\overline{P_3P_4}$, $\overline{P_4P_1}$ are $A(-\frac{1}{2}, \frac{7}{2})$, $B(-\frac{11}{2}, 2)$, $C(-\frac{7}{2}, -2)$, $D(\frac{3}{2}, -\frac{1}{2})$, respectively. The perimeter of the quadrilateral $ABCD$ is

$$AB + BC + CD + DA = \sqrt{(-\tfrac{1}{2} + \tfrac{11}{2})^2 + (\tfrac{7}{2} - 2)^2} + \sqrt{(-\tfrac{11}{2} + \tfrac{7}{2})^2 + (2 + 2)^2}$$
$$+ \sqrt{(-\tfrac{7}{2} - \tfrac{3}{2})^2 + (-2 + \tfrac{1}{2})^2} + \sqrt{(\tfrac{3}{2} + \tfrac{1}{2})^2 + (-\tfrac{1}{2} - \tfrac{7}{2})^2}$$
$$= \frac{\sqrt{109}}{2} + 2\sqrt{5} + \frac{\sqrt{109}}{2} + 2\sqrt{5} = \sqrt{109} + 4\sqrt{5}$$

(Note that the quadrilateral is a parallelogram since the opposite sides are equal.)
The sum of the diagonals of the given quadrilateral is

$$P_1P_3 + P_2P_4 = \sqrt{(3 + 7)^2 + (2 + 1)^2} + \sqrt{(-4 - 0)^2 + (5 + 3)^2} = \sqrt{109} + \sqrt{80} = \sqrt{109} + 4\sqrt{5}$$

44.14   Prove: The sum of the squares of the lengths of the medians of a triangle is equal to $\frac{3}{4}$ the sum of the square of the lengths of its sides.

Place the triangle so that its vertices are $O(0, 0)$, $A(a, 0)$, and $B(b, c)$. The midpoints of the sides OA, AB, BO are respectively $R(\frac{1}{2}a, 0)$, $S[\frac{1}{2}(a + b), \frac{1}{2}c]$, $T(\frac{1}{2}b, \frac{1}{2}c)$. See Fig. 44-8. Then

$$(OS)^2 + (AT)^2 + (BR)^2 = \{[\tfrac{1}{2}(a + b)]^2 + (\tfrac{1}{2}c)^2\} + [(\tfrac{1}{2}b - a)^2 + (\tfrac{1}{2}c)^2] + [(b - \tfrac{1}{2}a)^2 + c^2]$$
$$= \tfrac{3}{2}(a^2 - ab + b^2 + c^2)$$

$$(OA)^2 + (AB)^2 + (BO)^2 = (a^2) + [(b - a)^2 + c^2] + (b^2 + c^2) = 2(a^2 - ab + b^2 + c^2)$$

Thus, $(OS)^2 + (AT)^2 + (BR)^2 = \frac{3}{4}[(OA)^2 + (AB)^2 + (BO)^2]$.

Fig. 44-8

44.15 Show that the area of the triangle whose vertices are $P_1(x_1, y_1)$, $P_2(x_2, y_2)$, $P_3(x_3, y_3)$ is given, apart from sign, by $A = \frac{1}{2}(x_1y_2 + x_2y_3 + x_3y_1 - x_1y_3 - x_2y_1 - x_3y_2)$.

In Fig. 44-9,

$$\text{Area triangle } P_1P_2P_3 = \text{area trapezoid } M_3P_3P_2M_2 - \text{area trapezoid } M_1P_1P_2M_2$$
$$- \text{area trapezoid } M_3P_3P_1M_1$$
$$= \frac{1}{2}(y_3 + y_2)(x_2 - x_3) - \frac{1}{2}(y_1 + y_2)(x_2 - x_1) - \frac{1}{2}(y_3 + y_1)(x_1 - x_3)$$
$$= \frac{1}{2}(x_1y_2 + x_2y_3 + x_3y_1 - x_1y_3 - x_2y_1 - x_3y_2)$$

(NOTE.   In the derivation above, P_1, P_2, and P_3 have been assigned in counterclockwise order about the triangle; when so assigned, the formula yields a positive result.)

Fig. 44-9   Fig. 44-10

44.16 Find the area of the triangle whose vertices are $(-5, 1)$, $(4, 5)$, and $(7, -4)$.

To ensure a positive result, we locate the points and number them so that as we describe the path $P_1P_2P_3P_1$ the area lies on the left (see Fig. 44-10). Then

$$A = \frac{1}{2}\begin{vmatrix} 4 & 5 & 1 \\ -5 & 1 & 1 \\ 7 & -4 & 1 \end{vmatrix} = 46\frac{1}{2}$$

44.17 Find the equation of the locus of a point which is (a) equidistant from the points $A(-3, 2)$ and $B(5, 6)$, (b) at a distance 6 from the point $C(3, 4)$.

Let $P(x, y)$ be an arbitrary point on the required locus.

(a)   In Fig. 44-11, $AP = BP$; hence, $(AP)^2 = (BP)^2$ or $(x + 3)^2 + (y - 2)^2 = (x - 5)^2 + (y - 6)^2$. This reduces to $2x + y = 6$, the equation of the required locus.

(b)   In Fig. 44-12, $CP = 6$; hence, $(CP)^2 = 36$ or $(x - 3)^2 + (y - 4)^2 = 36$. The locus is a circle.

44.18 Write the equation of the circle having center at $C(-1, 2)$ and passing through $A(5, 4)$. See Fig. 44-13.

Let $P(x, y)$ be an arbitrary point on the circle. Then $CP = CA$; hence

$$(CP)^2 = (CA)^2 \qquad \text{or} \qquad (x + 1)^2 + (y - 2)^2 = (5 + 1)^2 + (4 - 2)^2 = 40$$

is the required equation.

Fig. 44-11   Fig. 44-12   Fig. 44-13

Supplementary Problems

44.19 The sides of a square are 12 units in length. Locate its vertices when it is placed in a coordinate system so that (*a*) one vertex is at the origin, two of its sides are along the coordinate axes, and the fourth vertex is in quadrant IV; (*b*) its center is at the origin and its sides are parallel to the coordinate axes; (*c*) its center is at the origin and its vertices are on the coordinate axes.

Ans. (*a*) $(0,0)$, $(12,0)$, $(0,-12)$, $(12,-12)$ (*b*) $(6,6)$, $(-6,6)$, $(-6,-6)$, $(6,-6)$
      (*c*) $(6\sqrt{2},0)$, $(0,6\sqrt{2})$, $(-6\sqrt{2},0)$, $(0,-6\sqrt{2})$

44.20 The radius of a circle is 8 units. (*a*) Locate its points of intersection with the coordinate axes if its center is at the origin. (*b*) Locate the center and the points of tangency if the circle is tangent to the coordinate axes with center in quadrant II.

Ans. (*a*) $(4,0)$, $(0,4)$, $(-4,0)$, $(0,-4)$ (*b*) $C(-8,8)$, $T_1(0,8)$, $T_2(-8,0)$

44.21 Find the area of the right triangle whose vertices are (*a*) $A(2,3)$, $B(8,7)$, $C(8,3)$ (*b*) $A(5,4)$, $B(-3,6)$, $C(-3,4)$.

Ans. (*a*) 12   (*b*) 8

44.22 Find the distance between each pair of points: (*a*) $(3,6)$ and $(7,3)$ (*b*) $(-4,2)$ and $(1,7)$ (*c*) $(10,4)$ and $(-2,-2)$ (*d*) $(0,0)$, and $(-3,-2)$.

Ans. (*a*) 5   (*b*) $5\sqrt{2}$ (*c*) $6\sqrt{5}$ (*d*) $\sqrt{13}$

44.23 Show that the triangle whose vertices are $(4,-1)$, $(-2,3)$, and $(-6,-3)$ is a right triangle. Show that the midpoint of its hypotenuse is equidistant from the vertices.

44.24 For the line segment P_1P_2 joining $P_1(-3,2)$ and $P_2(4,-4)$, locate (*a*) the midpoint (*b*) the points of trisection (*c*) the point dividing it in the ratio $3:2$ (*d*) the point dividing it in the ratio $-3:5$.

Ans. (*a*) $(\frac{1}{2},-1)$ (*b*) $(-\frac{2}{3},0)$, $(\frac{5}{3},-2)$ (*c*) $(\frac{6}{5},-\frac{8}{5})$ (*d*) $(-\frac{27}{2},11)$

44.25 Find the area of the triangle whose vertices are (*a*) $(4,-1)$, $(6,5)$, $(-2,9)$ (*b*) $(4,-3)$, $(9,4)$, $(-3,6)$.

Ans. (*a*) 28   (*b*) 47

44.26 Prove: The sum of the squares of the distances of any point in the plane to two opposite vertices of a rectangle is equal to the sum of the squares of the distances to the other two vertices.

Hint: Take the vertices at $(0,0)$, $(a,0)$, $(0,b)$, (a,b).

44.27 Prove: The segment joining the midpoints of two opposite sides of a quadrilateral and the segment joining the midpoints of its diagonals bisect each other.

Hint: Take the vertices at $(0,0)$, $(a,0)$, (b,c), (d,e).

44.28 Find the slope and inclination of the line through each pair of points of Problem 44.22.

Ans. (*a*) $-\frac{3}{4}$, $143°8'$ (*b*) $1, 45°$ (*c*) $\frac{1}{2}, 26°34'$ (*d*) $\frac{2}{3}, 33°41'$

44.29 Find the interior angles of the triangle whose vertices are $(3, 1)$, $(-3, -2)$, $(-4, 4)$.

　　　　Ans. $49°\,46'$, $72°\,54'$, $57°\,20'$

44.30 Find the area of the triangle of Problem 44.29.

　　　　Ans. $\frac{39}{2}$

44.31 Write the equation of the locus of a point $P(x, y)$ which moves so that (a) it is always 3 units to the left of the y axis, (b) it is always 4 units below the x axis, (c) its ordinate is always twice its abscissa, (d) its abscissa increased by 2 is equal to three times its ordinate.

　　　　Ans. (a) $x = -3$　　(b) $y = -4$　　(c) $y = 2x$　　(d) $x - 3y + 2 = 0$

44.32 Find the equation of the perpendicular bisector of the segment joining each of the following pairs of points: (a) $(2, -3)$ and $(-3, 2)$ (b) $(-4, -5)$ and $(3, -3)$.

　　　　Ans. (a) $x - y = 0$　　　(b) $14x + 4y + 23 = 0$

44.33 Find the equation of the perpendicular bisector of the base of the isosceles triangle whose vertices are $(1, 1)$, $(10, 2)$, $(6, -3)$. Show that the bisector passes through the third vertex.

　　　　Ans. $9x + y - 51 = 0$

44.34 Find the equation of the circle (a) having its center at $(-2, 4)$ and radius $= 6$ (b) having its center at $(4, 3)$ and passing through $(8, 0)$ (c) passing through the points $(4, 0)$, $(8, 0)$, $(0, 2)$, and $(0, 16)$ (d) circumscribing the right triangle whose vertices are $(2, -4)$, $(6, -4)$, and $(6, 6)$.

　　　　Ans. (a) $x^2 + y^2 + 4x - 8y - 16 = 0$　　(c) $x^2 + y^2 - 12x - 18y + 32 = 0$
　　　　　　　(b) $x^2 + y^2 - 8x - 6y = 0$　　　　　(d) $x^2 + y^2 - 8x - 2y - 12 = 0$

44.35 Find the equation of the locus of a point $P(x, y)$ which moves so that the sum of the squares of its distances from $A(2, 4)$ and $B(-3, 5)$ is 30.

　　　　Ans. $x^2 + y^2 + x - 9y + 12 = 0$

44.36 Find the equation of the locus of a point $P(x, y)$ which moves so that the difference of the squares of its distances from $A(2, 4)$ and $B(-3, 5)$ is 30.

　　　　Ans. $5x - y = -22$, $5x - y = 8$

Chapter 45

The Straight Line

THE EQUATION OF THE STRAIGHT LINE parallel to the y axis at a distance a from that axis is $x = a$. The equation of the straight line having slope m and passing through the point (x_1, y_1) is

$$y - y_1 = m(x - x_1) \qquad (\textit{Point-slope form})$$

(See Problem 45.1.)
   The equation of the line having slope m and y intercept b is

$$y = mx + b \qquad (\textit{Slope-intercept form})$$

(See Problem 45.2.)
   The equation of the line passing through the points (x_1, y_1) and (x_2, y_2), where $x_1 \neq x_2$, is

$$y - y_1 = \frac{y_2 - y_1}{x_2 - x_1}(x - x_1) \qquad (\textit{Two-point form})$$

(See Problem 45.3.)
   The equation of the line whose x intercept is a and whose y intercept is b, where $ab \neq 0$, is

$$\frac{x}{a} + \frac{y}{b} = 1 \qquad (\textit{Intercept form})$$

(See Problem 45.4.)

THE GENERAL EQUATION of the straight line is $Ax + By + C = 0$, where A, B, C are arbitrary constants except that not both A and B are zero.
   If $C = 0$, the line passes through the origin. If $B = 0$, the line is vertical; if $A = 0$, the line is horizontal. Otherwise, the line has slope $m = -A/B$ and y intercept $b = -C/B$.
   If two nonvertical lines are parallel, their slopes are equal. Thus the lines $Ax + By + C = 0$ and $Ax + By + D = 0$ are parallel.
   If two oblique lines are perpendicular, the slope of one is the negative reciprocal of the slope of the other. If m_1 and m_2 are the slopes of two perpendicular lines, then $m_1 = -1/m_2$ or $m_1 m_2 = -1$. Thus $Ax + By + C = 0$ and $Bx - Ay + D = 0$, where $AB \neq 0$, are perpendicular lines. (See Problems 45.5–45.8.)

Solved Problems

45.1   Construct and find the equation of the straight line which passes though the point $(-1, -2)$ with slope $(a)\ \frac{3}{4}$ $(b)\ -\frac{4}{5}$.

   (a) A point tracing the line *rises* (slope is positive) 3 units as it moves a horizontal distance of 4 units to the right. Thus, after locating the point $A(-1, -2)$, move 4 units to the right and 3 up to the point $B(3, 1)$. The required line is AB. [See Fig. 45-1(a).]
       Using $y - y_1 = m(x - x_1)$, the equation is $y + 2 = \frac{3}{4}(x + 1)$ or $3x - 4y - 5 = 0$.

313

Fig. 45-1

(b) A point tracing the line falls (slope is negative) 4 units as it moves a horizontal distance of 5 units to the right. Thus, after locating the point $A(-1, -2)$, move 5 units to the right and 4 units down to the point $B(4, -6)$. The required line is AB as shown in Fig. 45-1(b). Its equation is

$$y + 2 = -\tfrac{4}{5}(x + 1) \qquad \text{or} \qquad 4x + 5y + 14 = 0.$$

45.2 Determine the slope m and y intercept b of the following lines. Sketch each.

(a) $y = \tfrac{3}{2}x - 2$

    Ans. $m = \tfrac{3}{2};\ b = -2$

       To sketch the locus, locate the point $(0, -2)$. Then move 2 units to the right and 3 units up to another point on the required line. See Fig. 45-2(a).

(b) $y = -3x + \tfrac{5}{2}$

    Ans. $m = -3;\ b = \tfrac{5}{2}$

       To sketch the locus, locate the point $(0, \tfrac{5}{2})$. Then move 1 unit to the right and 3 units down to another point on the line. See Fig. 45-2(b).

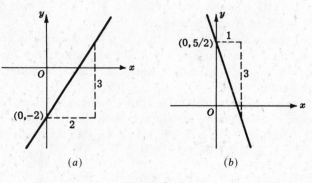

Fig. 45-2

45.3 Write the equations of the straight lines: (a) through $(2, 3)$ and $(-1, 4)$ (b) through $(-7, -2)$ and $(-2, -5)$ (c) through $(3, 3)$ and $(3, 6)$.

    We use $y - y_1 = \dfrac{y_2 - y_1}{x_2 - x_1}(x - x_1)$ and label each pair of points P_1 and P_2 in the order given.

(a) The equation is $y - 3 = \dfrac{4 - 3}{-1 - 2}(x - 2) = -\dfrac{1}{3}(x - 2) \qquad \text{or} \qquad x + 3y - 11 = 0.$

(b) The equation is $y + 2 = \dfrac{-5 + 2}{-2 + 7}(x + 7) = -\dfrac{3}{5}(x + 7) \qquad \text{or} \qquad 3x + 5y + 31 = 0.$

(c) Here $x_1 = x_2 = 3$. The required equation is $x - 3 = 0$.

45.4   Determine the x intercept a and the y intercept b of the following lines. Sketch each.

   (a) $3x - 2y - 4 = 0$
   When $y = 0$, $3x - 4 = 0$ and $x = \frac{4}{3}$; the x intercept is $a = \frac{4}{3}$.
   When $x = 0$, $-2y - 4 = 0$ and $y = -2$; the y intercept is $b = -2$.
   To obtain the locus, join the points $(\frac{4}{3}, 0)$ and $(0, -2)$ by a straight line. See Fig. 45-3(a).

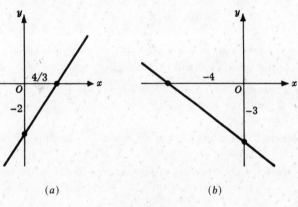

(a)   (b)

Fig. 45-3

   (b) $3x + 4y + 12 = 0$
   When $y = 0$, $3x + 12 = 0$ and $x = -4$; the x intercept is $a = -4$.
   When $x = 0$, $4y + 12 = 0$ and $y = -3$; the y intercept is $b = -3$.
   The locus is the straight line joining the points $(-4, 0)$ and $(0, -3)$. See Fig. 45-3(b).

45.5   Prove: If two oblique lines l_1 and l_2 of slope m_1 and m_2, respectively, are mutually perpendicular, then $m_1 = -1/m_2$.

   Let $m_2 = \tan \theta_2$, where θ_2 is the inclination of l_2. The inclination of l_1 is $\theta_1 = \theta_2 \pm 90°$ according as θ_2 is less than or greater than $90°$ [see Figs. 45-4(a) and (b)]. Then

$$m_1 = \tan \theta_1 = \tan (\theta_2 \pm 90°) = -\cot \theta_2 = -\frac{1}{\tan \theta_2} = -\frac{1}{m_2}$$

   (a) $\theta_2 < 90°$; $\theta_1 = \theta_2 + 90°$ (b) $\theta_2 > 90°$; $\theta_1 = \theta_2 - 90°$

Fig. 45-4

45.6   Find the equation of the straight line (a) through $(3, 1)$ and parallel to the line through $(3, -2)$ and $(-6, 5)$ (b) through $(-2, -4)$ and parallel to the line $8x - 2y + 3 = 0$.

   Two lines are parallel provided their slopes are equal.

(a) The slope of the line through $(3, -2)$ and $(-6, 5)$ is $m = \dfrac{y_2 - y_1}{x_2 - x_1} = \dfrac{5 + 2}{-6 - 3} = -\dfrac{7}{9}$. The equation of the line through $(3, 1)$ with slope $-\frac{7}{9}$ is $y - 1 = -\frac{7}{9}(x - 3)$ or $7x + 9y - 30 = 0$.

(b) **First Solution.** From $y = 4x + \frac{3}{2}$, the slope of the given line is $m = 4$. The equation of the line through $(-2, -4)$ with slope 4 is $y + 4 = 4(x + 2)$ or $4x - y + 4 = 0$.

Second Solution. The equation of the required line is of the form $8x - 2y + D = 0$. If $(-2, -4)$ is on the line, then $8(-2) - 2(-4) + D = 0$ and $D = 8$. The required equation is $8x - 2y + 8 = 0$ or $4x - y + 4 = 0$.

45.7 Find the equation of the straight line (a) through $(-1, -2)$ and perpendicular to the line through $(-2, 3)$ and $(-5, -6)$ (b) through $(2, -4)$ and perpendicular to the line $5x + 3y - 8 = 0$.

Two lines are perpendicular provided the slope of one is the negative reciprocal of the slope of the other.

(a) The slope of the line through $(-2, 3)$ and $(-5, -6)$ is $m = 3$; the slope of the required line is $-1/m = -\frac{1}{3}$. The required equation is $y + 2 = -\frac{1}{3}(x + 1)$ or $x + 3y + 7 = 0$.

(b) **First Solution.** The slope of the given line is $-\frac{5}{3}$; the slope of the required line is $\frac{3}{5}$. The required equation is $y + 4 = \frac{3}{5}(x - 2)$ or $3x - 5y - 26 = 0$.

Second Solution. The equation of the required line is of the form $3x - 5y + D = 0$. If $(2, -4)$ is on the line, then $3(2) - 5(-4) + D = 0$ and $D = -26$. The required equation is $3x - 5y - 26 = 0$.

45.8 Given the vertices $A(7, 9)$, $B(-5, -7)$, and $C(12, -3)$ of the triangle ABC (see Fig. 45-5), find

(a) The equation of the side AB

(b) The equation of the median through A

(c) The equation of the altitude through B

(d) The equation of the perpendicular bisector of the side AB

(e) The equation of the line through C with slope that of AB

(f) The equation of the line through C with slope the reciprocal of that of AB

(a) $\quad y + 7 = \dfrac{9 + 7}{7 + 5}(x + 5) = \dfrac{4}{3}(x + 5) \qquad$ or $\qquad 4x - 3y - 1 = 0$

(b) The median through a vertex bisects the opposite side. The midpoint of BC is $L(\frac{7}{2}, -5)$. The equation of the median through A is

$$y - 9 = \frac{-5 - 9}{\frac{7}{2} - 7}(x - 7) = 4(x - 7) \qquad \text{or} \qquad 4x - y - 19 = 0$$

(c) The altitude through B is perpendicular to CA. The slope of CA is $-\frac{12}{5}$ and its negative reciprocal is $\frac{5}{12}$. The equation is $y + 7 = \frac{5}{12}(x + 5)$ or $5x - 12y - 59 = 0$.

(d) The perpendicular bisector of AB passes through the midpoint $N(1, 1)$ and has slope $-\frac{3}{4}$. The equation is $y - 1 = -\frac{3}{4}(x - 1)$ or $3x + 4y - 7 = 0$.

(e) $\quad y + 3 = \frac{4}{3}(x - 12)$ or $4x - 3y - 57 = 0$. $\qquad$ (f) $\quad y + 3 = \frac{3}{4}(x - 12)$ or $3x - 4y - 48 = 0$.

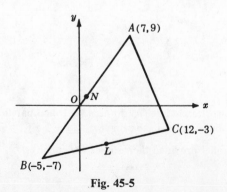

Fig. 45-5

Supplementary Problems

45.9 Determine the slope and the intercepts on the coordinate axes of each of the lines (a) $4x - 5y + 20 = 0$
(b) $2x + 3y - 12 = 0$.

> *Ans.* (a) $m = \frac{4}{5}$, $a = -5$, $b = 4$ (b) $m = -\frac{2}{3}$, $a = 6$, $b = 4$

45.10 Write the equation of each of the straight lines:

(a)	With slope 3 and y intercept 2.	*Ans.* $3x - y + 2 = 0$
(b)	Through $(5, 4)$ and parallel to $2x + 3y - 12 = 0$.	*Ans.* $2x + 3y - 22 = 0$
(c)	Through $(-3, 3)$ and with y intercept 6.	*Ans.* $x - y + 6 = 0$
(d)	Through $(-3, 3)$ and with x intercept 4.	*Ans.* $3x + 7y - 12 = 0$
(e)	Through $(5, 4)$ and $(-3, 3)$.	*Ans.* $x - 8y + 27 = 0$
(f)	With $a = 3$ and $b = -5$.	*Ans.* $5x - 3y - 15 = 0$
(g)	Through $(2, 3)$ and $(2, -5)$.	*Ans.* $x - 2 = 0$

45.11 Find the value of k such that the line $(k - 1)x + (k + 1)y - 7 = 0$ is parallel to the line $3x + 5y + 7 = 0$.

> *Ans.* $k = 4$

45.12 Given the triangle whose vertices are $A(-3, 2)$, $B(5, 6)$, $C(1, -4)$. Find the equations of

(a)	The sides.	*Ans.* $x - 2y + 7 = 0$, $5x - 2y - 13 = 0$, $3x + 2y + 5 = 0$
(b)	The medians.	*Ans.* $x + 6y - 9 = 0$, $7x - 6y + 1 = 0$, $x = 1$
(c)	The altitudes.	*Ans.* $2x + 5y - 4 = 0$, $2x - 3y + 8 = 0$, $2x + y + 2 = 0$
(d)	The perpendicular bisectors of the sides.	*Ans.* $2x + y - 6 = 0$, $2x + 5y - 11 = 0$, $2x - 3y - 1 = 0$

45.13 For the triangle of Problem 45.12: (a) find the coordinates of the centroid G (intersection of the medians), the orthocenter H (intersection of the altitudes), and circumcenter C (intersection of the perpendicular bisectors of the sides) (b) Show that G lies on the line joining H and C and divides the segment HC in the ratio $2:1$.

> *Ans.* (a) $G(1, \frac{4}{3})$, $H(-\frac{7}{4}, \frac{3}{2})$, $C(\frac{19}{8}, \frac{5}{4})$

45.14 Show that the line passing through the points $(5, -4)$ and $(-2, 7)$ is the perpendicular bisector of the line segment whose end points are $(-4, -2)$ and $(7, 5)$.

45.15 Use the determinant form of the formula for the area of a triangle (Chapter 47) to show

(a) The equation of the straight line through $P_1(x_1, y_1)$ and $P_2(x_2, y_2)$ is given by $\begin{vmatrix} x & y & 1 \\ x_1 & y_1 & 1 \\ x_2 & y_2 & 1 \end{vmatrix} = 0$.

(b) Three points $P_1(x_1, y_1)$, $P_2(x_2, y_2)$ and $P_3(x_3, y_3)$ are on a straight line provided $\begin{vmatrix} x_1 & y_1 & 1 \\ x_2 & y_2 & 1 \\ x_3 & y_3 & 1 \end{vmatrix} = 0$.

Chapter 46

Families of Straight Lines

THE EQUATION $y = 3x + b$ represents the set of all lines having slope 3. The quantity b is a variable over the set of lines, each value of b being associated with one and only one line of the set. To distinguish it from the variables x, y which vary over the points of each line, b is called a *parameter*. Such sets of lines are also called *one-parameter systems* or *families* of lines. See Fig. 46-1.

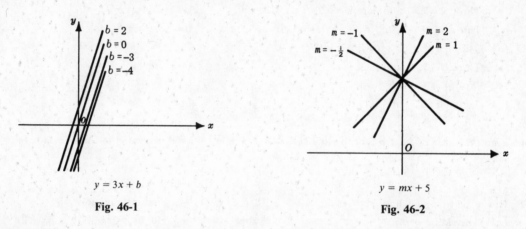

Fig. 46-1 Fig. 46-2

Similarly, the equation $y = mx + 5$, in which m is the parameter, represents the one-parameter family of lines having y intercept = 5 or passing through the point (0, 5). See Fig. 46-2. It is important to note that one line satisfying the geometric condition is not included since for no value of m does the equation $y = mx + 5$ yield the line $x = 0$. (See Problems 46.1–46.4.)

THE EQUATION $$A_1x + B_1y + C_1 + k(A_2x + B_2y + C_2) = 0 \tag{46.1}$$

represents the family of all lines, except for l_2, passing through the point of intersection of the lines

$$l_1: \quad A_1x + B_1y + C_1 = 0 \quad \text{and} \quad l_2: \quad A_2x + B_2y + C_2 = 0$$

(See Problem 46.5)

Equation (*46.1*) provides a means of finding the equation of a line which passes through the point of intersection of two given lines and satisfies one other condition, without having first to compute the coordinates of the point of intersection.

EXAMPLE 1. Write the equation of the line which passes through the point of intersection of the lines

$$l_1: \quad 3x + 5y + 29 = 0 \quad \text{and} \quad l_2: \quad 7x - 11y - 13 = 0$$

and through the point (2, 1).

Since the given point is not on l_2, the required line is among those given by

$$l_1 + kl_2: \quad 3x + 5y + 29 + k(7x - 11y - 13) = 0$$

For the line which passes through (2, 1), $3(2) + 5(1) + 29 + k[7(2) - 11(1) - 13] = 0$ and $k = 4$; hence, its equation is

$$l_1 + 4l_2: \quad 3x + 5y + 29 + 4(7x - 11y - 13) = 31x - 39y - 23 = 0$$

(See Problem 46.6.)

318

Solved Problems

46.1   Write the equation of the family of lines satisfying the given condition. Name the parameter and list any lines satisfying the condition but not obtained by assigning a value to the parameter.

(*a*)   Parallel to the *x* axis.

(*d*)   Perpendicular to $5x + 3y - 8 = 0$.

(*b*)   Through the point $(-3, 2)$.

(*e*)   The sum of whose intercepts is 10.

(*c*)   At a distance 5 from the origin.

(*a*)   The equation is $y = k$, *k* being the parameter.

(*b*)   The equation is $y - 2 = m(x + 3)$, *m* being the parameter. The line $x + 3 = 0$ is not included.

(*c*)   The equation is $x \cos \omega + y \sin \omega - 5 = 0$, ω being the parameter.

(*d*)   The slope of the given line is $-\frac{5}{3}$; hence the slope of the perpendicular is $\frac{3}{5}$. The equation of the family is $y = \frac{3}{5}x + c$ or $3x - 5y + 5c = 0$, *c* being the parameter. The latter is equivalent to $3x - 5y + k = 0$, with *k* as parameter.

(*e*)   Taking the *x* intercept as $a \neq 0$, the *y* intercept is $10 - a$ and the equation of the family is $\dfrac{x}{a} + \dfrac{y}{10 - a} = 1$.

   The parameter is *a*.

46.2   Describe each family of lines: (*a*) $x = k$ (*b*) $x \cos \omega + y \sin \omega - 10 = 0$ (*c*) $x/\cos \theta + y/\sin \theta = 1$ (*d*) $kx + \sqrt{1 - k^2}\,y - 10 = 0$.

(*a*)   This is the family of all vertical lines.

(*b*)   This is the family of all tangents to the circle with center at the origin and radius = 10.

(*c*)   This is the family of all lines the sum of the squares of whose intercepts is 1.

(*d*)   Since $\sqrt{1 - k^2}$ is to be assumed real, the range of $k = \cos \omega$ is $-1 \leq k \leq 1$, while the range of $\sqrt{1 - k^2} = \sin \omega$ is $0 \leq \sqrt{1 - k^2} \leq 1$. The equation is that of the family of tangents to the upper half circle of (*b*).

46.3   In each of the following write the equation of the family of lines satisfying the first condition and then obtain the equation of the line of the family satisfying the second condition: (*a*) parallel to $3x - 5y + 12 = 0$, through $P(1, -2)$ (*b*) perpendicular to $3x - 5y + 12 = 0$, through $P(1, -2)$ (*c*) through $P(-3, 2)$, at a distance 3 from the origin.

(*a*)   The equation of the family is $3x - 5y + k = 0$. Substituting $x = 1$, $y = -2$ in this equation and solving for *k*, we find $k = -13$. The required line has equation $3x - 5y - 13 = 0$.

(*b*)   The equation of the family is $5x + 3y + k = 0$ [see Problem 46.1(*d*)]. Proceeding as in (*a*) above, we find $k = 1$; the required line has equation $5x + 3y + 1 = 0$.

(*c*)   The equation of the family is $y - 2 = k(x + 3)$ or, in normal form, $\dfrac{kx - y + (3k + 2)}{\pm\sqrt{k^2 + 1}} = 0$. Setting the undirected distance of a line of the family from the origin equal to 3, we have $\left| \dfrac{3k + 2}{\pm\sqrt{k^2 + 1}} \right| = 3$. Then $\dfrac{(3k + 2)^2}{k^2 + 1} = 9$ and $k = \frac{5}{12}$. Thus, the required line has equation $y - 2 = \frac{5}{12}(x + 3)$ or $5x - 12y + 39 = 0$.

   Now there is a second line, having equation $x + 3 = 0$, satisfying the conditions but this line [see Problem 46.1(*b*)] was not included in the equation of the family.

46.4 Write the equation of the line which passes through the point of intersection of the lines $l_1: 3x + 5y + 26 = 0$ and $l_2: 7x - 11y - 13 = 0$ and satisfies the additional condition: (*a*) Passes through the origin.   (*b*) Is perpendicular to the line $7x + 3y - 8 = 0$.

Each of the required lines is a member of the family

$$l_1 + kl_2:   3x + 5y + 26 + k(7x - 11y - 13) = 0 \qquad (A)$$

(*a*) Substituting $x = 0, y = 0$ in (*A*), we find $k = 2$; the required line has equation

$$l_1 + 2l_2:   3x + 5y + 26 + 2(7x - 11y - 13) = 0 \qquad \text{or} \qquad x - y = 0$$

(*b*) The slope of the given line is $-\frac{7}{3}$ and the slope of a line of (*A*) is $-\dfrac{3 + 7k}{5 - 11k}$.

Setting one slope equal to the negative reciprocal of the other, we find $-\dfrac{7}{3} = \dfrac{5 - 11k}{3 + 7k}$ and $k = -\frac{9}{4}$. The required line has equation

$$l_1 - \tfrac{9}{4}l_2:   3x + 5y + 26 - \tfrac{9}{4}(7x - 11y - 13) = 0 \qquad \text{or} \qquad 3x - 7y - 13 = 0$$

46.5 Write the equation of the line which passes through the point of intersection of the lines $l_1: x + 4y - 18 = 0$ and $l_2: x + 2y - 2 = 0$, and satisfies the additional condition: (*a*) Is parallel to the line $3x + 8y + 1 = 0$. (*b*) Whose distance from the origin is 2.

The equation of the family of lines through the point of intersection of the given lines is

$$l_1 + kl_2:   x + 4y - 18 + k(x + 2y - 2) = 0 \quad \text{or} \quad (1 + k)x + (4 + 2k)y - (18 + 2k) = 0 \qquad (A)$$

(*a*) Since the required line is to have slope $-\frac{3}{8}$, we set $-\dfrac{1 + k}{4 + 2k} = -\dfrac{3}{8}$. Then $k = 2$ and the line has equation

$$l_1 + 2l_2:   3x + 8y - 22 = 0$$

(*b*) The normal form of (*A*) is $\dfrac{(1 + k)x + (4 + 2k)y - (18 + 2k)}{\pm\sqrt{17 + 18k + 5k^2}} = 0$. Setting $\left|\dfrac{-(18 + 2k)}{\pm\sqrt{17 + 18k + 5k^2}}\right| = 2$ and squaring both members, we find $k = \pm 4$. The required lines have equations

$$l_1 + 4l_2:   5x + 12y - 26 = 0 \qquad \text{and} \qquad l_1 - 4l_2:   3x + 4y + 10 = 0$$

Supplementary Problems

46.6 Write the equation of the family of lines satisfying the given condition. List any lines satisfying the condition but not obtained by assigning a value to the parameter.

(*a*) Perpendicular to the x axis.

(*b*) Through the point $(3, -1)$.

(*c*) At the distance 6 from the origin.

(*d*) Parallel to $2x + 5y - 8 = 0$.

(*e*) The product of whose intercepts on the coordinate axes is 10.

Ans.   (*a*) $x = k$ (*c*) $x \cos \omega + y \sin \omega - 6 = 0$ (*e*) $10x + k^2y - 10k = 0$

(*b*) $y = kx - 3k - 1, x = 3$ (*d*) $2x + 5y + k = 0$

46.7   Write the equation of the family of lines satisfying the first condition and then obtain the equation of that line of the family satisfying the second condition.

(a)   Parallel to $2x - 3y + 8 = 0$; passing through $P(2, -2)$.

(b)   Perpendicular to $2x - 3y + 8 = 0$; passing through $P(2, -2)$.

(c)   Sum of the intercepts on the coordinate axes is 2; forms with the coordinate axes a triangle of area 24.

(d)   At a distance 3 from the origin; passes through $P(1, 3)$.

(e)   Slope is $\frac{3}{2}$; product of the intercepts on the coordinate axes is -54.

(f)   The y intercept is 6; makes an angle of $135°$ with the line $7x - y - 23 = 0$.

(g)   Slope is $\frac{5}{12}$; 5 units from $(2, -3)$.

　　　　Ans.   (a) $2x - 3y - 10 = 0$　　　　　　　　　　(e) $3x - 2y \pm 18 = 0$

　　　　　　　　(b) $3x + 2y - 2 = 0$　　　　　　　　　　(f) $3x - 4y + 24 = 0$

　　　　　　　　(c) $3x - 4y - 24 = 0$,　$4x - 3y + 24 = 0$　　(g) $5x - 12y + 19 = 0$,　$5x - 12y - 111 = 0$

　　　　　　　　(d) $3x + 4y - 15 = 0$,　$y = 3$

46.8   Use the parametric method to find the equation of the line which passes through the point $P(3, -4)$ and has the additional property:

(a)   The sum of its intercepts on the coordinate axes is -5.

(b)   The product of its intercepts on the coordinate axes is the negative reciprocal of its slope.

(c)   It forms with the coordinate axes a triangle of area 1.

　　　　Ans.   (a) $2x - y - 10 = 0$,　$2x + 3y + 6 = 0$　　(b) $x + y + 1 = 0$,　$5x + 3y - 3 = 0$

　　　　　　　　(c) $2x + y - 2 = 0$,　$8x + 9y + 12 = 0$

46.9   Write the equation of the line which passes through the point of intersection of the lines $l_1: x - 2y - 4 = 0$ and $l_2: 4x - y - 4 = 0$ and satisfies the additional condition:

(a)   Passes through the origin

(b)   Passes through the point $P(4, -6)$

(c)   Is parallel to the line $16x - 11y + 3 = 0$

(d)   Is perpendicular to the line $9x + 22y - 8 = 0$

(e)   Is $\frac{12}{7}$ units from the origin

(f)   Makes an angle of $45°$ with the line $9x - 5y - 12 = 0$

(g)   Its y intercept is $-\frac{20}{11}$ its slope

(h)   Makes a triangle of area $\frac{12}{5}$ with the coordinate axes

　　　　Ans.   (a) $3x + y = 0$　　　　　　　　　　　(e) $7y + 12 = 0$,　$21x - 28y - 60 = 0$

　　　　　　　　(b) $5x + 4y + 4 = 0$　　　　　　　　　(f) $49x + 14y - 4 = 0$

　　　　　　　　(c) $16x - 11y - 28 = 0$　　　　　　　　(g) $11x - 8y - 20 = 0$

　　　　　　　　(d) $22x - 9y - 28 = 0$　　　　　　　　(h) $6x - 5y - 12 = 0$,　$15x - 2y - 12 = 0$,

　　　　　　　　　　　　　　　　　　　　　　　　　　　　　　$12(12 \mp \sqrt{89})x + (41 \pm 3\sqrt{89})y - 12(1 \mp \sqrt{89}) = 0$

Chapter 47

The Circle

A CIRCLE IS THE LOCUS of a point which moves in a plane so that it is always at a constant distance from a fixed point in the plane. The fixed point is called the *center* and the constant distance is the length of the *radius* of the circle.

THE STANDARD FORM of the equation of the circle whose center is at the point $C(h, k)$ and whose radius is the constant r is

$$(x - h)^2 + (y - k)^2 = r^2 \qquad (47.1)$$

See Fig. 47-1. (See Problem 47.1.)

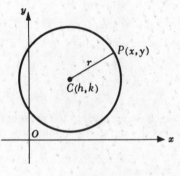

Fig. 47-1

THE GENERAL FORM of the equation of a circle is

$$x^2 + y^2 + 2Dx + 2Ey + F = 0 \qquad (47.2)$$

By completing the squares this may be put in the form

$$(x + D)^2 + (x + E)^2 = D^2 + E^2 - F$$

Thus, (47.2) represents a circle with center at $C(-D, -E)$ and radius $\sqrt{D^2 + E^2 - F}$ if $(D^2 + E^2 - F) > 0$, a point if $(D^2 + E^2 - F) = 0$, and an imaginary locus if $(D^2 + E^2 - F) < 0$. (See Problems 47.2–47.3.)

IN BOTH THE STANDARD AND GENERAL FORM the equation of a circle contains three independent arbitrary constants. It follows that a circle is uniquely determined by three independent conditions. (See Problems 47.4–47.6.)

THE EQUATION OF A TANGENT to a circle may be found by making use of the fact that a tangent and the radius drawn to the point of tangency are perpendicular lines. (See Problem 47.7.)

322

THE LENGTH OF A TANGENT to a circle from an external point P_1 is defined as the undirected distance from the point P_1 to the point of tangency. The two tangents from an external point are of equal length. See Fig. 47-2.

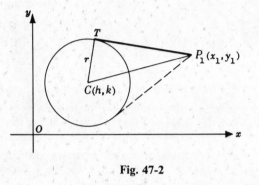

Fig. 47-2

The square of the length of a tangent from the external point $P_1(x_1, y_1)$ to a circle is obtained by substituting the coordinates of the point in the left member of the equation of the circle when written in the form $(x - h)^2 + (y - k)^2 - r^2 = 0$ or $x^2 + y^2 + 2Dx + 2Ey + F = 0$. (See Problems 47.8–47.9.)

THE EQUATION

$$x^2 + y^2 + 2D_1x + 2E_1y + F_1 + k(x^2 + y^2 + 2D_2x + 2E_2y + F_2) = 0 \qquad (47.3)$$

where

$$K_1: \quad x^2 + y^2 + 2D_1x + 2E_1y + F_1 = 0 \qquad \text{and} \qquad K_2: \quad x^2 + y^2 + 2D_2x + 2E_2y + F_2 = 0$$

are distinct circles and $k \neq -1$ is a parameter, represents a one-parameter family of circles.

If K_1 and K_2 are concentric, the circles of (47.3) are concentric with them.

If K_1 and K_2 are not concentric, the circles of (47.3) have a common line of centers with them and the centers of the circles (47.3) divide the segment joining the centers of K_1 and K_2 in the ratio $k : 1$.

If K_1 and K_2 intersect in two distinct points P_1 and P_2, (47.3) consists of all circles except K_2 which pass through these points. If K_1 and K_2 are tangent to each other at the point P_1, (47.3) consists of all circles except K_2 which are tangent to each other at P_1. If K_1 and K_2 have no point in common, any two circles of the family (47.3) have no point in common with each other. See Fig. 47-3. (See Problems 47.11–47.13.)

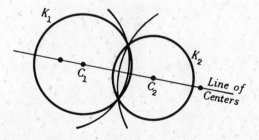

Fig. 47-3

Solved Problems

47.1   Write the equation of the circle satisfying the given conditions.

(a) $C(0, 0)$, $r = 5$ (d) $C(-5, 6)$ and tangent to the x axis

(b) $C(4, -2)$, $r = 8$ (e) $C(3, 4)$ and tangent to $2x - y + 5 = 0$

(c) $C(-4, -2)$ and passing through $P(1, 3)$ (f)   Center on $y = x$, tangent to both axes, $r = 4$

(a)   Using $(x - h)^2 + (y - k)^2 = r^2$, the equation is $(x - 0)^2 + (y - 0)^2 = 25$ or $x^2 + y^2 = 25$.

(b)   Using $(x - h)^2 + (y - k)^2 = r^2$, the equation is $(x - 4)^2 + (y + 2)^2 = 64$.

(c)   Since the center is at $C(-4, -2)$, the equation has the form $(x + 4)^2 + (y + 2)^2 = r^2$. The condition that $P(1, 3)$ lie on this circle is $(1 + 4)^2 + (3 + 2)^2 = r^2 = 50$. Hence the required equation is $(x + 4)^2 + (y + 2)^2 = 50$.

(d)   The tangent to a circle is perpendicular to the radius drawn to the point of tangency; hence $r = 6$. The equation of the circle is $(x + 5)^2 + (y - 6)^2 = 36$. See Fig. 47-4(a).

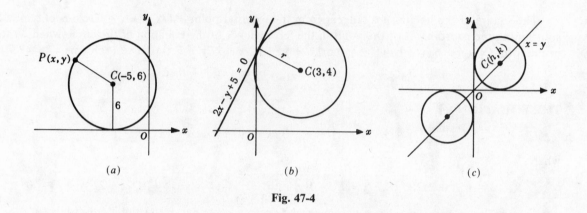

(a)   (b)   (c)

Fig. 47-4

(e)   The radius is the undirected distance of the point $C(3, 4)$ from the line $2x - y + 5 = 0$; thus $r = \left| \dfrac{2 \cdot 3 - 4 + 5}{-\sqrt{5}} \right| = \dfrac{7}{\sqrt{5}}$. The equation of the circle is $(x - 3)^2 + (y - 4)^2 = \frac{49}{5}$. See Fig. 47-4(b).

(f)   Since the center (h, k) lies on the line $x = y$, $h = k$; since the circle is tangent to both axes, $|h| = |k| = r$. Thus there are two circles satisfying the conditions, one with center $(4, 4)$ and equation $(x - 4)^2 + (y - 4)^2 = 16$, the other with center $(-4, -4)$ and equation $(x + 4)^2 + (y + 4)^2 = 16$. See Fig. 47-4(c).

47.2   Describe the locus represented by each of the following equations:

(a) $x^2 + y^2 - 10x + 8y + 5 = 0$ (c) $x^2 + y^2 + 4x - 6y + 24 = 0$

(b) $x^2 + y^2 - 6x - 8y + 25 = 0$ (d) $4x^2 + 4y^2 + 80x + 12y + 265 = 0$

(a)   From the standard form $(x - 5)^2 + (y + 4)^2 = 36$, the locus is a circle with center at $C(5, -4)$ and radius 6.

(b)   From the standard form $(x - 3)^2 + (y - 4)^2 = 0$, the locus is a point circle or the point $(3, 4)$.

(c)   Here we have $(x + 2)^2 + (y - 3)^2 = -11$; the locus is imaginary.

(d)   Dividing by 4, we have $x^2 + y^2 + 20x + 3y + \frac{265}{4} = 0$. From $(x + 10)^2 + (y + \frac{3}{2})^2 = 36$, the locus is a circle with center at $C(-10, -\frac{3}{2})$ and radius 6.

47.3 Show that the circles $x^2 + y^2 - 16x - 20y + 115 = 0$ and $x^2 + y^2 + 8x - 10y + 5 = 0$ are tangent and find the point of tangency.

The first circle has center $C_1(8, 10)$ and radius 7; the second has center $C_2(-4, 5)$ and radius 6. The two circles are tangent externally since the distance between their centers $C_1C_2 = \sqrt{144 + 25} = 13$ is equal to the *sum* of the radii.

The point of tangency $P(x, y)$ divides the segment C_2C_1 in the ratio $6:7$. Then

$$x = \frac{6 \cdot 8 + 7(-4)}{6 + 7} = \frac{20}{13}, \qquad y = \frac{6 \cdot 10 + 7 \cdot 5}{6 + 7} = \frac{95}{13}$$

and the point of tangency has coordinates $(\frac{20}{13}, \frac{95}{13})$.

47.4 Find the equation of the circle through the points $(5, 1)$, $(4, 6)$, and $(2, -2)$.

Take the equation in general form $x^2 + y^2 + 2Dx + 2Ey + F = 0$. Substituting successively the coordinates of the given points, we have

$$\begin{cases} 25 + 1 + 10D + 2E + F = 0 \\ 16 + 36 + 8D + 12E + F = 0 \\ 4 + 4 + 4D - 4E + F = 0 \end{cases} \quad \text{or} \quad \begin{cases} 10D + 2E + F = -26 \\ 8D + 12E + F = -52 \\ 4D - 4E + F = -8 \end{cases}$$

with solution $D = -\frac{1}{3}$, $E = -\frac{8}{3}$, $F = -\frac{52}{3}$. Thus the required equation is

$$x^2 + y^2 - \frac{2}{3}x - \frac{16}{3}y - \frac{52}{3} = 0 \quad \text{or} \quad 3x^2 + 3y^2 - 2x - 16y - 52 = 0$$

47.5 Write the equations of the circle having radius $\sqrt{13}$ and tangent to the line $2x - 3y + 1 = 0$ at $(1, 1)$.

Let the equation of the circle be $(x - h)^2 + (y - k)^2 = 13$. Since the coordinates $(1, 1)$ satisfy this equation, we have

$$(1 - h)^2 + (1 - k)^2 = 13 \tag{1}$$

The undirected distance from the tangent to the center of the circle is equal to the radius, that is,

$$\left| \frac{2h - 3k + 1}{-\sqrt{13}} \right| = \sqrt{13} \quad \text{and} \quad \frac{2h - 3k + 1}{\sqrt{13}} = \pm\sqrt{13} \tag{2}$$

Finally, the radius through $(1, 1)$ is perpendicular to the tangent there, that is,

$$\text{Slope of radius through } (1, 1) = -\frac{1}{\text{slope of given line}} \quad \text{or} \quad \frac{k - 1}{h - 1} = -\frac{3}{2} \tag{3}$$

Since there are only two unknowns, we may solve simultaneously any two of the three equations. Using (2) and (3), noting that there are two equations in (2), we find $h = 3$, $k = -2$ and $h = -1$, $k = 4$. The equations of the circles are $(x - 3)^2 + (y + 2)^2 = 13$ and $(x + 1)^2 + (y - 4)^2 = 13$.

47.6 Write the equations of the circles satisfying the following sets of conditions:

(a) Through $(2, 3)$ and $(-1, 6)$, with center on $2x + 5y + 1 = 0$.

(b) Tangent to $5x - y - 17 = 0$ at $(4, 3)$ and also tangent to $x - 5y - 5 = 0$.

(c) Tangent to $x - 2y + 2 = 0$ and to $2x - y - 17 = 0$, and passing through $(6, -1)$.

(d) With center in the first quadrant and tangent to lines $y = 0$, $5x - 12y = 0$, $12x + 5y - 39 = 0$.

Take the equation of the circle in standard form $(x - h)^2 + (y - k)^2 = r^2$.

(a) We obtain the following system of equations:

$$2h + 5k + 1 = 0 \qquad [\text{center } (h, k) \text{ on } 2x + 5y + 1 = 0] \tag{1}$$
$$(2 - h)^2 + (3 - k)^2 = r^2 \qquad [\text{point } (2, 3) \text{ on the circle}] \tag{2}$$
$$(-1 - h)^2 + (6 - k)^2 = r^2 \qquad [\text{point } (-1, 6) \text{ on the circle}] \tag{3}$$

The elimination of r between (2) and (3) yields $h - k + 4 = 0$ and when this is solved simultaneously with (1), we obtain $h = -3$, $k = 1$. By (2), $r^2 = (2 + 3)^2 + (3 - 1)^2 = 29$; the equation of the circle is $(x + 3)^2 + (y - 1)^2 = 29$.

(b) We obtain the following system of equations:

$$(4 - h)^2 + (3 - k)^2 = r^2 \qquad \text{[point (4, 3) on the circle]} \tag{1}$$

$$\left(\frac{5h - k - 17}{\sqrt{26}}\right)^2 = \left(\frac{h - 5k - 5}{\sqrt{26}}\right)^2 \qquad \begin{array}{l}\text{[the square of the directed distance from}\\ \text{each tangent to the center is } r^2]\end{array} \tag{2}$$

$$\frac{k - 3}{h - 4} = -\frac{1}{5} \qquad \begin{array}{l}\text{[the radius drawn to (4, 3) is perpendicular}\\ \text{to the tangent there]}\end{array} \tag{3}$$

The elimination of h between (2) and (3) yields

$$\left(\frac{-26k + 78}{\sqrt{26}}\right)^2 - \left(\frac{-10k + 14}{\sqrt{26}}\right)^2$$

or $9k^2 - 59k + 92 = (k - 4)(9k - 23) = 0$; then $k = 4, \frac{23}{9}$.

When $k = 4$, $h = -5k + 19 = -1$; then $r^2 = (4 + 1)^2 + (3 - 4)^2 = 26$ and the equation of the circle is $(x + 1)^2 + (y - 4)^2 = 26$. When $k = \frac{23}{9}$, $h = \frac{56}{9}$ and $r^2 = \frac{416}{81}$; the equation of the circle is $(x - \frac{56}{9})^2 + (y - \frac{23}{9})^2 = \frac{416}{81}$.

(c) Observing the directions indicated in Fig. 47-5(a), we obtain for each circle the following system of equations:

$$-\frac{h - 2k + 2}{-\sqrt{5}} = r \qquad (x - 2y + 2 = 0 \text{ is tangent to the circle}) \tag{1}$$

$$-\frac{2h - k - 17}{\sqrt{5}} = r \qquad (2x - y - 17 = 0 \text{ is tangent to the circle}) \tag{2}$$

$$(6 - h)^2 + (-1 - k)^2 = r^2 \qquad \text{[point (6, -1) is on the circle]} \tag{3}$$

The elimination of r between (1) and (2) yields

$$h - k - 5 = 0 \tag{4}$$

and the elimination of r between (1) and (3) yields

$$(6 - h)^2 + (-1 - k)^2 = \frac{(h - 2k + 2)^2}{5} \tag{5}$$

Eliminating h between (4) and (5), we have $(1 - k)^2 + (-1 - k)^2 = (-k + 7)^2/5$ or $9k^2 + 14k - 39 = (k + 3)(9k - 13) = 0$ and $k = -3, \frac{13}{9}$.

When $k = -3$, $h = k + 5 = 2$, $r^2 = (6 - 2)^2 + [-1 - (-3)]^2 = 20$ and the equation of the circle is $(x - 2)^2 + (y + 3)^2 = 20$. When $k = \frac{13}{9}$, $h = k + 5 = \frac{58}{9}$, $r^2 = (6 - \frac{58}{9})^2 + (-1 - \frac{13}{9})^2 = \frac{500}{81}$ and the circle has equation $(x - \frac{58}{9})^2 + (y - \frac{13}{9})^2 = \frac{500}{81}$.

(a)

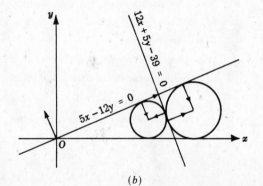

(b)

Fig. 47-5

(d) Observing the directions indicated in Fig. 47-5(b), we have for the inscribed circle the following system of equations:

$$k = r, \qquad -\frac{5h - 12k}{-13} = r \quad \text{and} \quad -\frac{12h + 5k - 39}{13} = r$$

Then $h = \frac{5}{2}$, $k = r = \frac{1}{2}$ and the equation of the circle is $(x - \frac{5}{2})^2 + (y - \frac{1}{2})^2 = \frac{1}{4}$.
For the other circle, we have the following system of equations:

$$k = r, \qquad -\frac{5h - 12k}{-13} = r \quad \text{and} \quad \frac{12h + 5k - 39}{13} = r$$

Here $h = \frac{15}{4}$, $k = r = \frac{3}{4}$, and the circle has equation $(x - \frac{15}{4})^2 + (y - \frac{3}{4})^2 = \frac{9}{16}$.

47.7 Find the equations of all tangents to the circle $x^2 + y^2 - 2x + 8y - 23 = 0$:

(a) at the point $(3, -10)$ on it   (b) having slope 3   (c) through the external point $(8, -3)$.

(a) The center of the given circle is $C(1, -4)$. Since the tangent at any point P on the circle is perpendicular to the radius through P, the slope of the tangent is $\frac{1}{3}$. The equation of the tangent is then $y + 10 = \frac{1}{3}(x - 3)$ or $x - 3y - 33 = 0$.

(b) Each tangent belongs to the family of lines $y = 3x + b$ or, in normal form, $\dfrac{3x - y + b}{\pm\sqrt{10}} = 0$.

The undirected distance from any tangent to the center of the circle is equal to the radius of the circle; thus

$$\left|\frac{3(1) - (-4) + b}{\pm\sqrt{10}}\right| = 2\sqrt{10} \quad \text{or} \quad \frac{7 + b}{\sqrt{10}} = \pm 2\sqrt{10} \quad \text{and} \quad b = 13, -27$$

There are then two tangents having equations $y = 3x + 13$ and $y = 3x - 27$.

(c) The tangents to a circle through the external point $(8, -3)$ belong to the family of lines $y + 3 = m(x - 8)$ or $y = mx - 8m - 3$. When this replacement for y is made in the equation of the circle, we have

$$x^2 + (mx - 8m - 3)^2 - 2x + 8(mx - 8m - 3) - 23 = (m^2 + 1)x^2$$
$$+ (-16m^2 + 2m - 2)x + (64m^2 - 16m - 38) = 0$$

Now this equation will have equal roots x provided the discriminant is zero, that is, provided $(-16m^2 + 2m - 2)^2 - 4(m^2 + 1)(64m^2 - 16m - 38) = -4(m - 3)(9m + 13) = 0$. Then $m = 3, -\frac{13}{9}$ and the equations of the tangents are $y + 3 = 3(x - 8)$ or $3x - y - 27 = 0$ and $y + 3 = -\frac{13}{9}(x - 8)$ or $13x + 9y - 77 = 0$.

47.8 Prove: The length t of the tangent from the point $P_1(x_1, y_1)$ to the circle $x^2 + y^2 + 2Dx + 2Ey + F = 0$, with center $C(-D, -E)$, is given by $t = \sqrt{x_1^2 + y_1^2 + 2Dx_1 + 2Ey_1 + F}$.

In Fig. 47-6, $(CP_1)^2 = (x_1 + D)^2 + (y_1 + E)^2$ and $(CR)^2 = D^2 + E^2 - F$.

Then $\qquad\qquad t^2 = (RP_1)^2 = (CP_1)^2 - (CR)^2 = x_1^2 + y_1^2 + 2Dx_1 + 2Ey_1 + F$

and $\qquad\qquad\qquad\quad t = \sqrt{x_1^2 + y_1^2 + 2Dx_1 + 2Ey_1 + F}$.

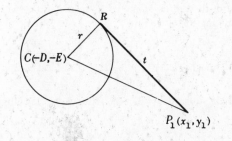

Fig. 47-6

47.9   Find the length of the tangent
   (a)   To the circle $x^2 + y^2 - 2x + 8y - 23 = 0$ from the point $(8, -3)$
   (b)   To the circle $4x^2 + 4y^2 - 2x + 5y - 8 = 0$ from the point $(-4, 4)$

   Denote the required length by t.

   (a)   Substituting the coordinates $(8, -3)$ in the left member of the equation of the circle, we have
   $$t^2 = (8)^2 + (-3)^2 - 2(8) + 8(-3) - 23 = 10 \qquad \text{and} \qquad t = \sqrt{10}$$

   (b)   From the general form of the equation $x^2 + y^2 - \tfrac{1}{2}x + \tfrac{5}{4}y - 2 = 0$, we find
   $$t^2 = (-4)^2 + (4)^2 - \tfrac{1}{2}(-4) + \tfrac{5}{4}(4) - 2 = 37 \qquad \text{and} \qquad t = \sqrt{37}$$

47.10   For the circle $x^2 + y^2 + 6x - 8y = 0$, find the values of m for which the lines of the family $y = mx - \tfrac{1}{3}$
   (a) intersect the circle in two distinct points   (b) are tangent to the circle   (c) do not meet the circle.

   Eliminating y between the two equations, we have
   $$x^2 + (mx - \tfrac{1}{3})^2 + 6x - 8(mx - \tfrac{1}{3}) = (m^2 + 1)x^2 + (6 - \tfrac{26}{3}m)x + \tfrac{25}{9} = 0$$

   This equation will have two real and distinct roots, two equal roots, or two imaginary roots according as its discriminant
   $$(6 - \tfrac{26}{3}m)^2 - 4(m^2 + 1)(\tfrac{25}{9}) = \tfrac{8}{9}(72m^2 - 117m + 28) = \tfrac{8}{9}(3m - 4)(24m - 7) >, =, \text{or} < 0$$

   (a)   The lines will intersect the circle in two distinct points when $m > \tfrac{4}{3}$ and $m < \tfrac{7}{24}$.
   (b)   The lines will be tangent to the circle when $m = \tfrac{4}{3}$ and $m = \tfrac{7}{24}$.
   (c)   The lines will not meet the circle when $\tfrac{7}{24} < m < \tfrac{4}{3}$.

47.11   Write the equation of the family of circles satisfying the given conditions: (a) having the common center $(-2, 3)$ (b) having radius $= 5$ (c) with center on the x axis.

   (a)   The equation is $(x + 2)^2 + (y - 3)^2 = r^2$, r being a parameter.
   (b)   The equation is $(x - h)^2 + (y - k)^2 = 25$, h and k being parameters.
   (c)   Let the center have coordinates $(k_1, 0)$ and denote the radius by k_2. The equation of the family is $(x - k_1)^2 + y^2 = k_2^2$, k_1 and k_2 being parameters.

47.12   Given the circles $K_1: x^2 + y^2 - 6x + 8y - 16 = 0$ and $K_2: x^2 + y^2 + 4x - 2y - 8 = 0$, write the equation of the family $K_1 + kK_2$ and determine the circle of the family having the further property: (a) passes through $P(1, -3)$ (b) tangent to the line $7x - 11y - 40 = 0$.

   The equation of the family is
   $$K_1 + kK_2: \quad x^2 + y^2 - 6x + 8y - 16 + k(x^2 + y^2 + 4x - 2y - 8) = 0 \tag{1}$$
   or
   $$(1 + k)x^2 + (1 + k)y^2 - (6 - 4k)x + (8 - 2k)y - (16 + 8k) = 0$$

   (a)   Substituting the coordinates of P in (1), we find $k = 3$. The circle of the family which passes through $(-1, 3)$ has equation $K_1 + 3K_2: 2x^2 + 2y^2 + 3x + y - 20 = 0$.

   (b)   Equating the square of the radius of $K_1 + kK_2$ to the square of the distance from the given tangent to the center and solving for k, we find
   $$\left(-\frac{3 - 2k}{1 + k}\right)^2 + \left(\frac{4 - k}{1 + k}\right)^2 + \frac{16 + 8k}{1 + k} = \left[\frac{7\left(\dfrac{3 - 2k}{1 + k}\right) - 11\left(-\dfrac{4 - k}{1 + k}\right) - 40}{\sqrt{170}}\right]^2$$

   or $403k^2 - 786k - 1269 = 0$ and $k = 3, -\tfrac{423}{403}$. The two circles tangent to the given line have equations
   $$K_1 + 3K_2: \quad 2x^2 + 2y^2 + 3x + y - 20 = 0$$
   and
   $$K_1 - \tfrac{423}{403}K_2: \quad 10x^2 + 10y^2 + 2055x - 2035y + 1532 = 0$$

47.13 Write the equation of the family of circles which are tangent to the circles of Problem 47.3 at their common point and determine the circle of the family having the property (*a*) center on $x + 4y + 16 = 0$ (*b*) radius is $\frac{1}{2}$.

The equation of the family is

$$K_1 + kK_2: \quad x^2 + y^2 - 16x - 20y + 115 + k(x^2 + y^2 + 8x - 10y + 5) = 0$$

(*a*) The centers of K_1 and K_2 are $(8, 10)$ and $(-4, 5)$, respectively; the equation of their line of centers is $5x - 12y + 80 = 0$. This line meets $x + 4y + 16 = 0$ in the point $(-16, 0)$, the required center.
Now $(-16, 0)$ divides the segment joining $(8, 10)$ and $(-4, 5)$ in the ratio $k : 1$; thus,

$$\frac{5k + 10}{k + 1} = 0 \quad \text{and} \quad k = -2$$

The equation of the circle is $K_1 - 2K_2 : x^2 + y^2 + 32x - 105 = 0$.

(*b*) Setting the square of the radius of $K_1 + kK_2$ equal to $(\frac{1}{2})^2$ and solving for k, we find

$$\left(\frac{4k - 8}{1 + k}\right)^2 + \left(-\frac{10 + 5k}{1 + k}\right)^2 - \frac{115 + 5k}{1 + k} = \frac{1}{4} \quad \text{and} \quad k = 1, \tfrac{15}{11}$$

The required circles have equations

$$K_1 + K_2 : x^2 + y^2 - 4x - 15y + 60 = 0 \quad \text{and} \quad K_1 + \tfrac{15}{11}K_2: \quad 13x^2 + 13y^2 - 28x - 185y + 670 = 0$$

47.14 For the intersecting circles $K_1: x^2 + y^2 - 6x - 8y = 0$ and $K_2: x^2 + y^2 - 4x + 6y - 12 = 0$, (*a*) find the equation of the radical axis (*b*) show that the radical axis is perpendicular to the line of centers (*c*) find the points of intersection of the circles.

(*a*) The radical axis has equation $K_1 - K_2 : x + 7y - 6 = 0$.

(*b*) The slope of the line of centers, joining $(3, 4)$ and $(2, -3)$, is 7; the slope of the radical axis is $-\frac{1}{7}$. These lines are mutually perpendicular.

(*c*) Solving simultaneously the equation of the radical axis and either of the equations of the circles, the points of intersection are $(6, 0)$ and $(-1, 1)$.

47.15 For each pair of circles

(*a*) $\begin{aligned} K_1: &\ x^2 + y^2 - 8x - 6y = 0 \\ K_2: &\ 4x^2 + 4y^2 - 10x - 10y - 13 = 0 \end{aligned}$ (*b*) $\begin{aligned} K_1: &\ x^2 + y^2 - 12x - 16y - 125 = 0 \\ K_2: &\ 3x^2 + 3y^2 - 60x - 16y + 113 = 0 \end{aligned}$

find the equation of the radical axis. Without finding the coordinates of their points of intersection, show that the circles (*a*) intersect in two distinct points while those of (*b*) are tangent internally.

(*a*) The equation of the radical axis is $K_1 - \frac{1}{4}K_2 : 22x + 14y - 13 = 0$. The undirected distance from the radical axis to the center $(4, 3)$ of K_1

$$\left| \frac{22(4) + 14(3) - 13}{2\sqrt{170}} \right| = \frac{117\sqrt{170}}{340}$$

is less than 5, the radius of K_1. Hence the radical axis and the circle K_2 intersect K_1 in two distinct points.

(*b*) The equation of the radical axis is $K_1 - \frac{1}{3}K_2 : 3x - 4y - 61 = 0$. The undirected distance from the radical axis to the center $(6, 8)$ of K_1

$$\left| \frac{3(6) - 4(8) - 61}{5} \right| = 15$$

is equal to the radius of K_1 and the circles are tangent to each other.
Since the distance between the centers of K_1 and K_2 $\sqrt{(10 - 6)^2 + (\tfrac{8}{3} - 8)^2} = \tfrac{20}{3}$ is equal to the difference between their radii, the circles are tangent internally.

Supplementary Problems

47.16 Write the equation of each of the following circles: (a) $C(0,0)$, radius 7 (b) $C(-4,8)$, radius 3 (c) $C(5,-4)$, through $(0,0)$ (d) $C(-4,-3)$, through $(2,1)$ (e) $C(-2,5)$, tangent to x axis (f) $C(-2,-5)$, tangent to $2x - y + 3 = 0$ (g) tangent to both axes, radius 5 (h) circumscribed about the right triangle whose vertices are $(3,4)$, $(-1,-4)$, $(5,-2)$ (i) circumscribed about the triangle of Problem 45.8.

> *Ans.* (a) $x^2 + y^2 = 49$
>
> (b) $x^2 + y^2 + 8x - 16y + 71 = 0$
>
> (c) $x^2 + y^2 - 10x + 8y = 0$
>
> (d) $x^2 + y^2 + 8x + 6y - 27 = 0$
>
> (e) $x^2 + y^2 + 4x - 10y + 4 = 0$
>
> (f) $5x^2 + 5y^2 + 20x + 50y + 129 = 0$
>
> (g) $x^2 + y^2 \pm 10x \pm 10y + 25 = 0$, $x^2 + y^2 \mp 10x \pm 10y + 25 = 0$
>
> (h) $x^2 + y^2 - 2x - 19 = 0$
>
> (i) $56x^2 + 56y^2 - 260x - y - 5451 = 0$

47.17 Find the center and radius of each of the circles.

(a) $x^2 + y^2 - 6x + 8y - 11 = 0$ *Ans.* $C(3,-4)$, $r = 6$

(b) $x^2 + y^2 - 4x - 6y - \frac{10}{3} = 0$ *Ans.* $C(2,3)$, $r = 7\sqrt{3}/3$

(c) $7x^2 + 7y^2 + 14x - 56y - 25 = 0$ *Ans.* $C(-1,4)$, $r = 12\sqrt{7}/7$

47.18 Explain why any line passing through $(4,-1)$ cannot be tangent to the circle $x^2 + y^2 - 4x + 6y - 12 = 0$.

47.19 (a) Show that the circles $x^2 + y^2 + 6x - 2y - 54 = 0$ and $x^2 + y^2 - 22x - 8y + 112 = 0$ do not intersect.

(b) Show that the circles $x^2 + y^2 + 2x - 6y + 9 = 0$ and $x^2 + y^2 + 8x - 6y + 9 = 0$ are tangent internally.

47.20 The equation of a given circle is $x^2 + y^2 = 36$. Find (a) the length of the chord which lies along the line $3x + 4y - 15 = 0$ (b) the equation of the chord whose midpoint is $(3,2)$.

Hint: In (a) draw the normal to the given line.   *Ans.* (a) $6\sqrt{3}$ (b) $3x + 2y - 13 = 0$

47.21 Find the equation of each circle satisfying the given conditions.

(a) Through $(6,0)$ and $(-2,-4)$, tangent to $4x + 3y - 25 = 0$.

(b) Tangent to $3x - 4y + 5 = 0$ at $(1,2)$, radius 5.

(c) Tangent to $x - 2y - 4 = 0$ and $2x - y - 6 = 0$, passes through $(-1,2)$.

(d) Tangent to $2x - 3y - 7 = 0$ at $(2,-1)$; passes through $(4,1)$.

(e) Tangent to $3x + y + 3 = 0$ at $(-3,6)$, tangent to $x + 3y - 7 = 0$.

> *Ans.* (a) $(x-3)^2 + (y+4)^2 = 25$, $(x - \frac{213}{121})^2 + (y + \frac{184}{121})^2 = 297\,025/14\,641$
>
> (b) $(x-4)^2 + (y+2)^2 = 25$, $(x+2)^2 + (y-6)^2 = 25$
>
> (c) $x^2 + y^2 - 2x - 2y - 3 = 0$, $x^2 + y^2 + 118x - 122y + 357 = 0$
>
> (d) $x^2 + y^2 + 4x - 10y - 23 = 0$
>
> (e) $x^2 + y^2 - 6x - 16y + 33 = 0$, $x^2 + y^2 + 9x - 11y + 48 = 0$

47.22 Find the equation of the tangent to the given circle at the given point on it.

   (a) $x^2 + y^2 = 169$, $(5, -12)$ (b) $x^2 + y^2 - 4x + 6y - 37 = 0$, $(3, 4)$

   Ans. (a) $5x - 12y - 169 = 0$ (b) $x + 7y - 31 = 0$

47.23 Find the equations of the tangent to each circle through the given external point.

   (a) $x^2 + y^2 = 25$, $(7, 1)$ (b) $x^2 + y^2 - 4x + 2y - 31 = 0$, $(-1, 5)$

   Ans. (a) $3x + 4y - 25 = 0, 4x - 3y - 25 = 0$ (b) $y - 5 = 0, 4x - 3y + 19 = 0$

47.24 Show that the circles $x^2 + y^2 + 4x - 6y = 0$ and $x^2 + y^2 + 6x + 4y = 0$ are *orthogonal*, that is, that the tangents to the two circles at a point of intersection are mutually perpendicular. Also, that the square of the distance between the centers of the circles is equal to the sum of the squares of the radii.

47.25 Determine the equation of the circle of the family of Problem 47.12:   (a) which has its center on the line $4x + 3y - 5 = 0$ (b) whose radius is $\sqrt{85}$.

   Ans. (a) $x^2 + y^2 - 16x + 18y - 24 = 0$,

   (b) $x^2 + y^2 + 14x - 12y = 0, 25x^2 + 25y^2 - 260x + 310y - 488 = 0$

47.26 Determine the equation of the circle of the family of Problem 47.13:   (a) which passes through the point $(0, 3)$ (b) which is tangent to the line $4x + 3y - 25 = 0$.

   Ans. (a) $5x^2 + 5y^2 + 16x - 60y + 135 = 0$

   (b) $x^2 + y^2 - 40x - 30y + 225 = 0, 8x^2 + 8y^2 - 20x - 115y + 425 = 0$

47.27 Prove: The radical axis of two distinct nonconcentric circles is perpendicular to their line of centers.

47.28 Find the equation of the locus of a point from which the tangents to the circle $x^2 + y^2 = 4$ are twice as long as the tangents to $x^2 + y^2 - 4x = 0$.

   Ans. $3x^2 + 3y^2 - 16x + 4 = 0$

Chapter 48

More Locus Problems

A PLANE CURVE IS THE LOCUS of all those points in the plane, and only those points, which satisfy a given geometric condition. To obtain the equation of the locus:

    (*1*)   Let $P(x, y)$ be any point on the locus.
    (*2*)   Express the given geometric condition by means of an equation in x and y.
    (*3*)   Simplify the equation obtained in (*2*).

EXAMPLE 1. Find the equation of the locus of a point which moves in the plane so that the sum of the squares of its distances from the two lines $l_1: 7x - 4y - 10 = 0$ and $l_2: 4x + 7y + 5 = 0$ is always equal to 3.

Let $P(x, y)$ be any point on the locus and drop the perpendiculars $\overline{AP}$ and $\overline{BP}$ to l_1 and l_2, respectively; then P must satisfy the condition $(AP)^2 + (BP)^2 = 3$. We obtain

$$\left(\frac{7x - 4y - 10}{\sqrt{65}}\right)^2 + \left(\frac{4x + 7y + 5}{-\sqrt{65}}\right)^2 = 3$$

or

$$13x^2 + 13y^2 - 20x + 30y - 14 = 0$$

a circle with center at $\left(\frac{10}{13}, -\frac{15}{13}\right)$ and radius $\sqrt{3}$. See Fig. 48-1. (See Problems 48.1–48.7.)

Fig. 48-1

THE PARAMETRIC METHOD. In expressing analytically the geometric condition which defines a locus, it is often more convenient to introduce one or more auxiliary variables or parameters in addition to the coordinates of the tracing point P. Then between the relations set up, the parameters are to be eliminated. If n parameters are introduced, $(n + 1)$ relations must be found in order to be able to effect their elimination. When parameters are so used to obtain the equation of the locus, we say that it has been obtained by the *parametric method*.

EXAMPLE 2. Find the equation of the locus of the midpoints of the ordinates (that is, the midpoints of the vertical line segments drawn from the x axis to points on the curve) of the circle $x^2 + y^2 = r^2$. See Fig. 48-2.

Let the vertical line drawn through $A(a, 0)$ on the x axis meet the circle in the point $B(a, b)$. If $P(x, y)$ is the tracing point of the required locus, its position on this line is at the midpoint of $\overline{AB}$. Then

$$(1)\quad x = a \qquad \text{and} \qquad (2)\quad y = \tfrac{1}{2}b$$

The third relation needed

$$(3)\quad a^2 + b^2 = r^2$$

expresses the condition that $B(a, b)$ be on the given circle.

332

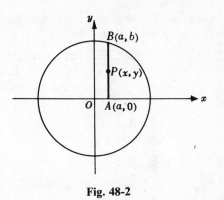

Fig. 48-2

To eliminate the parameters, we solve (1) and (2) for $a = x$ and $b = 2y$, respectively, and substitute in (3) to obtain $x^2 + 4y^2 = r^2$, the desired equation. (See Problems 48.8–48.9.)

Solved Problems

The following refer to loci in the plane.

48.1   Find the equation of the locus of a point which moves so that the slope of the line joining it to $A(3, -2)$ is twice that of the line joining it to the point $B(5, 8)$.

Let $P(x, y)$ be any point on the locus. Referring to Fig. 48-3,

$$\text{Slope of } \overleftrightarrow{AP} = 2(\text{slope of } \overleftrightarrow{BP}) \quad \text{ or } \quad \frac{y + 2}{x - 3} = 2\frac{y - 8}{x - 5}$$

The required equation is $xy - 18x - y + 58 = 0$. The locus is a hyperbola.

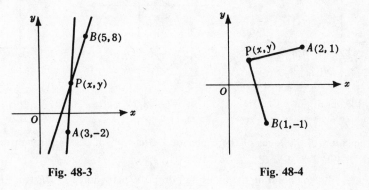

Fig. 48-3   **Fig. 48-4**

48.2   Find the equation of the locus of a point which moves so that the sum of the squares of its distances from the points $A(2, 1)$ and $B(1, -1)$ is always equal to 3.

Let $P(x, y)$ be any point on the locus. Using Fig. 48-4,

$$(AP)^2 + (BP)^2 = 3, \quad \left[\sqrt{(x - 2)^2 + (y - 1)^2}\right]^2 + \left[\sqrt{(x - 1)^2 + (y + 1)^2}\right]^2 = 3$$

Expanding and simplifying, we have $x^2 + y^2 - 3x + 2 = 0$, a circle with center at $(\frac{3}{2}, 0)$ and radius $\frac{1}{2}$.

48.3 Find the equation of the locus of a point which moves so that its distance from the point $F(p, 0)$ is equal to its distance from the line $d: x = -p$. (Assume $p > 0$.)

Let $P(x, y)$ be any point on the locus and draw PM perpendicular to d, as in Fig. 48-5. Then since $FP = MP$, $\sqrt{(x - p)^2 + y^2} = x + p$ or $(x - p)^2 + y^2 = (x + p)^2$. Expanding and simplifying, we have $y^2 = 4px$.

Fig. 48-5

48.4 Find the equation of the locus of a point which moves so that the sum of its distances from the points $A(5, 0)$ and $B(-5, 0)$ is 12.

Let $P(x, y)$ be any point on the locus, as in Fig. 48-6. Since $AP + BP = 12$,

$$\sqrt{(x - 5)^2 + y^2} + \sqrt{(x + 5)^2 + y^2} = 12 \quad \text{or} \quad \sqrt{(x - 5)^2 + y^2} = 12 - \sqrt{(x + 5)^2 + y^2}$$

Squaring and simplifying, we have

$$x^2 - 10x + 25 + y^2 = 144 - 24\sqrt{(x + 5)^2 + y^2} + x^2 + 10x + 25 + y^2 \quad \text{or} \quad 5x + 36 = 6\sqrt{(x + 5)^2 + y^2}$$

Squaring and simplifying, we have finally $11x^2 + 36y^2 - 396 = 0$ as the required equation. The locus is an ellipse.

Fig. 48-6   **Fig. 48-7**   **Fig. 48-8**

48.5 Find the equation of the locus of the center of a circle which passes through the point $A(2, -5)$ and is tangent to the line $y = 1$.

Let $P(x, y)$ be any point on the locus and drop the perpendicular BP from P to the line $y = 1$, as in Fig. 48-7. Since $AP = BP$, $\sqrt{(x - 2)^2 + (y + 5)^2} = |y - 1|$. Then $(x - 2)^2 + (y + 5)^2 = (y - 1)^2$ or $x^2 - 4x + 12y + 28 = 0$ is the desired equation. The locus is a parabola.

48.6 Find the equation of the locus of a point which moves so that its distance from the point $A(5, 0)$ is $e > 0$ times its distance from the line $l_1: x - \frac{36}{5} = 0$. Identify the locus when   (a) $e = \frac{5}{6}$ (b) $e = 1$ (c) $e = 2$.

Let $P(x, y)$ be any point on the locus and drop the perpendicular BP from P to l_1, as in Fig. 48-8. Since $AP = e|\overline{BP}|$, $\sqrt{(x - 5)^2 + y^2} = e|x - \frac{36}{5}|$, $(x - 5)^2 + y^2 = e^2(x - \frac{36}{5})^2$, and the required equation is $(1 - e^2)x^2 - (10 - 72e^2/5)x + y^2 + 25 - 1296e^2/25 = 0$.

(*a*) When $e = \frac{5}{6}$, the locus is the ellipse $11x^2 + 36y^2 - 396 = 0$.

(*b*) When $e = 1$, the locus is the parabola $25y^2 + 110x - 671 = 0$.

(*c*) When $e = 2$, the locus is the hyperbola $25y^2 - 75x^2 + 1190x - 4559 = 0$.

48.7 Find the equation of the locus of a point which moves so that the sum of its distances from the points $A(4, 4)$ and $B(-4, -2)$ is 12.

Let $P(x, y)$ be any point on the locus. See Fig. 48-9. Since $AP + BP = 12$, $\sqrt{(x - 4)^2 + (y - 4)^2} + \sqrt{(x + 4)^2 + (y + 2)^2} = 12$.

Proceeding as in Problem 48.3, we obtain $4x + 3y + 33 = 6\sqrt{(x + 4)^2 + (y + 2)^2}$, and finally $20x^2 - 24xy + 27y^2 + 24x - 54y - 369 = 0$.

Fig. 48-9

48.8 A line segment $\overline{AB}$ of constant length $2e$ moves with the end A always on the x axis and the end B always on the line $y = 6x$. Find the equation of the locus of the midpoint of $\overline{AB}$.

Let A have coordinates $(a, 0)$ and B have coordinates (b, c), as in Fig. 48-10. If $P(x, y)$ is the tracing point of the required locus, then

$$(1) \quad x = \tfrac{1}{2}(a + b) \quad \text{and} \quad (2) \quad y = \tfrac{1}{2}c$$

Also, since B is on the line $y = 6x$ and $AB = 2e$,

$$(3) \quad c = 6b \quad \text{and} \quad (4) \quad (a - b)^2 + c^2 = 4e^2$$

From (2), $c = 2y$; from (3), $b = c/6 = y/3$; from (1), $a = 2x - b = 2x - y/3$. Substituting in (4), we obtain the equation of the locus as

$$\left(2x - \frac{2y}{3}\right)^2 + (2y)^2 = 4e^2 \quad \text{or} \quad 9x^2 - 6xy + 10y^2 = 9e^2$$

Fig. 48-10

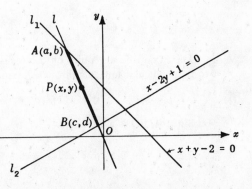

Fig. 48-11

48.9 A line l passing through the origin intersects the line $l_1: x + y - 2 = 0$ in A and the line $l_2: x - 2y + 1 = 0$ in B. Find the locus of the midpoint of $\overline{AB}$ as l revolves about the origin. Refer to Fig. 48-11.

Let A have coordinates (a, b); then A is on l_1 and

$$a + b - 2 = 0 \tag{1}$$

Let B have coordinates (c, d); then B is on l_2 and

$$c - 2d + 1 = 0 \tag{2}$$

If $P(x, y)$ is the tracing point of the locus it is the midpoint of AB; then

$$(3) \quad x = \tfrac{1}{2}(a + c) \qquad \text{and} \qquad (4) \quad y = \tfrac{1}{2}(b + d)$$

Also, since A, O, B, and P are collinear, we have

$$\frac{b}{a} = \frac{d}{c} = \frac{y}{x} \tag{5}$$

To eliminate the parameters we add (1) and (2), and substitute from (3) and (4) to get $(a + c) + (b + d) - 3d - 1 = 0$, $2x + 2y - 3d - 1 = 0$, or

$$2x + 2y - 1 = 3d \tag{6}$$

Next, we subtract (2) from twice (1) to get $2(a + c) - 3c + 2(b + d) - 5 = 0$ or

$$4x + 4y - 5 = 3c \tag{7}$$

Finally, dividing (6) by (7) and making use of (5) we have

$$\frac{2x + 2y - 1}{4x + 4y - 5} = \frac{d}{c} = \frac{y}{x}$$

The desired equation is $2x^2 - 2xy - 4y^2 - x + 5y = 0$.

48.10 Find the equation of the locus of the point of intersection of any two perpendicular tangents to the circle $x^2 + y^2 = 4$. See Fig. 48-12.

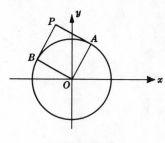

Fig. 48-12

Let $P(x, y)$ be any point on the locus and let $A(a, b)$ and $B(c, d)$ be the points of tangency of the two perpendicular tangents through P. Since A and B are on the circle, we have

$$(1) \quad a^2 + b^2 = 4 \qquad \text{and} \qquad (2) \quad c^2 + d^2 = 4$$

Since $APBO$ is a rectangle:

$$(3) \quad \frac{y - b}{x - a} = \frac{d}{c}, \qquad (4) \quad \frac{y - d}{x - c} = \frac{b}{a}, \qquad \text{and} \qquad (5) \quad \frac{b}{a} = -\frac{c}{d}.$$

Using (3), (5), (4):

$$\frac{y-b}{x-a} = \frac{d}{c} = -\frac{a}{b} = -\frac{x-c}{y-d} \quad \text{or} \quad (6) \quad x^2 + y^2 = (ax + by) + (cx + dy) - (ac + bd)$$

From $\dfrac{y-b}{x-a} = -\dfrac{a}{b}$, $ax + by = a^2 + b^2 = 4$; from $\dfrac{y-d}{x-c} = -\dfrac{c}{d}$, $cx + dy = c^2 + d^2 = 4$; from (5),
$ac + bd = 0$. Then (6) becomes $x^2 + y^2 = 8$, the equation of the locus.

Supplementary Problems

48.11 Find the equation of the locus of a point whose distance from the point $(3, 0)$ is equal to its distance from the line $x + 3 = 0$.

　　　Ans.　$y^2 - 12x = 0$

48.12 Find the equation of the locus of a point whose distance from the point $(-3, 4)$ is twice its distance from the line $y + 4 = 0$.

　　　Ans.　$x^2 - 3y^2 + 6x - 40y - 39 = 0$

48.13 Find the equation of the locus of the center of a circle which passes through $(-5, 2)$ and is tangent to the line $x + 4 = 0$.

　　　Ans.　$y^2 + 2x - 4y + 13 = 0$

48.14 Find the equation of the locus of a point the sum of whose distances from $(0, 4)$ and $(0, -4)$ is 10.

　　　Ans.　$25x^2 + 9y^2 - 225 = 0$

48.15 Find the equation of the locus of a point the difference of whose distances from $(4, 0)$ and $(-4, 0)$ is 6.

　　　Ans.　$7x^2 - 9y^2 - 63 = 0$

48.16 Find the equation of the locus of a point which is twice as far from the point $(3, 0)$ as from the line $12x + 5y - 13 = 0$.

　　　Ans.　$407x^2 + 480xy - 69y^2 - 234x - 520y - 845 = 0$

48.17 Find the equation of the locus of a point which is twice as far from the line $12x + 5y - 13 = 0$ as from the point $(3, 0)$.

　　　Ans.　$532x^2 - 120xy + 651y^2 - 3744x + 130y + 5915 = 0$

48.18 A line is drawn from the origin to a point on the ellipse $9x^2 + 25y^2 - 225 = 0$. Find the equation of the locus of the midpoint of the line.

　　　Ans.　$36x^2 + 100y^2 - 225 = 0$

48.19 A line is drawn from the point $(4, 0)$ to the hyperbola $25x^2 - 9y^2 - 225 = 0$. Find the equation of the locus of the midpoint of the line.

   Ans. $100x^2 - 36y^2 - 400x + 175 = 0$

48.20 A line through the origin intersects the line $2x + y - 2 = 0$ in A and the line $x - 2y + 2 = 0$ in B. Find the equation of the locus of the midpoint of AB.

   Ans. $2x^2 - 3xy - 2y^2 + x + 3y = 0$

48.21 A variable tangent is drawn to the circle $x^2 + y^2 - 4x = 0$. Find the equation of the locus of the foot of the perpendicular dropped from the origin to the tangent.

   Hint: Let (a, b) be the point of tangency and (x, y) be the foot of the perpendicular. Then $a^2 + b^2 - 4a = 0$, $(a - 2)x + by - 2a = 0$, and $b/(a - 2) = y/x$.

   Ans. $(x^2 + y^2 - 2x)^2 = 4(x^2 + y^2)$

48.22 Find the equation of the locus of the center of a circle which is tangent to the line $x - 8 = 0$ and to the circle $x^2 + y^2 = 36$.

   Hint: Let (x, y) be a point on the locus and (a, b) be the corresponding point of tangency of the two circles. Then $a^2 + b^2 = 36$, $(x - a)^2 + (y - b)^2 = (8 - x)^2$, and $y/x = b/a$.

   Ans. $(y^2 + 28x - 196)(y^2 + 4x - 4) = 0$

Chapter 49

The Parabola

THE LOCUS OF A POINT P which moves in a plane so that its distance from a fixed line of the plane and its distance from a fixed point of the plane, not on the line, are equal is called a *parabola*.

The fixed point F is called the *focus* and the fixed line d is called the *directrix* of the parabola. The line $\overleftrightarrow{FD}$ through the focus and perpendicular to the directrix is called the *axis* of the parabola. The axis intersects the parabola in the point V, the midpoint of $\overline{FD}$, called the *vertex*.

The line segment joining any two distinct points of the parabola is called a *chord*. A chord (as BB') which passes through the focus is called a *focal chord* while $\overleftrightarrow{FB}$ and $\overleftrightarrow{FB}'$ are called the *focal radii* of B and B', respectively. The focal chord LL' which is perpendicular to the axis is called the *latus rectum*. See Fig. 49-1.

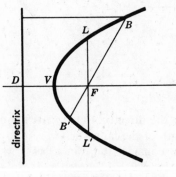

Fig. 49-1

THE EQUATION OF A PARABOLA assumes its simplest (*reduced*) form when its vertex is at the origin and its axis coincides with one of the coordinate axes.

When the vertex is at the origin and the axis coincides with the x axis, the equation of the parabola is

$$y^2 = 4px \qquad (49.1)$$

Then the focus is at $F(p, 0)$ and the equation of the directrix is $d: x = -p$. If $p > 0$, the parabola opens to the right; if $p < 0$, the parabola opens to the left. See Figs. 49-2(a) and (b). (See Problem 48.3.)

(a) $y^2 = 4px,\ p > 0$ (b) $y^2 = 4px,\ p < 0$

Fig. 49-2

When the vertex is at the origin and the axis coincides with the y axis, the equation of the parabola is

$$x^2 = 4py \qquad (49.2)$$

Then the focus is at $F(0, p)$ and the equation of the directrix is $d: y = -p$. If $p > 0$, the parabola opens upward; if $p < 0$, the parabola opens downward. See Figs. 49-3(a) and (b).

(a) $x^2 = 4py, p > 0$ (b) $x^2 = 4py, p < 0$

Fig. 49-3

In either case, the distance from the directrix to the vertex and the distance from the vertex to the focus are equal to $|p|$; also, the length of the latus rectum is $|4p|$.

[Some authors define $p > 0$ and consider the four cases $y^2 = 4px$, $y^2 = -4px$, $x^2 = 4py$, and $x^2 = -4py$. Other authors label the focus $F(\frac{1}{2}p, 0)$ and directrix $d: x = -\frac{1}{2}p$, and obtain $y^2 = 2px$, etc.]

THE EQUATION OF A PARABOLA assumes the *semireduced* form

$$(y - k)^2 = 4p(x - h) \qquad (49.1')$$

or

$$(x - h)^2 = 4p(y - k) \qquad (49.2')$$

when its vertex is at the point (h, k) and its axis is parallel respectively to the x axis or the y axis. The distance between the directrix and vertex, the distance between the vertex and focus, and the length of the latus rectum are the same as given in the section above.

EXAMPLE 1.   Sketch the locus and find the coordinates of the vertex and focus, the equations of the axis and directrix, and the length of the latus rectum of the parabola $y^2 - 6y + 8x + 41 = 0$.

We first put the equation in the form

$$(y - 3)^2 = -8(x + 4) \qquad (49.1')$$

and note, since $4p = -8$, that the parabola opens to the left. Having located the vertex at $V(-4, 3)$, we draw in the axis through V parallel to the x axis.

In locating the focus, we move from the vertex to the left (the parabola opens to the left) along the axis a distance $|p| = 2$ to the point $F(-6, 3)$. In locating the directrix, we move from the vertex to the right (away from the focus) along the axis a distance $|p| = 2$ to the point $D(-2, 3)$. The directrix passes through D and is perpendicular to the axis; its equation is $x + 2 = 0$.

The length of the latus rectum is $|4p| = 8$. We draw in the line segment $\overline{LL'}$, through F perpendicular to the axis FV so that $FL = L'F = |2p| = 4$.

Finally, using the points L, L', and V, we sketch the parabola. See Fig. 49-4.

$$(y - 3)^2 = -(8x + 4)$$

Fig. 49-4

Solved Problems

49.1	For each of the following parabolas, sketch the curve and find the coordinates of the vertex and focus, the equations of the axis and directrix, and the length of the latus rectum:

(a) $y^2 = 16x$	(b) $x^2 = -9y$	(c) $x^2 - 2x - 12y + 25 = 0$	(d) $y^2 + 4y + 20x + 4 = 0$.

(a)	The parabola opens to the right ($p > 0$) with vertex at $V(0, 0)$. The equation of its axis is $y = 0$. Moving from V to the right along the axis a distance $|p| = 4$, we locate the focus at $F(4, 0)$. Moving from V to the left along the axis a distance $|p| = 4$, we locate the point $D(-4, 0)$. Since the directrix passes through D perpendicular to the axis, its equation is $x + 4 = 0$. The length of the latus rectum LL' is $|4p| = 16$. See Fig. 49-5(a).

(a) $y^2 = 16x$	(b) $x^2 = -9y$

Fig. 49-5

(b)	The parabola opens downward ($p < 0$) with vertex at $V(0, 0)$. The equation of its axis is $x = 0$. Moving from V downward along the axis a distance $|p| = \frac{9}{4}$, we locate the focus at $F(0, -\frac{9}{4})$. Moving from V upward along the axis a distance $|p| = \frac{9}{4}$, we locate the point $D(0, \frac{9}{4})$; the equation of the directrix is $4y - 9 = 0$. The length of the latus rectum is $|4p| = 9$. See Fig. 49-5(b).

(c)	Here $(x - 1)^2 = 12(y - 2)$. The parabola opens upward ($p > 0$) with vertex at $V(1, 2)$. The equation of its axis is $x - 1 = 0$. Since $|p| = 3$, the focus is at $F(1, 5)$ and the equation of the directrix is $y + 1 = 0$. The length of the latus rectum is $|4p| = 12$. See Fig. 49-6(a).

(d) Here $(y + 2)^2 = -20x$. The parabola opens to the left ($p < 0$) with vertex at $V(0, -2)$. The equation of its axis is $y + 2 = 0$. Since $|p| = 5$, the focus is at $F(-5, -2)$ and the equation of the directrix is $x - 5 = 0$. The length of the latus rectum is $|4p| = 20$. See Fig. 49-6(b).

(a) $(x - 1)^2 = 12(y - 2)$ (b) $(y + 2)^2 = -20x$

Fig. 49-6

49.2   Find the equation of the parabola, given

(a) $V(0, 0)$; $F(0, -4)$

(b) $V(0, 0)$; directrix: $x = -5$

(c) $V(1, 4)$; $F(-2, 4)$

(d) $F(2, 3)$; directrix: $y = -1$

(e) $V(0, 0)$; axis: $y = 0$; passing through $(4, 5)$.

(f) Axis parallel to $y = 0$; passing through $(-2, 4)$, $(-3, 2)$, and $(-11, -2)$. Locate the vertex, focus, and ends of the latus rectum.

(a) Since the directed distance $p = VF = -4$, the parabola opens downward. Its equation is $x^2 = -16y$.

(b) The parabola opens to the right (away from the directrix). Since $p = DV = 5$, the equation is $y^2 = 20x$.

(c) Here the focus lies to the left of the vertex and the parabola opens to the left. The directed distance $p = VF = -3$ and the equation is $(y - 4)^2 = -12(x - 1)$.

(d) Here the focus lies above the directrix and the parabola opens upward. The axis of the parabola meets the directrix in $D(2, -1)$ and the vertex is at the midpoint $V(2, 1)$ of FD. Then $p = VF = 2$ and the equation is $(x - 2)^2 = 8(y - 1)$.

(e) The equation of this parabola is of the form $y^2 = 4px$. If $(4, 5)$ is a point on it then $(5)^2 = 4p(4)$, $4p = \frac{25}{4}$, and the equation is $y^2 = \frac{25}{4}x$.

(f) The equation of the parabola with axis parallel to the x axis is of the form $(y - k)^2 = 4p(x - h)$ or when expanded $x = ay^2 + by + c$. Substituting the coordinates of the given points into the latter form, we obtain the system of equations

$$-2 = 16a + 4b + c, \qquad -3 = 4a + 2b + c, \qquad -11 = 4a - 2b + c$$

having the simultaneous solution $a = -\frac{1}{4}$, $b = 2$, $c = -6$. The required equation is $x = -\frac{1}{4}y^2 + 2y - 6$ or $(y - 4)^2 = -4(x + 2)$. The vertex is at $V(-2, 4)$. Since the parabola opens to the left and $|p| = 1$, the focus is at $F(-3, 4)$. Since $|2p| = 2$, the ends of the latus rectum are at $L(-3, 6)$ and $L'(-3, 2)$.

49.3   Prove: The square of the length of the perpendicular dropped from any point of a parabola to its axis is equal to the product of the length of the latus rectum and the segment of the axis between the vertex and the foot of the perpendicular.

Consider the parabola $y^2 = 4px$, $p > 0$, and let $P(x, y)$ be any point on it.

The length of the perpendicular dropped from P to the axis (here, the x axis) is y and its square is y^2. The foot of the perpendicular is x units from the vertex (origin) and the length of the latus rectum is $4p$; their product is $4px$. The equation of the parabola assures us that these two quantities are equal. See Fig. 49-7.

Fig. 49-7

49.4   The cable of a suspension bridge has supporting towers which are 50 ft high and 400 ft apart. If the lowest point of the cable is 10 ft above the floor of the bridge, find the length of a supporting rod 100 ft from the center of the span.

Take the origin of coordinates at the lowest point of the cable and the positive y axis directed upward along the axis of symmetry of the parabola. Then the equation of the parabola has the form $x^2 = 4py$. Since $(200, 40)$ is a point on the parabola, $(200)^2 = 4p \cdot 40$ or $4p = 1000$, and the equation is $x^2 = 1000y$. When $x = 100$, $(100)^2 = 1000y$, and $y = 10$ ft. The length of the supporting rod is $10 + 10 = 20$ ft.

Supplementary Problems

49.5   For each of the following parabolas, sketch the curve and find the coordinates of the vertex and focus, the equations of the axis and directrix, and the length of the latus rectum:

(a) $x^2 = 12y$ *Ans.* $V(0, 0), F(0, 3); x = 0, y + 3 = 0; 12$

(b) $y^2 = -10x$ *Ans.* $V(0, 0), F(-\frac{5}{2}, 0); y = 0, 2x - 5 = 0; 10$

(c) $x^2 - 6x + 8y + 25 = 0$ *Ans.* $V(3, -2), F(3, -4); x - 3 = 0, y = 0; 8$

(d) $y^2 - 16x + 2y + 49 = 0$ *Ans.* $V(3, -1), F(7, -1); y + 1 = 0, x + 1 = 0; 16$

(e) $x^2 - 2x - 6y - 53 = 0$ *Ans.* $V(1, -9), F(1, -\frac{15}{2}); x - 1 = 0, 2y + 21 = 0; 6$

(f) $y^2 + 20x + 4y - 60 = 0$ *Ans.* $V(\frac{16}{5}, -2), F(-\frac{9}{5}, -2); y + 2 = 0, 5x - 41 = 0; 20$

49.6   Find the equation of the parabola, given

(a) $V(0, 0)$, opens to the right, latus rectum equal to 16.

(b) $V(0, 0), F(-2, 0)$

(c) $V(0, 0), F(0, 5)$

(d) $V(0, 0), d: y + 3 = 0$

(e) $V(0, 0), F$ on x axis, passes through $(-2, 6)$.

(f) $V(1, 3), F(-1, 3)$

(g) $F(3, 2)$, d: $y + 4 = 0$

(h) $V(3, -2)$, latus rectum equal to 20, axis: $y + 2 = 0$

(i)   ends of latus rectum $(2, -1)$ and $(2, 5)$

   Ans.   (a) $y^2 = 16x$

            (b) $y^2 = -8x$

            (c) $x^2 = 20y$

            (d) $x^2 = 12y$

            (e) $y^2 = -18x$

            (f) $y^2 + 8x - 6y + 1 = 0$

            (g) $x^2 - 6x - 12y - 3 = 0$

            (h) $y^2 - 20x + 4y + 64 = 0$, $y^2 + 20x + 4y - 56 = 0$

            (i) $y^2 - 6x - 4y + 7 = 0$, $y^2 + 6x - 4y - 17 = 0$

49.7   Prove: The ordinate at any point $P(x, y)$ on the parabola $y^2 = 4px$ is the mean proportional between the abscissa at P and the length of the latus rectum.

49.8   Find the length of the focal chord of $y^2 = 8x$ parallel to $y = 2\sqrt{2}x + 1$.

   Ans.   9

49.9   Verify the following geometric construction of the parabola when the focus and the directrix are given: Draw through F the perpendicular to d meeting it in D. Locate the vertex V on DF so that $DV = VF$. Take any point A on VF on the same side of V as F and draw BA parallel to d or perpendicular to DF. With F as a center and radius equal to DA draw an arc intersecting AB at P and P'. Then P and P' are two points of the parabola.

49.10   Two circles are drawn, each having a focal chord of a parabola as diameter. Show that their radical axis passes through the vertex of the parabola.

Chapter 50

The Ellipse

THE LOCUS OF A POINT P which moves in a plane so that the sum of its distances from two fixed points in the plane is constant is called an *ellipse*.

The fixed points F and F' are called the *foci* and their midpoint C is called the *center* of the ellipse. The line $\overleftrightarrow{FF'}$ joining the foci intersects the ellipse in the points V and V', called the *vertices*. The segment $\overline{V'V}$ intercepted on the line $\overleftrightarrow{FF'}$ by the ellipse is called its *major axis*; the segment $\overline{B'B}$ intercepted on the line through C perpendicular to $\overleftrightarrow{F'F}$ is called its *minor axis*.

A line segment whose extremities are any two points on the ellipse is called a *chord*. A chord which passes through a focus is called a *focal chord*; a focal chord perpendicular to the major axis is called a *latus rectum*. See Fig. 50-1.

$$FP + F'P = V'V$$

$$\frac{FP}{PD} = \frac{F'P}{PD'} = e = \frac{CF}{CV}$$

Fig. 50-1

The ellipse may also be defined as the locus of a point which moves so that the ratio of its distance from a fixed point to its distance from a fixed line is equal to $e < 1$. The fixed point is a focus F or F' and the fixed line d or d' is called a *directrix*. The ratio e is called the *eccentricity* of the ellipse.

Problems 48.4 and 48.6(a) illustrate the two definitions.

THE EQUATION OF AN ELLIPSE assume its simplest (*reduced*) form when its center is at the origin and its major axis lies along one of the coordinate axes.

When the center is at the origin and the major axis lies along the x axis, the equation of the ellipse is

$$\frac{x^2}{a^2} + \frac{y^2}{b^2} = 1 \tag{50.1}$$

Then the vertices are at $V(a, 0)$ and $V'(-a, 0)$ and the length of the major axis is $V'V = 2a$. The length of the minor axis is $B'B = 2b$. The foci are on the major axis at $F(c, 0)$ and $F'(-c, 0)$ where

$$c = \sqrt{a^2 - b^2}$$

345

When the center is at the origin and the major axis lies along the y axis, the equation of the ellipse is

$$\frac{x^2}{b^2} + \frac{y^2}{a^2} = 1 \qquad (50.2)$$

See Figs. 50-2(a) and (b).

$$(a) \quad \frac{x^2}{a^2} + \frac{y^2}{b^2} = 1, \ a > b \qquad\qquad\qquad (b) \quad \frac{x^2}{b^2} + \frac{y^2}{a^2} = 1, \ a > b$$

Fig. 50-2

Then the vertices are at $V(0, a)$ and $V'(0, -a)$ and the length of the major axis is $V'V = 2a$. The length of the minor axis is $B'B = 2b$. The foci are on the major axis at $F(0, c)$ and $F'(0, -c)$, where

$$c = \sqrt{a^2 - b^2}$$

In both cases, the length of a latus rectum is $2b^2/a$, the eccentricity is

$$e = \frac{c}{a} = \frac{\sqrt{a^2 - b^2}}{a}$$

and the directrices are perpendicular to the major axis at distances $\pm a^2/c = \pm a/e$ from the center.

THE EQUATION OF AN ELLIPSE assumes the *semireduced* form

$$\frac{(x - h)^2}{a^2} + \frac{(y - k)^2}{b^2} = 1 \qquad (50.1')$$

or

$$\frac{(x - h)^2}{b^2} + \frac{(y - k)^2}{a^2} = 1 \qquad (50.2')$$

Here the center is at the point (h, k) and the major axis is parallel respectively to the x axis and to the y axis. The lengths of the major and minor axes, the distance between the foci, the distance from the center to a directrix, the length of a latus rectum, and the eccentricity are as given in the section above.

EXAMPLE 1. Find the coordinates of the center, vertices, and foci; the lengths of the major and minor axes; the length of a latus rectum; the eccentricity; and the equations of the directrices of the ellipse

$$\frac{(x - 4)^2}{9} + \frac{(y + 2)^2}{25} = 1$$

Sketch the locus.

The center is at the point $C(4, -2)$.

Since $a^2 > b^2$, $a^2 = 25$ and $b^2 = 9$.

Since a^2 is under the term in y, the major axis is parallel to the y axis. To locate the vertices (the extremities of the major axis), we move from the center parallel to the y axis a distance $a = 5$ to the points $V(4, 3)$ and $V'(4, -7)$. To locate the extremities of the minor axis, we move from the center perpendicular to the major axis a distance $b = 3$ to the points $B'(1, -2)$ and $B(7, -2)$. The lengths of the major and minor axes are $2a = 10$ and $2b = 6$, respectively.

The distance from the center to a focus is $c = \sqrt{a^2 - b^2} = \sqrt{25 - 9} = 4$. To locate the foci, we move from the center along the major axis a distance $c = 4$ to the points $F(4, 2)$ and $F'(4, -6)$.

The length of a latus rectum is $2b^2/a = \frac{18}{5}$. The coordinates of the extremities of the latus rectum through F are $(\frac{11}{5}, 2)$ and $(\frac{29}{5}, 2)$, each being at a distance of one-half the length of the latus rectum from F; and the coordinates of the extremities of the latus rectum through F' are $(\frac{11}{5}, -6)$ and $(\frac{29}{5}, -6)$.

The eccentricity is $e = c/a = \frac{4}{5}$.

The distance from the center to a directrix is $a^2/c = \frac{25}{4}$. Since the directrices are perpendicular to the major axis, their equations are $d: y = -2 + \frac{25}{4} = \frac{17}{4}$ and $d': y = -2 - \frac{25}{4} = -\frac{33}{4}$. See Fig. 50-3.

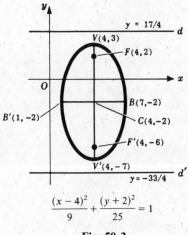

$$\frac{(x - 4)^2}{9} + \frac{(y + 2)^2}{25} = 1$$

Fig. 50-3

Solved Problems

50.1 For each of the following ellipses, find the coordinates of the center, vertices, and foci; the lengths of the major and minor axes; the length of a latus rectum; the eccentricity; and the equations of the directrices. Sketch the curve.

(a) $x^2/16 + y^2/4 = 1$ (b) $25x^2 + 9y^2 = 25$ (c) $x^2 + 9y^2 + 4x - 18y - 23 = 0$.

(a) Here $a^2 = 16$, $b^2 = 4$, and $c = \sqrt{a^2 - b^2} = \sqrt{16 - 4} = 2\sqrt{3}$.

The center is at the origin and the major axis is along the x axis (a^2 under x^2). The vertices are on the major axis at a distance $a = 4$ from the center; their coordinates are $V(4, 0)$ and $V'(-4, 0)$. The minor axis is along the y axis and its extremities, being at a distance $b = 2$ from the center, are at $B(0, 2)$ and $B'(0, -2)$. The foci are on the major axis at a distance $2\sqrt{3}$ from the center; their coordinates are $F(2\sqrt{3}, 0)$ and $F'(-2\sqrt{3}, 0)$.

The lengths of the major and minor axes are $2a = 8$ and $2b = 4$, respectively.

The length of a latus rectum is $2b^2/a = 2 \cdot 4/4 = 2$.

The eccentricity is $e = c/a = 2\sqrt{3}/4 = \frac{1}{2}\sqrt{3}$.

The directrices are perpendicular to the major axis and at a distance $a^2/c = 16/2\sqrt{3} = 8\sqrt{3}/3$ from the center; their equations are $x = \pm 8\sqrt{3}/3$.

See Fig. 50-4(a).

Fig. 50-4

(b) When the equation is put in the form

$$\frac{x^2}{1} + \frac{y^2}{\frac{25}{9}} = 1$$

we find $\qquad a^2 = \frac{25}{9}, \ b^2 = 1 \qquad$ and $\qquad c = \sqrt{a^2 - b^2} = \sqrt{\frac{25}{9} - 1} = \frac{4}{3}.$

The center is at the origin and the major axis is along the y axis (a^2 under y^2). The vertices are on the major axis at a distance $a = \frac{5}{3}$ from the center; their coordinates are $V(0, \frac{5}{3})$ and $V'(0, -\frac{5}{3})$. The extremities of the minor axis are on the x axis and at a distance $b = 1$ from the center; their coordinates are $B(1, 0)$ and $B'(-1, 0)$. The foci are on the major axis at a distance $c = \frac{4}{3}$ from the center; their coordinates are $F(0, \frac{4}{3})$ and $F'(0, -\frac{4}{3})$.

The lengths of the major and minor axes are $2a = \frac{10}{3}$ and $2b = 2$, respectively.

The length of a latus rectum is $2b^2/a = \frac{6}{5}$

The eccentricity is

$$e = \frac{c}{a} = \frac{\frac{4}{3}}{\frac{5}{3}} = \frac{4}{5}$$

The directrices are perpendicular to the major axis and at a distance $a^2/c = \frac{25}{12}$ from the center; their equations are $y = \pm\frac{25}{12}$.

See Fig. 50-4(b).

(c) When the equation is put in the form

$$\frac{(x + 2)^2}{36} + \frac{(y - 1)^2}{4} = 1$$

we have $a^2 = 36$, $b^2 = 4$, and $c = \sqrt{a^2 - b^2} = 4\sqrt{2}$.

The center is at the point $C(-2, 1)$ and the major axis is along the line $y = 1$. The vertices are on the major axis at a distance $a = 6$ from the center; their coordinates are $V(4, 1)$ and $V'(-8, 1)$. The extremities of the minor axis are on the line $x = -2$ and at a distance $b = 2$ from the center; their coordinates are $B(-2, 3)$ and $B'(-2, -1)$. The foci are on the major axis at a distance $c = 4\sqrt{2}$ from the center; their coordinates are $F(-2 + 4\sqrt{2}, 1)$ and $F'(-2 - 4\sqrt{2}, 1)$.

The lengths of the major and minor axes are $2a = 12$ and $2b = 4$, respectively.

The length of a latus rectum is $2b^2/a = \frac{4}{3}$.

The eccentricity is $c/a = 4\sqrt{2}/6 = 2\sqrt{2}/3$.

The directrices are perpendicular to the major axis and at a distance $a^2/c = 9\sqrt{2}/2$ from the center; their equations are $x = -2 \pm 9\sqrt{2}/2$.

See Fig. 50-5.

50.2 Find the equation of the ellipse, given

(a) Vertices $(\pm 8, 0)$, minor axis = 6.

(b) One vertex at $(0, 13)$, one focus at $(0, -12)$, center at $(0, 0)$.

Fig. 50-5

(c)   Foci $(\pm 10, 0)$, eccentricity $= \frac{5}{6}$.

(d)   Vertices $(8, 3)$ and $(-4, 3)$, one focus at $(6, 3)$.

(e)   Vertices $(3, -10)$ and $(3, 2)$, length of latus rectum $= 4$.

(f)   Directrices $4y - 33 = 0$, $4y + 17 = 0$; major axis on $x + 1 = 0$; eccentricity $= \frac{4}{5}$.

(a)   Here $2a = V'V = 16$, $2b = 6$, and the major axis is along the x axis. The equation of the ellipse is $x^2/a^2 + y^2/b^2 = x^2/64 + y^2/9 = 1$.

(b)   The major axis is along the y axis, $a = 13$, $c = 12$, and $b^2 = a^2 - c^2 = 25$. The equation of the ellipse is $x^2/b^2 + y^2/a^2 = x^2/25 + y^2/169 = 1$.

(c)   Here the major axis is along the x axis and $c = 10$. Since $e = c/a = 10/a = \frac{5}{6}$, $a = 12$ and $b^2 = a^2 - c^2 = 44$. The equation of the ellipse is $x^2/144 + y^2/44 = 1$.

(d)   The center is at the midpoint of $V'V$, that is, at $C(2, 3)$. Then $a = CV = 6$, $c = CF = 4$, and $b^2 = a^2 - c^2 = 20$. Since the major axis is parallel to the x axis, the equation of the ellipse is

$$\frac{(x-2)^2}{36} + \frac{(y-3)^2}{20} = 1$$

(e)   The center is at $C(3, -4)$ and $a = 6$. Since $2b^2/a = 2b^2/6 = 4$, $b^2 = 12$. The major axis is parallel to the y axis and the equation of the ellipse is $\dfrac{(x-3)^2}{12} + \dfrac{(y+4)^2}{36} = 1$.

(f)   The major axis intersects the directrices in $D(-1, \frac{33}{4})$ and $D'(-1, -\frac{17}{4})$. The center of the ellipse bisects $D'D$ and hence is at $C(-1, 2)$. Since $CD = a/e = \frac{25}{4}$ and $e = \frac{4}{5}$, $a = 5$ and $c = ae = 4$. Then $b^2 = a^2 - c^2 = 9$. Since the major axis is parallel to the y axis, the equation of the ellipse is $\dfrac{(x+1)^2}{9} + \dfrac{(y-2)^2}{25} = 1$.

50.3   Find the equation of the locus of the midpoints of a system of parallel chords of slope m of the ellipse $x^2/a^2 + y^2/b^2 = 1$.

Let $P(x, y)$ be any point on the locus and let $Q(p, q)$ and $R(r, s)$ be the extremities of the chord of which P is the midpoint. Then

(1) $b^2 p^2 + a^2 q^2 = a^2 b^2$, since $Q(p, q)$ is on the ellipse

(2) $b^2 r^2 + a^2 s^2 = a^2 b^2$, since $R(r, s)$ is on the ellipse

(3) $\dfrac{q - s}{p - r} = m$, since m is the slope of the chord

(4) $x = \frac{1}{2}(p + r)$, $y = \frac{1}{2}(q + s)$, since P is the midpoint of QR

Equating the left members of (1) and (2), we have $b^2 p^2 + a^2 q^2 = b^2 r^2 + a^2 s^2$. Then, using (3) and (4),

$$b^2(p^2 - r^2) = a^2(s^2 - q^2) \qquad \text{and} \qquad \frac{b^2}{a^2} = \frac{s^2 - q^2}{p^2 - r^2} = -\frac{q - s}{p - r} \cdot \frac{q + s}{p + r} = -m\frac{2y}{2x} = -m\frac{y}{x}$$

Thus the desired equation is $y = -\dfrac{b^2}{a^2 m} x$.

The locus passes through the center $C(0,0)$ and is called a *diameter* of the ellipse. See Fig. 50-6.

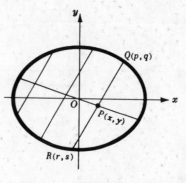

Fig. 50-6

Supplementary Problems

50.4   For each of the following ellipses, find the coordinates of the center, vertices, and foci; the lengths of the major and minor axes; the length of the latus rectum; the eccentricity; and the equations of the directrices. Sketch each curve.

(a) $4x^2 + 9y^2 = 36$ (c) $x^2 + 4y^2 - 6x + 32y + 69 = 0$

(b) $25x^2 + 16y^2 = 400$ (d) $16x^2 + 9y^2 + 32x - 36y - 92 = 0$

Ans.   (a) $C(0,0)$, $V(\pm 3, 0)$, $F(\pm\sqrt{5}, 0)$; 6, 4; $\frac{8}{3}$; $\sqrt{5}/3$; $x = \pm 9\sqrt{5}/5$

(b) $C(0,0)$, $V(0, \pm 5)$, $F(0, \pm 3)$; 10, 8; $\frac{32}{5}$; $\frac{3}{5}$; $y = \pm\frac{25}{3}$

(c) $C(3, -4)$, $V(5, -4)$, $V'(1, -4)$, $F(3 \pm \sqrt{3}, -4)$; 4, 2; 1; $\sqrt{3}/2$; $x = 3 \pm 4\sqrt{3}/3$

(d) $C(-1, 2)$, $V(-1, 6)$, $V'(-1, -2)$, $F(-1, 2 \pm \sqrt{7})$; 8, 6; $\frac{9}{2}$; $\sqrt{7}/4$; $y = 2 \pm 16\sqrt{7}/7$

50.5   Find the equation of the ellipse, given

(a) $V(\pm 13, 0)$, $F(12, 0)$ (d) $V(\pm 4, 0)$, latus rectum $= \frac{9}{2}$

(b) $C(0, 0)$, $a = 5$, $F(0, 4)$ (e) $V(7, 3)$, $V'(-3, 3)$, $F(6, 3)$

(c) $C(0, 0)$, $b = 2$, $d: x = -16\sqrt{7}/7$ (f) $F(5, 4)$, $F'(5, -2)$, $e = \sqrt{3}/3$

(g)   Ends of minor axis $(-2, 4)$, $(-2, 2)$; $d: x = 0$

(h)   Directrices: $y = \frac{11}{5}$, $y = -\frac{61}{5}$; major axis on $x = 3$, $e = \frac{5}{6}$

Ans.   (a) $25x^2 + 169y^2 = 4225$ (e) $9x^2 + 25y^2 - 36x - 150y + 36 = 0$

(b) $25x^2 + 9y^2 = 225$ (f) $3x^2 + 2y^2 - 30x - 4y + 23 = 0$

(c) $x^2 + 8y^2 = 32$, $7x^2 + 8y^2 = 32$ (g) $x^2 + 2y^2 + 4x - 12y + 20 = 0$

(d) $9x^2 + 16y^2 = 144$ (h) $36x^2 + 11y^2 - 216x + 110y + 203 = 0$

50.6   An arch in the form of a semi-ellipse is 50 ft wide at the base and 20 ft high. Find (a) its height 10 ft from the center of the base and (b) its width 10 ft above the base.

Ans.   (a) $4\sqrt{21}$ ft   (b) $25\sqrt{3}$ ft

50.7   Justify the following construction for the ellipse $b^2x^2 + a^2y^2 = a^2b^2$. Through the origin draw a line meeting the circle $x^2 + y^2 = a^2$ in $A(p, q)$ and the circle $x^2 + y^2 = b^2$ in $B(r, s)$; then the point $E(p, s)$ is on the ellipse.

50.8   Prove: The locus of the midpoints of the chords of an ellipse drawn through one end of the major axis is an ellipse.

50.9   Prove: If one diameter of an ellipse bisects all chords parallel to another diameter, then the second diameter bisects all chords parallel to the first. Two such diameters are called *conjugate diameters*.

50.10   An ellipse whose axes are parallel to the coordinate axes passes through the points $(12, -1)$, $(-4, 5)$, $(-2, -2)$, and $(10, 6)$. Find its equation.

     Ans. $x^2 + 4y^2 - 8x - 16y - 68 = 0$

50.11   Find the equation of the locus of the center of a circle which is tangent to the circles $x^2 + y^2 = 4$ and $x^2 + y^2 - 8x - 48 = 0$.

     Ans. $5x^2 + 9y^2 - 20x - 25 = 0$, $21x^2 + 25y^2 - 84x - 441 = 0$

Chapter 51

The Hyperbola

THE LOCUS OF A POINT P which moves in a plane so that the absolute value of the difference of its distances from two fixed points in the plane is constant is called a *hyperbola*. (Note that the locus consists of two distinct branches each of indefinite length.)

The fixed points F and F' are called the *foci* and their midpoint C is called the *center* of the hyperbola. The line $\overleftrightarrow{FF'}$ joining the foci intersects the hyperbola in the points V and V', called the *vertices*.

The segment $\overline{V'V}$ intercepted on the line $\overleftrightarrow{FF'}$ by the hyperbola is called its *transverse axis*. The line l through C and perpendicular to $\overleftrightarrow{FF'}$ does not intersect the curve but it will be found convenient to define a certain segment $\overline{B'B}$ on l, having C as midpoint, as the *conjugate axis*. See Fig. 51-1.

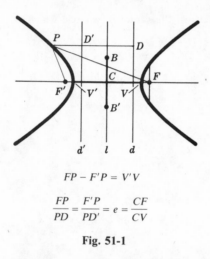

$$FP - F'P = V'V$$

$$\frac{FP}{PD} = \frac{F'P}{PD'} = e = \frac{CF}{CV}$$

Fig. 51-1

A line segment whose extremities are any two points (both on the same branch or one on each branch) on the hyperbola is called a *chord*. A chord which passes through a focus is called a *focal chord*, a focal chord perpendicular to the transverse axis is called a *latus rectum*.

The hyperbola may also be defined as the locus of a point which moves so that the ratio of its distance from a fixed point and its distance from a fixed line is equal to $e > 1$. The fixed point is a focus F or F' and the fixed line d or d' is called a *directrix*. The ratio e is called the *eccentricity* of the hyperbola.

THE EQUATION OF A HYPERBOLA assumes its simplest (*reduced*) form when its center is at the origin and its transverse axis lies along one of the coordinate axes.

When the center is at the origin and the transverse axis lies along the x axis, the equation of the hyperbola is

$$\frac{x^2}{a^2} - \frac{y^2}{b^2} = 1 \qquad (51.1)$$

Then the vertices are at $V(a, 0)$ and $V'(-a, 0)$ and the length of the transverse axis is $V'V = 2a$. The extremities of the conjugate axis are $B'(0, -b)$ and $B(0, b)$ and its length is $B'B = 2b$. The foci are on the transverse axis at $F'(-c, 0)$ and $F(c, 0)$, where

$$c = \sqrt{a^2 + b^2}$$

When the center is at the origin and the transverse axis lies along the y axis, the equation of the hyperbola is

$$\frac{y^2}{a^2} - \frac{x^2}{b^2} = 1 \tag{51.2}$$

Then the vertices are at $V(0, a)$ and $V'(0, -a)$ and the length of the transverse axis is $V'V = 2a$. The extremities of the conjugate axis are $B'(-b, 0)$ and $B(b, 0)$, and its length is $B'B = 2b$. The foci are on the transverse axis at $F(0, c)$ and $F'(0, -c)$, where

$$c = \sqrt{a^2 + b^2}$$

In both cases, the length of a latus rectum is $2b^2/a$, the eccentricity is $e = c/a = \sqrt{a^2 + b^2}/a$, and the directrices are perpendicular to the transverse axis at distances $\pm a^2/c = \pm a/e$ from the center. See Figs. 51-2(a) and (b).

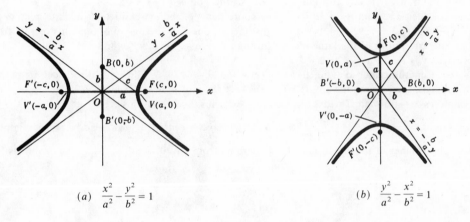

$$(a) \quad \frac{x^2}{a^2} - \frac{y^2}{b^2} = 1 \qquad\qquad\qquad (b) \quad \frac{y^2}{a^2} - \frac{x^2}{b^2} = 1$$

Fig. 51-2

THE STRAIGHT LINES with equations $y = \pm \dfrac{b}{a} x$ are called the *asymptotes* of the hyperbola (*51.1*),

and the lines with equations $x = \pm \dfrac{b}{a} y$ are called the *asymptotes* of the hyperbola (*51.2*).

These lines have the property that the perpendicular distance from a point on a hyperbola to one of them approaches zero as the point moves indefinitely far from the center. (For a proof, see Problem 51.4.)

THE EQUATION OF A HYPERBOLA assumes the *semireduced* form

$$\frac{(x - h)^2}{a^2} - \frac{(y - k)^2}{b^2} = 1 \tag{51.1'}$$

or

$$\frac{(y - k)^2}{a^2} - \frac{(x - h)^2}{b^2} = 1 \tag{51.2'}$$

Here the center is at the point $C(h, k)$ and the transverse axis is parallel respectively to the x axis or to the y axis. The lengths of the transverse and conjugate axes, the distance between the foci, the distance from the center to a directrix, the length of a latus rectum, the slope of the asymptotes, and the eccentricity are as given in the sections above.

EXAMPLE 1. Find the coordinates of the center, vertices, and foci; the lengths of the transverse and conjugate axes; the length of a latus rectum; the eccentricity; and the equations of the directrices and asymptotes of the hyperbola

$$\frac{(x+3)^2}{4} - \frac{(y-1)^2}{25} = 1$$

Sketch the locus.

The center is at the point $C(-3, 1)$. Since, when the equation is put in the reduced or semireduced form, a^2 is always in the positive term on the left, $a^2 = 4$ and $b^2 = 25$; then $a = 2$ and $b = 5$.

The transverse axis is parallel to the x axis (the positive term contains x). To locate the vertices, we move from the center along the transverse axis a distance $a = 2$ to the points $V(-1, 1)$ and $V'(-5, 1)$. To locate the extremities of the conjugate axis, we move from the center perpendicular to the transverse axis a distance $b = 5$ to the points $B'(-3, -4)$ and $B(-3, 6)$. The lengths of the transverse and conjugate axes are $2a = 4$ and $2b = 10$, respectively.

The distance from the center to a focus is $c = \sqrt{a^2 + b^2} = \sqrt{4 + 25} = \sqrt{29}$. To locate the foci, we move from the center along the transverse axis a distance $c = \sqrt{29}$ to the points $F(-3 + \sqrt{29}, 1)$ and $F'(-3 - \sqrt{29}, 1)$.

The length of a latus rectum is $2b^2/a = 25$. The coordinates of the extremities of the latus rectum through F are $(-3 + \sqrt{29}, \frac{27}{2})$ and $(-3 + \sqrt{29}, -\frac{23}{2})$, being at a distance of one-half the length of the latus rectum from F. The coordinates of the extremities of the latus rectum through F' are $(-3 - \sqrt{29}, \frac{27}{2})$ and $(-3 - \sqrt{29}, -\frac{23}{2})$.

The eccentricity is $e = c/a = \sqrt{29}/2$.

The distance from the center to a directrix is $a^2/c = 4/\sqrt{29} = 4\sqrt{29}/29$. Since the directrices are perpendicular to the transverse axis, their equations are $d: x = -3 + 4\sqrt{29}/29$ and $d': x = -3 - 4\sqrt{29}/29$.

The asymptotes pass through C with slopes $\pm b/a = \pm \frac{5}{2}$; thus their equations are $y - 1 = \pm \frac{5}{2}(x + 3)$. Combining these two equations, $\frac{(x+3)^2}{4} - \frac{(y-1)^2}{25} = 0$. Hence they may be obtained most readily by the simple trick of changing the right member of the equation of the hyperbola from 1 to 0. See Fig. 51-3. (See Problems 51.1-51.3.)

Fig. 51-3

THE HYPERBOLAS OF EQUATION $x^2 - y^2 = a^2$ and $y^2 - x^2 = a^2$ whose transverse and conjugate axes are of equal length $2a$ are called *equilateral hyperbolas*. Since their asymptotes $x \pm y = 0$ are mutually perpendicular, the equilateral hyperbola is also called the *rectangular hyperbola*. See Figs. 51-4(a) and (b).

(a) $x^2 - y^2 = a^2$ (b) $y^2 - x^2 = a^2$

Fig. 51-4

TWO HYPERBOLAS such that the transverse axis of each is the conjugate axis of the other, as

$$\frac{x^2}{16} - \frac{y^2}{9} = 1 \quad \text{and} \quad \frac{y^2}{9} - \frac{x^2}{16} = 1$$

are called *conjugate hyperbolas*, each being the conjugate of the other. See Fig. 51-5.

A pair of conjugate hyperbolas have the same center and the same asymptotes. Their foci lie on a circle whose center is the common center of the hyperbolas. (See Problem 51.5.)

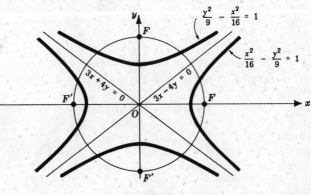

Fig. 51-5

Solved Problems

51.1   For each of the following hyperbolas find the coordinates of the center, the vertices, and the foci; the lengths of the transverse and conjugate axes; the length of a latus rectum; the eccentricity; and the equations of the directrices and asymptotes. Sketch each locus.

(a) $x^2/16 - y^2/4 = 1$ (b) $25y^2 - 9x^2 = 225$ (c) $9x^2 - 4y^2 - 36x + 32y + 8 = 0$.

(a)   Here $a^2 = 16$, $b^2 = 4$, and $c = \sqrt{a^2 + b^2} = 2\sqrt{5}$.

The center is at the origin and the transverse axis is along the x axis (a^2 under x^2). The vertices are on the transverse axis at a distance $a = 4$ from the center; their coordinates are $V(4, 0)$ and $V'(-4, 0)$. The extremities of the conjugate axis are on the y axis at a distance $b = 2$ from the center; their coordinates are $B(0, 2)$ and $B'(0, -2)$.

The foci are on the transverse axis at a distance $c = 2\sqrt{5}$ from the center; their coordinates are $F(2\sqrt{5}, 0)$ and $F'(-2\sqrt{5}, 0)$.

The lengths of the transverse and conjugate axes are $2a = 8$ and $2b = 4$, respectively.

The length of a latus rectum is $2b^2/a = 2$. The eccentricity is $e = c/a = 2\sqrt{5}/4 = \frac{1}{2}\sqrt{5}$.

The directrices are perpendicular to the transverse axis and at a distance $a^2/c = 16/2\sqrt{5} = 8\sqrt{5}/5$ from the center; their equations are $x = \pm 8\sqrt{5}/5$.

The equations of the asymptotes are $x^2/16 - y^2/4 = 0$ or $x = \pm 2y$. See Fig. 51-6.

Fig. 51-6

(*b*) When the equation is put in the form $y^2/9 - x^2/25 = 1$, we find $a^2 = 9$, $b^2 = 25$, and $c = \sqrt{a^2 + b^2} = \sqrt{34}$.

The center is at the origin and the transverse axis is along the y axis. The vertices are on the transverse axis at a distance $a = 3$ from the center; their coordinates are $V(0, 3)$ and $V'(0, -3)$. The extremities of the conjugate axis are on the x axis at a distance $b = 5$ from the center; their coordinates are $B(5, 0)$ and $B'(-5, 0)$.

The foci are on the transverse axis at a distance $c = \sqrt{34}$ from the center; their coordinates are $F(0, \sqrt{34})$ and $F'(0, -\sqrt{34})$.

The lengths of the transverse and conjugate axes are $2a = 6$ and $2b = 10$, respectively.

The length of a latus rectum is $2b^2/a = \frac{50}{3}$. The eccentricity is $e = c/a = \sqrt{34}/3$.

The directrices are perpendicular to the transverse axis and at a distance $a^2/c = 9/\sqrt{34} = 9\sqrt{34}/34$ from the center; their equations are $y = \pm 9\sqrt{34}/34$.

The equations of the asymptotes are $y^2/9 - x^2/25 = 0$ or $5y = \pm 3x$. See Fig. 51-7(*a*).

(*a*)

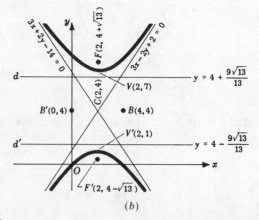

(*b*)

Fig. 51-7

(c) Putting the equation in the form $\dfrac{(y-4)^2}{9} - \dfrac{(x-2)^2}{4} = 1$, we have $a^2 = 9$, $b^2 = 4$, and $c = \sqrt{a^2 + b^2} = \sqrt{13}$.

The center is at the point $C(2,4)$ and the transverse axis is parallel to the y axis along the line $x = 2$. The vertices are on the transverse axis at a distance $a = 3$ from the center; their coordinates are $V(2,7)$ and $V'(2,1)$. The extremities of the conjugate axis are on the line $y = 4$ at a distance $b = 2$ from the center; their coordinates are $B(4,4)$ and $B'(0,4)$.

The foci are on the transverse axis at a distance $c = \sqrt{13}$ from the center; their coordinates are $F(2, 4 + \sqrt{13})$ and $F'(2, 4 - \sqrt{13})$.

The lengths of the transverse and conjugate axes are $2a = 6$ and $2b = 4$, respectively.

The length of a latus rectum is $2b^2/a = \frac{8}{3}$. The eccentricity is $e = c/a = \sqrt{13}/3$.

The directrices are perpendicular to the transverse axis and at a distance $a^2/c = 9/\sqrt{13} = 9\sqrt{13}/13$ from the center; their equations are $y = 4 \pm 9\sqrt{13}/13$.

The equations of the asymptotes are $\dfrac{(y-4)^2}{9} - \dfrac{(x-2)^2}{4} = 0$ or $3x - 2y + 2 = 0$ and $3x + 2y - 14 = 0$. See Fig. 51-7(b).

51.2 Find the equation of the hyperbola, given

(a) Center $(0,0)$, vertex $(4,0)$, focus $(5,0)$.

(b) Center $(0,0)$, focus $(0,-4)$, eccentricity $= 2$.

(c) Center $(0,0)$, vertex $(5,0)$, one asymptote $5y + 3x = 0$.

(d) Center $(-5,4)$, vertex $(-11,4)$, eccentricity $= \frac{5}{3}$.

(e) Vertices $(-11,1)$ and $(5,1)$, one asymptote $x - 4y + 7 = 0$.

(f) Transverse axis parallel to the x axis, asymptotes $3x + y - 7 = 0$ and $3x - y - 5 = 0$, passes through $(4,4)$.

(a) Here $a = CV = 4$, $c = CF = 5$, and $b^2 = c^2 - a^2 = 25 - 16 = 9$. The transverse axis is along the x axis and the equation of the hyperbola is $x^2/16 - y^2/9 = 1$.

(b) Since $c = F'C = 4$ and $e = c/a = 2$, $a = 2$ and $b^2 = c^2 - a^2 = 12$. The transverse axis is along the y axis and the equation of the hyperbola is $y^2/4 - x^2/12 = 1$.

(c) The slope of the asymptote is $-b/a = -\frac{3}{5}$ and, since $a = CV = 5$, $b = 3$. The transverse axis is along the x axis and the equation of the hyperbola is $x^2/25 - y^2/9 = 1$.

(d) Here $a = V'C = 6$ and $e = c/a = \frac{5}{3} = \frac{10}{6}$; then $c = 10$ and $b^2 = c^2 - a^2 = 64$. The transverse axis is parallel to the x axis and the equation of the hyperbola is $(x+5)^2/36 - (y-4)^2/64 = 1$.

(e) The center is at $(-3,1)$, the midpoint of VV'. The slope of the asymptote is $b/a = \frac{1}{4}$ and, since $a = CV = 8$, $b = 2$. The transverse axis is parallel to the x axis and the equation of the hyperbola is $(x+3)^2/64 - (y-1)^2/4 = 1$.

(f) The asymptotes intersect in the center $C(2,1)$. Since the slope of the asymptote $3x - y - 5 = 0$ is $b/a = \frac{3}{1}$, we take $a = m$ and $b = 3m$. The equation of the hyperbola may be written as $(x-2)^2/m^2 - (y-1)^2/9m^2 = 1$.

In order that the hyperbola pass through $(4,4)$, $4/m^2 - 9/9m^2 = 1$ and $m = \sqrt{3}$. Then $a = m = \sqrt{3}$, $b = 3m = 3\sqrt{3}$, and the required equation is $(x-2)^2/3 - (y-1)^2/27 = 1$.

51.3 Find the distance from the right-hand focus of $9x^2 - 4y^2 + 54x + 16y - 79 = 0$ to one of its asymptotes.

Putting the equation in the form $\dfrac{(x+3)^2}{16} - \dfrac{(y-2)^2}{36} = 1$, we find $a^2 = 16$, $b^2 = 36$, and $c = \sqrt{a^2 + b^2} = 2\sqrt{13}$. The right-hand focus is at $F(-3 + 2\sqrt{13}, 2)$ and the equations of the asymptotes are

$$\frac{(x+3)^2}{16} - \frac{(y-2)^2}{36} = 0 \qquad \text{or} \qquad 3x + 2y + 5 = 0 \qquad \text{and} \qquad 3x - 2y + 13 = 0$$

The distance from F to the first asymptote is $\left| \dfrac{3(-3 + 2\sqrt{13}) + 2 \cdot 2 + 5}{-\sqrt{13}} \right| = 6$.

51.4   Prove: The perpendicular distance from a point on the hyperbola $x^2/a^2 - y^2/b^2 = 1$ to an asymptote $y = \pm \dfrac{b}{a} x$ approaches zero as the point moves indefinitely far from the center.

In view of the symmetry of the curve, it will suffice to prove this for the first quadrant. Let $P_1(x_1, y_1)$ be a point in the first quadrant on the hyperbola so that $x_1^2/a^2 - y_1^2/b^2 = 1$. The undirected distance of P_1 from the asymptote $y = \dfrac{b}{a} x$ is

$$\left| \frac{bx_1 - ay_1}{-\sqrt{a^2 + b^2}} \right| = \frac{a^2 b^2}{\sqrt{a^2 + b^2}} \left| \frac{1}{bx_1 + ay_1} \right|$$

since $b^2 x_1^2 - a^2 y_1^2 = (bx_1 - ay_1)(bx_1 + ay_1) = a^2 b^2$ and

$$bx_1 - ay_1 = \frac{a^2 b^2}{bx_1 + ay_1}$$

Now as P_1 moves out along the curve in the first quadrant, both x_1 and y_1 increase indefinitely and $1/(bx_1 + ay_1)$ approaches zero as a limit.

The asymptotes may be located by the following simple device: Through the vertices V and V' draw lines perpendicular to the transverse axis and through the extremities B and B' of the conjugate axis draw lines parallel to the transverse axis forming the rectangle $RR'S'S$. The diagonals of this rectangle are the asymptotes of the hyperbola; for example, the diagonal RS' passes through the center $C(0,0)$ with slope $\dfrac{VR}{CV} = \dfrac{CB}{CV} = \dfrac{b}{a}$. With the asymptotes in position as guide lines, the hyperbola may be sketched readily. See Fig. 51-8.

Fig. 51-8

51.5   Write the equation of the conjugate of the hyperbola $25x^2 - 16y^2 = 400$ and sketch both curves.

The equation of the conjugate hyperbola is $16y^2 - 25x^2 = 400$. The common asymptotes have equations $y = \pm 5x/4$. The vertices of $25x^2 - 16y^2 = 400$ are at $(\pm 4, 0)$. The vertices of $16y^2 - 25x^2 = 400$ are at $(0, \pm 5)$. The curves are shown in Fig. 51-9.

Fig. 51-9

Supplementary Problems

51.6 For each of the following hyperbolas, find the coordinates of the center, the vertices, and the foci; the lengths of the transverse and conjugate axes; the length of the latus rectum; the eccentricity; the equations of the directrices; and the equations of the asymptotes. Sketch each curve.

(a) $4x^2 - 9y^2 = 36$ (c) $x^2 - 4y^2 + 6x + 16y - 11 = 0$

(b) $16y^2 - 9x^2 = 144$ (d) $144x^2 - 25y^2 - 576x + 200y + 3776 = 0$

Ans. (a) $C(0, 0)$, $V(\pm 3, 0)$, $F(\pm\sqrt{13}, 0)$; 6, 4; $\frac{8}{3}$; $\sqrt{13}/3$; $x = \pm 9\sqrt{13}/13$; $2x \pm 3y = 0$.

(b) $C(0, 0)$, $V(0, \pm 3)$, $F(0, \pm 5)$; 6, 8; $\frac{32}{3}$; $\frac{5}{3}$; $y = \pm\frac{9}{5}$; $3x \pm 4y = 0$.

(c) $C(-3, 2)$, $V(-1, 2)$, $V'(-5, 2)$, $F(-3 \pm \sqrt{5}, 2)$; 4, 2; 1; $\sqrt{5}/2$; $x = -3 \pm 4\sqrt{5}/5$;
$x + 2y - 1 = 0$, $x - 2y + 7 = 0$.

(d) $C(2, 4)$, $V(2, 16)$, $V'(2, -8)$, $F(2, 17)$, $F'(2, -9)$; 24, 10; $\frac{25}{6}$; $\frac{13}{12}$; $y = 4 \pm \frac{144}{13}$;
$12x - 5y - 4 = 0$, $12x + 5y - 44 = 0$.

51.7 Find the equation of the hyperbola, given

(a) $V(\pm 5, 0)$, $F(13, 0)$

(b) $C(0, 0)$, $a = 5$, $F(0, 6)$

(c) $V(\pm 12, 0)$; latus rectum equal to $\frac{32}{3}$

(d) $C(0, 0)$; $b = 5$; $d: y = \pm 16\sqrt{41}/41$

(e) $C(2, -3)$, $V(7, -3)$, asymptote $3x - 5y - 21 = 0$

(f) $C(-3, 1)$, $F(-3, 5)$, eccentricity equal to 2

(g) $C(2, 4)$; asymptotes $x + 2y - 10 = 0$, $x - 2y + 6 = 0$; passes through $(2, 0)$

(h) $C(-3, 2)$, $F(2, 2)$, asymptote $4x + 3y + 6 = 0$

Ans. (a) $144x^2 - 25y^2 = 3600$

(b) $25x^2 - 11y^2 + 275 = 0$

(c) $4x^2 - 9y^2 = 576$

(d) $16x^2 - 25y^2 + 400 = 0$

(e) $9x^2 - 25y^2 - 36x - 150y - 414 = 0$

(f) $x^2 - 3y^2 + 6x + 6y + 18 = 0$

(g) $x^2 - 4y^2 - 4x + 32y + 4 = 0$

(h) $16x^2 - 9y^2 + 96x + 36y - 36 = 0$

51.8 Prove: The length of the conjugate axis of a hyperbola is the mean proportional between the length of the transverse axis and the latus rectum.

51.9 Prove: The eccentricity of every equilateral hyperbola is equal to $\sqrt{2}$.

51.10 Prove: The foci of a pair of conjugate hyperbolas are concyclic.

Chapter 52

Introduction to Transformational Geometry

INTRODUCTION TO TRANSFORMATIONS. If you look back at the previous chapters on geometry, you will notice that, while we have concentrated on different topics from chapter to chapter, all the material had one very important thing in common: The *positions* of all the geometric figures were *fixed*. In other words, when we considered a triangle such as $\triangle ABC$ in Fig. 52-1, we did not move it. In this chapter, we consider objects in geometry as they change position. These objects (such as triangles, lines, points, and circles) will move as a result of transformations of the plane.

Fig. 52-1

DEFINITION: By a *transformation of the plane*, we mean a rule that assigns to each point in the plane a different point or the point itself.

Note that *each* point in the plane is assigned to *exactly one point*. Points that are assigned to themselves are called *fixed* points. If point P is assigned to point Q, then we say that *image* of P is Q, and the image of Q is P.

REFLECTIONS. Imagine that a mirror is placed along line m in Fig. 52-2. What would be the image of point S in the mirror? How would you describe S', the image of S? If we actually placed a mirror along m, we would see that the image of S lies on l, on the other side of m, and that the distance from S to O is equal to the distance from O to S' (see Fig. 52-3). We say that S' is the image of S under a reflection in line m. Notice that, under this reflection, O is the image of O.

Fig. 52-2   **Fig. 52-3**

DEFINITION: A *reflection in line m* is a transformation of the plane having the property that the image of any point S not on m is S', where m is the perpendicular bisector of SS'; the image of any point O on m is O itself.

We write $R_m(S) = S'$ to mean S' is the image of S under the reflection in line m.

EXAMPLE 1. Image of a Point. Find the image of (*a*) A, (*b*) B, (*c*) C, (*d*) $\overline{AC}$, and (*e*) $\angle DAC$ under the reflection in line t indicated in Fig. 52-4.

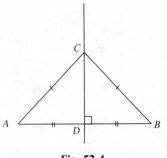

Fig. 52-4

SOLUTIONS

(*a*) B, because t is the perpendicular bisector of $\overline{AB}$

(*b*) A

(*c*) C, because C is on t

(*d*) $\overline{BC}$ (Why?)

(*e*) $\angle DBC$, because D and C are fixed, and $R_t(B) = A$

EXAMPLE 2. Image of a Triangle. What is the image of $\triangle ABC$ in Fig. 52-4 under a reflection in line t?

SOLUTION

We saw that $R_t(A) = B$, $R_t(B) = A$, and $R_t(C) = C$; thus $\triangle ABC$ is its own image. See Problems 52.1–52.3.

LINE SYMMETRY. Notice that the images of angles are angles and the images of segments are segments under a reflection in a line. When a figure is its own image under a reflection in a line (like $\triangle ABC$ in Fig. 52-4), we say the figure has *line symmetry*.

DEFINITION: A figure F exhibits *line symmetry* if there exists a line l such that the image of F under a reflection in line l is F itself. In this case, l is called a *line of symmetry* or an *axis of symmetry*.

Notice that when a figure exhibits line symmetry, all its points are not necessarily fixed. In Fig. 52-4, only points C and D are fixed in triangle ABC. See Problems 52.4–52.6.

EXAMPLE 3. Finding the Axis of Symmetry. In Fig. 52-5, find all axes of symmetry for the regular hexagon $ABCDEF$.

SOLUTION

$\overline{AD}$, $\overline{FC}$, $\overline{BE}$, and the indicated line l are all axes of symmetry. Find two others.

Fig. 52-5

EXAMPLE 4.   Discovering Line Symmetry.   Which of the objects in Fig. 52-6 exhibit line symmetry?

(a)   (b)   (c)   (d)   (e)

Fig. 52-6

SOLUTION

All except (c)

POINT SYMMETRY.   Not only can we transform the plane by reflections in a line, but we can also reflect in a point P. In Fig. 52-7, for example, we can reflect Q in the point P by finding the point Q' such that $QP = PQ'$.

DEFINITION: A *reflection in the point P* is a transformation of the plane such that the image of any point Q except P is Q', where $QP = PQ'$, and the image of P is P (i.e., P is fixed). If point F is its own image under such a transformation, then we say F exhibits *point symmetry.*

Fig. 52-7   **Fig. 52-8**

Figure 52-8 shows a regular hexagon $ABCDEF$, with $AO = OD$. Notice that A is the image of D under the reflection in O. We use the notation $R_O(A) = D$ and $R_O(D) = A$ to indicate that A and D are each other's images under a reflection in point O. See Problem 52.7.

EXAMPLE 5.   Finding Images under a Reflection in a Point.   Referring to Fig. 52-8, find (a) $R_O(B)$ (b) $R_O(C)$ (c) $R_O(\overline{AD})$ (d) $R_O(\angle AOB)$ and (e) $R_O(ABCDEF)$.

ANSWERS

(*a*) *E*   (*b*) *F*   (*c*) $\overline{AD}$ (*d*) *DOE*

(*e*) Hexagon *ABCDEF*. (Thus, *ABCDEF* exhibits point symmetry.)

EXAMPLE 6. Finding Point Symmetry. Which of the following exhibit point symmetry?

(*a*) Squares   (*b*) Rhombuses   (*c*) Scalene triangles   (*d*) *S*

ANSWERS

All except (*c*)

REFLECTIONS AND ANALYTIC GEOMETRY. Since points can change position in transformational geometry, analytic geometry is a particularly useful tool for these transformations. Recall that in analytic geometry, we deal extensively with the positions of points; being able to locate points and determine distances is of great help in exploring the properties of transformations.

EXAMPLE 7. Images under Reflections (Fig. 52-9).

(*a*) What is the image of point *A* under a reflection in the *x* axis? The *y* axis?

(*b*) What is the image of *B* under a reflection in the *y* axis?

(*c*) What is the image of *O* under a reflection in the point *O*?

(*d*) What is the image of *B* under a reflection in the line *y* = *x*?

(*e*) What is the image of *A* under a reflection in the line *x* = −1?

(*f*) What is the image of △*AOB* under a reflection in the *y* axis? Under a reflection in *O*?

Fig. 52-9

SOLUTIONS

(*a*) Point *A'* in Fig. 52-10 is the image of *A* under a reflection in the *x* axis; the coordinates of *A'* are (1, −1). Point *A''* is the image of *A* under a reflection in the *y* axis; *A''* = (−1, 1).

(*b*) Point *B'* in Fig. 52-11 is the image of *B* under a reflection in the *y* axis. Its coordinates are (−1, 0).

(*c*) Point *O* is a fixed point. The point in which we reflect is always fixed.

(*d*) $R_1(B) = B'(0, 1)$ in Fig. 52-12. Notice that line *l* is the perpendicular bisector of $\overline{BB'}$.

(*e*) $R_m(A) = A'(-3, 1)$ in Fig. 52-13. Note that *m* is the perpendicular bisector of *AA'*.

(*f*) The image of △*AOB* under a reflection in the *y* axis is △*A'B'O* in Fig. 52-14(*a*), where *A'* = (−1, 1), *B'* = (−1, 0), and *O* = (0, 0). The image under a reflection in the origin is △*A''B''O* in Fig. 52-14(*b*), where *A''* = (−1, −1), *B''* = (−1, 0), and *O* = (0, 0).

Fig. 52-10

Fig. 52-11

Fig. 52-12

Fig. 52-13

Fig. 52-14

PATTERNS IN REFLECTIONS. We can observe several patterns in the results of Example $7(f)$:

1. The distance from A' to B' in Fig. 52-14(a) equals the distance from A to B. In other words, distance is *preserved* under a reflection. Observe that measures of angles are also preserved. In other words, $m\angle BAO = m\angle B'A'O$ in Fig. 52-14(a), and that property appears to hold for other reflections. As you will see, other properties are preserved as well.

2. Under a reflection in the x axis, the point (a, b) moves to $(a, -b)$; under a reflection in the y axis, (a, b) moves to $(-a, b)$; and under a reflection in the origin, (a, b) moves to $(-a, -b)$. These patterns hold only for these reflections.

EXAMPLE 8.   More Images under Reflections.   In Fig. 52-15, find

(a)   The reflection of C in the y axis

(b)   The reflection of B in the origin

(c)   The reflection of $\triangle CAB$ in the x axis

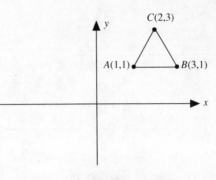

Fig. 52-15

ANSWERS

    (*a*) $(-2, 3)$ (*b*) $(-3, -1)$ (*c*) $\triangle C'A'B'$, where $C' = (2, -3)$, $A' = (1, -1)$, and $B' = (3, -1)$

TRANSLATIONS. Let us transform $\triangle ABC$ in Fig. 52-16(*a*) by adding 1 to each *x* coordinate and 2 to each *y* coordinate. The result is shown in Fig. 52-16(*b*). Notice that $\triangle ABC$ does not change shape, but it does move in the plane, in the direction of ray $\overrightarrow{OD}$, where $D = (1, 2)$. The *x* coordinate of *D* is the "amount" by which the *x* coordinates of the triangle are shifted, and the *y* coordinate of *D* is the "amount" by which the *y* coordinates are shifted. We call this kind of transformation a *translation*.

        (*a*)   (*b*)

Fig. 52-16

DEFINITION: A *translation* is a transformation of the plane such that the image of every point (a, b) is the point $(a + h, b + k)$, where *h* and *k* are given.

    A translation has the effect of moving every point the *same distance* in the *same direction*. We use the notation $T_{(h,k)}(a, b)$ to mean the image of (a, b) under a translation of *h* units in the *x* direction and *k* units in the *y* direction.

    As in a reflection, distance and angle measure are preserved in a translation.

EXAMPLE 9. **Finding the Image of a Point.** Find $T_{(-1,1)}(1, 4)$ and $T_{(-1,1)}(-1, 2)$.

SOLUTION

$$T_{(-1,1)}(1, 4) = (1 + (-1), 4 + 1) = (0, 5) \qquad T_{(-1,1)}(-1, 2) = (-1 + (-1), 2 + 1) = (-2, 3)$$

Notice in Fig. 52-17 that $(1, 4)$ and $(-1, 2)$ are translated the same number of units in the same dirction by the same translation *T*.

Fig. 52-17

EXAMPLE 10. Finding the Image of a Triangle.   Find the image of $\triangle ABC$ under the translation $T_{(1,2)}$, where $A = (0, 0)$, $B = (1, 1)$, and $C = (1, 0)$.

SOLUTION

$T_{(1,2)}(0, 0) = (1, 2)$, $T_{(1,2)}(1, 1) = (2, 3)$, and $T_{(1,2)}(1, 0) = (2, 2)$. Hence the image of $\triangle ABC$ is $\triangle A'B'C'$ in Fig. 52-18. All points are translated along ray $\overline{AA'} = \overline{OA'}$, where $A'(1, 2)$ has the coordinates of the translation.

Fig. 52-18

EXAMPLE 11. Finding the Image from Another Image.   Under a certain translation, $T(5, 2) = (7, 1)$. Find $T(-3, 6)$ under the same translation.

SOLUTION

We have $T_{(h,k)}(5, 2) = (7, 1)$. Thus $5 + h = 7$, or $h = 2$; and $2 + k = 1$, or $k = -1$. Then $T_{(2,-1)}(-3, 6) = (2 + (-3), -1 + 6) = (-1, 5)$.

EXAMPLE 12. Finding Various Images under Translation.

(a) Find $T_{(-1,0)}(6, 2)$.

(b) Find h and k if $T_{(h,k)}(1, 7) = (0, 0)$.

(c) Find the image of square $ABCD$ under the translation $T_{(1,1)}$, where $A = (0, 0)$, $B = (1, 0)$, $C = (0, 1)$, and $D = (1, 1)$.

(d) Find $T_{(h,k)}(1, 6)$ if $T_{(h,k)}(4, 1) = (0, -7)$.

(e) Find all fixed points under $T_{(-1,4)}$.

SOLUTIONS

(a) $T(6, 2) = (6 + (-1), 2 + 0) = (5, 2)$ (b) $h = 0 - 1 = -1; k = 0 - 7 = -7$

(c) $A'B'C'D'$, where $A' = (1, 1)$, $B' = (2, 1)$, $C' = (1, 2)$, and $D' = (2, 2)$

(d) $h = 0 - 4 = -4$ and $k = -7 - 1 = -8$, so $T(1, 6) = (-3, -2)$.

(e)   Only $T_{(0,0)}$ has fixed points. Any other translation, including $T_{(-1,4)}$, has none.

EXAMPLE 13.   Finding Images of Figures.   Let $A = (1, 1)$, $B = (2, 2)$, and $C = (3, 1)$. Find the image under $T_{(2,-1)}$ of (a) $\overline{AB}$, (b) $\triangle ABC$, and (c) $\angle CBA$.

SOLUTIONS

(a) $\overline{A'B'}$, where $A' = (3, 0)$ and $B' = (4, 1)$ (b) $A'B'C'$, with $C' = (5, 0)$ (c) $\angle C'B'A'$

ROTATIONS.   Consider square $ABCD$ in Fig. 52-19(a). Suppose we were to rotate that square counterclockwise 90° about P, as shown by the arrow. (Imagine that the square is separate from the page, but held to it by a pin through point P.) Then

The image of B would be A.
The image of D would be C.
The image of C would be B.
The image of A would be D.

Now consider point S in Fig. 52-19(b). We can rotate it counterclockwise by, say, 50° about P, as if it were one end of a ruler that was nailed to the page at P. The image of S' is in the diagram.

In both these rotations, the segment from P to the point being rotated is congruent to the segment from P to the image of that point.

(a)

(b)

Fig. 52-19

DEFINITION: A *rotation* through an angle of measure θ degrees about a point P is a transformation of the plane such that the image of P is P and, for any point $B \neq P$, the image of B is B', where $m\triangle BPB' = \theta$ and $BP = B'P$.

Figure 52-20 shows P, B, B', and θ. If $\theta > 0$, the rotation is counterclockwise. If $\theta < 0$, the rotation is clockwise.

We use the notation $\mathrm{Rot}_{(P,\theta)}(B)$ to mean the rotation of point B about point P through $\theta°$. Segment lengths and measures of angles are preserved under rotations.

Let us rotate point B in Fig. 52-21 through 180° about O. The image of B is $B'(-2, 2)$. Notice that this is also the image of B under the reflection in O.

Fig. 52-20   Fig. 52-21

Now let us rotate point B through $90°$ about O. The image here is $B''(-2, 2)$. Notice that this is the image of B under the reflection in the y axis.

Be Careful. These two similarities of transformations are coincidences. They come about because we are rotating $90°$ and $180°$ *about the origin*. Do not generalize beyond these cases! Note, though, that these coincidences yield the following formulas:

$$\mathrm{Rot}_{(P, 90°)}(a, b) = (-b, a) \qquad \text{and} \qquad \mathrm{Rot}_{(O, 180°)}(a, b) = (-a, -b)$$

EXAMPLE 14.   Finding the Rotation of a Point.   Let $A = (1, 3)$ and $B = (2, 1)$, and find (a) $\mathrm{Rot}_{(O, 90°)}(A)$, (b) $\mathrm{Rot}_{(O, 90°)}(B)$, and (c) the image of AB under a rotation of $90°$ about O.

SOLUTIONS

(a) $\mathrm{Rot}_{(O, 90°)}(a, b) = (-b, a) = (-3, 1)$ (b) $\mathrm{Rot}_{(O, 90°)}(2, 1) = (-1, 2)$

(c)   The image of $\overline{A'B'}$, where $A' = (-3, 1)$ and $B' = (-1, 2)$, as shown in Fig. 52-22

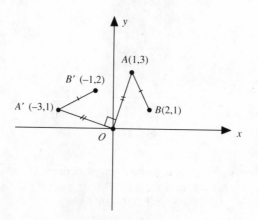

Fig. 52-22

EXAMPLE 15.   Finding the Image of a Triangle

(a)   Find the image of $\triangle ABC$ under a rotation of $180°$ about O, if $A = (1, 3)$, $B = (2, 1)$, and $C = (1, 1)$.

(b)   Find the image of $\angle BAC$.

SOLUTIONS

(a) $\text{Rot}_{(O,180°)}(A) = (-1, -3) = A'$ (b)   The image of $\angle BAC$ is $\angle B'A'C'$.

$\text{Rot}_{(O,180°)}(B) = (-2, -1) = B'$

$\text{Rot}_{(O,180°)}(C) = (-1, -1) = C'$

The image of $\triangle ABC$ is $\triangle A'B'C'$.

SYMMETRY OF ROTATION.   The image of the square in Fig. 52-19 under a rotation of 90° is the square itself. This is also true for a rotation of −90°, or 180°, and so on.

EXAMPLE 16.   Determining Rotational Symmetry (Fig. 52-23)

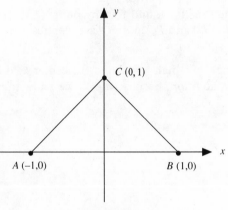

Fig. 52-23

(a)   Find $\text{Rot}_{(O,45°)}(O)$.

(b)   Find $\text{Rot}_{(O,90°)}$ of A, B, and C.

(c)   Find $\text{Rot}_{(O,-90°)}$ of A, B, and C.

(d)   Find the image of $\triangle ABC$ under $\text{Rot}_{(O,-90°)}$.

(e)   Does $\triangle ABC$ exhibit rotational symmetry?

SOLUTIONS

(a) $\text{Rot}(O) = 0$

(b) $\text{Rot}(A) = A'(0, 1)$; $\text{Rot}(B) = B'(0, 1)$; $\text{Rot}(C) = C'(-1, 0)$

(c) $\text{Rot}(A) = A''(0, 1)$; $\text{Rot}(B) = B''(0, -1)$; $\text{Rot}(C) = C''(1, 0)$

(d) $\text{Rot}(\triangle ABC) = A''B''C''$

(e)   No, because it is not its own image for any rotation except one of 360°.

DILATIONS.   Suppose we blew a balloon up most of the way and traced its outline, and then blew it up all the way and traced it again. The outlines might look like those in Fig. 52-24. Although the balloon has changed size from (a) to (b) its shape has not changed. Notice that if C is on AB, then its image C' is on $\overline{A'B'}$. Such a transformation in the plane is called a *dilation* (or dilatation). The "reverse" transformation is also a dilation: The balloon could be reduced in size in a transformation from that in (b) to that in (a).

Fig. 52-24

DEFINITION: Given a point P in the plane and a positive number n, a transformation of the plane having the following properties is called a *dilation of n*, and P is called the *center of dilation*: Point P is fixed, and for any point Q, the image of Q is the point Q' such that $PQ' = (n)(PQ)$ and $\overrightarrow{PQ}$ and $\overrightarrow{PQ'}$ are identical rays. The point Q' is usually denoted $D_n(Q)$.

Figure 52-25 shows a dilation in which $n = 2$ and the center of dilation is P. Hence, $D_2(A) = A'$, $D_2(B) = B'$, and $D_2(P) = P$. In addition, because $n = 2$, we know that $PA' = 2PA$ and $PB' = 2PB$.

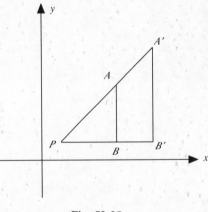

Fig. 52-25

Several properties of dilations are evident in Fig. 52-25:

1. Dilations *do not* preserve distance.
2. The image of a figure is similar to the figure under a dilation. In Fig. 52-25, $\triangle PAB \sim PB'B'$.
3. Angles *are* preserved under dilations (because of item 2 above).

When the center of a dilation is $O = (0, 0)$, we can find the images of points very easily: $D_n(x, y) = (nx, ny)$.

EXAMPLE 17.   Finding the Dilation of a Triangle.   Find the image of $\triangle ABC$ in Fig. 52-26 under a dilation of $n = \frac{1}{2}$ with center of dilation at $(0, 0)$.

SOLUTION

$D_{1/2}(1, 1) = (\frac{1}{2}, \frac{1}{2}) = B'$ as shown in Fig. 52-26. Also, $D_{1/2}(1, 0) = (\frac{1}{2}, 0) = A'$ and $D_{1/2}(2, 1) = (1, \frac{1}{2}) = C'$. Then $\triangle A'B'C'$ is the image of $\triangle BAC$, and $\triangle B'A'C' \sim \triangle BAC$. Note that the image here is smaller than the original triangle because $n = \frac{1}{2}$.

Fig. 52-26

EXAMPLE 18.   Finding an Unknown.   Given that $D_n(8, 0) = (1, 0)$, find n for a dilation in which $(0, 0)$ is the center of dilation.

SOLUTION

Since the origin is the center of dilation, $(1, 0) = (n8, n0)$. Therefore, $8n = 1$ and $n = \frac{1}{8}$.

EXAMPLE 19.   Dilating a Square.   Draw a square $ABCD$ in the coordinate plane such that $A = (1, 1)$, $B = (1, 2)$, $C = (2, 1)$, and $D = (2, 2)$. Then,

(*a*)   With O as the center of dilation, find the image of $ABCD$ under a dilation with $n = \frac{1}{3}$.

(*b*)   Find the midpoint M of $\overline{AB}$ and the midpoint M' of $\overline{A'B'}$.

(*c*)   Find $D_{1/3}(M)$.

SOLUTIONS

(*a*)   For a dilation with center $(0, 0)$ and $n = \frac{1}{3}$, we have $D(x, y) = (\frac{1}{3}x, \frac{1}{3}y)$. The image of $ABCD$ is $A'B'C'D'$, where $A' = (\frac{1}{3}, \frac{1}{3})$, $B' = (\frac{1}{3}, \frac{2}{3})$, $C' = (\frac{2}{3}, \frac{1}{3})$, and $D' = (\frac{2}{3}, \frac{2}{3})$.

(*b*) $M = (\frac{1}{2}(1 + 1), \frac{1}{2}(1 + 2)) = (1, \frac{3}{2})$ and $M' = (\frac{1}{2}(1), \frac{1}{2}(\frac{2}{3})) = (\frac{1}{3}, \frac{1}{2})$.

(*c*) $D(M) = M'$

PROPERTIES OF TRANSFORMATIONS.   We are now in a position to summarize the properties of transformations. In particular, we are interested in what is preserved under each kind of transformation.

1.   Reflections preserve (*a*) distance, (*b*) angle measure, (*c*) midpoints, (*d*) parallelism, and (*e*) collinearity.
2.   Translations preserve these same five properties, (*a*) through (*e*).
3.   Rotations preserve all five properties as well.
4.   Dilations preserve all except distance, that is, (*b*) through (*e*).

Solved Problems

52.1   Find the image of each of the following under the reflection in line *t* in Fig. 52-27: (*a*) point D (*b*) point C (*c*) point B (*d*) $\overline{AC}$.

   Ans.   (*a*) C (*b*) D (*c*) B (*d*) $\overline{AD}$

Fig. 52-27(*a*)   Fig. 52-27(*b*)

52.2   Find the image of rectangle *ABCD* under the reflection in line *t* in Fig. 52-27.

     Ans.   Rectangle *ABCD*

52.3   Is it true or false that every circle is its own image under a reflection in a diameter?

     Ans.   True

52.4   Find all axes of symmetry for the rectangle in Fig. 52-27.

     Ans.   *t* is one such axis. Are there any which are not diagonals?

52.5   Give (or draw) an example of a five-sided polygon that does not exhibit line symmetry.

     Ans.   There are many, Fig. 52-27(*b*) is one.

52.6   Explain why each figure in Fig. 52-28 exhibits line symmetry.

     Hint: For each, find an axis of symmetry.

(*a*)   (*b*)   (*c*)   (*d*)

Fig. 52-28

52.7   In Fig. 52-29, find (*a*) $R_O(B)$ (*b*) $R_O(A)$ (*c*) $R_O(O)$ (*d*) $R_O(\triangle AOB)$.

     Ans.   (*a*) $(-2, 0)$ (*b*) $(-2, -2)$ (*c*) O'

            (*d*) $\triangle A'OB'$ where $A' = (-2, -2)$ and $B' = (-2, 0)$

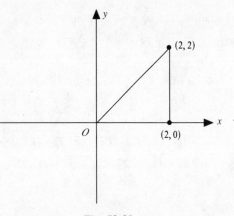

Fig. 52-29

Supplementary Problems

52.8 In Fig. 52-30, find:

(a) The image of E under a reflection in the y axis   *Ans.* E

(b) The image of B under a reflection in the y axis   *Ans.* C

(c) The image of $\overline{AB}$ under a reflection in the y axis   *Ans.* BA

(d) The image of BC under a reflection in the x axis   *Ans.* $B'C'$ where $B' = (-1, -1)$; $C' = (1, -1)$

(e) The image of $ABCD$ under a reflection in the x axis   *Ans.* $A'B'C'D'$ where $A' = A$, $D' = D$

Fig. 52-30

52.9 Find (a) $T_{(1,3)}(2, 8)$ (b) $T_{(1,3)}(6, 5)$ (c) $T_{(1,3)}(0, 0)$ (d) $T_{(1,3)}(1, 1)$.

  Ans.   (a) (3, 11)   (b) (7, 8)   (c) (1, 3)   (d) (2, 4)

52.10 Find the image of rectangle $ABCD$ in Fig. 52-30 under the translation $T_{(3,6)}$.

  Ans. $A'B'C'D'$ where $A' = (2, 6)$, etc.

52.11 Under a particular translation, $T(3, 4) = (0, 0)$. Find $T(-8, -6)$ under that same translation.

 Ans. Then T, here, $= T_{(-3,-4)}$; $T(-8, -6) = (-11, -10)$

52.12 Find (a) $T_{(4,3)}(0, -6)$; (b) $T_{(h,k)}(3, 7)$; (c) $T_{(h,k)}(e, f)$; (d) $T_{(h,k)}(4, 1)$ if $T_{(h,k)}(1, 1) = (2, 2)$.

 Ans. (a) $(4, -3)$;   (b) $(4 + h, 7 + k)$;   (c) $(e + h, f + k)$;   (d) $(5, 2)$

52.13 In Fig. 52-31, find (a) $T_{(0,0)}(EF)$; (b) $T_{(1,0)}(EF)$; (c) $T_{(0,1)}(EF)$; (d) $T_{(0,0)}(\triangle OEF)$.

 Ans. (a) EF;   (b) $E' = (1, 1)$; $F' = (2, 0)$;   (c) $E' = (0, 2)$; $F' = (1, 1)$;   (d) $O' = O$, etc.

Fig. 52-31

52.14 Let $x = (4, 1)$ and $y = (0, 3)$, and find (a) $\text{Rot}_{(O,90°)}(x)$; (b) $\text{Rot}_{(O,90°)}(y)$; (c) the image of yx under a rotation of 90° about O.

 Ans. (a) $(-1, 4)$;   (b) $(-3, 0)$;   (c) $y'x'$, where $y' = (-3, 0)$, $x' = (-1, 4)$

52.15 Find the image of $\triangle EOF$ in Fig. 52-31 under a rotation of 180° about O.

 Ans. $O' = 0$, $E' = (0, -1)$, $F' = (-1, 0)$; $\triangle E'O'F'$ is the image

52.16 In Fig. 52-32, find

 (a) $\text{Rot}_{(O,90°)}(A)$ and $\text{Rot}_{(O,90°)}(B)$

 (b) $\text{Rot}_{(O,-90°)}(A)$ and $\text{Rot}_{(O,90°)}(C)$

 (c)   the image of $OABC$ under $\text{Rot}_{(O,-90°)}$.

 Ans. (a) $(0, -1)$ and $(-1, 1)$ (b) $(0, 1)$ and $(-1, 0)$ (c) $O'A'B'C'$, where $O' = 0$,
     $A' = (0, 1)$, $B' = (-1, 1)$, $C' = (-1, 0)$

Fig. 52-32

52.17 Find (a) $D_{1/3}(-1, 3)$, (b) $D_{1/2}(5, -3)$, (c) $D_4(0, 0)$, and (d) $D_5(1, 6)$, where the center of dilation in each case is the origin.

>   *Ans.*   (a) $(-\frac{1}{3}, 1)$;   (b) $(\frac{5}{2}, -\frac{3}{2})$;   (c) $(0, 0)$;   (d) $(5, 30)$

52.18 If the center of a dilation D is $(0, 0)$ and $D_n(3, 6) = (5, 10)$, find n and $D_n(0, -7)$.

>   *Ans.* $n = \frac{5}{3}$; $(0, -\frac{35}{3})$

52.19 For A and B as given in Fig. 52-33 and dilations with centers at $(0, 0)$, find

(a)   The image of $\triangle OAB$ under a dilation with $n = \frac{1}{2}$

(b)   The image of the midpoint of $\overline{AB}$ under a dilation with $n = 3$

>   *Ans.*   (a) $O' = 0$, $A' = (0, \frac{1}{2})$, $B' = (\frac{1}{2}, 0)$; image is $\triangle OA'B'$ (b)   midpoint: $(\frac{1}{2}, \frac{1}{2})$; image is $(\frac{3}{2}, \frac{3}{2})$

Fig. 52-33

Chapter 53

Transformation of Coordinates

THE MOST GENERAL EQUATION of the second degree in x and y has the form

$$Ax^2 + 2Bxy + Cy^2 + 2Dx + 2Ey + F = 0 \qquad (53.1)$$

If (53.1) can be factored so that we have $(ax + by + c)(dx + ey + f) = 0$, the locus consists of two straight lines; if $B = 0$, $A = C$, the locus of (53.1) is a circle; otherwise, the locus is one of the conics of Chapters 52–54.

The locus of the equation

$$20x^2 - 24xy + 27y^2 + 24x - 54y - 369 = 0 \qquad (1)$$

obtained in Problem 48.7, and the locus of the equation

$$11x^2 + 36y^2 - 369 = 0 \qquad (2)$$

obtained in Problem 48.4, are identical ellipses. The difference in equations is due to their positions with respect to the coordinate axes.

In order to make a detailed study of the loci represented by (53.1), say (1), it will be necessary to introduce some device to change (1) into (2). The operations by which (1) is eventually replaced by (2) are two *transformations*. The general effect of these transformations may be interpreted as follows: Each point (x, y) of the plane remains fixed but changes its name, i.e., its coordinates, in accordance with a stated law, called the *equations of the transformation*.

TRANSLATION OF THE COORDINATE AXES. Recall that the transformation which moves the coordinate axes to a new position while keeping them always parallel to their original position is called a *translation*. In Fig. 53-1, $\overleftrightarrow{Ox}$ and $\overleftrightarrow{Oy}$ are the axes and O is the origin of the original system of coordinates while $\overleftrightarrow{O'x'}$ and $\overleftrightarrow{O'y'}$ are the axes and O' is the origin of the new (translated) system.

Fig. 53-1

Each point in the plane will now have two sets of coordinates, the original set being the directed distances in proper order of the point from the original axes and the new set being the directed distances from the new axes. In order to avoid errors we propose to write the coordinates of a point when referred to the original system as, for example, $A(a, b)$ and the coordinates when referred to the new system as $A(c, d)'$. Also, we shall find it convenient at times to speak of the unprimed and primed systems.

If the axes with the origin O are translated to a new position with origin O' having coordinates (h, k) when referred to the original system and if the coordinates of any point are (x, y) before and $(x', y')'$ after the translation, then the equations of transformation are

$$x = x' + h \qquad y = y' + k \qquad\qquad (53.2)$$

EXAMPLE 1.   By means of a translation, transform $3x^2 + 4y^2 - 12x + 16y - 8 = 0$ into another equation which lacks terms of the first degree.

> **First Solution.**   When the values of x and y from (53.2) are substituted in the given equation, we obtain

$$3(x' + h)^2 + 4(y' + k)^2 - 12(x' + h) + 16(y' + k) - 8 = 0$$

or $\qquad 3x'^2 + 4y'^2 + (6h - 12)x' + (8k + 16)y' + 3h^2 + 4k^2 - 12h + 16k - 8 = 0 \qquad (3)$

The equation will lack terms of the first degree provided $6h - 12 = 0$ and $8k + 16 = 0$, that is, provided $h = 2$ and $k = -2$. Thus, the translation $x = x' + 2$, $y = y' - 2$ reduces the given equation to $3x'^2 + 4y'^2 - 36 = 0$. The locus, an ellipse, together with the original and new system of coordinates are shown in Fig. 53-2.

Fig. 53-2

> **Second Solution.**   We put the given equation in the form

$$3(x^2 - 4x) + 4(y^2 + 4y) = 8$$

and complete the squares to obtain

$$3(x^2 - 4x + 4) + 4(y^2 + 4y + 4) = 8 + 3(4) + 4(4) = 36$$

or $\qquad\qquad 3(x - 2)^2 + 4(y + 2)^2 = 36 \qquad\qquad (4)$

The transformation $x - 2 = x'$, $y + 2 = y'$ or $x = x' + 2$, $y = y' - 2$ reduces (4) to $3x'^2 + 4y'^2 = 36$ as before. (See Problems 53.1–53.3.)

ROTATION OF THE COORDINATE AXES.   Recall that the transformation which holds the origin fixed while rotating the coordinate axes through a given angle is called a *rotation*.

If, while the origin remains fixed, the coordinate axes are rotated counterclockwise through an angle of measure θ, and if the coordinates of any point P are (x, y) before and $(x', y')'$ after the rotation, the equations of transformation are

$$x = x' \cos \theta - y' \sin \theta \qquad y = x' \sin \theta + y' \cos \theta \qquad\qquad (53.3)$$

since, from Fig. 53-3,

$$x = OM = ON - MN = ON - RQ = OQ \cos \theta - QP \sin \theta = x' \cos \theta - y' \sin \theta$$

and $\qquad y = MP = MR + RP = NQ + RP = OQ \sin \theta + QP \cos \theta = x' \sin \theta + y' \cos \theta$

Fig. 53-3

EXAMPLE 2. Transform the equation $x^2 + \sqrt{3}\,xy + 2y^2 - 5 = 0$ by rotating the coordinate axes through the angle 60°.

The equations of transformation are

$$x = x' \cos 60° - y' \sin 60° = \tfrac{1}{2}(x' - \sqrt{3}y'), \qquad y = x' \sin 60° + y' \cos 60° = \tfrac{1}{2}(\sqrt{3}x' + y').$$

Substituting for x and y in the given equation, we obtain

$$\tfrac{1}{4}(x' - \sqrt{3}y')^2 + \tfrac{1}{4}\sqrt{3}(x' - \sqrt{3}y')(\sqrt{3}x' + y') + \tfrac{1}{2}(\sqrt{3}x' + y')^2 - 5 = 0$$

$$\tfrac{5}{2}x'^2 + \tfrac{1}{2}y'^2 - 5 = 0 \qquad \text{or} \qquad 5x'^2 + y'^2 = 10$$

The locus, an ellipse, together with the original and new systems of coordinates are shown in Fig. 53-4. (See Problems 53.4–53.6.)

Fig. 53-4

THE SEMIREDUCED FORM OF THE SECOND-DEGREE EQUATION.

Under a rotation of the coordinate axes with equations of transformation (*53.3*), the general equation of the second degree

$$Ax^2 + 2Bxy + Cy^2 + 2Dx + 2Ey + F = 0 \tag{53.1}$$

becomes

$$A'x'^2 + 2B'x'y' + C'y'^2 + 2D'x' + 2E'y' + F' = 0 \tag{53.1'}$$

where

$$A' = A \cos^2 \theta + 2B \sin \theta \cos \theta + C \sin^2 \theta$$

$$B' = (C - A) \sin \theta \cos \theta + B(\cos^2 \theta - \sin^2 \theta)$$

$$= \tfrac{1}{2}(C - A) \sin 2\theta + B \cos 2\theta \tag{53.4}$$

$$C' = A \sin^2 \theta - 2B \sin \theta \cos \theta + C \cos^2 \theta$$

$$D' = D \cos \theta + E \sin \theta \qquad E' = E \cos \theta - D \sin \theta \qquad F' = F$$

If $B \neq 0$, $(53.1')$ will lack the term in $x'y'$ if θ is such that $0 < m\measuredangle\theta < 90°$ and $\tan 2\theta = \dfrac{2B}{A - C}$ when $A \neq C$, and $m\measuredangle\theta = 45°$ when $A = C$. Under this transformation, the general equation (53.1), takes the form

$$A'x'^2 + C'y'^2 + 2D'x' + 2E'y' + F' = 0 \qquad (53.1'')$$

which will be called the *semireduced form* of the second-degree equation.

Under *any* rotation of the coordinate axes, the quantities $A + C$ and $B^2 - AC$ are unchanged or *invariant*; that is, $A + C = A' + C'$ and $B^2 - AC = B'^2 - A'C'$. When (53.1) is transformed into $(53.1'')$, $B' = 0$ and $B^2 - AC = -A'C'$. Then

If $B^2 - AC < 0$, A' and C' agree in sign and (53.1) represents a real ellipse, a point ellipse, or an imaginary ellipse.

If $B^2 - AC = 0$, either $A' = 0$ or $C' = 0$; now $(53.1'')$ contains either a term in x'^2 or y'^2 (but not both), and (53.1) represents a parabola or a pair of parallel lines.

If $B^2 - AC > 0$, A' and C' differ in sign and (53.1) represents a hyperbola or a pair of intersecting lines.

THE REDUCED FORM OF THE SECOND-DEGREE EQUATION. Under a suitable translation the semireduced form $(53.1'')$ of the second-degree equation takes the *reduced form*

$$A'x''^2 + C'y''^2 = F'' \qquad \text{when } A'C' \neq 0$$

and
$$x''^2 = Gy'' \quad \text{or} \quad y''^2 = Hx'' \qquad \text{when } A'C' = 0$$

If $B^2 - AC \neq 0$, it is a matter of individual preference whether the rotation is performed before or after the translation of axes; however, if $B^2 - AC = 0$, the axes *must* be rotated first. (See Problems 53.7–53.9.)

Solved Problems

53.1   If the equations of translation are $x = x' - 3$, $y = y' + 4$, find

    (a)   The coordinates of $O(0, 0)$ when referred to the primed system of coordinates

    (b)   The coordinates of $O'(0, 0)'$ when referred to the unprimed system

    (c)   The coordinates of $P(5, -3)$ when referred to the primed system

    (d)   The coordinates of $P(5, -3)'$ when referred to the unprimed system

    (e)   The equation of l: $2x - 3y + 18 = 0$ when referred to the primed system

        (a)   For $x = 0$, $y = 0$ the equations of transformation yield $x' = 3$, $y' = -4$; thus, in the primed system we have $O(3, -4)'$.

        (b)   For $x' = 0$, $y' = 0$ the equations of transformation yield $x = -3$, $y = 4$; thus, in the unprimed system we have $O'(-3, 4)$.

        (c)   Here $x = 5$, $y = -3$; then $x' = 8$, $y' = -7$. Thus, we have $P(8, -7)'$.

        (d)   Here $x' = 5$, $y' = -3$; then $x = 2$, $y = 1$. Thus, we have $P(2, 1)$.

        (e)   When the values for x and y are substituted in the given equation, we have $2(x' - 3) - 3(y' + 4) + 18 = 2x' - 3y' = 0$ as the equation of l in the primed system. Note that the new origin was chosen on the line l.

53.2   Transform each of the following equations into another lacking terms of the first degree:

(a) $x^2 + 4y^2 - 2x - 12y + 1 = 0$,   ($b$) $9x^2 - 16y^2 - 36x - 96y - 252 = 0$,   ($c$) $xy + 4x - y - 8 = 0$.

(a)   Since the given equation lacks a term in xy, we use the second method of Example 1. We have

$$(x^2 - 2x) + 4(y^2 - 3y) = -1 \qquad (x^2 - 2x + 1) + 4(y^2 - 3y + \tfrac{9}{4}) = -1 + 1 + 4(\tfrac{9}{4}) = 9$$

and $\qquad\qquad\qquad\qquad (x - 1)^2 + 4(y - \tfrac{3}{2})^2 = 9$

This equation takes the form $x'^2 + 4y'^2 = 9$ under the transformation $x - 1 = x'$, $y - \tfrac{3}{2} = y'$ or $x = x' + 1$, $y = y' + \tfrac{3}{2}$.

(b)   We have

$$9(x^2 - 4x) - 16(y^2 + 6y) = 252 \qquad 9(x^2 - 4x + 4) - 16(y^2 + 6y + 9) = 252 + 36 - 144 = 144$$

and $\qquad\qquad\qquad\qquad 9(x - 2)^2 - 16(y + 3)^2 = 144$

This equation takes the form $9x'^2 - 16y'^2 = 144$ under the transformation $x - 2 = x'$, $y + 3 = y'$ or $x = x' + 2$, $y = y' - 3$.

(c)   Since the given equation contains a term in xy, we must use the first method of Example 1. We have, using Equations (*53.2*),

$$(x' + h)(y' + k) + 4(x' + h) - (y' + k) - 8 = x'y' + (k + 4)x' + (h - 1)y' + hk + 4h - k - 8 = 0$$

The first-degree terms will disappear provided $k + 4 = 0$ and $h - 1 = 0$; that is, provided we take $h = 1$ and $k = -4$. For this choice, the equation becomes $x'y' - 4 = 0$.

53.3   By a translation of the axes, simplify each of the following:

(a) $x^2 + 6x - 4y + 1 = 0$,   (b) $y^2 + 4y + 8x - 2 = 0$.

(a)   Since the given equation lacks a term in y^2, it is not possible to use the second method of Example 1. Using the transformation (*53.2*), we find

$$(x' + h)^2 + 6(x' + h) - 4(y' + k) + 1 = x'^2 + 2(h + 3)x' - 4y' + h^2 + 6h - 4k + 1 = 0$$

If we take $h = -3$, the term in x' disappears but it is clear that we cannot make the term in y' disappear. However, in this case, we may make the constant term

$$h^2 + 6h - 4k + 1 = (-3)^2 + 6(-3) - 4k + 1 = -8 - 4k$$

disappear by taking $k = -2$. Thus, the transformed equation becomes

$$x'^2 - 4y' = 0 \qquad \text{or} \qquad x'^2 = 4y'$$

It is now clear that this simplification may be effected by the following variation of the second method of Example 1:

$$x^2 + 6x = 4y - 1 \qquad x^2 + 6x + 9 = 4y - 1 + 9 = 4y + 8$$

or $\qquad\qquad\qquad\qquad (x + 3)^2 = 4(y + 2)$

Then the transformation $x + 3 = x'$, $y + 2 = y'$ or $x = x' - 3$, $y = y' - 2$ reduces the equation to $x'^2 = 4y'$.

(b)   We have $y^2 + 4y = -8x + 2$, $y^2 + 4y + 4 = -8x + 2 + 4 = -8x + 6$, $(y + 2)^2 = -8(x - \tfrac{3}{4})$, and finally $y'^2 = -8x'$ under the transformation $x = x' + \tfrac{3}{4}$, $y = y' - 2$.

53.4   Write the equations of transformation for a rotation of the coordinate axes through an angle of 45° and use them to find

(a)   The coordinates of $P(\sqrt{2}, 3\sqrt{2})'$ when referred to the original (unprimed) system

(b)   The coordinates of $O(0, 0)$ when referred to the new (primed) system

(c)   The coordinates of $P(\sqrt{2}, 3\sqrt{2})$ when referred to the primed system (*d*)   The equation of the line *l*: $x + y + 3\sqrt{2} = 0$ when referred to the primed system

(*e*)   The equation of the line *l*: $3x - 3y + 4 = 0$ when referred to the primed system

The equations of transformation are

$$x = x' \cos 45° - y' \sin 45° = \frac{1}{\sqrt{2}}(x' - y')$$

$$y = x' \sin 45° + y' \cos 45° = \frac{1}{\sqrt{2}}(x' + y')$$

(*a*)   For $x' = \sqrt{2}$, $y' = 3\sqrt{2}$, the equations of transformation yield

$$x = \frac{1}{\sqrt{2}}(\sqrt{2} - 3\sqrt{2}) = -2 \qquad y = \frac{1}{\sqrt{2}}(\sqrt{2} + 3\sqrt{2}) = 4$$

Thus, in the unprimed system the coordinates are $P(-2, 4)$.

(*b*)   When the equations of transformation are solved for x' and y', we have

$$x' = \frac{1}{\sqrt{2}}(x + y) \qquad y' = -\frac{1}{\sqrt{2}}(x - y)$$

For $x = 0$, $y = 0$ these equations yield $x' = 0$, $y' = 0$; thus, in the primed system, we have $O(0, 0)'$. Since the coordinates are unchanged, the origin is called an *invariant point* of the transformation.

(*c*)   For $x = \sqrt{2}$, $y = 3\sqrt{2}$ the equations of (*b*) yield $x' = 4$, $y' = 2$. In the primed system, we have $P(4, 2)'$.

(*d*)   When the values for x and y from the equations of transformation are substituted in the given equation of the line, we have

$$\frac{1}{\sqrt{2}}(x' - y') + \frac{1}{\sqrt{2}}(x' + y') + 3\sqrt{2} = \frac{2}{\sqrt{2}}x' + 3\sqrt{2} = 0 \qquad \text{or} \qquad x' + 3 = 0$$

Note that the x' axis is perpendicular to the given line.

(*e*)   Here

$$\frac{3}{\sqrt{2}}(x' - y') - \frac{3}{\sqrt{2}}(x' + y') + 4 = -\frac{6}{\sqrt{2}}y' + 4 = 0 \qquad \text{or} \qquad 3y' - 2\sqrt{2} = 0$$

Note that the x' axis is parallel to the given line.

53.5   Transform the equation $2x^2 - 4xy + 5y^2 - 18x + 12y - 24 = 0$ by rotating the coordinate axes through the angle θ where $\sin \theta = 1/\sqrt{5}$ and $\cos \theta = 2/\sqrt{5}$.

The equations of transformation are $x = \frac{1}{\sqrt{5}}(2x' - y')$, $y = \frac{1}{\sqrt{5}}(x' + 2y')$ and when the values of x and y are substituted in the given equation, we find

$$x'^2 + 6y'^2 - \frac{24}{\sqrt{5}}x' + \frac{42}{\sqrt{5}}y' - 24 = 0$$

Thus, the effect of the transformation is to produce an equation in which the cross-product term $x'y'$ is missing.

53.6   After a rotation of axes with equations of transformation

$$x = \tfrac{1}{13}(12x' - 5y') \qquad y = \tfrac{1}{13}(5x' + 12y')$$

followed by a translation with equations of transformation

$$x' = x'' + \tfrac{16}{13} \qquad y' = y'' - \tfrac{63}{13}$$

a certain equation of the second degree is reduced to $y''^2 = -8x''$. Sketch the locus, showing each set of coordinate axes.

In order to distinguish between the three coordinate systems, we shall use the term *unprimed* for the original system, *primed* for the system after the rotation, and *double-primed* for the system after the translation. We begin with the original (unprimed) axes in the usual position.

Now any pair of numbers, as $(12, 5)$, which are proportional respectively to $\cos \theta = \frac{12}{13}$ and $\sin \theta = \frac{5}{13}$ are the coordinates of a point on the x' axis. With the x' and y' axes in position, we next seek the origin O'' of the double-primed system. Using first the equations of translation and then the equations of rotation, we find $O''(0,0)'' = O''(\frac{16}{13}, -\frac{63}{13})' = O''(3, -4)$. Locating $O''(3, -4)$, with reference to the original system, we draw the x'' and y'' axes through O'' parallel to the x' and y' axes. Finally, on this latter set of axes, we sketch the parabola $y''^2 = -8x''$. See Fig. 53-5.

Fig. 53-5

In Problems 53.7–53.9 determine the nature of the locus, obtain the reduced form of the equation, and sketch the locus showing all sets of coordinate axes.

53.7 $20x^2 - 24xy + 27y^2 + 24x - 54y - 369 = 0$

Since $B^2 - AC = (-12)^2 - 20 \cdot 27 < 0$, the locus is an ellipse.

First Solution.   The angle θ through which the axes must be rotated to eliminate the term in xy is given by

$$\tan 2\theta = \frac{2B}{A - C} = \frac{-24}{20 - 27} = \frac{24}{7}$$

Then $\cos 2\theta = \dfrac{7}{\sqrt{(24)^2 + (7)^2}} = \dfrac{7}{25}$, $\sin \theta = \sqrt{\dfrac{1 - \cos 2\theta}{2}} = \dfrac{3}{5}$, $\cos \theta = \sqrt{\dfrac{1 + \cos 2\theta}{2}} = \dfrac{4}{5}$,

and the equations of rotation are $x = \frac{1}{5}(4x' - 3y')$, $y = \frac{1}{5}(3x' + 4y')$. When this transformation is applied to the given equation, we find

$$11x'^2 + 36y'^2 - \tfrac{66}{5}x' - \tfrac{288}{5}y' - 369 = 0 \qquad \text{as the semireduced form}$$

Completing squares, we have

$$11(x'^2 - \tfrac{6}{5}x' + \tfrac{9}{25}) + 36(y'^2 - \tfrac{8}{5}y' + \tfrac{16}{25}) = 369 + 11(\tfrac{9}{25}) + 36(\tfrac{16}{25}) = 396$$

or $11(x' - \tfrac{3}{5})^2 + 36(y' - \tfrac{4}{5})^2 = 396$

The translation $x' = x'' + \frac{3}{5}$, $y' = y'' + \frac{4}{5}$ gives the reduced form $11x''^2 + 36y''^2 = 396$.

The x' axis passes through the point $(4, 3)$ and the coordinates of the new origin are $O''(0, 0)'' = O''(\frac{3}{5}, \frac{4}{5})' = O''(0, 1)$. See Fig. 53-6.

Fig. 53-6

Second Solution.   Since some may prefer to eliminate the first-degree terms before rotating the axes, we give the details for this locus.

Applying the transformation (*53.2*) to the given equation, we obtain

$$20(x' + h)^2 - 24(x' + h)(y' + k) + 27(y' + k)^2 + 24(x' + h) - 54(y' + k) - 369$$

$$= 20x'^2 - 24x'y' + 27y'^2 + (40h - 24k + 24)x' - (24h - 54k + 54)y'$$

$$+ 20h^2 - 24hk + 27k^2 + 24h - 54k - 369 = 0.$$

If the terms of first degree are to disappear, h and k must be chosen so that $40h - 24k + 24 = 0$, $24h - 54k + 54 = 0$. Then $h = 0, k = 1$ and the equation of the locus becomes $20x'^2 - 24x'y' + 27y'^2 - 396 = 0$.

As in the first solution, the equations of rotation to eliminate the term in $x'y'$ are $x' = \frac{1}{5}(4x'' - 3y'')$, $y' = \frac{1}{5}(3x'' + 4y'')$. When this transformation is made, we have

$$\frac{20}{25}(4x'' - 3y'')^2 - \frac{24}{25}(4x'' - 3y'')(3x'' + 4y'') + \frac{27}{25}(3x'' + 4y'')^2 - 396 = 0$$

or $11x''^2 + 36y''^2 = 396,$ as before.

53.8 $x^2 + 2xy + y^2 + 10\sqrt{2}x - 2\sqrt{2}y + 8 = 0.$

Here $B^2 - AC = 1 - 1 \cdot 1 = 0$; the locus is either a parabola or a pair of parallel lines. Since $A = C = 1$, we rotate the axes through the angle $\theta = 45°$ to obtain the semireduced form. When the transformation

$$x = \frac{1}{\sqrt{2}}(x' - y'), \qquad y = \frac{1}{\sqrt{2}}(x' + y')$$

is applied to the given equation, we find $x'^2 + 4x' - 6y' + 4 = (x' + 2)^2 - 6y' = 0$. Then the translation $x' = x'' - 2$, $y' = y''$ produces the reduced form $x''^2 - 6y'' = 0$.

The locus is a parabola. See Fig. 53-7. The x' axis passes through the point $(1, 1)$ and $O''(0, 0)'' = O''(-2, 0)' = O''(-\sqrt{2}, -\sqrt{2})$.

Fig. 53-7

53.9　　$27x^2 + 120xy + 77y^2 + 234x + 858y + 117 = 0$.

Here $B^2 - AC = (60)^2 - 27 \cdot 77 > 0$; the locus is a hyperbola or a pair of intersecting lines. From $\tan 2\theta = 120/(27 - 77) = -\frac{12}{5}$, we have $\cos 2\theta = -\frac{5}{13}$; then $\sin \theta = \sqrt{\frac{1}{2}(1 + \frac{5}{13})} = 3/\sqrt{13}$, $\cos \theta = \sqrt{\frac{1}{2}(1 - \frac{5}{13})} = 2/\sqrt{13}$, and the equations of rotation are

$$x = \frac{1}{\sqrt{13}}(2x' - 3y'), \qquad y = \frac{1}{\sqrt{13}}(3x' + 2y')$$

Applying the transformation, we obtain $9x'^2 - y'^2 + 18\sqrt{13}x' + 6\sqrt{13}y' + 9 = 0$.

Completing the squares, we have $9(x' + \sqrt{13})^2 - (y' - 3\sqrt{13})^2 + 9 = 0$ which after the translation $x' = x'' - \sqrt{13}$, $y' = y'' + 3\sqrt{13}$ becomes $9x''^2 - y''^2 + 9 = 0$ or $y''^2 - 9x''^2 = 9$.

The locus is a hyperbola. See Fig. 53-8. The x' axis passes through the point $(2, 3)$ and $O''(0, 0)'' = O''(-\sqrt{13}, 3\sqrt{13})' = O''(-11, 3)$.

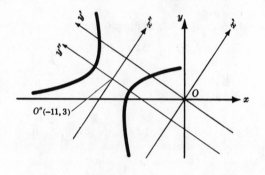

Fig. 53-8

Supplementary Problems

53.10　If the equations of translation are $x = x' + 2$, $y = y' - 5$, find

(a)　The coordinates of $O(0, 0)$ when referred to the primed system

(b)　The coordinates of $P(-2, 4)$ when referred to the primed system

(c)　The coordinates of $P(-2, 4)'$ when referred to the unprimed system

(d)　The equation of l: $5x + 2y = 0$ when referred to the primed system

(e)　The equation of l: $x - 2y + 4 = 0$ when referred to the primed system

　　　Ans.　(a)　$(-2, 5)'$　(b)　$(-4, 9)'$　(c)　$(0, -1)$　(d)　$5x' + 2y' = 0$　(e)　$x' - 2y' + 16 = 0$

53.11　Simplify each of the following equations by a suitable translation. Draw the figure showing both sets of axes.

(a)　$4x^2 + y^2 - 16x + 6y - 11 = 0$　　　　(d)　$x^2 - 12x - 8y - 4 = 0$

(b)　$9x^2 - 4y^2 - 36x + 48y - 144 = 0$　　(e)　$16y^2 + 5x + 32y + 6 = 0$

(c)　$9x^2 - 4y^2 - 36x + 48y - 72 = 0$　　(f)　$xy - 4x + 3y + 24 = 0$

　　　Ans.　(a)　$4x'^2 + y'^2 = 36$　　　(c)　$4y'^2 - 9x'^2 = 36$　　(e)　$16y'^2 = -5x'$

　　　　　　　(b)　$9x'^2 - 4y'^2 = 36$　　(d)　$x'^2 - 8y' = 0$　　　(f)　$x'y' + 36 = 0$

53.12 Simplify each of the following equations by rotating the axes through the indicated angle. Draw the figure showing both sets of axes.

(a) $x^2 - y^2 = 16$; 45°   (c) $16x^2 + 24xy + 9y^2 + 60x - 80y = 0$; Arccos $\frac{4}{5}$

(b) $9x^2 + 24xy + 16y^2 = 25$; Arccos $\frac{3}{5}$ (d) $31x^2 - 24xy + 21y^2 = 39$; Arccos $2/\sqrt{13}$

   Ans. (a) $x'y' + 8 = 0$ (b) $x'^2 = 1$ (c) $x'^2 = 4y'$ (d) $x'^2 + 3y'^2 = 3$

53.13 Simplify each equation by suitable transformations and draw a figure showing all sets of axes.

(a) $3x^2 + 2xy + 3y^2 - 8x + 16y + 30 = 0$

(b) $13x^2 + 12xy - 3y^2 - 15x - 15y = 0$

(c) $25x^2 - 120xy + 144y^2 + 1300x + 1274y - 2704 = 0$

(d) $108x^2 - 312xy + 17y^2 + 750y + 225 = 0$

(e) $16x^2 + 24xy + 9y^2 - 60x - 170y - 175 = 0$

(f) $37x^2 + 32xy + 13y^2 - 42\sqrt{5}x - 6\sqrt{5}y = 0$

   Ans. (a) $2x''^2 + y''^2 = 4$ (c) $y''^2 = -10x''$ (e) $x''^2 = 4y''$

   (b) $30x''^2 - 10y''^2 = 3$ (d) $4x''^2 - 9y''^2 = 36$ (f) $9x''^2 + y''^2 = 18$

53.14 Apply the equations of transformation (53.2) directly to (53.1) and show that the first-degree terms may be made to disappear provided $B^2 - AC \neq 0$.

53.15 Use (53.3) to show: (a) $A' + C' = A + C$,   (b) $B'^2 - A'C' = B^2 - AC$.

53.16 Solve (53.1) to obtain $x = \dfrac{-(By + D) \pm \sqrt{H}}{A}$, where $H = (B^2 - AC)y^2 + 2(BD - AE)y + D^2 - AF$. Show

that H is a perfect square when $\Delta = \begin{vmatrix} A & B & D \\ B & C & E \\ D & E & F \end{vmatrix} = 0$. Thus, prove that (53.1) represents a degenerate

locus if and only if $\Delta = 0$.

53.17 Prove that Δ of Problem 53.16 is invariant under translation and rotation of the axes.

Chapter 54

Polar Coordinates

IN THE POLAR COORDINATE SYSTEM a point in the plane is located by giving its position relative to a fixed point and a fixed half line (direction) through the fixed point. The fixed point O (see Fig. 54-1) is called the *pole* and the fixed half line OA is called the *polar axis*.

Let θ denote the smallest positive angle measured counterclockwise in degrees or radians from OA to OB, and let r denote the (positively) directed distance OP. Then P is uniquely determined when r and θ are known. These two measures constitute the *polar coordinates* of P and we write $P(r, \theta)$. The quantity r is called the *radius vector* and θ is called the *vectorial angle* of P. Note that a positive direction, indicated by the arrow, has been assigned on the half line $\overrightarrow{OB}$.

Fig. 54-1   Fig. 54-2   Fig. 54-3

EXAMPLE 1. Locate the point $P(3, 240°)$ or $P(3, 4\pi/3)$. Refer to Fig. 54-2.

Lay off the vectorial angle $\theta = m \angle AOB = 240°$, measured counterclockwise from $\overrightarrow{OA}$, and on $\overrightarrow{OB}$ locate P such that $r = OP = 3$.

In the paragraph above we have restricted r and θ so that $r \geq 0$ and $0° \leq \theta < 360°$. In general, these restrictions will be observed; however, at times it will be more convenient to permit r and θ to have positive or negative values. If θ is negative and r is positive, we lay off the angle $\theta = \angle AOB$, measured clockwise from $\overrightarrow{OA}$, and locate P on $\overrightarrow{OB}$ so that $OP = r$. If r is negative, we lay off $\theta = \angle AOB$, extend $\overrightarrow{OB}$ through the pole to B', and locate P on $\overrightarrow{OB'}$ a distance $|r|$ from O.

EXAMPLE 2. Locate the point $P(-2, -60°)$ or $P(-2, -\pi/3)$. Refer to Fig. 54-3.

Lay off the vectorial angle $\theta = \angle AOB = 60°$, measured clockwise from OA, extend OB through the pole to B', and on OB' locate P a distance 2 units from O. (See Problems 54.1–54.2.)

Although not a part of the polar system, it will be helpful at times to make use of a half line, called the 90° axis, which issues from the pole perpendicular to the polar axis.

TRANSFORMATIONS BETWEEN POLAR AND RECTANGULAR COORDINATES. If the pole and polar axis of the polar system coincide respectively with the origin and positive x axis of the rectangular system, and if P has rectangular coordinates (x, y) and polar coordinates (r, θ), then the following relations hold:

    (1) $x = r \cos \theta$ (4) $\theta = \arctan y/x$
    (2) $y = r \sin \theta$ (5) $\sin \theta = y/r$ and $\cos \theta = x/r$
    (3) $r = \sqrt{x^2 + y^2}$

If relations (3)–(5) are to yield the restricted set of coordinates of the section above, θ is to be taken as the smallest positive angle satisfying (5) or, what is equivalent, θ is the smallest positive angle satisfying (4) and terminating in the quadrant in which $P(x, y)$ lies.

EXAMPLE 3. Find the rectangular coordinates of $P(3, 300°)$.

Here $r = 3$ and $\theta = 300°$; then $x = r \cos \theta = 3 \cos 300° = 3(\frac{1}{2}) = \frac{3}{2}$, $y = r \sin \theta = 3 \sin 300° = 3(-\frac{1}{2}\sqrt{3}) = -3\sqrt{3}/2$, and the rectangular coordinates are $(\frac{3}{2}, -3\sqrt{3}/2)$.

EXAMPLE 4. Find the polar equation of the circle whose rectangular equation is $x^2 + y^2 - 8x + 6y - 2 = 0$.

Since $x = r \cos \theta$, $y = r \sin \theta$, and $x^2 + y^2 = r^2$, the polar equation is $r^2 - 8r \cos \theta + 6r \sin \theta - 2 = 0$. (See Problems 54.3–54.5.)

CURVE SKETCHING IN POLAR COORDINATES.

Preliminary to sketching the locus of a polar equation, we discuss symmetry, extent, etc., as in the case of rectangular equations. However, there are certain complications at times due to the fact that in polar coordinates a given curve may have more than one equation.

EXAMPLE 5. Let $P(r, \theta)$ be an arbitrary point on the curve $r = 4 \cos \theta - 2$. Now P has other representations: $(-r, \theta + \pi)$, $(-r, \theta - \pi)$, $(r, \theta - 2\pi)$, ...

Since (r, θ) satisfies the equation $r = 4 \cos \theta - 2$, $(-r, \theta + \pi)$ satisfies the equation $-r = 4 \cos (\theta + \pi) - 2 = -4 \cos \theta - 2$ or $r = 4 \cos \theta + 2$. Thus, $r = 4 \cos \theta - 2$ and $r = 4 \cos \theta + 2$ are equations of the same curve. Such equations are called *equivalent*. The reader will show that $(-r, \theta - \pi)$ satisfies $r = 4 \cos \theta + 2$ and $(r, \theta - 2\pi)$ satisfies $r = 4 \cos \theta - 2$.

EXAMPLE 6. Show that point $A(-1, \pi/6)$ is on the ellipse $r = \dfrac{3}{4 + 2 \sin \theta}$.

Note that the given coordinates do not satisfy the given equation.

First Solution. Another set of coordinates for A is $(1, 7\pi/6)$. Since these coordinates satisfy the equation, A is on the ellipse.

Second Solution. An equivalent equation for the ellipse is

$$-r = \frac{3}{4 + 2 \sin (\theta - \pi)} \qquad \text{or} \qquad r = \frac{-3}{4 - 2 \sin \theta}$$

Since the given coordinates satisfy this equation, A is on the ellipse.

SYMMETRY.

A locus is symmetric with respect to the polar axis if an equivalent equation is obtained when

(a) θ is replaced by $-\theta$, or
(b) θ is replaced by $\pi - \theta$ and r by $-r$ in the given equation.

A locus is symmetric with respect to the 90° axis if an equivalent equation is obtained when

(a) θ is replaced by $\pi - \theta$, or
(b) θ is replaced by $-\theta$ and r by $-r$ in the given equation.

A locus is symmetric with respect to the pole if an equivalent equation is obtained when

(a) θ is replaced by $\pi + \theta$, or
(b) r is replaced by $-r$ in the given equation.

EXTENT. The locus whose polar equation is $r = f(\theta)$ is a closed curve if r is real and finite for all values of θ, but is not a closed curve if there are values of one variable which make the other become infinite.

The equation should also be examined for values of one variable which make the other imaginary.

At times, as in the equation $r = a(1 + \sin \theta)$, the values of θ which give r its maximum values can be readily determined. Since the maximum value of $\sin \theta$ is 1, the maximum value of r is $2a$ which it assumes when $\theta = \frac{1}{2}\pi$.

DIRECTIONS AT THE POLE. Unlike all other points, the pole has infinitely many pairs of coordinates $(0, \theta)$ when θ is restricted to $0° \le \theta < 360°$. While two such pairs $(0, \theta_1)$ and $(0, \theta_2)$ define the pole, they indicate different directions (measured from the polar axis) there. Thus, the values of θ for which $r = f(\theta) = 0$ give the directions of the tangents to the locus $r = f(\theta)$ at the pole.

POINTS ON THE LOCUS. We may find as many points on a locus as desired by assigning values to θ in the given equation and solving for the corresponding values of r.

EXAMPLE 7. Discuss and sketch the locus of the cardiod $r = a(1 - \sin \theta)$.

Symmetry. An equivalent equation is obtained when θ is replaced by $\pi - \theta$; the locus is symmetric with respect to the 90° axis.

Extent. Since r is real and $\le 2a$ for all values of θ, the locus is a closed curve, lying within a circle of radius $2a$ with center at the pole. Since $\sin \theta$ is of period 2π, the complete locus is described as θ varies from 0 to 2π.

Directions at the pole. When $r = 0$, $\sin \theta = 1$ and $\theta = \frac{1}{2}\pi$. Thus, the locus is tangent to the 90° axis at the pole.

After locating the points in Table 54.1 and making use of symmetry of the locus with respect to the 90° axis, we obtain the required curve as shown in Fig. 54-4. (See Problems 54.11–54.17.)

Fig. 54-4

Table 54.1

θ	$\frac{1}{2}\pi$	$2\pi/3$	$3\pi/4$	$5\pi/6$	π	$7\pi/6$	$5\pi/4$	$4\pi/3$	$3\pi/2$
r	0	0.13a	0.29a	0.5a	a	1.5a	1.71a	1.87a	2a

INTERSECTIONS OF POLAR CURVES. It is to be expected that in finding the points of intersection of two curves with polar equations $r = f_1(\theta)$ and $r = f_2(\theta)$, we set $f_1(\theta) = f_2(\theta)$ and solve for θ. However, because of the multiplicity of representations both of the coordinates of a point and the equation of a curve, this procedure will fail at times to account for all of the intersections. Thus, it is a better policy to determine from a figure the exact number of intersections before attempting to find them.

EXAMPLE 8. Since each of the circles $r = 2 \sin \theta$ and $r = 2 \cos \theta$ passes through the pole, the circles intersect in the pole and in one other point. See Fig. 54-5. Since each locus is completely described on the interval 0 to π, we set $2 \sin \theta = 2 \cos \theta$ and solve for θ on this interval. The solution $\theta = \frac{1}{4}\pi$ yields the point $(\sqrt{2}, \frac{1}{4}\pi)$.

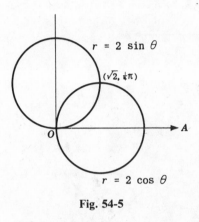

Fig. 54-5

Analytically we may determine whether or not the pole is a point of intersection by setting $r = 0$ in each of the equations and solving for θ. Setting $\sin \theta = 0$ we find $\theta = 0$, and setting $\cos \theta = 0$ we find $\theta = \frac{1}{2}\pi$. Since both equations have solutions, the pole is a point of intersection. The procedure above did not yield this solution since the coordinates of the pole $(0, 0)$ satisfy $r = 2 \sin \theta$ while the coordinates $(0, \frac{1}{2}\pi)$ satisfy $r = 2 \cos \theta$. (See Problems 54.18–54.19.)

Solved Problems

54.1 Locate the following points and determine which coincide with $P(2, 150°)$ and which with $Q(2, 30°)$:

(a) $A(2, 750°)$ (b) $B(-2, -30°)$ (c) $C(-2, 330°)$ (d) $D(-2, -150°)$ (e) $E(2, -210°)$

The points B, C, and E coincide with P; the points A and D coincide with Q. See Figs. 54-6(a)-(e).

54.2 Find the distance between the points

(a) $P_1(5, 20°)$ and $P_2(3, 140°)$ (b) $P_1(4, 50°)$ and $P_2(3, 140°)$ (c) $P_1(r_1, \theta_1)$ and $P_2(r_2, \theta_2)$.

In any triangle OP_1P_2, $(P_1P_2)^2 = (OP_1)^2 + (OP_2)^2 - 2(OP_1)(OP_2) \cos \angle P_1OP_2$.

(a) From Fig. 54-7(a), $(P_1P_2)^2 = (5)^2 + (3)^2 - 2 \cdot 5 \cdot 3 \cos 120° = 49$; hence, $P_1P_2 = 7$.

(b) From Fig. 54-7(b), $P_1P_2 = \sqrt{(4)^2 + (3)^2 - 2 \cdot 4 \cdot 3 \cos 90°} = 5$.

(c) From Fig. 54-7(c), $P_1P_2 = \sqrt{r_1^2 + r_2^2 - 2r_1r_2 \cos (\theta_1 - \theta_2)}$.

54.3 Find the set of polar coordinates, satisfying $r \geq 0, 0° \leq \theta < 360°$, of P whose rectangular coordinates are
(a) $(2, -2\sqrt{3})$ (b) (a, a) (c) $(-3, 0)$ (d) $(0, 2)$. Find two other sets of polar coordinates for each point.

Fig. 54-6

Fig. 54-7

(a) We have $r = \sqrt{x^2 + y^2} = \sqrt{(2)^2 + (-2\sqrt{3})^2} = 4$ and $\theta = \arctan y/x = \arctan(-\sqrt{3})$. Since the point is in the fourth quadrant, we take $\theta = 300°$. The polar coordinates are $(4, 300°)$ or $(4, 5\pi/3)$. Equivalent sets of polar coordinates are $(4, -60°)$ and $(-4, 2\pi/3)$.

(b) Here $r = \sqrt{a^2 + a^2} = a\sqrt{2}$ and $\theta = \arctan 1$, when $a > 0$. Since the point is in the first quadrant, we take $\theta = \frac{1}{4}\pi$. The polar coordinates are $(a\sqrt{2}, \frac{1}{4}\pi)$. Equivalent sets are $(a\sqrt{2}, -7\pi/4)$ and $(-a\sqrt{2}, -3\pi/4)$.

(c) Here $r = \sqrt{(-3)^2 + (0)^2} = 3$. Since the point is on the negative x axis, we take $\theta = \pi$ and the polar coordinates are $(3, \pi)$. Equivalent sets are $(-3, 0)$ and $(3, -\pi)$.

(d) Here $r = \sqrt{(0)^2 + (2)^2} = 2$. Since the point is on the positive y axis, we take $\theta = \pi/2$ and the polar coordinates are $(2, \pi/2)$. Equivalent sets are $(2, -3\pi/2)$ and $(-2, 3\pi/2)$.

54.4 Transform each of the following rectangular equations into their polar form:

(a) $x^2 + y^2 = 25$ (c) $3x - y = 0$ (e) $(x^2 + y^2 - ax)^2 = a^2(x^2 + y^2)$

(b) $x^2 - y^2 = 4$ (d) $x^2 + y^2 = 4x$ (f) $x^3 + xy^2 + 6x^2 - 2y^2 = 0$

We make use of the transformation: $x = r\cos\theta$, $y = r\sin\theta$, $x^2 + y^2 = r^2$.

(a) By direct substitution we obtain $r^2 = 25$ or $r = \pm 5$. Now $r = 5$ and $r = -5$ are equivalent equations since they represent the same locus, a circle with center at the origin and radius 5.

(b) We have $(r\cos\theta)^2 - (r\sin\theta)^2 = r^2(\cos^2\theta - \sin^2\theta) = r^2\cos 2\theta = 4$.

(c) Here $3r\cos\theta - r\sin\theta = 0$ or $\tan\theta = 3$. The polar equation is $\theta = \arctan 3$.

(d)　We have $r^2 = 4r \cos \theta$ or $r = 4 \cos \theta$ as the equation of the circle of radius 2 which passes through the origin and has its center on the polar axis.

(e)　Here $(r^2 - ar \cos \theta)^2 = a^2 r^2$; then $(r - a \cos \theta)^2 = a^2$ and $r - a \cos \theta = \pm a$. Thus we may take $r = a(1 + \cos \theta)$ or $r = -a(1 - \cos \theta)$ as the polar equation of the locus.

(f)　Writing it as $x(x^2 + y^2) + 6x^2 - 2y^2 = 0$, we have $r^3 \cos \theta + 6r^2 \cos^2 \theta - 2r^2 \sin^2 \theta = 0$. Then $r \cos \theta = 2 \sin^2 \theta - 6 \cos^2 \theta = 2(\sin^2 \theta + \cos^2 \theta) - 8 \cos^2 \theta = 2 - 8 \cos^2 \theta$ and $r = 2(\sec \theta - 4 \cos \theta)$ is the polar equation.

54.5　Transform each of the following equations into its rectangular form:

(a)　$r = -2$　　　(c)　$r \cos \theta = -6$　　　(e)　$r = 4(1 + \sin \theta)$　　　(f)　$r = \dfrac{4}{2 - \cos \theta}$

(b)　$\theta = 3\pi/4$　　　(d)　$r = 2 \sin \theta$

　　In general, we attempt to put the polar equation in a form so that the substitutions $x^2 + y^2$ for r^2, x for $r \cos \theta$ and y for $r \sin \theta$ can be made.

(a)　Squaring, we have $r^2 = 4$; the rectangular equation is $x^2 + y^2 = 4$.

(b)　Here $\theta = \arctan y/x = 3\pi/4$; then $y/x = \tan 3\pi/4 = -1$ and the rectangular equation is $x + y = 0$.

(c)　The rectangular form is $x = -6$.

(d)　We first multiply the given equation by r to obtain $r^2 = 2r \sin \theta$. The rectangular form is $x^2 + y^2 = 2y$.

(e)　After multiplying by r, we have $r^2 = 4r + 4r \sin \theta$ or $r^2 - 4r \sin \theta = 4r$; then $(r^2 - 4r \sin \theta)^2 = 16r^2$ and the rectangular equation is $(x^2 + y^2 - 4y)^2 = 16(x^2 + y^2)$.

(f)　Here $2r - r \cos \theta = 4$ or $2r = r \cos \theta + 4$; then $4r^2 = (r \cos \theta + 4)^2$ and the rectangular form of the ellipse is $4(x^2 + y^2) = (x + 4)^2$ or $3x^2 + 4y^2 - 8x - 16 = 0$.

54.6　Derive the polar equation of the straight line:

(a)　Passing through the pole with vectorial angle k

(b)　Perpendicular to the polar axis and $p > 0$ units from the pole

(c)　Parallel to the polar axis and $p > 0$ units from the pole

　　Let $P(r, \theta)$ be an arbitrary point on the line.

(a)　From Fig. 54-8(a) the required equation is $\theta = k$.

(b)　From Fig. 54-8(b) the equation is $r \cos \theta = p$ or $r \cos \theta = -p$ according as the line is to the right or left of the pole.

(c)　From Fig. 54-8(c) the equation is $r \sin \theta = p$ or $r \sin \theta = -p$ according as the line is above or below the pole.

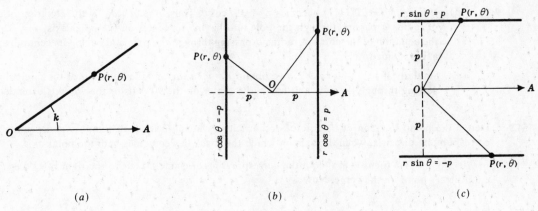

(a)　　　　　　　　　　　(b)　　　　　　　　　　　(c)

Fig. 54-8

54.7   Derive the polar equivalent of the normal form of the rectangular equation of the straight line not passing through the pole.

Let $P(r, \theta)$ be an arbitrary point on the line. Then the foot of the normal from the pole has coordinates $N(p, \omega)$. Using triangle ONP, the required equation is $r \cos(\theta - \omega) = p$. See Fig. 54-9.

Fig. 54-9

54.8   Derive the polar equation of the circle of radius a whose center is at (c, γ).

Let $P(r, \theta)$ be an arbitrary point on the circle. See Fig. 54-10. Then [see Problem 54.2(c)]

$$r^2 + c^2 - 2rc \cos(\gamma - \theta) = a^2$$

or
$$r^2 - 2rc \cos(\gamma - \theta) + c^2 - a^2 = 0 \tag{1}$$

is the required equation.

Fig. 54-10

The following special cases are of interest:

(a)   If the center is at the pole, (1) becomes $r^2 = a^2$. Then $r = a$ or $r = -a$ is the equation of the circle of radius a with center at the pole.

(b)   If $(c, \gamma) = (\pm a, 0°)$, (1) becomes $r = \pm 2a \cos \theta$. Thus, $r = 2a \cos \theta$ is the equation of the circle of radius a passing through the pole and having its center on the polar axis; $r = -2a \cos \theta$ is the equation of the circle of radius a passing through the pole and having its center on the polar axis extended.

(c)   Similarly if $(c, \gamma) = (\pm a, 90°)$, we obtain $r = \pm 2a \sin \theta$ as the equation of the circle of radius a passing through the pole and having its center on the 90° axis or the 90° axis extended.

54.9   Derive the polar equation of a conic of eccentricity e, having a focus at the pole and p units from the corresponding directrix, when the axis on which the focus lies coincides with the polar axis.

In Fig. 54-11 a focus is at O and the corresponding directrix DD' is to the right of O. Let $P(r, \theta)$ be an arbitrary point on the conic. Now

$$\frac{OP}{PM} = e$$

Fig. 54-11

where $OP = r$ and $PM = PB + BM = r \sin(\theta - 90°) + p = p - r \cos \theta$. Thus

$$\frac{r}{p - r \cos \theta} = e, \qquad r(1 + e \cos \theta) = ep, \qquad \text{and} \qquad r = \frac{ep}{1 + e \cos \theta}.$$

It is left for the reader to derive the equation $r = \dfrac{ep}{1 - e \cos \theta}$ when the directrix DD' lies to the left of O.

Similarly it may be shown that the polar equation of a conic of eccentricity e, having a focus at the pole and p units from the corresponding directrix, is

$$r = \frac{ep}{1 \pm e \sin \theta}$$

where the positive sign (negative sign) is used when the directrix lies above (below) the pole.

54.10   Find the locus of the third vertex of a triangle whose base is a fixed line of length $2a$ and the product of the other two sides is the constant b^2.

Take the base of the triangle along the polar axis with the midpoint of the base at the pole. The coordinates of the end points of the base are $B(a, 0)$ and $C(a, \pi)$. Denote the third (variable) vertex by $P(r, \theta)$. See Fig. 54-12.

Fig. 54-12

From the triangle BOP, $(BP)^2 = r^2 + a^2 - 2ar \cos \theta$ and from the triangle COP, $(CP)^2 = r^2 + a^2 - 2ar \cos(\pi - \theta) = r^2 + a^2 + 2ar \cos \theta$. Now $(BP)(CP) = b^2$; hence

$$(r^2 + a^2 - 2ar \cos \theta)(r^2 + a^2 + 2ar \cos \theta) = (b^2)^2 = b^4$$

Then
$$(r^2 + a^2)^2 - 4a^2r^2 \cos^2 \theta = b^4$$

$$r^4 + 2a^2r^2(1 - 2\cos^2 \theta) = r^4 - 2a^2r^2 \cos 2\theta = b^4 - a^4$$

$$r^4 - 2a^2r^2 \cos 2\theta + a^4 \cos^2 2\theta = b^4 - a^4 + a^4 \cos^2 2\theta = b^4 - a^4 \sin^2 2\theta$$

and the required equation is $r^2 = a^2 \cos 2\theta \pm \sqrt{b^4 - a^4 \sin^2 2\theta}$.

54.11 Sketch the conic $r = \dfrac{3}{2 - 2\sin\theta}$.

To put the equation in standard form, in which the first term in the denominator is 1, divide numerator and denominator by 2 and obtain $r = \dfrac{\frac{3}{2}}{1 - \sin\theta}$. The locus is a parabola ($e = 1$) with focus at the pole. It opens upward ($\theta = \frac{1}{2}\pi$ makes r infinite).

When $\theta = 0$, $r = \frac{3}{2}$; the length of the latus rectum is 3. When $\theta = 3\pi/2$, $r = \frac{3}{4}$; the vertex is on the 90° axis extended $\frac{3}{4}$ unit below the pole. With these facts the parabola may be sketched readily as in Fig. 54-13. The equation in rectangular coordinates is $4x^2 = 12y + 9$.

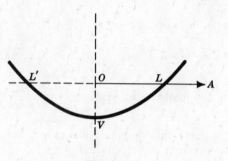

Fig. 54-13

54.12 Sketch the conic $r = \dfrac{18}{5 + 4\sin\theta}$.

After dividing numerator and denominator by 5, we have

$$r = \dfrac{\frac{18}{5}}{1 + \frac{4}{5}\sin\theta}$$

The locus is an ellipse ($e = \frac{4}{5}$) with a focus at the pole.

Since an equivalent equation is obtained when θ is replaced by $\pi - \theta$, the ellipse is symmetric with respect to the 90° axis; thus, the major axis is along the 90° axis. Since $ep = \frac{18}{5}$ and $e = \frac{4}{5}$, $p = \frac{9}{2}$; the directrix is $\frac{9}{2}$ units above the pole. When $\theta = \frac{1}{2}\pi$, $r = 2$; when $\theta = 3\pi/2$, $r = 18$. Thus the vertices are 2 units above and 18 units below the pole. Since $a = \frac{1}{2}(2 + 18) = 10$, $b = \sqrt{a^2(1 - e^2)} = 6$. With these facts the ellipse may be readily sketched as in Fig. 54-14.

Fig. 54-14

In rectangular coordinates, the equation is $25x^2 + 9y^2 + 144y - 324 = 0$.

54.13 Sketch the conic $r = \dfrac{8}{3 - 5\cos\theta}$.

After dividing the numerator and denominator by 3, we have $r = \dfrac{\frac{8}{3}}{1 - \frac{5}{3}\cos\theta}$. The locus is a hyperbola ($e = \frac{5}{3}$) with a focus at the pole.

An equivalent equation is obtained when θ is replaced by $-\theta$; hence the hyperbola is symmetric with respect to the polar axis and its transverse axis is on the polar axis. When $\theta = 0$, $r = -4$ and when $\theta = \pi$, $r = 1$; the vertices are respectively 4 units and 1 unit to the left of the pole. Then $a = \frac{1}{2}(4 - 1) = \frac{3}{2}$ and $b = \sqrt{a^2(e^2 - 1)} = 2$. The asymptotes, having slopes $\pm b/a = \pm\frac{4}{3}$, intersect at the center $\frac{1}{2}(1 + 4) = \frac{5}{2}$ units to the left of the pole. Since $ep = \frac{8}{3}$ and $e = \frac{5}{3}$, $p = \frac{8}{5}$; the directrix is $\frac{8}{5}$ units to the left of the pole.

In rectangular coordinates, the equation is $16x^2 - 9y^2 + 80x + 64 = 0$. See Fig. 54-15.

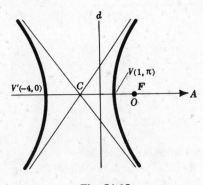

Fig. 54-15

54.14 Sketch the limacon $r = 2a\cos\theta + b$ when (a) $a = 2, b = 5$ (b) $a = 2, b = 4$ (c) $a = 2, b = 3$.

(a) The equation is $r = 4\cos\theta + 5$.

   Symmetry. An equivalent equation is obtained when θ is replaced by $-\theta$; the locus is symmetric with respect to the polar axis.

   Extent. Since r is real and finite for all values of θ, the locus is a closed curve. Since $\cos\theta$ is of period 2π, the complete locus is described as θ varies from 0 to 2π.

   Directions at the Pole. When $r = 0$, $\cos\theta = -\frac{5}{4}$; the locus does not pass through the pole. After locating the points in Table 54.2 and making use of symmetry with respect to the polar axis, we obtain the required curve shown in Fig. 54-16(a). The equation in rectangular coordinates is $(x^2 + y^2 - 4x)^2 = 25(x^2 + y^2)$.

Table 54.2

θ	0	$\pi/6$	$\pi/4$	$\pi/5$	$\pi/2$	$2\pi/3$	$3\pi/4$	$5\pi/6$	π
r	9.00	8.48	7.84	7.00	5.00	3.00	2.16	1.52	1.00

(b) The equation is $r = 4(1 + \cos\theta)$.

   The locus is a closed curve, symmetric with respect to the polar axis, and is completely described as θ varies from 0 to 2π.

   When $r = 0$, $\cos\theta = -1$ and $\theta = \pi$. The locus passes through the pole and is tangent to the polar axis there.

   After locating the points in Table 54.3 and making use of symmetry, we obtain the required curve shown in Fig. 54-16(b).

   In rectangular coordinates the equation of the cardiod is $(x^2 + y^2 - 4x)^2 = 16(x^2 + y^2)$.

Fig. 54-16

Table 54.3

θ	0	$\pi/6$	$\pi/4$	$\pi/3$	$\pi/2$	$2\pi/3$	$3\pi/4$	$5\pi/6$	π
r	8.00	7.48	6.84	6.00	4.00	2.00	1.16	0.52	0

(c)   The equation is $r = 4\cos\theta + 3$.

The locus is a closed curve, symmetric with respect to the polar axis, and is completely described as θ varies from 0 to 2π.

When $r = 0$, $\cos\theta = -\frac{3}{4} = -0.750$ and $\theta = 138°\,40'$, $221°\,20'$. The locus passes through the pole with tangents $\theta = 138°\,40'$ and $\theta = 221°\,20'$.

After putting in these tangents as guide lines, locating the points in Table 54.4, and making use of symmetry, we obtain the required curve shown in Fig. 54-16(c).

The equation in rectangular coordinates is $(x^2 + y^2 - 4x)^2 = 9(x^2 + y^2)$.

Table 54.4

θ	0	$\pi/6$	$\pi/4$	$\pi/3$	$\pi/2$	$2\pi/3$	$3\pi/4$	$5\pi/6$	π
r	7.00	6.48	5.84	5.00	3.00	1.00	0.16	-0.48	-1.00

54.15   Sketch the rose $r = a\cos 3\theta$.

The locus is a closed curve, symmetric with respect to the polar axis. When $r = 0$, $\cos 3\theta = 0$ and $\theta = \pi/6$, $\pi/2$, $5\pi/6$, $7\pi/6$, $\ldots$; the locus passes through the pole with tangent lines $\theta = \pi/6$, $\theta = \pi/2$, and $\theta = 5\pi/6$ there.

The variation of r as θ changes is shown in Table 54.5.

Table 54.5

θ	3θ	r
0 to $\pi/6$	0 to $\pi/2$	a to 0
$\pi/6$ to $\pi/3$	$\pi/2$ to π	0 to $-a$
$\pi/3$ to $\pi/2$	π to $3\pi/2$	$-a$ to 0
$\pi/2$ to $2\pi/3$	$3\pi/2$ to 2π	0 to a
$2\pi/3$ to $5\pi/6$	2π to $5\pi/2$	a to 0
$5\pi/6$ to π	$5\pi/2$ to 3π	0 to $-a$

Caution. The values plotted are (r, θ) not $(r, 3\theta)$. The curve starts at a distance a to the right of the pole on the polar axis, passes through the pole tangent to the line $\theta = \pi/6$, reaches the tip of a loop when $\theta = \pi/3$, passes through the pole tangent to the line $\theta = \pi/2$, and so on. The locus is known as a three-leaved rose.

The rectangular equation is $(x^2 + y^2)^2 = ax(x^2 + 3y^2)$. See Fig. 54-17.

Fig. 54-17

In general, the roses $r = a \sin n\theta$ and $r = a \cos n\theta$ consist of n leaves when n is an *odd* integer.

54.16   Sketch the rose $r = a \sin 4\theta$.

The locus is a closed curve, symmetric with respect to the polar axis (an equivalent equation is obtained when θ is replaced by $\pi - \theta$ and r by $-r$), with respect to the 90° axis (an equivalent equation is obtained when θ is replaced by $-\theta$ and r by $-r$), and with respect to the pole (an equivalent equation is obtained when θ is replaced by $\pi + \theta$).

When $r = 0$, $\sin 4\theta = 0$ and $\theta = 0, \pi/4, \pi/2, 3\pi/4, \ldots$; the locus passes through the pole with tangent lines $\theta = 0$, $\theta = \pi/4$, $\theta = \pi/2$, and $\theta = 3\pi/4$ there.

The variation of r as θ changes from 0 to $\pi/2$ is shown in Table 54.6.

Table 54.6

θ	4θ	r
0 to $\pi/8$	0 to $\pi/2$	0 to a
$\pi/8$ to $\pi/4$	$\pi/2$ to π	a to 0
$\pi/4$ to $3\pi/8$	π to $3\pi/2$	0 to $-a$
$3\pi/8$ to $\pi/2$	$3\pi/2$ to 2π	$-a$ to 0

The complete curve, consisting of 8 leaves, can be traced by making use of the symmetry. See Fig. 54-18.

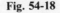

Fig. 54-18

In rectangular coordinates, the equation of the locus is $(x^2 + y^2)^5 = 16a^2(x^3 y - xy^3)^2$.

In general, the roses $r = a \sin n\theta$ and $r = a \cos n\theta$ consist of $2n$ leaves when n is an *even* integer.

54.17   Sketch the locus of $r = \cos \frac{1}{2}\theta$.

Other equations of the locus are $-r = \cos \frac{1}{2}(\theta + \pi) = -\sin \frac{1}{2}\theta$ or $r = \sin \frac{1}{2}\theta$, $-r = \cos \frac{1}{2}(\theta - \pi) = \sin \frac{1}{2}\theta$ or $r = -\sin \frac{1}{2}\theta$, and $r = \cos \frac{1}{2}(\theta - 2\pi) = -\cos \frac{1}{2}\theta$.

The locus is a closed curve, symmetric with respect to the polar axis, the 90° axis, and the pole. It is completely described as θ varies from 0 to 4π.

When $r = 0$, $\theta = \pi, 3\pi, \ldots$; the line $\theta = \pi$ is tangent to the locus at the pole.

The curve is traced by locating the points (Table 54.7) and making use of symmetry. See Fig. 54-19.

Table 54.7

θ	$\frac{1}{2}\theta$	r
0	0	1.00
$\pi/6$	$\pi/12$	0.97
$\pi/3$	$\pi/6$	0.87
$\pi/2$	$\pi/4$	0.71
$2\pi/3$	$\pi/3$	0.50
$5\pi/6$	$5\pi/12$	0.26
π	$\pi/2$	0.00

Fig. 54-19

54.18   Find the points of intersection of the limacon $r = 2\cos\theta + 4$ and the circle $r = 8\cos\theta$.

From Fig. 54-20 there are two points of intersection.

Setting $2\cos\theta + 4 = 8\cos\theta$, we obtain $\cos\theta = \frac{2}{3}$; then $\theta = 48°\,10'$ and $311°\,50'$. (We solve for θ on the range $0 \le \theta < 2\pi$ since the limacon is completely described on this range.) The points of intersection are $(\frac{16}{3}, 48°\,10')$ and $(\frac{16}{3}, 311°\,50')$.

Fig. 54-20   **Fig. 54-21**

54.19   Find the points of intersection of the ellipse $r = \dfrac{4}{2 + \cos\theta}$ and the limacon $r = 4\cos\theta - 2$.

From Fig. 54-21 there are four points of intersection.

Setting $\dfrac{4}{2 + \cos\theta} = 4\cos\theta - 2$, we have $2\cos^2\theta + 3\cos\theta - 4 = 0$. Then $\cos\theta = \dfrac{-3 \pm \sqrt{41}}{4} = 0.851$ or -2.351, and $\theta = 31°\,40'$ and $328°\,20'$. The corresponding points are $E(-5 + \sqrt{41}, 31°\,40')$ and $B(-5 + \sqrt{41}, 328°\,20')$.

To obtain the other two points, we solve the equation of the ellipse with another equation $r = 4\cos\theta + 2$

(see Example 5) of the limacon. From $4 \cos \theta + 2 = \dfrac{4}{2 + \cos \theta}$, we obtain $2 \cos^2 \theta + 5 \cos \theta = (\cos \theta)$ $(2 \cos \theta + 5) = 0$. Then $\cos \theta = 0$ and $\theta = \frac{1}{2}\pi$ and $3\pi/2$. The corresponding points are $C(2, \pi/2)$ and $D(2, 3\pi/2)$.

[NOTE.   When sketching $r = 4 \cos \theta - 2$, the coordinates of C were found as $(-2, 3\pi/2)$ and those of D as $(-2, \pi/2)$.]

Supplementary Problems

54.20 Find the rectangular coordinates of P whose polar coordinates are

(a) $(-2, 45°)$ (b) $(3, \pi)$ (c) $(2, \pi/2)$ (d) $(4, 2\pi/3)$

   Ans. (a) $(-\sqrt{2}, -\sqrt{2})$ (b) $(-3, 0)$ (c) $(0, 2)$ d, $(-2, 2\sqrt{3})$

54.21 Find a set of polar coordinates of P whose rectangular coordinates are

(a) $(1, \sqrt{3})$ (b) $(0, -5)$ (c) $(1, -1)$ (d) $(-12, 5)$

   Ans. (a) $(2, \pi/3)$ (b) $(5, 3\pi/2)$ (c) $(\sqrt{2}, 7\pi/4)$ (d) $(13, \pi - \text{Arctan}\,\frac{5}{12})$

54.22 Transform each of the following rectangular equations into polar form:

(a) $x^2 + y^2 = 16$ (b) $y^2 - x^2 = 9$ (c) $x = 4$ (d) $y = \sqrt{3}x$ (e) $xy = 12$
(f) $(x^2 + y^2)x = 4y^2$

   Ans. (a) $r = 4$ (c) $r \cos \theta = 4$ (e) $r^2 \sin 2\theta = 24$
          (b) $r^2 \cos 2\theta + 9 = 0$ (d) $\theta = \pi/3$ (f) $r = 4 \tan \theta \sin \theta$

54.23 Transform each of the following polar equations into rectangular form:

(a) $r \sin \theta = -4$ (c) $r = 2 \cos \theta$ (e) $r = 1 - 2 \cos \theta$ (f) $r = \dfrac{4}{1 - 2 \sin \theta}$
(b) $r = -4$ (d) $r = \sin 2\theta$

   Ans. (a) $y = -4$ (c) $x^2 + y^2 - 2x = 0$ (e) $(x^2 + y^2 + 2x)^2 = x^2 + y^2$
          (b) $x^2 + y^2 = 16$ (d) $(x^2 + y^2)^3 = 4x^2 y^2$ (f) $x^2 - 3y^2 - 16y - 16 = 0$

54.24 Derive the polar equation $r = \dfrac{ep}{1 \pm e \sin \theta}$ of the conic of eccentricity e with a focus at the pole and with corresponding directrix p units from the focus.

54.25 Write the polar equation of each of the following:

(a) Straight line bisecting the second and fourth quadrants

(b) Straight line through $(4, 2\pi/3)$ and perpendicular to the polar axis

(c) Straight line through $N(3, \pi/6)$ and perpendicular to the radius vector of N

(d) Circle with center at $C(4, 3\pi/2)$ and radius $= 4$

(e) Circle with center at $C(-4, 0)$ and radius $= 4$

(f) Circle with center at $C(4, \pi/3)$ and radius $= 4$

(g) Parabola with focus at the pole and directrix $r = -4 \sec \theta$

(h) Parabola with focus at the pole and vertex at $V(3, \pi/2)$

(i) Ellipse with eccentricity $\frac{3}{4}$, one focus at the pole and the corresponding directrix 5 units above the polar axis

(j) Ellipse with one focus at the pole, the other focus at $(8, \pi)$ and eccentricity $= \frac{2}{3}$

(k) Hyperbola with eccentricity $= \frac{3}{2}$, one focus at the pole, and the corresponding directrix 5 units to the left of the 90° axis

(l) Hyperbola, conjugate axis $= 24$ parallel to and below the polar axis, transverse axis $= 10$, and one focus at the pole

Ans. (a) $\theta = 3\pi/4$ (c) $r \cos(\theta - \pi/6) = 3$ (e) $r = -8 \cos \theta$

(b) $r \cos \theta = -2$ (d) $r = -8 \sin \theta$ (f) $r = 8 \cos(\theta - \pi/3)$

(g) $r = \dfrac{4}{1 - \cos \theta}$ (i) $r = \dfrac{15}{4 + 3 \sin \theta}$ (k) $r = \dfrac{15}{2 - 3 \cos \theta}$

(h) $r = \dfrac{6}{1 + \sin \theta}$ (j) $r = \dfrac{10}{3 + 2 \cos \theta}$ (l) $r = \dfrac{144}{5 - 13 \sin \theta}$

54.26 Discuss and sketch:

(a) $r = \sin(\theta - 45°) = -2$ (f) $r = \dfrac{2}{1 - \cos \theta}$ (l) $r = 4 \sin 2\theta$

(b) $r = 10 \sin \theta$ (m) $r = 2a \tan \theta \sin \theta$

(c) $r = -6 \cos \theta$ (g) $r = 2 - 4 \cos \theta$ (n) $r = 4 \tan^2 \theta \sec \theta$

(h) $r = 4 - 2 \cos \theta$ (o) $r = \cos \frac{3}{2}\theta$

(d) $r = \dfrac{8}{2 - \sin \theta}$ (i) $r^2 = 9 \cos 2\theta$

(j) $r^2 = 16 \sin 2\theta$ (p) $r = 2\theta$

(e) $r = \dfrac{6}{1 + 2 \cos \theta}$ (k) $r = 2 \cos 2\theta$ (q) $r = a/\theta$

54.27 Find the complete intersection of

(a) $r = 2 \cos \theta, \ r = 1$ (b) $r^2 = 4 \cos 2\theta, \ r = 2\sqrt{2} \sin \theta$ (c) $r = 1 + \sin \theta, \ r = \sqrt{3} \cos \theta$

Ans. (a) $(1, \pi/3), (1, 5\pi/3)$ (b) $(0, 0), (\sqrt{2}, \pi/6), (\sqrt{2}, 5\pi/6)$ (c) $(0, 0), (\frac{3}{2}, \pi/6)$

54.28 If each radius vector through the pole to the line $r = a \sec \theta$ is increased and decreased by the constant b, the equation of the curve (conchoid of Nicomedes) thus generated is $r = a \sec \theta \pm b$. Sketch the three types corresponding to $b > a$, $b = a$, and $b < a$.

54.29 Let $\alpha = \angle XOM$ be any acute angle. Lay off $a = OA$ on OX and construct the perpendicular at A to OX meeting OM in C. Construct the conchoid $r = a \sec \theta + b$ with $a = OA = c \cos \alpha$ and $b = 2OC = 2c$, that is, the conchoid $r = c \cos \alpha \sec \theta + 2c$. Let the line $r = c \sin \alpha \csc \theta$ through C parallel to OX meet the conchoid in P. Show that $\theta = \angle XOP = \alpha/3$. See Fig. 54-22. This solution of the famous trisection problem was given by Nicomedes (circa 180 B.C.).

Fig. 54-22

Chapter 55

Parametric Equations

IN THIS CHAPTER we consider the analytic representation of a plane curve by means of a pair of equations, as $x = t$, $y = 2t + 3$, in which the coordinates of a variable point (x, y) on the curve are expressed as a function of a third variable or *parameter*. Such equations are called *parametric equations* of the curve.

A table of values of x and y is readily obtained from the given parametric equations by assigning values to the parameter. After plotting the several points (x, y), the locus may be sketched in the usual manner.

EXAMPLE. Sketch the locus of $x = t$, $y = 2t + 3$.

We form the table of values (Table 55.1), plot the points (x, y), and join these points to obtain the straight line shown in Fig. 55-1. (See Problem 55-1.)

Table 55.1

t	2	0	-3
x	2	0	-3
y	7	3	-3

Fig. 55-1

At times it will be possible to eliminate the parameter between the two equations and thus obtain the rectangular equation of the curve. In the example above, the elimination of t is easy and results in $y = 2x + 3$. At other times, however, it will be impractical or impossible to eliminate the parameter. (See Problem 55.2.)

Parametric representation of a curve is *not* unique. For example, $x = t$, $y = 2t + 3$; $x = \frac{1}{2}u$, $y = u + 3$; $x = v - 1$, $y = 2v + 1$ are parametric representations with parameters t, u, v, respectively, of the straight line whose rectangular equation is $y = 2x + 3$. (See Problems 55.3–55.4.)

PATH OF A PROJECTILE. If a body is projected from the origin with initial velocity v_0 ft/s at an angle α with the positive x axis and if all forces acting on the body after projection, excepting the force of gravity, are ignored, the coordinates of the body t seconds thereafter are given by

$$x = v_0 t \cos \alpha, \qquad y = v_0 t \sin \alpha - \tfrac{1}{2}gt^2$$

where g is the acceleration due to gravity. For convenience, we take $g = 32$ ft/s². (See Problems 55.5–55.7.)

401

Solved Problems

55.1   Sketch the locus of each of the following:

(a) $x = t,\ y = t^2$ (b) $x = 4t,\ y = 1/t$ (c) $x = 5\cos\theta,\ y = 5\sin\theta$ (d) $x = 2 + \cos\theta,\ y = \cos 2\theta$

   The table of values and sketch of each are given in Figs. 55-2 to 55-5. In Fig. 55-3, $t = 0$; hence, we must examine the locus for values of t near 0. In Fig. 55-4 the complete locus is described on the interval $0 \le \theta \le 2\pi$. In Fig. 55-5 the complete locus is described on the interval $0 \le \theta \le \pi$. Note that only that part of the parabola below $y = 1$ is obtained; thus the complete parabola is not defined by the parametric equations.

t	2	1	0	−1	−2
x	2	1	0	−1	−2
y	4	1	0	1	4

Fig. 55-2

t	2	1	1/2	1/10	−1/10	−1/2	−1	−2
x	8	4	2	2/5	−2/5	−2	−4	−8
y	1/2	1	2	10	−10	−2	−1	−1/2

Fig. 55-3

θ	0	$\pi/4$	$\pi/2$	$3\pi/4$	π	$5\pi/4$	$3\pi/2$	$7\pi/4$	2π
x	5.00	3.55	0	−3.55	−5.00	−3.55	0	3.55	5.00
y	0	3.55	5.00	3.55	0	−3.55	−5.00	−3.55	0

Fig. 55-4

θ	0	$\pi/4$	$\pi/2$	$3\pi/4$	π
x	3.00	2.71	2.00	1.29	1.00
y	1.00	0	−1.00	0	1.00

Fig. 55-5

55.2   In each of the following, eliminate the parameter to obtain the equation of the locus in rectangular coordinates: (a)–(d), Problem 55.1 (e) $x = t^2 + t$, $y = t^2 - t$ (f) $x = 3 \sec \phi$, $y = 2 \tan \phi$ (g) $x = v_0(\cos \alpha)t$, $y = v_0(\sin \alpha)t - \frac{1}{2}gt^2$, t being the parameter.

   (a) Here $t = x$ and $y = t^2 = x^2$. The required equation is $y = x^2$.

   (b) Since $t = 1/y$, $x = 4t = 4/y$ and the equation is $xy = 4$.

   (c) $x^2 + y^2 = (5 \cos \theta)^2 + (5 \sin \theta)^2 = 25(\cos^2 \theta + \sin^2 \theta)$. The required equation is $x^2 + y^2 = 25$.

   (d) $\cos \theta = x - 2$ and $y = \cos 2\theta = 2 \cos^2 \theta - 1 = 2(x - 2)^2 - 1$. The equation is $(x - 2)^2 = \frac{1}{2}(y + 1)$.

   (e) Subtracting one of the equations from the other, $t = \frac{1}{2}(x - y)$. Then $x = \frac{1}{4}(x - y)^2 + \frac{1}{2}(x - y)$ and the required equation is $x^2 - 2xy + y^2 - 2x - 2y = 0$.

   (f) $\tan \phi = \frac{1}{2}y$ and $x^2 = 9 \sec^2 \phi = 9(1 + \tan^2 \phi) = 9(1 + \frac{1}{4}y^2)$. The equation is $4x^2 - 9y^2 = 36$.

   (g) $t = \dfrac{x}{v_0 \cos \alpha}$ and $y = \dfrac{v_0(\sin \alpha)x}{v_0 \cos \alpha} - \dfrac{1}{2} \cdot \dfrac{gx^2}{(v_0 \cos \alpha)^2}$. Then $y = x \tan \alpha - \dfrac{g}{2v_0^2 \cos^2 \alpha}x^2$.

55.3   Find parametric equations for each of the following, making use of the suggested substitution:

   (a) $y^2 - 2y + 2x - 5 = 0$, $x = t + 3$ (c) $9x^2 + 16y^2 = 144$, $x = 4 \cos \theta$

   (b) $y^2 - 2y + 2x = 5$, $y = 1 + t$ (d) $x^3 + y^3 - 3axy = 0$, $y = mx$

   (a) First write the equation as $(y - 1)^2 = -2(x - 3)$. Upon making the suggested substitution, we have $(y - 1)^2 = -2t$ or $y = 1 \pm \sqrt{-2t}$. We may take as parametric equations $x = t + 3$, $y = 1 + \sqrt{-2t}$ or $x = t + 3$, $y = 1 - \sqrt{-2t}$.

   (b) From $(y - 1)^2 = -2(x - 3)$, we obtain $t^2 = -2(x - 3)$ or $x = 3 - \frac{1}{2}t^2$. The parametric equations are $x = 3 - \frac{1}{2}t^2$, $y = 1 + t$.

   (c) We have $9(16 \cos^2 \theta) + 16y^2 = 144$ or $y^2 = 9(1 - \cos^2 \theta) = 9 \sin^2 \theta$. The parametric equations are $x = 4 \cos \theta$, $y = 3 \sin \theta$ or $x = 4 \cos \theta$, $y = -3 \sin \theta$.

   (d) Substituting, we have $x^3 + m^3x^3 - 3amx^2 = 0$. Dividing by x^2, we obtain $x = \dfrac{3am}{1 + m^3}$. Then $y = mx = \dfrac{3am^2}{1 + m^3}$ and the parametric equations are $x = \dfrac{3am}{1 + m^3}$, $y = \dfrac{3am^2}{1 + m^3}$.

55.4 (a) Show that $x = a \cos \theta$, $y = b \sin \theta$, $a > b$, are parametric equations of an ellipse.

   (b) Show that these equations indicate the following method for constructing an ellipse whose major and minor axes $2a$ and $2b$ are given: With the origin as common center draw two circles having radii a and b. Through the origin draw a half line l meeting the smaller circle in B and the larger circle in A. From A drop a perpendicular to the x axis meeting it in R; from B drop a perpendicular to the x axis meeting it in S and a perpendicular to RA meeting it in P. Then as l revolves about O, P describes the ellipse.

   (a) We have $\cos \theta = x/a$ and $\sin \theta = y/b$; then $x^2/a^2 + y^2/b^2 = \cos^2 \theta + \sin^2 \theta = 1$ is the equation of an ellipse.

   (b) In Fig. 55-6, let P have coordinates (x, y) and denote by θ the angle which l makes with the positive x axis. Then $x = OR = OA \cos \theta = a \cos \theta$ and $y = RP = SB = OB \sin \theta = b \sin \theta$ are parametric equations of the ellipse whose major and minor axes are a and b.

55.5   A body is projected from the origin with initial velocity v_0 ft/s at an angle α with the positive x axis. Assuming that the only force acting upon the body after projection is the attraction of the earth, obtain parametric equations of the path of the body with t (the number of seconds after projection) as parameter.

   If a body is released near the surface of the earth and all forces acting upon it other than gravity are neglected, the distance s ft through which it will fall in t seconds is given by $s = \frac{1}{2}gt^2$, where $g = 32$ ft/s^2 approximately.

   If a body is given motion as stated in the problem and if after projection *no other force* acts on it, the motion is in a straight line (Newton's first law of motion) and after t seconds the body has coordinates $(v_0t \cos \alpha, v_0t \sin \alpha)$.

Fig. 55-6

Since, when small distances are involved, the force of gravity may be assumed to act vertically, the coordinates of the projected body after t seconds of motion are given by

$$x = v_0 t \cos \alpha, \qquad y = v_0 t \sin \alpha - \tfrac{1}{2}gt^2. \tag{1}$$

In rectangular coordinates, the equation of the path is [see Problem 55.2(g)]

$$y = x \tan \alpha - \frac{g}{2v_0^2 \cos^2 \alpha} x^2 \tag{2}$$

55.6 A bird is shot when flying horizontally 120 ft directly above the hunter. If its speed is 30 mi/hr, find the time during which it falls and the distance it will be from the hunter when it strikes the earth.

As in Fig. 55-7, take the bird to be at the origin when shot. Since 30 mi/hr $= 30 \cdot 5280/(60 \cdot 60) = 44$ ft/s $= v_0$ and $\alpha = 0°$, the equations of motion are

(a) $x = v_0 t \cos \alpha = 44t \cos 0° = 44t$
(b) $y = v_0 t \sin \alpha - \tfrac{1}{2}gt^2 = -16t^2$

When the bird reaches the ground, its coordinates are $(x, -120)$.
From (b), $-120 = -16t^2$ and $t = \tfrac{1}{2}\sqrt{30}$; from ($a$), $x = 44(\tfrac{1}{2}\sqrt{30}) = 22\sqrt{30}$. Thus the bird will fall for $\tfrac{1}{2}\sqrt{30}$ s and will reach the ground $22\sqrt{30}$ ft from the hunter.

Fig. 55-7   Fig. 55-8

55.7 A ball is projected upward from the ground at an angle 60° from the horizontal with initial velocity 60 ft/s. Find (a) the time it will be in the air (b) its range, that is, the horizontal distance it will travel, and (c) its maximum height attained. Refer to Fig. 55-8.

Let the ball be projected from the origin; then the equations of motion are

$$x = v_0 t \cos \alpha = 60t \cos 60° = 30t, \qquad y = v_0 t \sin \alpha - \tfrac{1}{2}gt^2 = 30t\sqrt{3} - 16t^2.$$

(a)   When $y = 0$, $16t^2 - 30t\sqrt{3} = 0$ and $t = 0$, $15\sqrt{3}/8$. Now $t = 0$ is the time when the ball was projected and $t = 15\sqrt{3}/8$ is the time when it reaches the ground again. Thus the ball was in the air for $15\sqrt{3}/8$ s.

(b)   When $t = 15\sqrt{3}/8$, $x = 30 \cdot 15\sqrt{3}/8 = 225\sqrt{3}/4$. The range is $225\sqrt{3}/4$ ft.

(c)   **First Solution.** The ball will attain its maximum height when $t = \frac{1}{2}(15\sqrt{3}/8) = 15\sqrt{3}/16$. Then
$$y = 30t\sqrt{3} - 16t^2 = 30(15\sqrt{3}/16)\sqrt{3} - 16(15\sqrt{3}/16)^2 = \tfrac{675}{16} \text{ ft, the maximum height.}$$

Second Solution. The maximum height is attained where the horizontal distance of the ball is one-half the range, i.e., when $x = 225\sqrt{3}/8$. Using the rectangular equation $y = -\frac{4}{225}x^2 + x\sqrt{3}$, we obtain $\frac{675}{16}$ ft as before.

55.8   The locus of a fixed point P on the circumference of a circle of radius a as the circle rolls without slipping along a straight line is called a *cycloid*. Obtain parametric equations of this locus.

Take the x axis to be the line along which the circle is to roll and place the circle initially with its center C on the y axis and P at the origin. Figure 55-9 shows the position of P after the circle has rolled through an angle θ. Drop perpendiculars $\overline{PR}$ and $\overline{CS}$ to the x axis and PA to SC. Let P have coordinates (x, y). Then

$$x = OR = OS - RS = \text{arc } PS - PA = a\theta - a\sin\theta$$

and
$$y = RP = SA = SC - AC = a - a\cos\theta$$

Thus the equations of the locus are $x = a(\theta - \sin\theta)$, $y = a(1 - \cos\theta)$.

The maximum height of an arch is $2a$, the diameter of the circle, and the span of an arch or the distance between two consecutive cusps is $2\pi a$, the circumference of the circle.

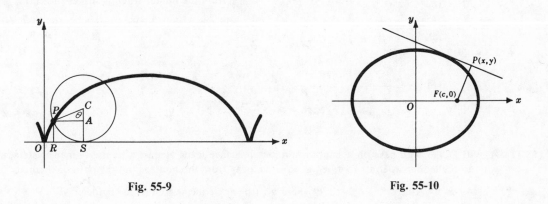

Fig. 55-9   Fig. 55-10

55.9   Find the equation of the locus of the feet of perpendiculars drawn from a focus to the tangents of the ellipse $b^2x^2 + a^2y^2 = a^2b^2$. This curve is called the *pedal curve* of the ellipse with respect to the focus. See Fig. 55-10.

Let $P(x, y)$ be any point on the locus. The equations of the tangents of slope m to the ellipse are

$$y = mx \pm \sqrt{a^2m^2 + b^2} \qquad \text{or} \qquad (1') \quad y - mx = \pm\sqrt{a^2m^2 + b^2}$$

The perpendicular to these tangents through $F(c, 0)$ has equation

$$y = -\frac{1}{m}x + \frac{c}{m} \qquad \text{or} \qquad (2') \quad my + x = \sqrt{a^2 - b^2}$$

Squaring ($1'$) and ($2'$), and adding, we obtain

$$(1 + m^2)y^2 + (1 + m^2)x^2 = a^2(1 + m^2) \qquad \text{or, since } 1 + m^2 \neq 0, \qquad x^2 + y^2 = a^2$$

as the desired equation.

Supplementary Problems

55.10 Sketch the locus and find the rectangular equation of the curve whose parametric equations are

  (a) $x = t + 2, y = 3t + 5$ Ans. $y = 3x - 1$

  (b) $x = \tan \theta, y = 4 \cot \theta$ Ans. $xy = 4$

  (c) $x = 2t^2 + 3, y = 3t + 2$ Ans. $2y^2 - 8y - 9x + 35 = 0$

  (d) $x = 3 + 2 \tan \theta, y = -1 + 5 \sec \theta$ Ans. $4y^2 - 25x^2 + 8y + 150x - 321 = 0$

  (e) $x = 2 \sin^4 \theta, y = 2 \cos^4 \theta$ Ans. $(x - y)^2 - 4(x + y) + 4 = 0$

  (f) $x = \tan \theta, y = \tan 2\theta$ Ans. $x^2 y - y + 2x = 0$

  (g) $x = \sqrt{\cos t}, y = \tan \frac{1}{2} t$ Ans. $y^2(1 + x^2) + x^2 - 1 = 0$

  (h) $x = a \cos^3 \theta, y = a \sin^3 \theta$ Ans. $x^{2/3} + y^{2/3} = a^{2/3}$

  (i) $x = \dfrac{2t}{1 + t^2}, y = \dfrac{1 - t^2}{1 + t^2}$ Ans. $x^2 + y^2 = 1$

  (j) $x = \dfrac{2at^2}{1 + t^2}, y = \dfrac{2at^3}{1 + t^2}$ Ans. $y^2(2a - x) = x^3$

55.11 Find parametric equations for each of the following, using the suggested value for x or y:

  (a) $x^3 = 4y^2, x = t^2$ (c) $4x^2 - 9y^2 = 36, x = 3 \sec \theta$ (e) $x(x^2 + y^2) = x^2 - y^2, y = tx$

  (b) $y = x^2 + x - 6, x = t + 2$ (d) $y = 2x^2 - 1, x = \cos t$ (f) $(x^2 + 16)y = 64, x = 4 \tan \theta$

  Ans. (a) $x = t^2, y = \frac{1}{2} t^3$

      (b) $x = t + 2, y = t^2 + 5t$

      (c) $x = 3 \sec \theta, y = 2 \tan \theta$

      (d) $x = \cos t, y = \cos 2t$

      (e) $x = \dfrac{1 - t^2}{1 + t^2}, y = \dfrac{t(1 - t^2)}{1 + t^2}$

      (f) $x = 4 \tan \theta, y = 4 \cos^2 \theta$

55.12 A 30-ft ladder with base on a smooth horizontal surface leans against a house. A man is standing $\frac{2}{3}$ the way up the ladder when its foot begins to slide away from the house. Find the path of the man.

  Ans. $x = 10 \cos \theta, y = 20 \sin \theta$ where θ is the angle at the foot of the ladder.

55.13 A stone is thrown upward with initial speed 48 ft/s at an angle measuring 60° with the horizontal from the top of a cliff 100 ft above the surface of a lake. Find (a) its greatest distance above the lake (b) when it will strike the surface of the lake, and (c) its horizontal distance from the point where thrown when it strikes the surface.

  Hint: Take the origin at the top of the cliff.

  Ans. (a) 127 ft   (b) $\dfrac{3\sqrt{3} + \sqrt{127}}{4}$ s later   (c) $6(3\sqrt{3} + \sqrt{127})$ ft

55.14 Find the locus of the vertices of all right triangles having hypotenuse of length $2a$.

  Hint: Take the hypotenuse along the x axis with its midpoint at the origin and let θ be an acute angle of the triangle.

  Ans. $x = a \cos 2\theta, y = a \sin 2\theta$ or $x = a \sin 2\theta, y = a \cos 2\theta$

55.15 From the two-point form of the equation of a straight line derive the parametric equations $x = x_1 + t(x_2 - x_1)$, $y = y_1 + t(y_2 - y_1)$. What values of the parameter t give the points on the segment $P_1 P_2$? Identify the points corresponding to $t = \frac{1}{2}, \frac{1}{3}, \frac{2}{3}$.

55.16 Verify the following method for constructing a hyperbola with transverse axis $2a$ and conjugate axis $2b$, where $a \neq b$. With the origin as common center draw two circles having radii a and b. Through the origin pass a half line l making an angle θ with the positive x axis and intersecting the larger circle in A. Let the tangent at A to the circle meet the x axis in B. Through C, the intersection of the smaller circle and the positive x axis, erect a perpendicular meeting l in D. Through D pass a line parallel to the x axis and through B a line perpendicular to the x axis, and denote their intersection by P. Then P is a point on the hyperbola.

55.17 Obtain the rectangular equation $x = a \arccos \dfrac{a - y}{a} \mp \sqrt{2ay - y^2}$ of the cycloid of Problem 55.8.

Chapter 56

Points in Space

RECTANGULAR COORDINATES IN SPACE. Consider the three mutually perpendicular planes of Fig. 56-1. These three planes (the xy plane, the xz plane, the yz plane) are called the *coordinate planes*; their three lines of intersection are called the *coordinate axes* (the x axis, the y axis, the z axis); and their common point O is called the *origin*. Positive direction is indicated on each axis by an arrow-tip.

(NOTE. The coordinate system of Fig. 56-1 is called a left-handed system. When the x and y axes are interchanged, the system becomes right-handed.)

The coordinate planes divide the space into eight regions, called *octants*. The octant whose edges are $\vec{Ox}, \vec{Oy}, \vec{Oz}$ is called the *first octant*; the other octants are not numbered.

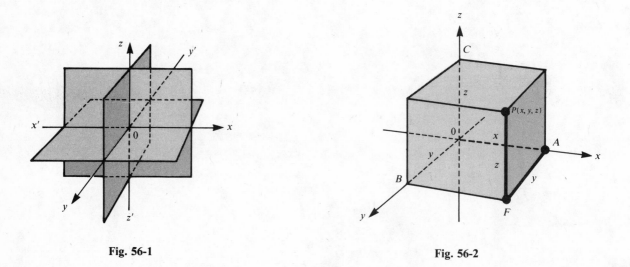

Fig. 56-1	**Fig. 56-2**

Let P be any point in space, not in a coordinate plane, and through P pass planes parallel to the coordinate planes meeting the coordinate axes in the points A, B, C and forming the rectangular parallelepiped of Fig. 56-2. The directed distances $x = OA$, $y = OB$, $z = OC$ are called, respectively, the x coordinate, the y coordinate, the z coordinate of P and we write $P(x, y, z)$.

Since $\overline{AF} \cong \overline{OB}$ and $\overline{FP} \cong \overline{OC}$, it is preferable to use the three edges $\overline{OA}$, $\overline{AF}$, $\overline{FP}$ instead of the complete parallelepiped in locating a given point.

EXAMPLE 1. Locate the points:

(*a*) $(2, 3, 4)$ (*b*) $(-2, -2, 3)$ (*c*) $(2, -2, -3)$

As standard procedure in representing on paper the left-handed system, we shall draw $\angle xOz$ measuring 90° and $\angle xOy$ measuring 135°. Then distances on parallels to the x and z axes will be drawn to full scale while distances parallel to the y axis will be drawn about $\frac{7}{10}$ of full scale.

(*a*) From the origin move 2 units to the right along the x axis to $A(2, 0, 0)$, from A move 3 units forward parallel to the y axis to $F(2, 3, 0)$, and from F move 4 units upward parallel to the z axis to $P(2, 3, 4)$. See Fig. 56-3(*a*).

(*b*) From the origin move 2 units to the left along the x axis to $A(-2, 0, 0)$, from A move 2 units backward parallel to the y axis to $F(-2, -2, 0)$, and from F move 3 units upward parallel to the z axis to $P(-2, -2, 3)$. See Fig. 56-3(*b*).

408

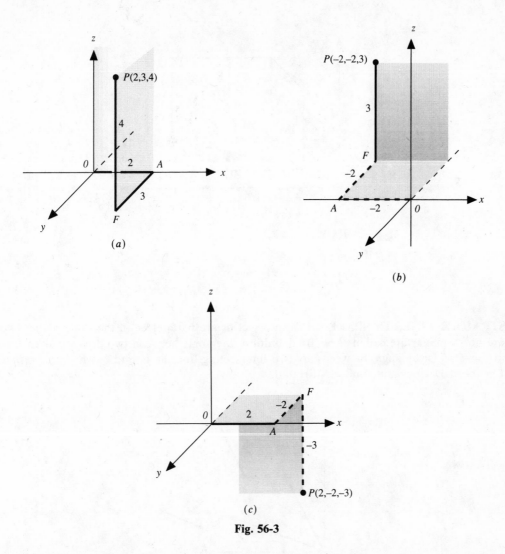

Fig. 56-3

(c)   From the origin move 2 units to the right along the x axis to $A(2, 0, 0)$, from A move 2 units backward
      parallel to the y axis to $F(2, -2, 0)$, and from F move 3 units downward parallel to the z axis to $P(2, -2, -3)$.
      See Fig. 56-3(c).

(See Problem 56.1.)

THE DISTANCE BETWEEN TWO POINTS $P_1(x_1, y_1, z_1)$ and $P_2(x_2, y_2, z_2)$ is from Fig. 56-4

$$d = P_1P_2 = \sqrt{(P_1R)^2 + (RP_2)^2} = \sqrt{(P_1S)^2 + (SR)^2 + (RP_2)^2} = \sqrt{(x_2 - x_1)^2 + (y_2 - y_1)^2 + (z_2 - z_1)^2}$$

$$(56.1)$$

(See Problem 56.2.)

IF $P_1(x_1, y_1, z_1)$ **AND** $P_2(x_1, y_2, z_2)$ are the end points of a line segment and if $P(x, y, z)$ divides the
segment in the ratio $P_1P/PP_2 = r_1/r_2$, then

$$x = \frac{r_2 x_1 + r_1 x_2}{r_1 + r_2}, \qquad y = \frac{r_2 y_1 + r_1 y_2}{r_1 + r_2}, \qquad z = \frac{r_2 z_1 + r_1 z_2}{r_1 + r_2}. \qquad (56.2)$$

The coordinates of the midpoint of $\overline{P_1P_2}$ are $(\frac{1}{2}(x_1 + x_2), \frac{1}{2}(y_1 + y_2), \frac{1}{2}(z_1 + z_2))$. (See Problems 56.3–56.4.)

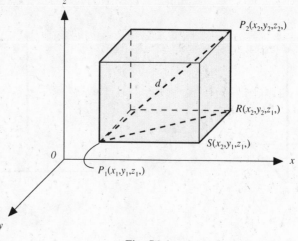

Fig. 56-4

TWO STRAIGHT LINES IN SPACE which intersect or are parallel lie in the same plane; two lines which are not coplanar are called *skew*. By definition, the angle between two directed skew lines as b and c in Fig. 56-5 is the angle between any two intersecting lines as b' and c' which are respectively parallel to the skew lines and similarly directed.

Fig. 56-5

DIRECTION COSINES OF A LINE. In the plane a directed line l [positive direction upward in Figs. 56-6(a) and (b)] forms the angles α and β with the positive directions on the x and y axes. However, in our study of the line in the plane we have favored the angle α over the angle β, calling it the angle of inclination of the line and its tangent the slope of the line.

In Fig. 56-6(a), $\alpha + \beta = \frac{1}{2}\pi$ and $m = \tan \alpha = \dfrac{\sin \alpha}{\cos \alpha} = \dfrac{\sin(\frac{1}{2}\pi - \beta)}{\cos \alpha} = \dfrac{\cos \beta}{\cos \alpha}$; and in Fig. 56-6($b$),

$\alpha = \frac{1}{2}\pi + \beta$ and $\tan \alpha = \dfrac{\sin(\frac{1}{2}\pi + \beta)}{\cos \alpha} = \dfrac{\cos \beta}{\cos \alpha}$. Now the angles α and β, called *direction angles* of the

line, or their cosines, $\cos \alpha$ and $\cos \beta$, called *direction cosines* of the line might have been used instead of the slope to give the direction of the line l. Indeed, it will be the direction cosines which will be generalized in our study of the straight line in space.

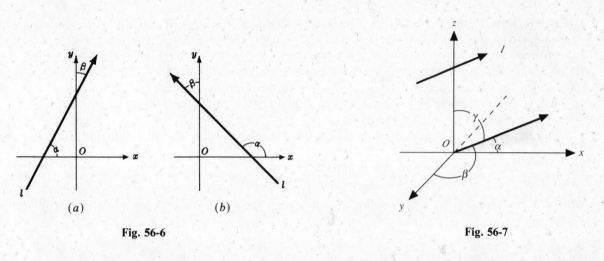

(a)   (b)

Fig. 56-6   **Fig. 56-7**

The direction of a line in space will be given by the three angles, called *direction angles* of the line, which it or that line through the origin parallel to it makes with the coordinate axes. If, as in Fig. 56-7, the direction angles α, β, γ, where $O \le \alpha, \beta, \gamma < \pi$, are the respective angles between the positive directions on the x, y, z axis and the directed line l (positive direction upward), the direction angles of this line when oppositely directed are $\alpha' = \pi - \alpha$, $\beta' = \pi - \beta$, $\gamma' = \pi - \gamma$. Thus, an undirected line in space has two sets of direction angles α, β, γ and $\pi - \alpha$, $\pi - \beta$, $\pi - \gamma$, and two sets of direction cosines $[\cos \alpha, \cos \beta, \cos \gamma]$ and $[-\cos \alpha, -\cos \beta, -\cos \gamma]$ since $\cos(\pi - \phi) = -\cos \phi$. To avoid confusion with the coordinates of a point, the triples of direction cosines of a line will be enclosed in a bracket. Thus we shall write $l: [A, B, C]$ to indicate the line whose direction cosines are the triple A, B, C.

The direction cosines of line l determined by points $P_1(x_1, y_1, z_1)$ and $P_2(x_2, y_2, z_2)$, and directed from P_1 to P_2 is (see Fig. 56-4)

$$\cos \alpha = \cos \angle P_2 P_1 S = \frac{P_1 S}{P_1 P_2} = \frac{x_2 - x_1}{d}, \qquad \cos \beta = \frac{y_2 - y_1}{d}, \qquad \cos \gamma = \frac{z_2 - z_1}{d}.$$

When l is directed from P_2 to P_1 the direction cosines are

$$\left[\frac{x_1 - x_2}{d}, \frac{y_1 - y_2}{d}, \frac{z_1 - z_2}{d} \right]$$

Except for the natural preference for $[\frac{1}{3}, \frac{2}{3}, \frac{2}{3}]$ over $[-\frac{1}{3}, -\frac{2}{3}, -\frac{2}{3}]$, it is immaterial which set of direction cosines is used when dealing with an undirected line.

EXAMPLE 2.   Find the two sets of direction cosines and indicate the positive direction along the line passing through the points $P_1(3, -1, 2)$ and $P_2(5, 2, -4)$.

We have $d = \sqrt{(2)^2 + (3)^2 + (-6)^2} = 7$. One set of direction cosines is

$$\left[\frac{x_2 - x_1}{d}, \frac{y_2 - y_1}{d}, \frac{z_2 - z_1}{d} \right] = \left[\frac{2}{7}, \frac{3}{7}, -\frac{6}{7} \right]$$

the positive direction being from P_1 to P_2. When the line is directed from P_2 to P_1, the direction cosines are $[-\frac{2}{7}, -\frac{3}{7}, \frac{6}{7}]$.

The sum of the squares of the direction cosines of any line is equal to 1, i.e.,

$$\cos^2 \alpha + \cos^2 \beta + \cos^2 \gamma = 1$$

It follows immediately that at least one of the direction cosines of any line is different from 0.

DIRECTION NUMBERS OF A LINE. Instead of the direction cosines of a line, it is frequently more convenient to use any triple of numbers, preferably small integers when possible, which are proportional to the direction cosines. Any such triple is called a set of *direction numbers* of the line. For example, if the direction cosines are $[\frac{2}{3}, -\frac{2}{3}, -\frac{1}{3}]$, sets of direction numbers are $[2, -2, -1]$, $[-2, 2, 1]$, $[4, -4, -2]$, etc.; if the direction cosines are $[\frac{1}{2}, 1/\sqrt{2}, -\frac{1}{2}]$, a set of direction numbers is $[1, \sqrt{2}, -1]$.

Sets of direction numbers for the line through points $P_1(x_1, y_1, z_1)$ and $P_2(x_2, y_2, z_2)$ are $[x_2 - x_1, y_2 - y_1, z_2 - z_1]$ and $[x_1 - x_2, y_1 - y_2, z_1 - z_2]$.

If $[a, b, c]$ is a set of direction numbers of a line, then the direction cosines of the line are given by

$$\cos \alpha = \pm \frac{a}{\sqrt{a^2 + b^2 + c^2}}, \qquad \cos \beta = \pm \frac{b}{\sqrt{a^2 + b^2 + c^2}}, \qquad \cos \gamma = \pm \frac{c}{\sqrt{a^2 + b^2 + c^2}} \qquad (56.3)$$

where the usual convention of first reading the upper signs and then the lower signs holds. (See Problems 56.5–56.8.)

THE ANGLE θ BETWEEN TWO DIRECTED LINES

$$l_1: \quad [\cos \alpha_1, \cos \beta_1, \cos \gamma_1] \qquad \text{and} \qquad l_2: \quad [\cos \alpha_2, \cos \beta_2, \cos \gamma_2]$$

is given by

$$\cos \theta = \cos \alpha_1 \cos \alpha_2 + \cos \beta_1 \cos \beta_2 + \cos \gamma_1 \cos \gamma_2 \qquad (56.4)$$

(For a proof see Problem 56.9.)

If the two lines are parallel then $\theta = 0$ or π, according as the lines are similarly or oppositely directed, and $\cos \alpha_1 \cos \alpha_2 + \cos \beta_1 \cos \beta_2 + \cos \gamma_1 \cos \gamma_2 = \pm 1$. If the sign is $+$, then $\cos \alpha_1 = \cos \alpha_2$, $\cos \beta_1 = \cos \beta_2$, $\cos \gamma_1 = \cos \gamma_2$; if the sign is $-$, then $\cos \alpha_1 = -\cos \alpha_2$, $\cos \beta_1 = -\cos \beta_2$, $\cos \gamma_1 = -\cos \gamma_2$. Thus two undirected lines are parallel if and only if their direction cosines are the same or differ only in sign. In terms of direction numbers, *two lines are parallel if and only if corresponding direction numbers are proportional.*

If the two lines are perpendicular, then $\theta = \frac{1}{2}\pi$ or $3\pi/2$, according as the lines are similarly or oppositely directed, and

$$\cos \alpha_1 \cos \alpha_2 + \cos \beta_1 \cos \beta_2 + \cos \gamma_1 \cos \gamma_2 = 0 \qquad (56.5)$$

In terms of direction numbers, *two lines with direction numbers $[a_1, b_1, c_1]$ and $[a_2, b_2, c_2]$, respectively, are perpendicular if and only if*

$$a_1 \cdot a_2 + b_1 \cdot b_2 + c_1 \cdot c_2 = 0. \qquad (56.5')$$

(See Problems 56.10–56.12.)

THE DIRECTION NUMBER DEVICE.

If $l_1: [a_1, b_1, c_1]$ and $l_2: [a_2, b_2, c_2]$ are two nonparallel lines, then a set of direction numbers $[a, b, c]$ of any line perpendicular to both l_1 and l_2 is given by

$$a = \begin{vmatrix} b_1 & c_1 \\ b_2 & c_2 \end{vmatrix}, \qquad b = \begin{vmatrix} c_1 & a_1 \\ c_2 & a_2 \end{vmatrix}, \qquad c = \begin{vmatrix} a_1 & b_1 \\ a_2 & b_2 \end{vmatrix}.$$

These three determinants can be obtained readily as follows:

(1) Write the two sets of direction numbers in three columns $\begin{matrix} a_1 & b_1 & c_1 \\ a_2 & b_2 & c_2 \end{matrix}$.

(2) Repeat the first two columns to obtain $\begin{matrix} a_1 & b_1 & c_1 & a_1 & b_1 \\ a_2 & b_2 & c_2 & a_2 & b_2 \end{matrix}$ and strike out the first column to

have $\begin{matrix} a_1 & b_1 & c_1 & a_1 & b_1 \\ a_2 & b_2 & c_2 & a_2 & b_2 \end{matrix}$.

Then a is the determinant of the first and second columns remaining, b is the determinant of the second and third columns, and c is the determinant of the third and fourth columns. This procedure will be called the *direction number device*. Note, however, that it is a mechanical procedure for obtaining one solution of two homogeneous equations in three unknowns and thus has other applications.

EXAMPLE 3.   Find a set of direction numbers $[a, b, c]$ of any line perpendicular to l_1: $[2, 3, 4]$ and l_2: $[1, -2, -3]$.

Using the direction number device, we write $\begin{matrix} 2 & 3 & 4 & 2 & 3 \\ 1 & -2 & -3 & 1 & -2 \end{matrix}$. Then

$$a = \begin{vmatrix} 3 & 4 \\ -2 & -3 \end{vmatrix} = -1, \qquad b = \begin{vmatrix} 4 & 2 \\ -3 & 1 \end{vmatrix} = 10, \qquad c = \begin{vmatrix} 2 & 3 \\ 1 & -2 \end{vmatrix} = -7.$$

A set of direction numbers is $[-1, 10, -7]$ or, if preferred, $[1, -10, 7]$. (See Problem 56.13.)

Solved Problems

56.1 What is the locus of a point:

  (*a*) Whose z coordinate is always 0?

  (*b*) Whose z coordinate is always 3?

  (*c*) Whose x coordinate is always -5?

  (*d*) Whose x and y coordinates are always 0?

  (*e*) Whose x coordinate is always 2 and whose y coordinate is always 3?

  (*a*) All points $(a, b, 0)$ lie in the xy plane; the locus is that plane.

  (*b*) Every point is 3 units above the xy plane; the locus is the plane parallel to the xy plane and 3 units above it.

  (*c*) A plane parallel to the yz plane and 5 units to the left of it.

  (*d*) All points $(0, 0, c)$ lie on the z axis; the locus is that line.

  (*e*) In locating the point $P(2, 3, c)$, the x and y coordinates are used to locate the point $F(2, 3, 0)$ in the xy plane and then a distance $|c|$ is measured from F parallel to the z axis. The locus is the line parallel to the z axis passing through the point $(2, 3, 0)$ in the xy plane.

56.2 (*a*) Find the distance between the points $P_1(-1, -3, 3)$ and $P_2(2, -4, 1)$.

  (*b*) Find the perimeter of the triangle whose vertices are $A(-2, -4, -3)$, $B(1, 0, 9)$, $C(2, 0, 9)$.

  (*c*) Show that the points $A(1, 2, 4)$, $B(4, 1, 6)$, and $C(-5, 4, 0)$ are collinear.

  (*a*) Here $d = \sqrt{(x_2 - x_1)^2 + (y_2 - y_1)^2 + (z_2 - z_1)^2} = \sqrt{[2 - (-1)]^2 + [-4 - (-3)]^2 + (1 - 3)^2} = \sqrt{14}$.

  (*b*) We find $AB = \sqrt{[1 - (-2)]^2 + [0 - (-4)]^2 + [9 - (-3)]^2} = 13$, $BC = \sqrt{(2 - 1)^2 + (0 - 0)^2 + (9 - 9)^2} = 1$, and $CA = \sqrt{(-2 - 2)^2 + (-4 - 0)^2 + (-3 - 9)^2} = 4\sqrt{11}$. The perimeter is $13 + 1 + 4\sqrt{11} = 14 + 4\sqrt{11}$.

  (*c*) Here $AB = \sqrt{(3)^2 + (-1)^2 + (2)^2} = \sqrt{14}$, $BC = \sqrt{(-9)^2 + (3)^2 + (-6)^2} = 3\sqrt{14}$, and $CA = \sqrt{(6)^2 + (-2)^2 + (4)^2} = 2\sqrt{14}$. Since $BC = CA + AB$, the points are collinear.

56.3   Find the coordinates of the point P of division for each pair of points and given ratio. Find also the midpoint of the segment. (a) $P_1(3, 2, -4)$, $P_2(6, -1, 2)$; $1:2$ (b) $P_1(2, 5, 4)$, $P_2(-6, 3, 8)$; $-3:5$.

(a)   Here $r_1 = 1$ and $r_2 = 2$. Then

$$x = \frac{r_2 x_1 + r_1 x_2}{r_1 + r_2} = \frac{2 \cdot 3 + 1 \cdot 6}{1 + 2} = 4, \qquad y = \frac{r_2 y_1 + r_1 y_2}{r_1 + r_2} = \frac{2 \cdot 2 + 1(-1)}{1 + 2} = 1, \qquad z = \frac{r_2 z_1 + r_1 z_2}{r_1 + r_2} = -2$$

and the required point is $P(4, 1, -2)$. The midpoint has coordinates $(\frac{1}{2}(x_1 + x_2), \frac{1}{2}(y_1 + y_2), \frac{1}{2}(z_1 + z_2)) = (\frac{9}{2}, \frac{1}{2}, -1)$.

(b)   Here $r_1 = -3$ and $r_2 = 5$. Then

$$x = \frac{5 \cdot 2 + (-3)(-6)}{-3 + 5} = 14, \qquad y = \frac{5 \cdot 5 + (-3)3}{-3 + 5} = 8, \qquad z = \frac{5 \cdot 4 + (-3)8}{-3 + 5} = -2$$

and the required point is $P(14, 8, -2)$. The midpoint has coordinates $(-2, 4, 6)$.

56.4   Prove: The three lines joining the midpoints of the opposite edges of a tetrahedron pass through a point P which bisects each of them.

Let the tetrahedron, shown in Fig. 56-8, have vertices $O(0, 0, 0)$, $A(a, 0, 0)$, $B(b, c, 0)$, and $C(d, e, f)$.

The midpoints of $\overline{OB}$ and $\overline{AC}$ are, respectively, $D(\frac{1}{2}b, \frac{1}{2}c, 0)$ and $E(\frac{1}{2}(a + d), \frac{1}{2}e, \frac{1}{2}f)$, and the midpoint of DE is $P(\frac{1}{4}(a + b + d), \frac{1}{4}(c + e), \frac{1}{4}f)$. The midpoints of $\overline{OA}$ and $\overline{BC}$ are, respectively, $F(\frac{1}{2}a, 0, 0)$ and $G(\frac{1}{2}(b + d), \frac{1}{2}(c + e), \frac{1}{2}f)$, and the midpoint of FG is P. It is left for the reader to find the midpoints H and I of $\overline{OC}$ and $\overline{AB}$, and show that P is the midpoint of $\overline{HI}$.

Fig. 56-8

56.5   Find the direction cosines of the line:

(a)   Passing through $P_1(3, 4, 5)$ and $P_2(-1, 2, 3)$ and directed from P_1 to P_2

(b)   Passing through $P_1(2, -1, -3)$ and $P_2(-4, 2, 1)$ and directed from P_2 to P_1

(c)   Passing through $O(0, 0, 0)$ and $P(a, b, c)$ and directed from O to P

(d)   Passing through $P_1(4, -1, 2)$ and $P_2(2, 1, 3)$ and directed so that γ is acute

(a)   We have

$$\cos \alpha = \frac{x_2 - x_1}{d} = \frac{-4}{2\sqrt{6}}, \qquad \cos \beta = \frac{y_2 - y_1}{d} = \frac{-2}{2\sqrt{6}}, \qquad \cos \gamma = \frac{z_2 - z_1}{d} = \frac{-2}{2\sqrt{6}}.$$

The direction cosines are $\left[-\dfrac{2}{\sqrt{6}}, -\dfrac{1}{\sqrt{6}}, -\dfrac{1}{\sqrt{6}} \right]$.

(b) $$\cos \alpha = \frac{x_1 - x_2}{d} = \frac{6}{\sqrt{61}}, \qquad \cos \beta = \frac{y_1 - y_2}{d} = \frac{-3}{\sqrt{61}}, \qquad \cos \gamma = \frac{z_1 - z_2}{d} = \frac{-4}{\sqrt{61}}.$$

The direction cosines are $\left[\dfrac{6}{\sqrt{61}}, -\dfrac{3}{\sqrt{61}}, -\dfrac{4}{\sqrt{61}} \right]$.

(c) $$\cos \alpha = \frac{a - 0}{d} = \frac{a}{\sqrt{a^2 + b^2 + c^2}}, \qquad \cos \beta = \frac{b - 0}{d} = \frac{b}{\sqrt{a^2 + b^2 + c^2}}, \qquad \cos \gamma = \frac{c - 0}{d} = \frac{c}{\sqrt{a^2 + b^2 + c^2}}$$

The direction cosines are $\left[\dfrac{a}{\sqrt{a^2 + b^2 + c^2}}, \dfrac{b}{\sqrt{a^2 + b^2 + c^2}}, \dfrac{c}{\sqrt{a^2 + b^2 + c^2}} \right]$.

(d)   The two sets of direction cosines of the undirected line are

$$\cos \alpha = \pm \tfrac{2}{3}, \qquad \cos \beta = \mp \tfrac{2}{3}, \qquad \cos \gamma = \mp \tfrac{1}{3}$$

one set being given by the upper signs and the other by the lower signs. When γ is acute, $\cos \gamma > 0$; hence the required set is $[-\tfrac{2}{3}, \tfrac{2}{3}, \tfrac{1}{3}]$.

56.6   Given the direction angles α measuring $120°$ and β measuring $45°$, find γ if the line is directed upward.

$\cos^2 \alpha + \cos^2 \beta + \cos^2 \gamma = \cos^2 120° + \cos^2 45° + \cos^2 \gamma = (-\tfrac{1}{2})^2 + (1\sqrt{2})^2 + \cos^2 \gamma = 1$. Then $\cos^2 \gamma = \tfrac{1}{4}$ and $\cos \gamma = \pm \tfrac{1}{2}$. When the line is directed upward, $\cos \gamma = \tfrac{1}{2}$ and $\gamma = 60°$.

56.7   The direction numbers of a line l are given as $[2, -3, 6]$. Find the direction cosines of l when directed upward.

The direction cosines of l are given by

$$\cos \alpha = \pm \frac{a}{\sqrt{a^2 + b^2 + c^2}} = \pm \frac{2}{7}, \qquad \cos \beta = \pm \left(\frac{-3}{7} \right), \qquad \cos \gamma = \pm \frac{6}{7}.$$

When γ is acute, $\cos \gamma > 0$, and the direction cosines are $[\tfrac{2}{7}, -\tfrac{3}{7}, \tfrac{6}{7}]$.

56.8   Use direction numbers to show that the points $A(1, 2, 4)$, $B(4, 1, 6)$, and $C(-5, 4, 0)$ are collinear. [See Problem 56.2(c).]

A set of direction numbers for the line $\overleftrightarrow{AB}$ is $[3, -1, 2]$, for BC is $[-9, 3, -6]$. Since the two sets are proportional, the lines are parallel; since the lines have a point in common they are coincident and the points are collinear.

56.9   Prove: the angle θ between two directed lines l_1: $[\cos \alpha_1, \cos \beta_1, \cos \gamma_1]$ and l_2: $[\cos \alpha_2, \cos \beta_2, \cos \gamma_2]$ is given by $\cos \theta = \cos \alpha_1 \cos \alpha_2 + \cos \beta_1 \cos \beta_2 + \cos \gamma_1 \cos \gamma_2$.

The angle θ is by definition the angle between two lines issuing from the origin parallel, respectively, to the given lines l_1 and l_2 and similarly directed.

Consider the triangle OP_1P_2, in Fig. 56-9, whose vertices are the origin and the points $P_1(\cos \alpha_1,$ $\cos \beta_1, \cos \gamma_1)$ and $P_2(\cos \alpha_2, \cos \beta_2, \cos \gamma_2)$. The line segment OP_1 is of length 1 (why?) and is parallel to l_1; similarly, OP_2 is of length 1 and is parallel to l_2. Thus, $\angle P_1OP_2 = \theta$. By the Law of Cosines, $(P_1P_2)^2 = (OP_1)^2 + (OP_2)^2 - 2(OP_1)(OP_2) \cos \theta$ and $\cos \theta = \cos \alpha_1 \cos \alpha_2 + \cos \beta_1 \cos \beta_2 + \cos \gamma_1 \cos \gamma_2$.

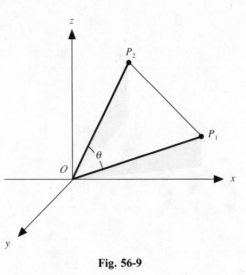

Fig. 56-9

56.10 (a) Find the angle between the directed lines $l_1 : [\frac{2}{7}, \frac{3}{7}, \frac{6}{7}]$ and $l_2 : [\frac{2}{3}, -\frac{1}{3}, \frac{2}{3}]$.

      (b) Find the acute angle between the lines $l_1 : [-2, 1, 2]$ and $l_2 : [2, -6, -3]$.

      (c) The line l_1 passes through $A(5, -2, 3)$ and $B(2, 1, -4)$, and the line l_2 passes through $C(-4, 1, -2)$ and $D(-3, 2, 3)$. Find the acute angle between them.

        (a) We have $\cos \theta = \cos \alpha_1 \cos \alpha_2 + \cos \beta_1 \cos \beta_2 + \cos \gamma_1 \cos \gamma_2 = \frac{2}{7} \cdot \frac{2}{3} + \frac{3}{7}(-\frac{1}{3}) + \frac{6}{7} \cdot \frac{2}{3} = \frac{13}{21} = 0.619$ and $\theta = 51° 50'$.

        (b) Since $\sqrt{(-2)^2 + (1)^2 + (2)^2} = 3$, we take $[-\frac{2}{3}, \frac{1}{3}, \frac{2}{3}]$ as direction cosines of l_1. Since $\sqrt{(2)^2 + (-6)^2 + (-3)^2} = 7$, we take $[\frac{2}{7}, -\frac{6}{7}, -\frac{3}{7}]$ as direction cosines of l_2. Then $\cos \theta = -\frac{2}{3} \cdot \frac{2}{7} + \frac{1}{3}(-\frac{6}{7}) + \frac{2}{3}(-\frac{3}{7}) = -\frac{16}{21} = -0.762$ and $\theta = 139° 40'$. The required angle is $40° 20'$.

        (c) Take $[3\sqrt{67}, -3\sqrt{67}, 7/\sqrt{67}]$ as direction cosines of l_1 and $[1/3\sqrt{3}, 1/3\sqrt{3}, 5/3\sqrt{3}]$ as direction cosines of l_2. Then

$$\cos \theta = \frac{1}{\sqrt{3 \cdot 67}} - \frac{1}{\sqrt{3 \cdot 67}} + \frac{35}{3\sqrt{3 \cdot 67}} = \frac{35}{3\sqrt{201}} = 0.823 \quad \text{and} \quad \theta = 34° 40'$$

56.11 (a) Show that the line joining $A(9, 2, 6)$ and $B(5, -3, 2)$ and the line joining $C(-1, -5, -2)$ and $D(7, 5, 6)$ are parallel.

      (b) Show that the line joining $A(7, 2, 3)$ and $B(-2, 5, 2)$ and the line joining $C(4, 10, 1)$ and $D(1, 2, 4)$ are mutually perpendicular.

        (a) Here $[9 - 5, 2 - (-3), 6 - 2] = [4, 5, 4]$ is a set of direction numbers of AB and $[-1 - 7, -5 - 5, -2 - 6] = [-8, -10, -8]$ is a set of direction numbers of CD. Since the two sets are proportional, the two lines are parallel.

        (b) Here $[9, -3, 1]$ is a set of direction numbers of AB and $[3, 8, -3]$ is a set of direction numbers of CD. Since [see Equation $(56.5')$] $9 \cdot 3 + (-3)8 + 1(-3) = 0$, the lines are perpendicular.

56.12 Find the area of the triangle whose vertices are $A(4, 2, 3)$, $B(7, -2, 4)$ and $C(3, -4, 6)$.

      The area of triangle ABC is given by $\frac{1}{2}(AB)(AC) \sin A$. We have $AB = \sqrt{26}$ and $AC = \sqrt{46}$.

      To find $\sin A$, we direct the sides $\overline{AB}$ and $\overline{AC}$ away from the origin as in Fig. 56-10. Then $\overline{AB}$ has direction cosines $[3/\sqrt{26}, -4/\sqrt{26}, 1/\sqrt{26}]$, $\overline{AC}$ has direction cosines $[-1/\sqrt{46}, -6/\sqrt{46}, 3/\sqrt{46}]$,

$$\cos A = \frac{3}{\sqrt{26}} \cdot \frac{-1}{\sqrt{46}} + \frac{-4}{\sqrt{26}} \cdot \frac{-6}{\sqrt{46}} + \frac{1}{\sqrt{26}} \cdot \frac{3}{\sqrt{46}} = \frac{24}{\sqrt{26}\sqrt{46}}$$

and

$$\sin A = \sqrt{1 - \cos^2 A} = \frac{2\sqrt{155}}{\sqrt{26}\sqrt{46}}$$

      The required area is $\frac{1}{2}\sqrt{26} \cdot \sqrt{46} \cdot \dfrac{2\sqrt{155}}{\sqrt{26}\sqrt{46}} = \sqrt{155}$.

Fig. 56-10

56.13 Find a set of direction numbers for any line which is perpendicular to

(a) $l_1: [1, -2, -3]$ and $l_2: [4, -1, -5]$

(b) The triangle whose vertices are $A(4, 2, 3)$, $B(7, -2, 4)$, and $C(3, -4, 6)$

    (a) Using the direction number device

$$\begin{matrix} 1 \\ 4 \end{matrix} \begin{matrix} -2 & -3 & 1 & -2 \\ -1 & -5 & 4 & -1 \end{matrix}, \quad \text{we obtain} \quad a = \begin{vmatrix} -2 & -3 \\ -1 & -5 \end{vmatrix} = 7, \quad b = \begin{vmatrix} -3 & 1 \\ -5 & 4 \end{vmatrix} = -7, \quad c = \begin{vmatrix} 1 & -2 \\ 4 & -1 \end{vmatrix} = 7.$$

    Thus a set of direction numbers is $[7, -7, 7]$; a simpler set is $[1, -1, 1]$.

    (b) Since the triangle lies in a plane determined by the lines AB and AC, we seek direction numbers for any line perpendicular to these lines. For AB and AC, respective sets of direction numbers are $[3, -4, 1]$ and $[-1, -6, 3]$. Using the direction number device

$$\begin{matrix} 3 \\ -1 \end{matrix} \begin{matrix} -4 & 1 & 3 & -4 \\ -6 & 3 & -1 & -6 \end{matrix}, \quad a = \begin{vmatrix} -4 & 1 \\ -6 & 3 \end{vmatrix} = -6, \quad b = \begin{vmatrix} 1 & 3 \\ 3 & -1 \end{vmatrix} = -10, \quad c = \begin{vmatrix} 3 & -4 \\ -1 & -6 \end{vmatrix} = -22.$$

    Then $[-6, -10, -22]$ is a set of direction numbers and $[3, 5, 11]$ is a simpler set.

Supplementary Problems

56.14 Find the undirected distance between each pair of points:

(a) $(4, 1, 5)$ and $(2, -1, 4)$ (b) $(9, 7, -2)$ and $(6, 5, 4)$ (c) $(9, -2, -3)$ and $(-3, 4, 0)$.

    Ans. (a) 3   (b) 7   (c) $3\sqrt{21}$

56.15 Find the undirected distance of each of the following points from (i) the origin (ii) the x axis (iii) the y axis and (iv) the z axis: (a) $(2, 6, -3)$ (b) $(2, -\sqrt{3}, 3)$.

    Ans. (a) $7, 2, 6, 3$ (b) $4, 2, \sqrt{3}, 3$

56.16 For each pair of points, find the coordinates of the point dividing $\overline{P_1 P_2}$ in the given ratio; find also the coordinates of the midpoint.

(a) $P_1(4, 1, 5)$, $P_2(2, -1, 4)$, $3:2$ *Ans.* $\left(\frac{14}{5}, -\frac{1}{5}, \frac{22}{5}\right)$, $\left(3, 0, \frac{9}{2}\right)$

(b) $P_1(9, 7, -2)$, $P_2(6, 5, 4)$, $1:4$ *Ans.* $\left(\frac{42}{5}, \frac{33}{5}, -\frac{4}{5}\right)$, $\left(\frac{15}{2}, 6, 1\right)$

(c) $P_1(9, -2, -3)$, $P_2(-3, 4, 0)$, $-1:3$ *Ans.* $\left(15, -5, -\frac{9}{2}\right)$, $\left(3, 1, -\frac{3}{2}\right)$

(d) $P_1(0, 0, 0)$, $P_2(2, 3, 4)$, $2:-3$ *Ans.* $(-4, -6, -8)$, $\left(1, \frac{3}{2}, 2\right)$

56.17 Find the equation of the locus of a point which is (*a*) always equidistant from the points $(4, 1, 5)$ and $(2, -1, 4)$ (*b*) always at a distance 6 units from $(4, 1, 5)$ (*c*) always two-thirds as far from the *y* axis as from the origin.

   Ans. (*a*) $4x + 4y + 2z - 21 = 0$ (*b*) $x^2 + y^2 + z^2 - 8x - 2y - 10z + 6 = 0$
   (*c*) $5x^2 - 4y^2 + 5z^2 = 0$

56.18 Find a set of direction cosines and a set of direction numbers for the line joining P_1 and P_2, given

   (*a*) $P_1(0, 0, 0)$, $P_2(4, 8, -8)$ *Ans.* $[\frac{1}{3}, \frac{2}{3}, -\frac{2}{3}]$, $[1, 2, -2]$

   (*b*) $P_1(1, 3, 5)$, $P_2(-1, 0, -1)$ *Ans.* $[-\frac{2}{7}, -\frac{3}{7}, -\frac{6}{7}]$, $[2, 3, 6]$

   (*c*) $P_1(5, 6, -3)$, $P_2(1, -6, 3)$ *Ans.* $[-\frac{2}{7}, -\frac{6}{7}, \frac{3}{7}]$, $[2, 6, -3]$

   (*d*) $P_1(4, 2, -6)$, $P_2(-2, 1, 3)$ *Ans.* $\left[\dfrac{6}{\sqrt{118}}, \dfrac{1}{\sqrt{118}}, \dfrac{-9}{\sqrt{118}}\right]$, $[6, 1, -9]$

56.19 Find $\cos \gamma$, given (*a*) $\cos \alpha = \frac{14}{15}$, $\cos \beta = -\frac{1}{3}$ (*b*) $\alpha = 60°$, $\beta = 135°$

   Ans. (*a*) $\pm\frac{2}{15}$ (*b*) $\pm\frac{1}{2}$

56.20 Find

   (*a*) The acute angle between the line having direction numbers $[-4, -1, -8]$ and the line joining the points $(6, 4, -1)$ and $(4, 0, 3)$.

   (*b*) The interior angles of the triangle whose vertices are $A(2, -1, 0)$, $B(4, 1, -1)$, $C(5, -1, -4)$.

   Ans. (*a*) $68° 20'$ (*b*) $A = 48° 10'$, $B = 95° 10'$, $C = 36° 40'$

56.21 Find the coordinates of the point P in which the line joining $A(5, -1, 4)$ and $B(-5, 7, 0)$ pierces the *yz* plane.

   Hint: Let P have coordinates $(0, b, c)$ and express the condition (see Problem 56.8) that A, B, P be collinear.

   Ans. $P(0, 3, 2)$

56.22 Find relations which the coordinates of $P(x, y, z)$ must satisfy if P is to be collinear with $(2, 3, 1)$ and $(1, -2, -5)$.

   Ans. $x - 2 : y - 3 : z - 1 = 1 : 5 : 6$ or $\dfrac{x - 2}{1} = \dfrac{y - 3}{5} = \dfrac{z - 1}{6}$

56.23 Find a set of direction numbers for any line perpendicular to

   (*a*) Each of the lines $l_1 : [1, 2, -4]$ and $l_2 : [2, -1, 3]$

   (*b*) Each of the lines joining $A(2, -1, 5)$ to $B(-1, 3, 4)$ and $C(0, -5, 4)$

   Ans. (*a*) $[2, -11, -5]$ (*b*) $[8, 1, -20]$

56.24 Find the coordinates of the point P in which the line joining the points $A(4, 11, 18)$ and $B(-1, -4, -7)$ intersects the line joining the points $C(3, 1, 5)$ and $D(5, 0, 7)$.

   Hint: Let P divide $\overline{AB}$ in the ratio $1 : r$ and $\overline{CD}$ in the ratio $1 : s$, and obtain relations $rs - r - 4s - 6 = 0$, etc.

   Ans. $(1, 2, 3)$

56.25 Prove that the four line segments joining each vertex of a tetrahedron to the point of intersection of the medians of the opposite face have a point G in common. Prove that each of the four line segments is divided in the ratio $1 : 3$ by G.

   (NOTE. G, the point P of Problem 56.4, is called the centroid of the tetrahedron.)

Part V

INTRODUCTION TO CALCULUS

Chapter 57

The Derivative

IN THIS AND SUBSEQUENT CHAPTERS, it will be understood that number refers always to a real number, that the range of any variable (as x) is a set of real numbers, and that a function of one variable [as $f(x)$] is a single-valued function.

In this chapter a procedure is given by which from a given function $y = f(x)$ another function, denoted by y' or $f'(x)$ and called the derivative of y or of $f(x)$ with respect to x, is obtained. Depending upon the quantities denoted by x and $y = f(x)$, the derivative may be interpreted as the slope of a tangent line to a curve, as velocity, as acceleration, etc.

LIMIT OF A FUNCTION. A given function $f(x)$ is said to have a *limit M* as x approaches c [in symbols, $\lim_{x \to c} f(x) = M$] if $f(x)$ can be made as close to M as we please for all values of $x \neq c$ but sufficiently near to c, by having x get sufficiently close to c (approaching both from the left and right).

EXAMPLE 1. Consider $f(x) = x^2 - 2$ for x near 3.

If x is near to 3, say $2.99 < x < 3.01$, then $(2.99)^2 - 2 < f(x) < (3.01)^2 - 2$ or $6.9401 < f(x) < 7.0601$.

If x is nearer to 3, say $2.999 < x < 3.001$, then $(2.999)^2 - 2 < f(x) < (3.001)^2 - 2$ or $6.994001 < f(x) < 7.006001$.

If x is still nearer to 3, say $2.9999 < x < 3.0001$, then $(2.9999)^2 - 2 < f(x) < (3.0001)^2 - 2$ or $6.99940001 < f(x) < 7.00060001$.

It appears reasonable to conclude that as x is taken in a smaller and smaller interval about 3, the corresponding $f(x)$ will lie in a smaller and smaller interval about 7. Conversely, it seems reasonable to conclude that if we demand that $f(x)$ have values lying in smaller and smaller intervals about 7, we need only to choose x in sufficiently smaller and smaller intervals about 3. Thus we conclude

$$\lim_{x \to 3} (x^2 - 2) = 7$$

EXAMPLE 2. Consider $f(x) = \dfrac{x^2 - x - 6}{x - 3}$ for x near 3.

When $x \neq 3$, $f(x) = \dfrac{x^2 - x - 6}{x - 3} = x + 2$. Thus, for x near 3, $x + 2$ is near to 5 and

$$\lim_{x \to 3} \frac{x^2 - x - 6}{x - 3} = 5$$

ONE-SIDED LIMITS. We say that the limit of $f(x)$ as x approaches a from the left is L [$\lim_{x \to a^-} f(x) = L$] when $f(x)$ gets arbitrarily close to L as x approaches a from the left-hand-side of a. [Similarly, $\lim_{x \to b^+} f(x) = M$ is the right-hand limit.] Clearly, if $\lim_{x \to a^-} f(x) = \lim_{x \to a^+} f(x) = L$, then $\lim_{x \to a} f(x) = L$.

THEOREMS ON LIMITS. If $\lim_{x \to c} f(x) = M$ and $\lim_{x \to c} g(x) = N$, then

I. $\lim_{x \to c} [f(x) \pm g(x)] = \lim_{x \to c} f(x) \pm \lim_{x \to c} g(x) = M \pm N$.

II. $\lim_{x \to c} [kf(x)] = k \lim_{x \to c} f(x) = kM$, where k is a constant.

III. $\lim_{x \to c} [f(x) \cdot g(x)] = \lim_{x \to c} f(x) \cdot \lim_{x \to c} g(x) = MN$.

$$\text{IV.} \quad \lim_{x \to c} \frac{f(x)}{g(x)} = \frac{\lim\limits_{x \to c} f(x)}{\lim\limits_{x \to c} g(x)} = \frac{M}{N}, \text{ provided } N \neq 0.$$

(See Problem 57.2.)

CONTINUOUS FUNCTIONS. A function $f(x)$ is called continuous at $x = c$, provided

   (1) $f(c)$ is defined,
   (2) $\lim\limits_{x \to c} f(x)$ exists,
   (3) $\lim\limits_{x \to c} f(x) = f(c)$.

EXAMPLE 3
(a) The function $f(x) = x^2 - 2$ of Example 1 is continuous at $x = 3$ since (1) $f(3) = 7$, (2) $\lim\limits_{x \to 3} (x^2 - 2) = 7$,

   (3) $\lim\limits_{x \to 3} (x^2 - 2) = f(3)$.

(b) The function $f(x) = \dfrac{x^2 - x - 6}{x - 3}$ of Example 2 is not continuous at $x = 3$ since $f(3)$ is not defined. (See
   Problem 57.3.)

   A function $f(x)$ is said to be *continuous* on the interval (a, b) if it is continuous for every value of x of the interval. A polynomial in x is continuous since it is continuous for all values of x. A rational function in x, $f(x) = P(x)/Q(x)$, where $P(x)$ and $Q(x)$ are polynomials, is continuous for all values of x except those for which $Q(x) = 0$. Thus $f(x) = \dfrac{x^2 + x + 1}{(x - 1)(x^2 + 2)}$ is continuous for all values of x, except $x = 1$.

CONTINUITY ON A CLOSED INTERVAL. If a function $y = g(x)$ is continuous for all values of x in $[a, b]$, then it is continuous on (a, b) and also at a and b.
   However, in this case, $g(x)$ is continuous at a means $\lim\limits_{x \to a^+} g(x) = g(a)$; similarly $\lim\limits_{x \to b^-} g(x) = g(b)$.

(See Problem 57.4.)

INCREMENTS. Let x_0 and x_1 be two distinct values of x. It is customary to denote their difference $x_1 - x_0$ by Δx (read, delta x) and to write $x_0 + \Delta x$ for x_1.
   Now if $y = f(x)$ and if x changes in value from $x = x_0$ to $x = x_0 + \Delta x$, y will change in value from $y_0 = f(x_0)$ to $y_0 + \Delta y = f(x_0 + \Delta x)$. The change in y due to a change in x from $x = x_0$ to $x = x_0 + \Delta x$ is $\Delta y = f(x_0 + \Delta x) - f(x_0)$.

EXAMPLE 4. Compute the change in $y = f(x) = x^2 - 2x + 5$ when x changes in value from (a) $x = 3$ to $x = 3.2$, (b) $x = 3$ to $x = 2.9$.

   (a) Take $x_0 = 3$ and $\Delta x = 0.2$. Then $y_0 = f(x_0) = f(3) = 8$, $y_0 + \Delta y = f(x_0 + \Delta x) = f(3.2) = 8.84$, and $\Delta y = 8.84 - 8 = 0.84$.

   (b) Take $x_0 = 3$ and $\Delta x = -0.1$. Then $y_0 = f(3) = 8$, $y_0 + \Delta y = f(2.9) = 7.61$, and $\Delta y = 7.61 - 8 = -0.39$.

(See Problems 57.5-57.6.)

THE DERIVATIVE.   The *derivative* of $y = f(x)$ at $x = x_0$ is

$$\lim_{\Delta x \to 0} \frac{\Delta y}{\Delta x} = \lim_{\Delta x \to 0} \frac{f(x_0 + \Delta x) - f(x_0)}{\Delta x}$$

provided the limit exists. $\lim_{\Delta x \to 0} \dfrac{\Delta y}{\Delta x}$ is called the *difference quotient*.

In finding derivatives, we shall use the following five-step rule:

(*1*)   Write $y_0 = f(x_0)$.
(*2*)   Write $y_0 + \Delta y = f(x_0 + \Delta x)$.
(*3*)   Obtain $\Delta y = f(x_0 + \Delta x) - f(x_0)$.
(*4*)   Obtain $\Delta y / \Delta x$.
(*5*)   Evaluate $\lim_{\Delta x \to 0} \dfrac{\Delta y}{\Delta x}$.

EXAMPLE 5.   Find the derivative of $y = f(x) = 2x^2 - 3x + 5$ at $x = x_0$.

(*1*) $$y_0 = f(x_0) = 2x_0^2 - 3x_0 + 5$$

(*2*) $$\begin{aligned} y_0 + \Delta y = f(x_0 + \Delta x) &= 2(x_0 + \Delta x)^2 - 3(x_0 + \Delta x) + 5 \\ &= 2x_0^2 + 4x_0 \cdot \Delta x + 2(\Delta x)^2 - 3x_0 - 3 \cdot \Delta x + 5 \end{aligned}$$

(*3*) $$\Delta y = f(x_0 + \Delta x) - f(x_0) = 4x_0 \cdot \Delta x - 3 \cdot \Delta x + 2(\Delta x)^2$$

(*4*) $$\frac{\Delta y}{\Delta x} = 4x_0 - 3 + 2 \cdot \Delta x$$

(*5*) $$\lim_{\Delta x \to 0} \frac{\Delta y}{\Delta x} = \lim_{\Delta x \to 0} (4x_0 - 3 + 2 \cdot \Delta x) = 4x_0 - 3.$$

If in the example above the subscript 0 is deleted, the five-step rule yields a function of x (here, $4x - 3$) called the derivative with respect to x of the given function. The derivative with respect to x of the function $y = f(x)$ is denoted by one of the symbols y', $\dfrac{dy}{dx}$, $f'(x)$, or $D_x y$.

Provided it exists, the value of the derivative for any given value of x, say x_0, will be denoted by

$y'|_{x=x_0}$, $\dfrac{dy}{dx}\Big|_{x=x_0}$, or $f'(x_0)$. (See Problems 57.8–57.15.)

HIGHER-ORDER DERIVATIVES.   The process of finding the derivative of a given function is called *differentiation*.

By differentiation, we obtain from a given function $y = f(x)$ another function $y' = f'(x)$ which will now be called the *first derivative* of y or of $f(x)$ with respect to x. If, in turn, $y' = f'(x)$ is differentiated with respect to x, another function $y'' = f''(x)$, called the *second derivative* of y or of $f(x)$ is obtained. Similarly, a third derivative may be found, and so on.

EXAMPLE 6.   Let $y = f(x) = x^4 - 3x^2 + 8x + 6$. Then $y' = f'(x) = 4x^3 - 6x + 8$, $y'' = f''(x) = 12x^2 - 6$, and $y''' = f'''(x) = 24x$. (See Problem 57.16.)

Solved Problems

57.1   Investigate $f(x) = 1/x$ for values of x near $x = 0$.

If x is near 0, say $-.01 < x < .01$, then $\dfrac{1}{-.01} < \dfrac{1}{x} < \dfrac{1}{.01}$ or $-100 < \dfrac{1}{x} < 100$.

If x is nearer to 0, say $-.0001 < x < .0001$, then $\dfrac{1}{-.0001} < \dfrac{1}{x} < \dfrac{1}{.0001}$ or $-10\,000 < \dfrac{1}{x} < 10\,000$.

It is clear that as x is taken in smaller and smaller intervals about 0, the corresponding $f(x)$ does *not* lie in smaller and smaller intervals about any number M. Hence, $\lim\limits_{x \to 0}(1/x)$ does not exist.

57.2   Evaluate when possible:

(*a*) $\lim\limits_{x \to 2}(4x^2 - 5x)$,   (*b*) $\lim\limits_{x \to 1}(x^2 - 4x + 10)$,   (*c*) $\lim\limits_{x \to 2}\dfrac{x^2 + 6x + 5}{x^2 - 2x - 3}$,

(*d*) $\lim\limits_{x \to 3}\dfrac{x^2 + 6x + 5}{x^2 - 2x - 3}$,   (*e*) $\lim\limits_{x \to -1}\dfrac{x^2 + 6x + 5}{x^2 - 2x - 3}$.

(*a*) $\lim\limits_{x \to 2}(4x^2 - 5x) = \lim\limits_{x \to 2}4x^2 - \lim\limits_{x \to 2}5x = 4\lim\limits_{x \to 2}x^2 - 5\lim\limits_{x \to 2}x = 4 \cdot 4 - 5 \cdot 2 = 6.$

(*b*) $\lim\limits_{x \to 1}(x^2 - 4x + 10) = (1)^2 - 4 \cdot 1 + 10 = 7.$

(*c*) $\lim\limits_{x \to 2}(x^2 + 6x + 5) = 21$ and $\lim\limits_{x \to 2}(x^2 - 2x - 3) = -3$; hence

$$\lim_{x \to 2}\frac{x^2 + 6x + 5}{x^2 - 2x - 3} = \frac{\lim\limits_{x \to 2}(x^2 + 6x + 5)}{\lim\limits_{x \to 2}(x^2 - 2x - 3)} = \frac{21}{-3} = -7$$

(*d*) $\lim\limits_{x \to 3}(x^2 + 6x + 5) = 32$ and $\lim\limits_{x \to 3}(x^2 - 2x - 3) = 0$; hence $\lim\limits_{x \to 3}\dfrac{x^2 + 6x + 5}{x^2 - 2x - 3}$ does not exist.

(*e*) $\lim\limits_{x \to -1}(x^2 + 6x + 5) = 0$ and $\lim\limits_{x \to -1}(x^2 - 2x - 3) = 0$. Then, when $x \neq -1$,

$$\frac{x^2 + 6x + 5}{x^2 - 2x - 3} = \frac{(x + 5)(x + 1)}{(x - 3)(x + 1)} = \frac{x + 5}{x - 3} \quad \text{and} \quad \lim_{x \to -1}\frac{x^2 + 6x + 5}{x^2 - 2x - 3} = \lim_{x \to -1}\frac{x + 5}{x - 3} = \frac{4}{-4} = -1$$

57.3   Tell why each graph in Fig. 57-1 is not continuous at a.

(*a*) $y = g(x)$ exists at a and $\lim\limits_{x \to a}g(x)$ exists, but $\lim\limits_{x \to a}g(x) \neq g(a)$.

(*b*) $\lim\limits_{x \to a}h(x)$ does not exist since $\lim\limits_{x \to a^+}h(x) \neq \lim\limits_{x \to a^-}h(x)$.

(*c*) $f(a)$ is not defused.

57.4   Discuss the continuity of $y = \sqrt{x - 1}$. See Fig. 57-2.

Here, $f(1) = \sqrt{1 - 1} = 0$; $\lim\limits_{x \to 1^+}\sqrt{x - 1} = 0$, thus $f(x)$ is continuous on $[1, \infty)$.

(a)

(b)

(c)

Fig. 57-1

Fig. 57-2

57.5   Let $P(x_0, y_0)$ and $Q(x_0 + \Delta x, y_0 + \Delta y)$ be two distinct points on the parabola $y = x^2 - 3$. Compute $\Delta y / \Delta x$ and interpret.

Here

$$y_0 = x_0^2 - 3$$

$$y_0 + \Delta y = (x_0 + \Delta x)^2 - 3 = x_0^2 + 2x_0 \cdot \Delta x + (\Delta x)^2 - 3$$

$$\Delta y = [x_0^2 + 2x_0 \cdot \Delta x + (\Delta x)^2 - 3] - [x_0^2 - 3] = 2x_0 \cdot \Delta x + (\Delta x)^2$$

and

$$\frac{\Delta y}{\Delta x} = 2x_0 + \Delta x$$

In Fig. 57-3, PR is parallel to the x axis and QR is parallel to the y axis. If α denotes the inclination of the secant line PQ, $\tan \alpha = \Delta y / \Delta x$; thus, $\Delta y / \Delta x$ is the slope of the secant line PQ.

Fig. 57-3

57.6 If $s = 3t^2 + 10$ is the distance a body moving in a straight line is from a fixed point O of the line at time t, (a) find the change Δs in s when t changes from $t = t_0$ to $t = t_0 + \Delta t$, (b) find $\Delta s / \Delta t$ and interpret.

(a) Here $s_0 = 3t_0^2 + 10$. Then $s_0 + \Delta s = 3(t_0 + \Delta t)^2 + 10 = 3t_0^2 + 6t_0 \cdot \Delta t + 3(\Delta t)^2 + 10$ and $\Delta s = 6t_0 \cdot \Delta t + 3(\Delta t)^2$.

(b) $\dfrac{\Delta s}{\Delta t} = \dfrac{6t_0 \cdot \Delta t + 3(\Delta t)^2}{\Delta t} = 6t_0 + 3\Delta t$. Since Δs is the distance the body moves in time Δt, $\dfrac{\Delta s}{\Delta t}$ is the average rate of change of distance with respect to time or the average velocity of the body in the interval t_0 to $t_0 + \Delta t$.

57.7 Find

(a) $g'(x)$, given $g(x) = 5$ (c) $k'(x)$, given $k(x) = 4x^2$

(b) $h'(x)$, given $h(x) = 3x$ (d) $f'(x)$, given $f(x) = 4x^2 + 3x + 5$

Thus verify: If $f(x) = k(x) + h(x) + g(x)$, then $f'(x) = k'(x) + h'(x) + g'(x)$.

(a) $y = g(x) = 5$ (b) $y = h(x) = 3x$
   $y + \Delta y = g(x + \Delta x) = 5$ $y + \Delta y = 3(x + \Delta x) = 3x + 3\Delta x$
   $\Delta y = 0$ $\Delta y = 3\Delta x$
   $\dfrac{\Delta y}{\Delta x} = 0$ $\dfrac{\Delta y}{\Delta x} = 3$
   $g'(x) = \lim\limits_{\Delta x \to 0} 0 = 0$ $h'(x) = \lim\limits_{\Delta x \to 0} 3 = 3$

(c) $y = k(x) = 4x^2$
   $y + \Delta y = 4(x + \Delta x)^2 = 4x^2 + 8x \cdot \Delta x + 4(\Delta x)^2$
   $\Delta y = 8x \cdot \Delta x + 4(\Delta x)^2$
   $\dfrac{\Delta y}{\Delta x} = 8x + 4\Delta x$
   $k'(x) = \lim\limits_{\Delta x \to 0} (8x + 4\Delta x) = 8x$

(d) $y = f(x) = 4x^2 + 3x + 5$
   $y + \Delta y = 4(x + \Delta x)^2 + 3(x + \Delta x) + 5 = 4x^2 + 8x \cdot \Delta x + 4(\Delta x)^2 + 3x + 3\Delta x + 5$
   $\Delta y = 8x \cdot \Delta x + 3\Delta x + 4(\Delta x)^2$
   $\dfrac{\Delta y}{\Delta x} = 8x + 3 + 4\Delta x$
   $f'(x) = \lim\limits_{\Delta x \to 0} (8x + 3 + 4\Delta x) = 8x + 3$

Thus $f'(x) = k'(x) + h'(x) + g'(x) = 8x + 3 + 0$.

57.8   Place a straight edge along PQ in Fig. 57-3. Keeping P fixed, let Q move along the curve toward P and thus verify that the straight edge approaches the tangent line PT as limiting position.

Now as Q moves toward P, $\Delta x \to 0$ and $\displaystyle \lim_{\Delta x \to 0} \frac{\Delta y}{\Delta x} = \lim_{\Delta x \to 0} (2x_0 + \Delta x) = 2x_0$. Thus the slope of the tangent line to $y = f(x) = x^2 - 3$ at the point $P(x_0, y_0)$ is $m = f'(x_0) = 2x_0$.

57.9   Find the slope and equation of the tangent line to the given curve $y = f(x)$ at the given point:

(a) $y = 2x^3$ at $(1, 2)$, (b) $y = -3x^2 + 4x + 5$ at $(3, -10)$, (c) $y = x^2 - 4x + 3$ at $(2, -1)$.

(a)   By the five-step rule, $f'(x) = 6x^2$; then the slope $m = f'(1) = 6$. The equation of the tangent line at $(1, 2)$ is $y - 2 = 6(x - 1)$ or $6x - y - 4 = 0$.

(b)   Here $f'(x) = -6x + 4$ and $m = f'(3) = -14$. The equation of the tangent line at $(3, -10)$ is $14x + y - 32 = 0$.

(c)   Here $f'(x) = 2x - 4$ and $m = f'(2) = 0$. The equation of the tangent line at $(2, -1)$ is $y + 1 = 0$. Identify the given point on the parabola.

57.10   Find the equation of the tangent line to the parabola $y^2 = 8x$ at (a) the point $(2, 4)$ (b) the point $(2, -4)$.

Let $P(x, y)$ and $Q(x + \Delta x, y + \Delta y)$ be two nearby points on the parabola. Then

$$y^2 = 8x \tag{1}$$

$$(y + \Delta y)^2 = 8(x + \Delta x) \tag{2}$$

or $\qquad\qquad y^2 + 2y \cdot \Delta y + (\Delta y)^2 = 8x + 8 \cdot \Delta x$

Subtracting (1) from (2),

$$2y \cdot \Delta y + (\Delta y)^2 = 8 \cdot \Delta x, \qquad \Delta y(2y + \Delta y) = 8 \cdot \Delta x, \qquad \text{and} \qquad \frac{\Delta y}{\Delta x} = \frac{8}{2y + \Delta y}.$$

Now as Q moves along the curve toward P, $\Delta x \to 0$ and $\Delta y \to 0$. Thus

$$m = \lim_{\Delta x \to 0} \frac{\Delta y}{\Delta x} = \lim_{\Delta y \to 0} \frac{8}{2y + \Delta y} = \frac{8}{2y} = \frac{4}{y}$$

(a)   At point $(2, 4)$ the slope of the tangent line (also called the slope of the curve) is $m = \frac{4}{4} = 1$ and the equation of the tangent line is $x - y + 2 = 0$.

(b)   At point $(2, -4)$ the slope of the tangent line is $m = -1$ and the equation is $x + y + 2 = 0$.

57.11   Find the equation of the tangent line to the ellipse $4x^2 + 9y^2 = 25$ at (a) the point $(2, 1)$ (b) the point $(0, \frac{5}{3})$.

Let $P(x, y)$ and $Q(x + \Delta x, y + \Delta y)$ be two nearby points on the ellipse. (Why should P not be taken at an extremity of the major axis?) Then

$$4x^2 + 9y^2 = 25 \tag{1}$$

$$4x^2 + 8x \cdot \Delta x + 4(\Delta x)^2 + 9y^2 + 18y \cdot \Delta y + 9(\Delta y)^2 = 25 \tag{2}$$

Subtracting (1) from (2), $8x \cdot \Delta x + 4(\Delta x)^2 + 18y \cdot \Delta y + 9(\Delta y)^2 = 0$. Then

$$\Delta y(18y + 9\Delta y) = -\Delta x(8x + 4 \cdot \Delta x) \qquad \text{and} \qquad \frac{\Delta y}{\Delta x} = -\frac{8x + 4\Delta y}{18y + 9\Delta y}$$

When Q moves along the curve toward P, $\Delta x \to 0$ and $\Delta y \to 0$. Then $m = \displaystyle \lim_{\Delta x \to 0} \frac{\Delta y}{\Delta x} = -\frac{4x}{9y}$.

(a)   At point $(2, 1)$, $m = -\frac{8}{9}$ and the equation of the tangent line is $8x + 9y - 25 = 0$.

(b)   At point $(0, \frac{5}{3})$, $m = 0$ and the equation of the tangent line is $y - \frac{5}{3} = 0$.

57.12 The normal line to a curve at a point P on it is perpendicular to the tangent line at P. Find the equation of the normal line to the given curve at the given point of (a) Problem 57.9(b) (b) Problem 57.11.

  (a) The slope of the tangent line is -14; the slope of the normal line is $\frac{1}{14}$. The equation of the normal line is $y + 10 = \frac{1}{14}(x - 3)$ or $x - 14y - 143 = 0$.

  (b) The slope of the tangent line at $(2, 1)$ is $-\frac{8}{9}$; the slope of the normal line is $\frac{9}{8}$. The equation of the normal line is $9x - 8y - 10 = 0$.
  At the point $(0, \frac{5}{3})$ the normal line is vertical. Its equation is $x = 0$.

57.13 If $s = f(t)$ is the distance of a body, moving in a straight line, from a fixed point O of the line at time t, then (see Problem 57.6) $\dfrac{\Delta s}{\Delta t} = \dfrac{f(t + \Delta t) - f(t)}{\Delta t}$ is the average velocity of the body in the interval of time t to $t + \Delta t$ and

$$v = s' = \lim_{\Delta t \to 0} \frac{\Delta s}{\Delta t} = \lim_{\Delta t \to 0} \frac{f(t + \Delta t) - f(t)}{\Delta t}$$

is the *instantaneous velocity of the body* at time t. For $s = 3t^2 + 10$ of Problem 57.6, find the (instantaneous) velocity of the body at time (a) $t = 0$, (b) $t = 4$.

  Here $v = s' = 6t$. (a) When $t = 0$, $v = 0$. (b) When $t = 4$, $v = 24$.

57.14 The height above the ground of a bullet shot vertically upward with initial velocity of 1152 ft/s is given by $s = 1152t - 16t^2$. Find (a) the velocity of the bullet 20 s after it was fired (b) the time required for the bullet to reach its maximum height and the maximum height attained.

  Here $v = 1152 - 32t$.

  (a) When $t = 20$, $v = 1152 - 32(20) = 512$ ft/s.

  (b) At its maximum height, the velocity of the bullet is 0 ft/s. When $v = 1152 - 32t = 0$, $t = 36$ s. When $t = 36$, $s = 1152(36) - 16(36)^2 = 20\,736$ ft, the maximum height.

57.15 Find the derivative of each of the following polynomials:

  (a) $f(x) = 3x^2 - 6x + 5$, (b) $f(x) = 2x^3 - 8x + 4$, (c) $f(x) = (x - 2)^2(x - 3)^2$.

  (a) $f'(x) = 3 \cdot 2x^{2-1} - 6x^{1-1} + 0 = 6x - 6$.

  (b) $f'(x) = 2 \cdot 3x^{3-1} - 8x^{1-1} + 0 = 6x^2 - 8$.

  (c) Here $f(x) = x^4 - 10x^3 + 37x^2 - 60x + 36$. Then $f'(x) = 4x^3 - 30x^2 + 74x - 60 = (x - 2)(x - 3)(4x - 10)$.

57.16 For each of the following functions, find $f'(x)$, $f''(x)$, and $f'''(x)$:

  (a) $f(x) = 2x^2 + 7x - 5$ (b) $f(x) = x^3 - 6x^2$ (c) $f(x) = x^5 - x^3 + 3x$.

  (a) $f'(x) = 4x + 7$, $f''(x) = 4$, $f'''(x) = 0$

  (b) $f'(x) = 3x^2 - 12x$, $f''(x) = 6x - 12$, $f'''(x) = 6$

  (c) $f'(x) = 5x^4 - 3x^2 + 3$, $f''(x) = 20x^3 - 6x$, $f'''(x) = 60x^2 - 6$

Supplementary Problems

57.17 Find all (real) values of x for which each of the following is defined:

(a) $x^2 - 3x + 4$ (d) $\dfrac{1}{(x-2)(x+3)}$ (g) $\dfrac{1}{x^2+4}$

(b) $\dfrac{1}{x^2}$ (e) $\dfrac{1}{x^2-4x+3}$ (h) $\dfrac{x^2-9}{x-3}$

(c) $\dfrac{1}{x-2}$ (f) $\dfrac{1}{x^2-4}$ (i) $\dfrac{x-3}{x^2-9}$

Ans. (a) all x (c) $x \neq 2$ (e) $x \neq 1, 3$ (g) all x (i) $x \neq \pm 3$

(b) $x \neq 0$ (d) $x \neq 2, -3$ (f) $x \neq \pm 2$ (h) $x \neq 3$

57.18 For each function $f(x)$ of Problem 57.17 evaluate $\lim\limits_{x \to 1} f(x)$, when it exists.

Ans. (a) 2   (b) 1   (c) -1 (d) $-\frac{1}{4}$ (f) $-\frac{1}{3}$ (g) $\frac{1}{5}$ (h) 4   (i) $\frac{1}{4}$

57.19 For each function $f(x)$ of Problem 57.17 evaluate $\lim\limits_{x \to 3} f(x)$, when it exists.

Ans. (a) 4   (b) $\frac{1}{9}$ (c) 1   (d) $\frac{1}{6}$ (f) $\frac{1}{5}$ (g) $\frac{1}{13}$ (h) 6   (i) $\frac{1}{6}$

57.20 Use the five-step rule to obtain $f'(x)$ or $f'(t)$, given   (a) $f(x) = 3x + 5$ (b) $f(x) = x^2 - 3x$

(c) $f(t) = 2t^2 + 8t + 9$ (d) $f(t) = 2t^3 - 12t^2 + 20t + 3$

Ans. (a) 3   (b) $2x - 3$ (c) $4t + 8$ (d) $6t^2 - 24t + 20$

57.21 Find the equations of the tangent and normal to each curve at the given point on it.

(a) $y = x^2 + 2$, $P(1, 3)$ Ans. $2x - y + 1 = 0$, $x + 2y - 7 = 0$

(b) $y = 2x^2 - 3x$, $P(1, -1)$ Ans. $x - y - 2 = 0$, $x + y = 0$

(c) $y = x^2 - 4x + 5$, $P(1, 2)$ Ans. $2x + y - 4 = 0$, $x - 2y + 3 = 0$

(d) $y = x^2 + 3x - 10$, $P(2, 0)$ Ans. $7x - y - 14 = 0$, $x + 7y - 2 = 0$

(e) $x^2 + y^2 = 25$, $P(4, 3)$ Ans. $4x + 3y - 25 = 0$, $3x - 4y = 0$

(f) $y^2 = 4x - 8$, $P(3, -2)$ Ans. $x + y - 1 = 0$, $x - y - 5 = 0$

(g) $x^2 + 4y^2 = 8$, $P(-2, -1)$ Ans. $x + 2y + 4 = 0$, $2x - y + 3 = 0$

(h) $2x^2 - y^2 = 9$, $P(-3, 3)$ Ans. $2x + y + 3 = 0$, $x - 2y + 9 = 0$

57.22 A particle moves along the x axis according to the law $s = 2t^2 + 8t + 9$ [see Problem 57.20(c)], where s (ft) is the directed distance of the particle from the origin O at time t (seconds). Locate the particle and find its velocity when (a) $t = 0$, (b) $t = 1$.

Ans. (a) 9 ft to the right of O, $v = 8$ ft/s   (b) 19 ft to the right of O, $v = 12$ ft/s

57.23 A particle moves along the x axis according to the law $s = 2t^3 - 12t^2 + 20t + 3$ [see Problem 57.20(d)], where s is defined as in Problem 57.22.

(a) Locate the particle and find its velocity when $t = 2$.   (b) Locate the particle when $v = 2$ ft/s.

Ans. (a) 11 ft to the right of O, $v = -4$ ft/s

(b) $t = 1$, 13 ft to the right of O; $t = 3$, 9 ft to the right of O

57.24 The height (s m) of a bullet shot vertically upwards is given by $s = 1280t - 16t^2$, with t measured in seconds. (a) What is the initial velocity? (b) For how long will it rise? (c) How high will it rise?

   Ans. (a) 1280 m/s   (b) 40 s   (c) 25 600 m

57.25 Find the coordinates of the points for which the slope of the tangent to $y = x^3 - 12x + 1$ is 0.

   Ans. $(2, -15)$, $(-2, 17)$

57.26 At what point on $y = \frac{1}{2}x^2 - 2x + 3$ is the tangent perpendicular to that at the point $(1, 0)$?

   Ans. $(3, \frac{3}{2})$

57.27 Show that the equation of the tangent to the conic $Ax^2 + 2Bxy + Cy^2 + 2Dx + 2Ey + F = 0$ at the point $P_1(x_1, y_1)$ on it is given by $Ax_1x + B(x_1y + y_1x) + Cy_1y + D(x_1 + x) + E(y_1 + y) + F = 0$. Use this as a formula to solve Problem 57.21.

57.28 Show that the tangents at the extremities of the latus rectum of the parabola $y^2 = 4px$ (a) are mutually perpendicular (b) intersect on the directrix.

57.29 Show that the tangent of slope $m \neq 0$ to the parabola $y^2 = 4px$ has equation $y = mx + p/m$.

57.30 Show that the slope of the tangent at either end of either latus rectum of the ellipse $b^2x^2 + a^2y^2 = a^2b^2$ is equal numerically to its eccentricity. Investigate the case of the hyperbola.

Chapter 58

Differentiation of Algebraic Expressions

DIFFERENTIATION FORMULAS

I. If $y = f(x) = kx^n$, where k and n are constants, then $y' = f'(x) = knx^{n-1}$. (For a verification, see Problem 58.1.)

II. If $y = k \cdot u^n$, where k and n are constants and u is a function of x, then $y' = knu^{n-1} \cdot u'$, provided u' exists. This is the chain rule. (For a verification, see Problem 58.2.)

EXAMPLE 1. Find y', given (a) $y = 8x^{5/4}$, (b) $y = (x^2 + 4x - 1)^{3/2}$.

(a) Here $k = 8$, $n = \frac{5}{4}$. Then $y' = knx^{n-1} = 8 \cdot \frac{5}{4}x^{5/4-1} = 10x^{1/4}$.

(b) Let $u = x^2 + 4x - 1$ so that $y = u^{3/2}$. Then differentiating with respect to x, $u' = 2x + 4$ and

$$y' = \tfrac{3}{2}u^{1/2} \cdot u' = \tfrac{3}{2}\sqrt{x^2 + 4x - 1}(2x + 4) = 3(x + 2)\sqrt{x^2 + 4x - 1}$$

(See Problem 58.3.)

III. If $y = f(x) \cdot g(x)$, then $y' = f(x) \cdot g'(x) + g(x) \cdot f'(x)$, provided $f'(x)$ and $g'(x)$ exist. (For the derivation, see Problem 58.4.)

EXAMPLE 2. Find y' when $y = (x^3 + 3x^2 + 1)(x^2 + 2)$.

Take $f(x) = x^3 + 3x^2 + 1$ and $g(x) = x^2 + 2$. Then $f'(x) = 3x^2 + 6x$, $g'(x) = 2x$, and

$$\begin{aligned}
y' &= f(x) \cdot g'(x) + g(x) \cdot f'(x) \\
&= (x^3 + 3x^2 + 1)(2x) + (x^2 + 2)(3x^2 + 6x) \\
&= 5x^4 + 12x^3 + 6x^2 + 14x
\end{aligned}$$

IV. If $y = \dfrac{f(x)}{g(x)}$, then $y' = \dfrac{g(x) \cdot f'(x) - f(x) \cdot g'(x)}{[g(x)]^2}$, when $f'(x)$ and $g'(x)$ exist and $g(x) \neq 0$. (For a derivation, see Problem 58.6.)

EXAMPLE 3. Find y', given $y = \dfrac{x + 1}{x^2 + 1}$.

Take $f(x) = x + 1$ and $g(x) = x^2 + 1$. Then

$$y' = \frac{g(x) \cdot f'(x) - f(x) \cdot g'(x)}{[g(x)]^2} = \frac{(x^2 + 1)(1) - (x + 1)(2x)}{(x^2 + 1)^2} = \frac{1 - 2x - x^2}{(x^2 + 1)^2}.$$

Solved Problems

58.1 Use the five-step rule to obtain y' when $y = 6x^{3/2}$.

We have
$$y = 6x^{3/2}$$

$$y + \Delta y = 6(x + \Delta x)^{3/2}$$

$$\Delta y = 6(x + \Delta x)^{3/2} - 6x^{3/2} = 6[(x + \Delta x)^{3/2} - x^{3/2}]$$

and
$$\frac{\Delta y}{\Delta x} = 6 \cdot \frac{(x+\Delta x)^{3/2} - x^{3/2}}{\Delta x} = 6 \cdot \frac{(x+\Delta x)^{3/2} - x^{3/2}}{\Delta x} \cdot \frac{(x+\Delta x)^{3/2} + x^{3/2}}{(x+\Delta x)^{3/2} + x^{3/2}}$$

$$= 6 \cdot \frac{(x+\Delta x)^3 - x^3}{\Delta x[(x+\Delta x)^{3/2} + x^{3/2}]} = 6 \cdot \frac{3x^2 + 3x \cdot \Delta x + (\Delta x)^2}{(x+\Delta x)^{3/2} + x^{3/2}}$$

Then
$$y' = \lim_{\Delta x \to 0} 6 \cdot \frac{3x^2 + 3x \cdot \Delta x + (\Delta x)^2}{(x+\Delta x)^{3/2} + x^{3/2}} = 6 \cdot \frac{3x^2}{2x^{3/2}} = 9x^{1/2}$$

(NOTE. By Formula I, with $k = 6$ and $n = \frac{3}{2}$, we find $y' = knx^{n-1} = 6 \cdot \frac{3}{2}x^{1/2} = 9x^{1/2}$.)

58.2   Use the five-step rule to find y' when $y = (x^2 + 4)^{1/2}$. Solve also by using Formula II.

We have
$$y = (x^2 + 4)^{1/2}$$

$$y + \Delta y = [(x+\Delta x)^2 + 4]^{1/2}$$

$$\Delta y = [(x+\Delta x)^2 + 4]^{1/2} - (x^2 + 4)^{1/2}$$

and
$$\frac{\Delta y}{\Delta x} = \frac{[(x+\Delta x)^2 + 4]^{1/2} - (x^2 + 4)^{1/2}}{\Delta x} \cdot \frac{[(x+\Delta x)^2 + 4]^{1/2} + (x^2 + 4)^{1/2}}{[(x+\Delta x)^2 + 4]^{1/2} + (x^2 + 4)^{1/2}}$$

$$= \frac{(x+\Delta x)^2 + 4 - (x^2 + 4)}{\Delta x\{[(x+\Delta x)^2 + 4]^{1/2} + (x^2 + 4)^{1/2}\}} = \frac{2x + \Delta x}{[(x+\Delta x)^2 + 4]^{1/2} + (x^2 + 4)^{1/2}}$$

Then
$$y' = \lim_{\Delta x \to 0} \frac{2x + \Delta x}{[(x+\Delta x)^2 + 4]^{1/2} + (x^2 + 4)^{1/2}} = \frac{2x}{2(x^2 + 4)^{1/2}} = \frac{x}{(x^2 + 4)^{1/2}}$$

Let $u = x^2 + 4$ so that $y = u^{1/2}$. Then $u' = 2x$ and $y' = \frac{1}{2}u^{-1/2} \cdot u' = \frac{1}{2}(x^2 + 4)^{-1/2} \cdot 2x = \frac{x}{\sqrt{x^2 + 4}}$.

58.3   Find y', given (a) $y = (2x - 5)^3$ (b) $y = \frac{2}{3}(x^6 + 4x^3 + 5)^2$ (c) $y = \frac{4}{x^2}$ (d) $y = \frac{1}{\sqrt{x}}$ (e) $y = 2(3x^2 + 2)^{1/2}$

(a)   Let $u = 2x - 5$ so that $y = u^3$. Then, differentiating with respect to x, $u' = 2$ and $y' = 3u^2 \cdot u' = 3(2x - 5)^2 \cdot 2 = 6(2x - 5)^2$.

(b)   Let $u = x^6 + 4x^3 + 5$ so that $y = \frac{2}{3}u^2$. Differentiating with respect to x, $u' = 6x^5 + 12x^2$ and $y' = \frac{4}{3}u \cdot u' = \frac{4}{3}(x^6 + 4x^3 + 5)(6x^5 + 12x^2) = 8(x^6 + 4x^3 + 5)(x^5 + 2x^2)$.

(c)   Here $y = 4x^{-2}$ and $y' = 4(-2)x^{-3} = -8/x^3$.

(d)   Here $y = x^{-1/2}$ and $y' = (-\frac{1}{2})x^{-3/2} = -1/(2x\sqrt{x})$.

(e)   Since $y = 2(3x^2 + 2)^{1/2}$, $y' = 2(\frac{1}{2})(3x^2 + 2)^{-1/2}(6x) = 6x/(3x^2 + 2)^{1/2}$.

58.4   Derive: If $y = f(x) \cdot g(x)$, then $y' = f(x) \cdot g'(x) + g(x) \cdot f'(x)$, provided $f'(x)$ and $g'(x)$ exist.

Let $u = f(x)$ and $v = g(x)$ so that $y = u \cdot v$.
As x changes to $x + \Delta x$, let u change to $u + \Delta u$, v change to $v + \Delta v$, and y change to $y + \Delta y$. Then

$$y + \Delta y = (u + \Delta u)(v + \Delta v) = uv + u \cdot \Delta v + v \cdot \Delta u + \Delta u \cdot \Delta v$$

$$\Delta y = u \cdot \Delta v + v \cdot \Delta u + \Delta u \cdot \Delta v$$

and
$$\frac{\Delta y}{\Delta x} = u \cdot \frac{\Delta v}{\Delta x} + v \cdot \frac{\Delta u}{\Delta x} + \Delta u \cdot \frac{\Delta v}{\Delta x}$$

Then
$$y' = \lim_{\Delta x \to 0} \left(u \cdot \frac{\Delta v}{\Delta x} + v \cdot \frac{\Delta u}{\Delta x} + \Delta u \cdot \frac{\Delta v}{\Delta x} \right)$$

$$= u \cdot v' + v \cdot u' + 0 \cdot v' = u \cdot v' + v \cdot u' = f(x) \cdot g'(x) + g(x) \cdot f'(x)$$

58.5 Find y', given (a) $y = x^5(1 - x^2)^4$, (b) $y = x^2\sqrt{x^2 + 4}$, (c) $y = (3x + 1)^2(2x^3 - 3)^{1/3}$.

(a) Set $f(x) = x^5$ and $g(x) = (1 - x^2)^4$. Then $f'(x) = 5x^4$, $g'(x) = 4(1 - x^2)^3(-2x)$, and

$$y' = f(x) \cdot g'(x) + g(x) \cdot f'(x) = x^5 \cdot 4(1 - x^2)^3(-2x) + (1 - x^2)^4 \cdot 5x^4$$

$$= x^4(1 - x^2)^3[-8x^2 + 5(1 - x^2)] = x^4(1 - x^2)^3(5 - 13x^2)$$

(b) $$y = x^2 \cdot \tfrac{1}{2}(x^2 + 4)^{-1/2} \cdot 2x + (x^2 + 4)^{1/2} \cdot 2x$$

$$= x^3(x^2 + 4)^{-1/2} + 2x(x^2 + 4)^{1/2} = \frac{x^3 + 2x(x^2 + 4)}{(x^2 + 4)^{1/2}} = \frac{3x^3 + 8x}{\sqrt{x^2 + 4}}$$

(c) Here $y = (3x + 1)^2(2x^3 - 3)^{1/3}$ and

$$y' = (3x + 1)^2 \cdot \tfrac{1}{3}(2x^3 - 3)^{-2/3} \cdot 6x^2 + (2x^3 - 3)^{1/3} \cdot 2(3x + 1) \cdot 3$$

$$= 2x^2(3x + 1)^2(2x^3 - 3)^{-2/3} + 6(3x + 1)(2x^3 - 3)^{1/3} = \frac{2x^2(3x + 1)^2 + 6(3x + 1)(2x^3 - 3)}{(2x^2 - 3)^{2/3}}$$

$$= \frac{2(3x + 1)[x^2(3x + 1) + 3(2x^3 - 3)]}{(2x^3 - 3)^{2/3}} = \frac{2(3x + 1)(9x^3 + x^2 - 9)}{(2x^3 - 3)^{2/3}}$$

58.6 Prove: If $y = \dfrac{f(x)}{g(x)}$, if $f'(x)$ and $g'(x)$ exist, and if $g(x) \neq 0$, then $y' = \dfrac{g(x) \cdot f'(x) - f(x) \cdot g'(x)}{[g(x)]^2}$.

Let $u = f(x)$ and $v = g(x)$ so that $y = u/v$. Then

$$y + \Delta y = \frac{u + \Delta u}{v + \Delta v} \qquad \Delta y = \frac{u + \Delta u}{v + \Delta v} - \frac{u}{v} = \frac{v \cdot \Delta u - u \cdot \Delta v}{v(v + \Delta v)}$$

and

$$\frac{\Delta y}{\Delta x} = \frac{v \cdot \Delta u - u \cdot \Delta v}{\Delta x \cdot v(v + \Delta v)} = \frac{v \cdot \dfrac{\Delta u}{\Delta x} - u \cdot \dfrac{\Delta v}{\Delta x}}{v(v + \Delta v)}$$

Then

$$y' = \lim_{\Delta x \to 0} \frac{v \cdot \dfrac{\Delta u}{\Delta x} - u \cdot \dfrac{\Delta v}{\Delta x}}{v(v + \Delta v)} = \frac{v \cdot u' - u \cdot v'}{v^2} = \frac{g(x) \cdot f'(x) - f(x) \cdot g'(x)}{[g(x)]^2}$$

58.7 Find y', given (a) $y = 1/x^3$, (b) $y = \dfrac{2x}{x - 3}$, (c) $y = \dfrac{x + 5}{x^2 - 1}$, (d) $y = \dfrac{x^3}{\sqrt{4 - x^2}}$, (e) $y = \dfrac{\sqrt{2x - 3x^2}}{x + 1}$.

(a) Take $f(x) = 1$ and $g(x) = x^3$; then

$$y' = \frac{g(x) \cdot f'(x) - f(x) \cdot g'(x)}{[g(x)]^2} = \frac{x^3 \cdot 0 - 1 \cdot 3x^2}{(x^3)^2} = -\frac{3}{x^4}$$

Note that it is simpler here to write $y = x^{-3}$ and $y' = -3x^{-4} = -3/x^4$.

(b) Take $f(x) = 2x$ and $g(x) = x - 3$; then

$$y' = \frac{g(x) \cdot f'(x) - f(x) \cdot g'(x)}{[g(x)]^2} = \frac{(x - 3) \cdot 2 - 2x \cdot 1}{(x - 3)^2} = \frac{-6}{(x - 3)^2}$$

Note that $y = \dfrac{2x}{x - 3} = 2 + \dfrac{6}{x - 3} = 2 + 6(x - 3)^{-1}$ and $y' = 6(-1)(x - 3)^{-2} = -6/(x - 3)^2$.

(c) $$y' = \frac{(x^2 - 1)(1) - (x + 5)(2x)}{(x^2 - 1)^2} = -\frac{1 + 10x + x^2}{(x^2 - 1)^2}$$

(d) $$y' = \frac{(4 - x^2)^{1/2} \cdot 3x^2 - x^3 \cdot \tfrac{1}{2}(4 - x^2)^{-1/2}(-2x)}{4 - x^2} = \frac{(4 - x^2)(3x^2) + x^3 \cdot x}{(4 - x^2)^{3/2}} = \frac{12x^2 - 2x^4}{(4 - x^2)^{3/2}}$$

The derivative exists for $-2 < x < 2$.

(e) $$y' = \frac{(x+1) \cdot \frac{1}{2}(2x - 3x^2)^{-1/2}(2 - 6x) - (2x - 3x^2)^{1/2}(1)}{(x+1)^2}$$

$$= \frac{(x+1)(1 - 3x) - (2x - 3x^2)}{(x+1)^2(2x - 3x^2)^{1/2}} = \frac{1 - 4x}{(x+1)^2(2x - 3x^2)^{1/2}}$$

The derivative exists for $0 < x < \frac{2}{3}$.

58.8 Find y', y'', y''', given (a) $y = 1/x$, (b) $y = \dfrac{1}{x-1} + \dfrac{1}{x+1}$, (c) $y = \dfrac{2}{x^2 - 1}$.

(a) Here $y = x^{-1}$; then $y' = -1 \cdot x^{-2} = -x^{-2}$, $y'' = 2x^{-3}$, $y''' = -6x^{-4}$ or $y' = -1/x^2$, $y'' = 2/x^3$, $y''' = -6/x^4$.

(b) Here $y = (x - 1)^{-1} + (x + 1)^{-1}$; then

$$y' = -1(x - 1)^{-2} + (-1)(x + 1)^{-2} = -\frac{1}{(x - 1)^2} - \frac{1}{(x + 1)^2}$$

$$y'' = 2(x - 1)^{-3} + 2(x + 1)^{-3} = \frac{2}{(x - 1)^3} + \frac{2}{(x + 1)^3}$$

$$y''' = -6(x - 1)^{-4} - 6(x + 1)^{-4} = -\frac{6}{(x - 1)^4} - \frac{6}{(x + 1)^4}$$

(c) Here $y = 2(x^2 - 1)^{-1}$; then

$$y' = 2(-1)(x^2 - 1)^{-2} \cdot (2x) = -4x(x^2 - 1)^{-2} = \frac{-4x}{(x^2 - 1)^2}$$

$$y'' = -4(x^2 - 1)^{-2} - 4x(-2)(x^2 - 1)^{-3}(2x) = -4(x^2 - 1)^{-2} + 16x^2(x^2 - 1)^{-3} = \frac{12x^2 + 4}{(x^2 - 1)^3}$$

$$y''' = \frac{(x^2 - 1)^3(24x) - (12x^2 + 4) \cdot 3(x^2 - 1)^2(2x)}{(x^2 - 1)^6} = \frac{(x^2 - 1)(24x) - 6x(12x^2 + 4)}{(x^2 - 1)^4} = \frac{-48x(x^2 + 1)}{(x^2 - 1)^4}$$

Supplementary Problems

58.9 Use the differentiation formulas to find y', given

(a) $y = 2x^3 + 4x^2 - 5x + 8$ *Ans.* $y' = 6x^2 + 8x - 5$

(b) $y = -5 + 3x - \frac{3}{2}x^2 - 7x^3$ *Ans.* $y' = 3 - 3x - 21x^2$

(c) $y = (x - 2)^4$ *Ans.* $y' = 4(x - 2)^3$

(d) $y = (x^2 + 2)^3$ *Ans.* $y' = 6x(x^2 + 2)^2$

(e) $y = (4 - x^2)^{10}$ *Ans.* $y' = -20x(4 - x^2)^9$

(f) $y = (2x^2 + 4x - 5)^6$ *Ans.* $y' = 24(x + 1)(2x^2 + 4x - 5)^5$

(g) $y = \frac{1}{5}x^{5/2} + \frac{1}{3}x^{3/2}$ *Ans.* $y' = \frac{1}{2}x^{1/2}(x + 1)$

(h) $y = (x^2 - 4)^{3/2}$ *Ans.* $y' = 3x(x^2 - 4)^{1/2}$

(i) $y = (1 - x^2)^{1/2}$ *Ans.* $y' = -\dfrac{x}{(1 - x^2)^{1/2}}$

(j) $y = \dfrac{6}{x} + \dfrac{4}{x^2} - \dfrac{3}{x^3}$ *Ans.* $y' = -\dfrac{6}{x^2} - \dfrac{8}{x^3} + \dfrac{9}{x^4}$

(k) $y = x^3(x + 1)^2$ *Ans.* $y' = x^2(x + 1)(5x + 3)$

(l) $y = (x + 1)^3(x - 3)^2$ *Ans.* $y' = (x + 1)^2(x - 3)(5x - 7)$

(m) $y = (x + 2)^2(2 - x)^3$ *Ans.* $y' = -(x + 2)(2 - x)^2(5x + 2)$

(n) $y = \dfrac{x + 1}{x - 1}$ *Ans.* $y' = -\dfrac{2}{(x - 1)^2}$

(o) $y = \dfrac{x^2 + 2x - 3}{x^2}$ *Ans.* $y' = \dfrac{6 - 2x}{x^3}$

(p) $y = \dfrac{x^2 + 1}{x^2 + 2}$ *Ans.* $y' = \dfrac{2x}{(x^2 + 2)^2}$

(q) $y = \dfrac{1}{(2x + 1)^3}$ *Ans.* $y' = -\dfrac{6}{(2x + 1)^4}$

(r) $y = \dfrac{1}{(x^2 - 9)^{1/2}}$ *Ans.* $y' = -\dfrac{x}{(x^2 - 9)^{3/2}}$

(s) $y = \dfrac{1}{(16 - x^2)^{1/2}}$ *Ans.* $y' = \dfrac{x}{(16 - x^2)^{3/2}}$

(t) $y = \dfrac{x}{(x + 1)^{1/2}}$ *Ans.* $y' = \dfrac{x + 2}{2(x + 1)^{3/2}}$

(u) $y = \dfrac{(x^2 + 2)^{1/2}}{x}$ *Ans.* $y' = \dfrac{-2}{x^2(x^2 + 2)^{1/2}}$

58.10   For each of the following, find $f'(x)$, $f''(x)$, and $f'''(x)$:

(a) $f(x) = 3x^4 - 8x^3 + 12x^2 + 5$ (c) $f(x) = \dfrac{1}{4 - x}$

(b) $f(x) = x^3 - 6x^2 + 9x + 18$ (d) $f(x) = (1 - x^2)^{3/2}$

Ans.   (a) $f'(x) = 12x(x^2 - 2x + 2)$, $f''(x) = 12(3x^2 - 4x + 2)$, $f'''(x) = 24(3x - 2)$

(b) $f'(x) = 3(x^2 - 4x + 3)$, $f''(x) = 6(x - 2)$, $f'''(x) = 6$

(c) $f'(x) = \dfrac{1}{(4 - x)^2}$, $f''(x) = \dfrac{2}{(4 - x)^3}$, $f'''(x) = \dfrac{6}{(4 - x)^4}$

(d) $f'(x) = -3x(1 - x^2)^{1/2}$, $f''(x) = \dfrac{3(2x^2 - 1)}{(1 - x^2)^{1/2}}$, $f'''(x) = \dfrac{3x(3 - 2x^2)}{(1 - x^2)^{3/2}}$

58.11   In each of the following state the values of x for which $f(x)$ is continuous; also find $f'(x)$ and state the values of x for which it is defined.

(a) $f(x) = \dfrac{1}{x^2}$ (b) $f(x) = \dfrac{1}{x - 2}$ (c) $f(x) = (x - 2)^{4/3}$ (d) $f(x) = (x - 2)^{1/3}$

Ans.   (a) $x \neq 0$; $f'(x) = \dfrac{-2}{x^3}$, $x \neq 0$ (c)   all x; $f'(x) = \frac{4}{3}(x - 2)^{1/3}$; all x

(b) $x \neq 2$; $f'(x) = \dfrac{-1}{(x - 2)^2}$, $x \neq 2$ (d)   all x; $f'(x) = \dfrac{1}{3(x - 2)^{2/3}}$, $x \neq 2$

[NOTE.   Parts (a) and (b) verify: If $f(x)$ is not continuous at $x = x_0$ then $f'(x)$ does not exist at $x = x_0$.
Parts (c) and (d) verify: If $f(x)$ is continuous at $x = x_0$, its derivative $f'(x)$ may or may not exist at $x = x_0$.]

Chapter 59

Applications of Derivatives

INCREASING AND DECREASING FUNCTIONS. A function $y = f(x)$ is said to be an *increasing function* if y *increases* as x increases, and a *decreasing function* if y *decreases* as x increases.

Let the graph of $y = f(x)$ be as shown in Fig. 59-1. Clearly $y = f(x)$ is an increasing function from A to B and from C to D, and is a decreasing function from B to C and from D to E. At any point of the curve between A and B (also between C and D), the inclination θ of the tangent line to the curve is acute; hence, $f'(x) = \tan \theta > 0$. At any point of the curve between B and C (also, between D and E), the inclination θ of the tangent line is obtuse; hence, $f'(x) = \tan \theta < 0$.

Fig. 59-1

Thus for values of x for which $f'(x) > 0$, the function $f(x)$ is an increasing function; for values of x for which $f'(x) < 0$, the function is a decreasing function.

When $x = b$, $x = c$, and $x = d$, the function is neither increasing nor decreasing since $f'(x) = 0$. Such values of x are called *critical values* for the function $f(x)$.

EXAMPLE 1. For the function $f(x) = x^2 - 6x + 8$, $f'(x) = 2x - 6$.

Setting $f'(x) = 0$, we find the critical value $x = 3$. Now $f'(x) < 0$ when $x < 3$, and $f'(x) > 0$ when $x > 3$. Thus, $f(x) = x^2 - 6x + 8$ is a decreasing function when $x < 3$ and an increasing function when $x > 3$. (See Problem 59.1.)

RELATIVE MAXIMUM AND MINIMUM VALUES. Let the curve of Fig. 59-1 be traced from left to right. Leaving A, the tracing point rises to B and then begins to fall. At B the ordinate $f(b)$ is greater than at any point of the curve near to B. We say that the point $B(b, f(b))$ is a *relative maximum point* of the curve or that the function $f(x)$ has a *relative maximum* $[=f(b)]$ when $x = b$. By the same argument $D(d, f(d))$ is also a relative maximum point of the curve or $f(x)$ has a relative maximum $[=f(d)]$ when $x = d$.

Leaving B, the tracing point falls to C and then begins to rise. At C the ordinate $f(c)$ is smaller than at any point of the curve near to C. We say that the point $C(c, f(c))$ is a *relative minimum point* of the curve or that $f(x)$ has a *relative minimum* $[=f(c)]$ when $x = c$.

Note that the relative maximum and minimum of this function occur at the critical values. While not true for all functions, the above statement is true for all of the functions considered in this chapter.

Test for relative maximum. If $x = a$ is a critical value for $y = f(x)$ and if $f'(x) > 0$ for all values of x less than but near to $x = a$ while $f'(x) < 0$ for all values of x greater than but near to $x = a$, then $f(a)$ is a relative maximum value of the function.

435

Test for relative minimum. If $x = a$ is a critical value for $y = f(x)$ and if $f'(x) < 0$ for all values of x less than but near to $x = a$ while $f'(x) > 0$ for all values of x greater than but near to $x = a$, then $f(a)$ is a relative minimum value of the function.

If as x increases in value through a critical value, $x = a$, $f'(x)$ does not change sign, then $f(a)$ is neither a relative maximum nor a relative minimum value of the function.

EXAMPLE 2. For the function of Example 1, the critical value is $x = 3$.

Since $f'(x) = 2(x - 3) < 0$ for $x < 3$ and $f'(x) > 0$ for $x > 3$, the given function has a relative minimum value $f(3) = -1$.

In geometric terms, the point $(3, -1)$ is a relative minimum point of the curve $y = x^2 - 6x + 8$. (See Problem 59.2.)

ANOTHER TEST FOR MAXIMUM AND MINIMUM VALUES.

At A on the curve of Fig. 59-1, the inclination θ of the tangent line is acute. As the tracing point moves from A to B, θ decreases; thus $f'(x) = \tan \theta$ is a decreasing function. At B, $f'(x) = 0$. As the tracing point moves from B to G, θ is obtuse and decreasing; thus $f'(x) = \tan \theta$ is a decreasing function. Hence from A to G, $f'(x)$ is a decreasing function and its derivative $f''(x) < 0$. In particular, $f''(b) < 0$. Similarly, $f''(d) < 0$.

As the tracing point moves from G to C, θ is obtuse and increasing; thus $f'(x)$ is an increasing function. At C, $f'(x) = 0$. As the tracing point moves from C to H, θ is acute and increasing; thus $f'(x)$ is an increasing function. Hence, from G to H, $f'(x)$ is an increasing function and $f''(x) > 0$. In particular, $f''(c) > 0$.

Test for relative maximum. If $x = a$ is a critical value for $y = f(x)$ and if $f''(a) < 0$, then $f(a)$ is a relative maximum value of the function $f(x)$.

Test for relative minimum. If $x = a$ is a critical value for $y = f(x)$ and if $f''(a) > 0$, then $f(a)$ is a relative minimum value of the function $f(x)$.

The test fails when $f''(a) = 0$. When this occurs, the tests of the preceding section must be used. (See Problem 59.3)

CONCAVITY.

Suppose that $f(x)$ is a differentiable function on (a, b). Then, if $f'(x)$ is increasing on (a, b), we call f concave upward on (a, b). See Fig. 59-2(a). If $f'(x)$ is decreasing on (a, b), we say f is concave downward on (a, b). See Fig. 59-2(b).

Fig. 59-2

INFLECTION POINT OF A CURVE.

If at $x = a$, not necessarily a critical value for $f(x)$, the concavity changes from downward to upward or upward to downward $(a, f(a))$ is an inflection point of $f(x)$. See Fig. 59-3.

In Fig. 59-1, G and H are inflection points of the curve. Note that at points between A and G the tangent lines to the curve lie above the curve, at points between G and H the tangent lines lie below *(inflection point)*   *(inflection point)*

Fig. 59-3

the curve, and at points between H and E the tangent lines lie above the curve. At G and H, the points of inflection, the tangent line *crosses* the curve. Thus, $f''(x)$ must be zero at an inflection point and change sign there.

EXAMPLE 3.   For the function $f(x) = x^2 - 6x + 8$ of Example 1, $f'(x) = 2x - 6$ and $f''(x) = 2$. At the critical value $x = 3$, $f''(x) > 0$; hence $f(3) = -1$ is a relative minimum value. Since $f''(x) = 2 \neq 0$, the parabola $y = x^2 - 6x + 8$ has no inflection point. (See Problem 59.4.)

EXAMPLE 4.   For the function $f(x) = x^3 + x^2 + x$, $f''(x) = 6x + 2 = 0$ when $x = -\frac{1}{3}$. Since $f'''(-\frac{1}{3}) \neq 0$, concavity must be changing when $x = -\frac{1}{3}$.

VELOCITY AND ACCELERATION.   Let a particle move along a horizontal line and let its distance (in feet) at time $t \geq 0$ (in seconds) from a fixed point O of the line be given by $s = f(t)$. Let the positive direction on the line be to the right (that is, the direction of increasing s). A complete description of the motion may be obtained by examining $f(t)$, $f'(t)$, and $f''(t)$. It was noted in Chapter 57 that $f'(t)$ gives the velocity v of the particle. The *acceleration* of the particle is given by $a = f''(t)$.

EXAMPLE 5.   Discuss the motion of a particle which moves along a horizontal line according to the equation $s = t^3 - 6t^2 + 9t + 2$.
   When $t = 0$, $s = f(0) = 2$. The particle begins its motion from $A(s = 2)$. See Fig. 59.4.

Fig. 59-4

Direction of Motion.   Here $v = f'(t) = 3t^2 - 12t + 9 = 3(t - 1)(t - 3)$. When $t = 0$, $v = f'(0) = 9$; the particle leaves A with initial velocity 9 ft/s.
   Now $v = 0$ when $t = 1$ and $t = 3$. Thus, the particle moves (from A) for 1 s, stops momentarily ($v = 0$, when $t = 1$), moves off for two more seconds, stops momentarily, and then moves off indefinitely.
   On the interval $0 < t < 1$, $v > 0$. Now $v > 0$ indicates that s is increasing; thus the body leaves A with initial velocity 9 ft/s and moves to the right for 1 s to B [$s = f(1) = 6$] where it stops momentarily.
   On the interval $1 < t < 3$, $v < 0$. Now $v < 0$ indicates that s is decreasing; thus the particle leaves B and moves to the left for 2 s to A [$s = f(3) = 2$] where it stops momentarily.
   On the interval $t > 3$, $v > 0$. The particle leaves A for the second time and moves to the right indefinitely.

Velocity and Speed.   We have $a = f''(t) = 6t - 12 = 6(t - 2)$. The acceleration is 0 when $t = 2$.

On the interval $0 < t < 2$, $a < 0$. Now $a < 0$ indicates that v is decreasing; thus the particle moves for the first 2 s with decreasing velocity. For the first second (from A to B) the velocity decreases from $v = 9$ to $v = 0$. The speed (numerical value of the velocity) decreases from 9 to 0, that is, the particle "slows up." When $t = 2$, $f(t) = 4$ (the particle is at C) and $f'(t) = -3$. Thus from B to C ($t = 1$ to $t = 2$), the velocity decreases from $v = 0$ to $v = -3$. On the other hand, the speed increases from 0 to 3, that is, the particle "speeds up."

On the interval $t > 2$, $a > 0$; thus the velocity is increasing. From C to A ($t = 2$ to $t = 3$) the velocity increases from $v = -3$ to $v = 0$ while the speed decreases from 3 to 0. Thereafter ($t > 3$) both the velocity and speed increase indefinitely. (See Problem 59.9.)

DIFFERENTIALS.   Let $y = f(x)$. Define dx (read, differential x) by the relation $dx = \Delta x$ and define dy (read differential y) by the relation $dy = f'(x) \cdot dx$. Note $dy \neq \Delta y$.

EXAMPLE 6.   If $y = f(x) = x^3$, then

$$\Delta y = (x + \Delta x)^3 - x^3 = 3x^2 \cdot \Delta x + 3x(\Delta x)^2 + (\Delta x)^3 = 3x^2 \cdot dx + 3x(dx)^2 + (dx)^3$$

while $dy = f'(x) \cdot dx = 3x^2 \cdot dx$. Thus if dx is small numerically, dy is a fairly good approximation of Δy and simpler to compute.

Suppose now that $x = 10$ and $dx = \Delta x = .01$. Then for the function above, $\Delta y = 3(10)^2(.01) + 3(10)(.01)^2 + (.01)^3 = 3.0031$ while $dy = 3(10)^2(.01) = 3$.

Solved Problems

59.1   Determine the intervals on which each of the following is an increasing function and the intervals on which it is a decreasing function:

(a) $f(x) = x^2 - 8x$ (c) $f(x) = -x^3 + 3x^2 + 9x + 5$ (e) $f(x) = (x - 2)^3$

(b) $f(x) = 2x^3 - 24x + 5$ (d) $f(x) = x^3 + 3x$ (f) $f(x) = (x - 1)^3(x - 2)$

(a)   Here $f'(x) = 2(x - 4)$. Setting this equal to 0 and solving, we find the critical value to be $x = 4$. We locate the point $x = 4$ on the x axis and find that $f'(x) < 0$ for $x < 4$, and $f'(x) > 0$ when $x > 4$. See Fig. 59-5. Thus, $f(x) = x^2 - 8x$ is an increasing function when $x > 4$, and is a decreasing function when $x < 4$.

Fig. 59-5

(b) $f'(x) = 6x^2 - 24 = 6(x + 2)(x - 2)$; the critical values are $x = -2$ and $x = 2$. Locating these points and determining the sign of $f'(x)$ on each of the intervals $x < -2$, $-2 < x < 2$, and $x > 2$ (see Fig. 59-6), we find that $f(x) = 2x^3 - 24x + 5$ is an increasing function on the intervals $x < -2$ and $x > 2$, and is a decreasing function on the interval $-2 < x < 2$.

Fig. 59-6

(c) $f'(x) = -3x^2 + 6x + 9 = -3(x+1)(x-3)$; the critical values are $x = -1$ and $x = 3$. See Fig. 59-7. Then $f(x)$ is an increasing function on the interval $-1 < x < 3$, and a decreasing function on the intervals $x < -1$ and $x > 3$.

Fig. 59-7

(d) $f'(x) = 3x^2 + 3 = 3(x^2 + 1)$; there are no critical values. Since $f'(x) > 0$ for all values of x, $f(x)$ is everywhere an increasing function.

(e) $f'(x) = 3(x - 2)^2$; the critical value is $x = 2$. See Fig. 59-8. Then $f(x)$ is an increasing function on the intervals $x < 2$ and $x > 2$.

Fig. 59-8

(f) $f'(x) = (x - 1)^2(4x - 7)$; the critical values are $x = 1$ and $x = \frac{7}{4}$. See Fig. 59-9. Then $f(x)$ is an increasing function on the interval $x > \frac{7}{4}$ and is a decreasing function on the intervals $x < 1$ and $1 < x < \frac{7}{4}$.

$$\begin{array}{c|c|c}
f'(x) < 0 & f'(x) < 0 & f'(x) > 0 \\
\hline
1 & 7/4 & \longrightarrow x
\end{array}$$

Fig. 59-9

59.2　Find the relative maximum and minimum values of the functions of Problem 59.1

(a)　The critical value is $x = 4$. Since $f'(x) < 0$ for $x < 4$ and $f'(x) > 0$ for $x > 4$, the function has a relative minimum value $f(4) = -16$.

(b)　The critical values are $x = -2$ and $x = 2$. Since $f'(x) > 0$ for $x < -2$ and $f'(x) < 0$ for $-2 < x < 2$, the function has a relative maximum value $f(-2) = 37$. Since $f'(x) < 0$ for $-2 < x < 2$ and $f'(x) > 0$ for $x > 2$, the function has a relative minimum value $f(2) = -27$.

(c)　The critical values are $x = -1$ and $x = 3$. Since $f'(x) < 0$ for $x < -1$ and $f'(x) > 0$ for $-1 < x < 3$, $f(x)$ has a relative minimum value $f(-1) = 0$. Since $f'(x) > 0$ for $-1 < x < 3$ and $f'(x) < 0$ for $x > 3$, the function has a relative maximum value $f(3) = 32$.

(d)　The function has neither a relative maximum nor a relative minimum value.

(e)　The critical value is $x = 2$. Since $f'(x) > 0$ for $x < 2$ and $f'(x) > 0$ for $x > 2$, the function has neither a relative maximum nor minimum value.

(f)　The critical values are $x = 1$ and $x = \frac{7}{4}$. The function has a relative minimum value $f(\frac{7}{4}) = -\frac{27}{256}$. The critical value $x = 1$ yields neither a relative maximum nor minimum value.

59.3　Find the relative maximum and minimum values of the functions of Problem 59.1, using the second derivative test.

(a)　$f(x) = x^2 - 8x, f'(x) = 2x - 8, f''(x) = 2$.
　　　The critical value is $x = 4$. Since $f''(4) = 2 \neq 0, f(4) = -16$ is a relative minimum value of the function.

(b)　$f(x) = 2x^3 - 24x + 5, f'(x) = 6x^2 - 24, f''(x) = 12x$.
　　　The critical values are $x = -2$ and $x = 2$. Since $f''(-2) = -24 < 0, f(-2) = 37$ is a relative maximum value of the function; since $f''(2) = 24 > 0, f(2) = -27$ is a relative minimum value.

(c) $f(x) = -x^3 + 3x^2 + 9x + 5, f'(x) = -3x^2 + 6x + 9, f''(x) = -6x + 6.$

The critical values are $x = -1$ and $x = 3$. Since $f''(-1) > 0, f(-1) = 0$ is a relative minimum value of the function; since $f''(3) < 0, f(3) = 32$ is a relative maximum value.

(d) $f(x) = x^3 + 3x, f'(x) = 3x^2 + 3, f''(x) = 6x.$

There are no critical values; hence the function has neither a relative minimum nor a relative maximum value.

(e) $f(x) = (x - 2)^3, f'(x) = 3(x - 2)^2, f''(x) = 6(x - 2).$

The critical value is $x = 2$. Since $f''(2) = 0$, the test fails. The test of Problem 59.2 shows that the function has neither a relative maximum nor a relative minimum value.

(f) $f(x) = (x - 1)^3(x - 2), f'(x) = (x - 1)^2(4x - 7), f''(x) = 6(2x - 3)(x - 1).$

The critical values are $x = 1$ and $x = \frac{7}{4}$. Since $f''(1) = 0$, the test fails; the test of Problem 59.2 shows that $f(1)$ is neither a relative maximum nor a relative minimum value of the function. Since $f''(\frac{7}{4}) > 0, f(\frac{7}{4}) = -\frac{27}{256}$ is a relative minimum value.

59.4 Find the inflection points and plot the graph of each of the given curves. In sketching the graph, locate the x and y intercepts when they can be found, the relative maximum and minimum points (see Problem 59.2), and the inflection points, if any. Additional points may be found if necessary.

(a) $y = f(x) = x^2 - 8x$ (c) $y = f(x) = -x^3 + 3x^2 + 9x + 5$ (e) $y = f(x) = (x - 2)^3$

(b) $y = f(x) = 2x^3 - 24x + 5$ (d) $y = f(x) = x^3 + 3x$ (f) $y = f(x) = (x - 1)^3(x - 2)$

(a) Since $f''(x) = 2$, the parabola does not have an inflection point. It is always concave upward.

The x and y intercepts are $x = 0$, $x = 8$, and $y = 0$; $(4, -16)$ is a relative minimum point. See Fig. 59-10(a).

Fig. 59-10

(b) $f''(x) = 12x$ and $f'''(x) = 12$. Since $f''(x) = 0$ when $x = 0$ and $f'''(0) = 12 \neq 0$, $(0, 5)$ is an inflection point. Notice the change in concavity.

The y intercept is $y = 5$, the x intercepts cannot be determined; $(-2, 37)$ is a relative maximum point, $(2, -27)$ is a relative minimum point; $(0, 5)$ is an inflection point. See Fig. 59-10(b).

(c) $f''(x) = -6x + 6$ and $f'''(x) = -6$. Since $f''(x) = 0$ when $x = 1$ and $f'''(1) = -6 \neq 0$, $(1, 16)$ is an inflection point.

The x and y intercepts are $x = -1$, $x = 5$, and $y = 5$; $(-1, 0)$ is a relative minimum point and $(3, 32)$ is a relative maximum point; $(1, 16)$ is an inflection point. See Fig. 54-10(c).

(d) $f''(x) = 6x$ and $f'''(x) = 6$. The point $(0, 0)$ is an inflection point.

The x and y intercepts are $x = 0$, $y = 0$; $(0, 0)$ is an inflection point. The curve can be sketched after locating the points $(1, 4)$, $(2, 14)$, $(-1, -4)$, and $(-2, -14)$. See Fig. 59-10(d).

(e) $f''(x) = 6(x - 2)$ and $f'''(x) = 6$. The point $(2, 0)$ is an inflection point.

The x and y intercepts are $x = 2$, $y = -8$; $(2, 0)$ is an inflection point. The curve can be sketched after locating the points $(3, 1)$, $(4, 8)$, and $(1, -1)$. See Fig. 59-10(e).

(f) $f''(x) = 6(2x - 3)(x - 1)$ and $f'''(x) = 6(4x - 5)$. The inflection points are $(1, 0)$ and $(\frac{3}{2}, -\frac{1}{16})$.

The x and y intercepts are $x = 1$, $x = 2$, and $y = 2$; $(\frac{7}{4}, -\frac{27}{256})$ is a relative minimum point; $(1, 0)$ and $(\frac{3}{2}, -\frac{1}{16})$ are inflection points. For the graph, see Fig. 59-10(f).

59.5 Find two integers whose sum is 12 and whose product is a maximum.

Let x and $12 - x$ be the integers; their product is $P = f(x) = x(12 - x) = 12x - x^2$.

Since $f'(x) = 12 - 2x = 2(6 - x)$, $x = 6$ is the critical value. Now $f''(x) = -2$; hence $f''(6) = -2 < 0$ and $x = 6$ yields a relative maximum. The integers are 6 and 6.

Note that we have, in effect, proved that the rectangle of given perimeter has maximum area when it is a square.

59.6 A farmer wishes to enclose a rectangular plot for a pasture, using a wire fence on three sides and a hedge row as the fourth side. If he has 2400 ft of wiring, what is the greatest area he can fence off?

Let x denote the length of the equal sides to be wired; then the length of the third side is $2400 - 2x$. See Fig. 59-11.

The area is $A = f(x) = x(2400 - 2x) = 2400x - 2x^2$. Now $f'(x) = 2400 - 4x = 4(600 - x)$ and the critical value is $x = 600$. Since $f''(x) = -4$, $x = 600$ yields a relative maximum $= f(600) = 720\,000$ ft^2.

Fig. 59-11   Fig. 59-12

59.7 A page is to contain 54 square inches of printed material. If the margins are 1 in. at top and bottom and $1\frac{1}{2}$ in. at the sides, find the most economical dimensions of the page. See Fig. 59-12.

Let the dimensions of the printed material be denoted by x and y; then $xy = 54$.

The dimensions of the page are $x + 3$ and $y + 2$; the area of the page is $A = (x + 3)(y + 2)$.

Since $y = 54/x$, $A = f(x) = (x + 3)(54/x + 2) = 60 + 162/x + 2x$. Then $f'(x) = -162/x^2 + 2$ and the critical values are $x = \pm 9$. Since $f''(x) = 324/x^3$, the relative minimum is given by $x = 9$. The required dimensions of the page are 12 in. wide and 8 in. high.

59.8 A cylindrical container with circular base is to hold 64 cubic centimeters. Find the dimensions so that the amount (surface area) of metal required is a minimum when (*a*) the container is an open cup and (*b*) a closed can.

Let r and h respectively be the radius of the base and height in centimeters, V be the volume of the container, and A be the surface area of the metal required.

(*a*) $V = \pi r^2 h = 64$ and $A = 2\pi rh + \pi r^2$. Solving for $h = 64/\pi r^2$ in the first relation and substituting in the second, we have $A = 2\pi r\left(\dfrac{64}{\pi r^2}\right) + \pi r^2 = \dfrac{128}{r} + \pi r^2$.

Then $\dfrac{dA}{dr} = -\dfrac{128}{r^2} + 2\pi r = \dfrac{2(\pi r^3 - 64)}{r^2}$ and the critical value is $r = \dfrac{4}{\sqrt[3]{\pi}}$.

Now $h = \dfrac{64}{\pi r^2} = \dfrac{4}{\sqrt[3]{\pi}}$; thus, $r = h = \dfrac{4}{\sqrt[3]{\pi}}$ cm.

(*b*) $V = \pi r^2 h = 64$ and $A = 2\pi rh + 2\pi r^2 = 2\pi r\left(\dfrac{64}{\pi r^2}\right) + 2\pi r^2 = \dfrac{128}{r} + 2\pi r^2$.

Then $\dfrac{dA}{dr} = -\dfrac{128}{r^2} + 4\pi r = \dfrac{4(\pi r^3 - 32)}{r^2}$ and the critical value is $r = 2\sqrt[3]{\dfrac{4}{\pi}}$.

Now $h = \dfrac{64}{\pi r^2} = 4\sqrt[3]{\dfrac{4}{\pi}}$; thus, $h = 2r = 4\sqrt[3]{\dfrac{4}{\pi}}$ cm.

59.9 Study the motion of a particle which moves along a horizontal line in accordance with

(*a*) $s = t^3 - 6t^2 + 3$,   (*b*) $s = t^3 - 5t^2 + 7t - 3$,   (*c*) $v = (t - 1)^2(t - 4)$,   (*d*) $v = (t - 1)^4$.

(*a*)   Here $v = 3t^2 - 12t = 3t(t - 4) = 0$ when $t = 0$ and $t = 4$; $a = 6t - 12 = 6(t - 2) = 0$ when $t = 2$. The particle leaves $A(s = 3)$ with velocity 0 and moves to $B(-29)$ where it stops momentarily; thereafter it moves to the right. See Fig. 59-13.

The intervals of increasing and decreasing speed are shown in Fig. 59-14.

Fig. 59-13

Fig. 59-14

(*b*)   Here $v = 3t^2 - 10t + 7 = (t - 1)(3t - 7) = 0$ when $t = 1$ and $t = \frac{7}{3}$; $a = 6t - 10 = 2(3t - 5) = 0$ when $t = \frac{5}{3}$. The particle leaves $A(s = -3)$ with velocity 7 ft/s and moves to 0 where it stops momentarily, then it moves to $B(-\frac{32}{27})$ where it stops momentarily. Thereafter it moves to the right. See Fig. 59-15.

The intervals of increasing and decreasing speed are shown in Fig. 59-16.

Fig. 59-15

Fig. 59-16

(c)   Here $v = 0$ when $t = 1$ and $t = 4$. Also $a = 3t^2 - 12t + 9 = 3(t-1)(t-3) = 0$ when $t = 1$ and $t = 3$. The intervals of increasing and decreasing speed are shown in Fig. 59-17.

   Note that the particle stops momentarily at the end of 1 s but does not then reverse its direction of motion as, for example, in (b).

(d)   Here $a = 4(t-1)^3$ and $v = a = 0$ when $t = 1$. The intervals of increasing and decreasing speed are shown in Fig. 59-18.

Fig. 59-17

Fig. 59-18

59.10 Find dy in terms of x and dx, given

(a) $y = f(x) = x^2 + 5x + 6$,   (b) $y = f(x) = x^4 - 4x^3 + 8$,   (c) $y = f(x) = x^2 + 1/x^2$.

(a)   Since $f'(x) = 2x + 5$, $dy = f'(x)\, dx = (2x + 5)\, dx$.

(b)   Since $f'(x) = 4x^3 - 12x^2$, $dy = (4x^3 - 12x^2)\, dx$.

(c)   Since $f'(x) = 2x - 2/x^3$, $dy = (2x - 2/x^3)\, dx$.

59.11 Find the approximate displacement of a particle moving along the x axis in accordance with the law $s = t^4 - t^2$, from the time $t = 1.99$ to $t = 2$.

   Here $ds = (4t^3 - 2t)\, dt$. We take $t = 2$ and $dt = -0.01$. Then $ds = (4 \cdot 8 - 2 \cdot 2)(-.01) = -0.28$ and the displacement is 0.28 unit.

59.12 Find using differentials the approximate area of a square whose side is 3.01 cm.

   Here $A = x^2$ and $dA = 2x\, dx$. Taking $x = 3$ and $dx = 0.01$, we find $dA = 2 \cdot 3(.01) = 0.06$ cm^2. Now the area (9 cm^2) of a square 3 cm. on a side is increased approximately 0.06 cm^2 when the side is increased to 3.01 cm. Hence the approximate area is 9.06 cm^2. The true area is 9.0601 cm^2.

Supplementary Problems

59.13 Determine the intervals on which each of the following is an increasing function and the intervals on which it is a decreasing function.

(a) $f(x) = x^2$ *Ans.*   Dec. for $x < 0$; inc. for $x > 0$

(b) $f(x) = 4 - x^2$ *Ans.*   Inc. for $x < 0$; dec. for $x > 0$

(c) $f(x) = x^2 + 6x - 5$ *Ans.*   Dec. for $x < -3$; inc. for $x > -3$

(d) $f(x) = 3x^2 + 6x + 18$ *Ans.*   Dec. for $x < -1$; inc. for $x > -1$

(e) $f(x) = (x-2)^4$ *Ans.*   Dec. for $x < 2$; inc. for $x > 2$

(f) $f(x) = (x-1)^3(x+2)^2$ *Ans.*   Inc. for $x < -2$; dec. for $-2 < x < -\frac{4}{5}$;
                                         inc. for $-\frac{4}{5} < x < 1$ and for $x > 1$

59.14   Find the relative maximum and minimum values of the functions of Problem 59.13.

      Ans.   (*a*) Min. = 0   (*c*) Min. = −14   (*e*) Min. = 0

              (*b*) Max. = 4   (*d*) Min. = 15   (*f*) Max. = 0, Min. = −26 244/3125

59.15   Investigate for relative maximum (minimum) points and points of inflection. Sketch each locus.

    (*a*) $y = x^2 - 4x + 8$ (*b*) $y = (x - 1)^3 + 5$ (*c*) $y = x^4 + 32x + 40$ (*d*) $y = x^3 - 3x^2 - 9x + 6$

      Ans. (*a*) Min. (2, 4)   (*c*) Min. (−2, −8)

            (*b*) I.P. (1, 5)   (*d*) Max. (−1, 11), Min. (3, −21), I.P. (1, −5)

59.16   The sum of two positive numbers is 12. Find the numbers

    (*a*)   If the sum of their squares is a minimum

    (*b*)   If the product of one and the square of the other is a maximum

    (*c*)   If the product of one and the cube of the other is a maximum

      Ans. (*a*) 6 and 6   (*b*) 4 and 8   (*c*) 3 and 9

59.17   Find the dimensions of the largest open box which can be made from a sheet of tin 24 in. square by cutting equal squares from the corners and turning up the sides.

      Ans. $16 \times 16 \times 4$ in.

59.18   Find the dimensions of the largest open box which can be made from a sheet of tin 60 in. by 28 in. by cutting equal squares from the corners and turning up the sides.

      Ans. $48 \times 16 \times 6$ in.

59.19   A rectangular field is to be enclosed by a fence and divided into two smaller plots by a fence parallel to one of the sides. Find the dimensions of the largest such field which can be enclosed by 1200 ft of fencing.

      Ans. 200×300 ft

59.20   If a farmer harvests his crop today he will have 1200 kg worth $2.00 per kg. Every week he waits, the crop increases by 100 kg but the price drops 10¢ per kg. When should he harvest the crop?

      Ans. 4 weeks from today

59.21   The base of an isosceles triangle is 20 ft and its altitude is 40 ft. Find the dimensions of the largest inscribed rectangle if two of the vertices are on the base of the triangle.

      Ans. 10×20 ft

59.22   For each of the following compute Δy, dy, and $\Delta y - dy$.

    (*a*) $y = \frac{1}{2}x^2 + x$; $x = 2, \Delta x = \frac{1}{4}$ *Ans.* $\Delta y = \frac{25}{32}$, $dy = \frac{3}{4}$, $\Delta y - dy = \frac{1}{32}$

    (*b*) $y = x^2 - x$; $x = 3, \Delta x = .01$ *Ans.* $\Delta y = .0501$, $dy = .05$, $\Delta y - dy = .0001$

59.23   Approximate using differentials the volume of a cube whose side is 3.005 in.

      Ans. 27.135 in^3

59.24   Approximate using differentials the area of a circular ring whose inner radius is 5 in. and whose width is $\frac{1}{8}$ in.

      Ans. 1.25π in^2

Chapter 60

Integration

IF $F(x)$ IS A FUNCTION whose derivative $F'(x) = f(x)$, then $F(x)$ is called an *integral* of $f(x)$.

For example, $F(x) = x^3$ is an integral of $f(x) = 3x^2$ since $F'(x) = 3x^2 = f(x)$. Also, $G(x) = x^3 + 5$ and $H(x) = x^3 - 6$ are integrals of $f(x) = 3x^2$. Why?

If $F(x)$ and $G(x)$ are two distinct integrals of $f(x)$, then $F(x) = G(x) + C$, where C is a constant. (See Problem 60.1)

THE INDEFINITE INTEGRAL or ANTIDERIVATIVE of $f(x)$, denoted by $\displaystyle\int f(x)\, dx$, is the most general integral of $f(x)$, that is,

$$\int f(x)\, dx = F(x) + C$$

where $F(x)$ is any function such that $F'(x) = f(x)$ and C is an arbitrary constant. Thus the indefinite integral of $f(x) = 3x^2$ is $\displaystyle\int 3x^2\, dx = x^3 + C$.

We shall use the following antidifferentiation formulas:

I. $\displaystyle\int x^n\, dx = \frac{x^{n+1}}{n+1} + C$, where $n \neq -1$

II. $\displaystyle\int cf(x)\, dx = c \int f(x)\, dx$, where c is a constant

III. $\displaystyle\int [f(x) + g(x)]\, dx = \int f(x)\, dx + \int g(x)\, dx$

EXAMPLE 1

(a) $\displaystyle\int x^5\, dx = \frac{x^{5+1}}{5+1} + C = \frac{x^6}{6} + C$

(b) $\displaystyle\int 4x^3\, dx = 4 \int x^3\, dx = 4 \cdot \frac{x^4}{4} + C = x^4 + C$

(c) $\displaystyle\int 3x\, dx = 3 \int x\, dx = 3 \cdot \frac{x^2}{2} + C = \frac{3}{2}x^2 + C$

(d) $\displaystyle\int \frac{dx}{x^3} = \int x^{-3}\, dx = \frac{x^{-2}}{-2} + C = -\frac{1}{2x^2} + C$

(e) $\displaystyle\int (x^5 + 4x^3 + 3x)\, dx = \int x^5\, dx + \int 4x^3\, dx + \int 3x\, dx = \frac{x^6}{6} + x^4 - \frac{3x^2}{2} + C$

(See Problems 60.2–60.8.)

AREA BY SUMMATION. Consider the area A bounded by the curve $y = f(x) \geq 0$, the x axis, and the ordinates $x = a$ and $x = b$, where $b > a$.

Let the interval $a \leq x \leq b$ be divided into n equal parts each of length Δx. At each point of subdivision construct the ordinate thus dividing the area into n strips, as in Fig. 60-1. Since the area

445

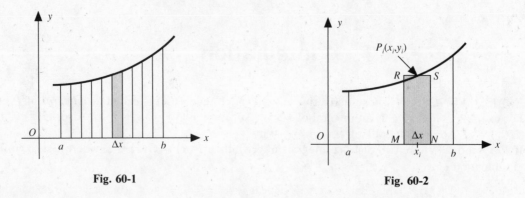

Fig. 60-1   **Fig. 60-2**

of the strips are unknown, we propose to approximate each strip by a rectangle whose area can be found. In Fig. 60-2 a representative strip and its approximating rectangle are shown.

Suppose the representative strip is the ith strip counting from the left, and let $x = x_i$ be the coordinate of the midpoint of its base. Denote by $y_i = f(x_i)$ the ordinate of the point P_i (on the curve) whose abscissa is x_i. Through P_i pass a line parallel to the x axis and complete the rectangle $MRSN$. This rectangle of area $y_i \Delta x$ is the approximating rectangle of the ith strip. When each strip is treated similarly, it seems reasonable to take

$$y_1 \Delta x + y_2 \Delta x + y_3 \Delta x + \cdots + y_n \Delta x = \sum_{i=1}^{n} y_i \Delta x$$

as an approximation of the area sought.

Now suppose that the number of strips (with approximating rectangles) is indefinitely increased so that $\Delta x \to 0$. It is evident from the figure that by so increasing the number of approximating rectangles the sum of their areas more nearly approximates the area sought, that is,

$$A = \lim_{n \to \infty} \sum_{i=1}^{n} y_i \Delta x$$

IF WE DEFINE $\displaystyle\int_a^b f(x)\,dx$ (read, the *definite integral* of $f(x)$ between $x = a$ and $x = b$) as follows

$$\int_a^b f(x)\,dx = F(x)\Big|_a^b = F(b) - F(a), \qquad (F'(x) = f(x))$$

then the area bounded by $y = f(x) \geq 0$, the x-axis, and the ordinates $x = a$ and $x = b$, $(b > a)$, is given by

$$A = \int_a^b f(x)\,dx$$

(See Problems 60.9–60.12.)

Solved Problems

60.1   Prove: If $F(x)$ and $G(x)$ are distinct integrals of $f(x)$ then $F(x) = G(x) + C$, where C is a constant.

Since $F(x)$ and $G(x)$ are integrals of $f(x)$, $F'(x) = G'(x) = f(x)$.
Suppose $F(x) - G(x) = H(x)$; differentiating with respect to x, $F'(x) - G'(x) = H'(x)$ and $H'(x) = 0$. Thus $H(x)$ is a constant, say C, and $F(x) = G(x) + C$.

60.2   (a) $\int \sqrt{x}\, dx = \int x^{1/2}\, dx = \dfrac{x^{3/2}}{\frac{3}{2}} + C = \dfrac{2}{3} x^{3/2} + C$

(b) $\int (3x^2 + 5)\, dx = x^3 + 5x + C$

(c) $\int (5x^6 + 2x^3 - 4x + 3)\, dx = \frac{5}{7} x^7 + \frac{1}{2} x^4 - 2x^2 + 3x + C$

(d) $\int (80x^{19} - 32x^{15} - 12x^{-3})\, dx = 4x^{20} - 2x^{16} + \dfrac{6}{x^2} + C$

60.3   At every point (x, y) of a certain curve, the slope is equal to 8 times the abscissa. Find the equation of the curve if it passes through $(1, 3)$.

Since $m = \dfrac{dy}{dx} = 8x$ we have $dy = 8x\, dx$. Then $y = \int 8x\, dx = 4x^2 + C$, a family of parabolas. We seek the equation of the parabola of this family which passes through the point $(1, 3)$. Then $3 = 4(1)^2 + C$ and $C = -1$. The curve has equation $y = 4x^2 - 1$.

60.4   For a certain curve $y'' = 6x - 10$. Find its equation if it passes through point $(1, 1)$ with slope -1.

Since $y'' = 6x - 10$, $y' = 3x^2 - 10x + C_1$; since $y' = -1$ when $x = 1$, we have $-1 = 3 - 10 + C_1$ and $C_1 = 6$. Then $y' = 3x^2 - 10x + 6$.

Now $y = x^3 - 5x^2 + 6x + C_2$ and since $y = 1$ when $x = 1$, $1 = 1 - 5 + 6 + C_2$ and $C_2 = -1$. Thus the equation of the curve is $y = x^3 - 5x^2 + 6x - 1$.

60.5   The velocity at time t of a particle moving along the x axis is given by $v = x' = 2t + 5$. Find the position of the particle at time t, if $x = 2$ when $t = 0$.

Select a point on the x axis as origin and assume positive direction to the right. Then at the beginning of the motion $(t = 0)$ the particle is 2 units to the right of the origin.

Since $v = \dfrac{dx}{dt} = 2t + 5$, $dx = (2t + 5)\, dt$. Then $x = \int (2t + 5)\, dt = t^2 + 5t + C$.

Substituting $x = 2$ and $t = 0$, we have $2 = 0 + 0 + C$ so that $C = 2$. Thus the position of the particle at time t is given by $x = t^2 + 5t + 2$.

60.6   A body moving in a straight line has an acceleration equal to $6t^2$, where time (t) is measured in seconds and distance s is measured in feet. If the body starts from rest, how far will it move during the first 2 s?

Let the body start from the origin; then it is given that when $t = 0$, $v = 0$ and $s = 0$.

Since $a = \dfrac{dv}{dt} = 6t^2$, $dv = 6t^2\, dt$. Then $v = \int 6t^2\, dt = 2t^3 + C_1$. When $t = 0$, $v = 0$; then $0 = 2 \cdot 0 + C_1$ and $C_1 = 0$. Thus $v = 2t^3$.

Now $v = \dfrac{ds}{dt} = 2t^3$; then $ds = 2t^3\, dt$ and $s = \int 2t^3\, dt = \frac{1}{2} t^4 + C_2$. When $t = 0$, $s = 0$; then $C_2 = 0$ and $s = \frac{1}{2} t^4$. When $t = 2$, $s = \frac{1}{2}(2)^4 = 8$. The body moves 8 ft during the first 2 s.

60.7   A ball is thrown upward from the top of a building 320 ft high with initial velocity 128 ft/s. Determine the velocity with which the ball will strike the street below. (Assume acceleration is 32 ft/s, directed downward.)

First we choose an origin from which all distances are to be measured and a direction (upward or downward) which will be called positive.

First Solution.   Take the origin at the top of the building and positive direction as upward.

Then $$a = \frac{dv}{dt} = -32 \quad \text{and} \quad v = -32t + C_1$$

When the ball is released, $t = 0$ and $v = 128$; then $128 = -32(0) + C_1$ and $C_1 = 128$.
  Now $v = ds/dt = -32t + 128$ and $s = -16t^2 + 128t + C_2$. When the ball is released, $t = 0$ and $s = 0$; then $C_2 = 0$ and $s = -16t^2 + 128t$.
  When the ball strikes the street it is 320 ft below the origin, that is, $s = -320$; hence $-320 = -16t^2 + 128t$, $t^2 - 8t - 20 = (t + 2)(t - 10) = 0$, and $t = 10$. Finally, when $t = 10$, $v = -32(10) + 128 = -192$ ft/s.

Second Solution.   Take the origin on the street and positive direction as before. Then $a = dv/dt = -32$ and $v = -32t + 128$ as in the first solution.
  Now $s = -16t^2 + 128t + C_2$ but when $t = 0$, $s = 320$. Thus $C_2 = 320$ and $s = -16t^2 + 128t + 320$. When the ball strikes the street $s = 0$; then $t = 10$ and $v = -192$ ft/s as before.

60.8   A ball was dropped from a balloon 640 ft above the ground. If the balloon was rising at the rate of 48 ft/s, find (*a*) the greatest distance above the ground attained by the ball (*b*) the time the ball was in the air (*c*) the speed of the ball when it struck the ground.

  Assume the origin at the point where the ball strikes the ground and positive distance to be directed upward. Then

$$a = \frac{dv}{dt} = -32 \quad \text{and} \quad v = -32t + C_1$$

When $t = 0$, $v = 48$; hence $C_1 = 48$. Then $v = ds/dt = -32t + 48$ and $s = -16t^2 + 48t + C_2$. When $t = 0$, $s = 640$; hence $C_2 = 640$ and $s = -16t^2 + 48t + 640$.

(*a*)   When $v = 0$, $t = \frac{3}{2}$ and $s = -16(\frac{3}{2})^2 + 48(\frac{3}{2}) + 640 = 676$. The greatest height attained by the ball was 676 ft.

(*b*)   When $s = 0$, $-16t^2 + 48t + 640 = 0$ and $t = -5, 8$. The ball was in the air for 8 s.

(*c*)   When $t = 8$, $v = -32(8) + 48 = -208$. The ball struck the ground with speed 208 ft/s.

60.9   Find the area bounded by the line $y = 4x$, the x axis, and the ordinates $x = 0$ and $x = 5$.

  Here $y \geq 0$ on the interval $0 \leq x \leq 5$. Then

$$A = \int_0^5 4x\, dx = 2x^2\big|_0^5 = 50 \text{ sq units}$$

  Note that we have found the area of a right triangle whose legs are 5 and 20 units. See Fig. 60-3. The area is $\frac{1}{2}(5)(20) = 50$ sq units.

Fig. 60-3

60.10 Find the area bounded by the parabola $y = 8 + 2x - x^2$ and the x axis.

The x intercepts are $x = -2$ and $x = 4$; $y \geq 0$ on the interval $-2 \leq x \leq 4$. See Fig. 60-4. Hence

$$A = \int_{-2}^{4} (8 + 2x - x^2)\, dx = \left(8x + x^2 - \frac{x^3}{3}\right)\Big|_{-2}^{4}$$

$$= \left(8 \cdot 4 + 4^2 - \frac{4^3}{3}\right) - \left[8(-2) + (-2)^2 - \frac{(-2)^3}{3}\right] = 36 \text{ sq units}$$

Fig. 60-4

60.11 Find the area bounded by the parabola $y = x^2 + 2x - 3$, the x axis, and the ordinates $c = -2$ and $x = 0$.
On the interval $-2 \leq x \leq 0$, $y \leq 0$. Here

$$A = \int_{-2}^{0} (x^2 + 2x - 3)\, dx = \left(\frac{x^3}{3} + x^2 - 3x\right)\Big|_{-2}^{0} = 0 - \left[\frac{(-2)^3}{3} + (-2)^2 - 3(-2)\right] = -\frac{22}{3}$$

The negative sign indicates that the area lies entirely below the x axis. The area is $\frac{22}{3}$ sq units. See Fig. 60-5.

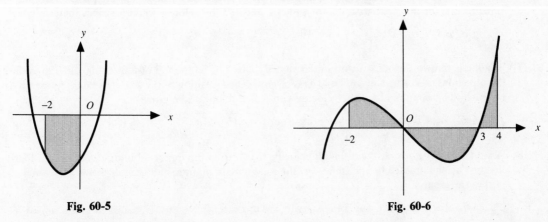

Fig. 60-5   **Fig. 60-6**

60.12 Find the area bounded by the curve $y = x^3 - 9x$, the x axis, and the ordinates $x = -2$ and $x = 4$.
The purpose of this problem is to show that the required area is *not* given by $\int_{-2}^{4} (x^3 - 9x)\, dx$.
From Fig. 60-6, we note that y changes sign at $x = 0$ and at $x = 3$. The required area consists of three pieces, the individual areas being given, apart from sign, by

$$A_1 = \int_{-2}^{0} (x^3 - 9x)\, dx = (\tfrac{1}{4}x^4 - \tfrac{9}{2}x^2)\big|_{2}^{0} = 0 - (4 - 18) = 14$$

$$A_2 = \int_{0}^{3} (x^3 - 9x)\, dx = (\tfrac{1}{4}x^4 - \tfrac{9}{2}x^2)\big|_{0}^{3} = (\tfrac{81}{4} - \tfrac{81}{2}) - 0 = -\tfrac{81}{4}$$

$$A_3 = \int_{3}^{4} (x^3 - 9x)\, dx = (\tfrac{1}{4}x^4 - \tfrac{9}{2}x^2)\big|_{3}^{4} = (64 - 72) - (\tfrac{81}{4} - \tfrac{81}{2}) = \tfrac{49}{4}$$

Thus $A = A_1 - A_2 + A_3 = 14 + \frac{81}{4} + \frac{49}{4} = \frac{93}{2}$ sq units.
Note that $\int_{-2}^{4} (x^3 - 9x)\, dx = 6 < A_1$, an absurd result.

Supplementary Problems

60.13 Work out the following indefinite integrals.

(a) $4 \int dx = 4x + C$

(b) $\int \frac{1}{2} x \, dx = \frac{1}{4} x^2 + C$

(c) $\int (3x^2 + 4x - 5) \, dx = x^3 + 2x^2 - 5x + C$

(d) $\int x(1 - x) \, dx = \frac{1}{2} x^2 - \frac{1}{3} x^3 + C$

(e) $\int 3(x + 1)^2 \, dx = (x + 1)^3 + C$

(f) $\int (x - 1)(x + 2) \, dx = \frac{1}{3} x^3 + \frac{1}{2} x^2 - 2x + C$

(g) $\int \dfrac{dx}{x^2} = -\dfrac{1}{x} + C$

(h) $\int \dfrac{x^2 - 2}{x^2} \, dx = x + \dfrac{2}{x} + C$

60.14 Find the equation of the family of curves whose slope is the given function of x. Find also the equation of the curve of the family passing through the given point.

(a) $m = 1, (1, -2)$ *Ans.* $y = x + C, y = x - 3$

(b) $m = -6x, (0, 0)$ *Ans.* $y = -3x^2 + C, y = -3x^2$

(c) $m = 3x^2 + 2x, (1, -3)$ *Ans.* $y = x^3 + x^2 + C, y = x^3 + x^2 - 5$

(d) $m = 6x^2, (0, 1)$ *Ans.* $y = 2x^3 + C, y = 2x^3 + 1$

60.15 For a certain curve $y'' = 6x + 8$. Find its equation if it passes through $(1, 2)$ with slope $m = 6$.

Ans. $y = x^3 + 4x^2 - 5x + 2$

60.16 A stone is dropped from the top of a building 400 ft high. Taking the origin at the top of the building and positive direction downward, find (a) the velocity of the stone at time t (b) the position at time t (c) the time it takes for the stone to reach the ground, and (d) the velocity when it strikes the ground.

Ans. (a) $v = 32t$ (b) $s = 16t^2$ (c) 5 s   (d) 160 ft/s

60.17 A stone is thrown downward with initial velocity 20 ft/s from the top of a building 336 ft high. Following the directions of Problem 60.16, find (a) the velocity and position of the stone 2 s later (b) the time it takes to reach the ground, and (c) the velocity with which it strikes the ground.

Ans. (a) 84 ft/s, 232 ft above the ground   (b) 4 s   (c) 148 ft/s

60.18 A stone is thrown upward with initial velocity 16 ft/s from the top of a building 192 ft high. Find (a) the greatest height attained by the stone, (b) the total time in motion, and (c) the speed with which the stone strikes the ground.

Ans. (a) 196 ft   (b) 4 s   (c) 112 ft/s

60.19 A boy on top of a building 192 ft high throws a rock straight down. What initial velocity did he give it if it strikes the ground after 3 s?

Ans. 16 ft/s

60.20 Find the area bounded by the x axis, the given curve, and the indicated ordinates.

(a) $y = x^2$ between $x = 2$ and $x = 4$ *Ans.* $\frac{56}{3}$ square units

(b) $y = 4 - 3x^2$ between $x = -1$ and $x = 1$ *Ans.* 6 square units

(c) $y = x^{1/2}$ between $x = 0$ and $x = 9$ *Ans.* 18 square units

(d) $y = x^2 - x - 6$ between $x = 0$ and $x = 2$ *Ans.* $\frac{34}{3}$ square units

(e) $y = x^3$ between $x = -2$ and $x = 4$ *Ans.* 68 square units

(f) $y = x^3 - x$ between $x = -1$ and $x = 1$ *Ans.* $\frac{1}{2}$ square unit

Appendix A

Four-Place Common Logarithms

N	0	1	2	3	4	5	6	7	8	9
10	0000	0043	0086	0128	0170	0212	0253	0294	0334	0374
11	0414	0453	0492	0531	0569	0607	0645	0682	0719	0755
12	0792	0828	0864	0899	0934	0969	1004	1038	1072	1106
13	1139	1173	1206	1239	1271	1303	1335	1367	1399	1430
14	1461	1492	1523	1553	1584	1614	1644	1673	1703	1732
15	1761	1790	1818	1847	1875	1903	1931	1959	1987	2014
16	2041	2068	2095	2122	2148	2175	2201	2227	2253	2279
17	2304	2330	2355	2380	2405	2430	2455	2480	2504	2529
18	2553	2577	2601	2625	2648	2672	2695	2718	2742	2765
19	2788	2810	2833	2856	2878	2900	2923	2945	2967	2989
20	3010	3032	3054	3075	3096	3118	3139	3160	3181	3201
21	3222	3243	3263	3284	3304	3324	3345	3365	3385	3404
22	3424	3444	3464	3483	3502	3522	3541	3560	3579	3598
23	3617	3636	3655	3674	3692	3711	3729	3747	3766	3784
24	3802	3820	3838	3856	3874	3892	3909	3927	3945	3962
25	3979	3997	4014	4031	4048	4065	4082	4099	4116	4133
26	4150	4166	4183	4200	4216	4232	4249	4265	4281	4298
27	4314	4330	4346	4362	4378	4393	4409	4425	4440	4456
28	4472	4487	4502	4518	4533	4548	4564	4579	4594	4609
29	4624	4639	4654	4669	4683	4698	4713	4728	4742	4757
30	4771	4786	4800	4814	4829	4843	4857	4871	4886	4900
31	4914	4928	4942	4955	4969	4983	4997	5011	5024	5038
32	5051	5065	5079	5092	5105	5119	5132	5145	5159	5172
33	5185	5198	5211	5224	5237	5250	5263	5276	5289	5302
34	5315	5328	5340	5353	5366	5378	5391	5403	5416	5428
35	5441	5453	5465	5478	5490	5502	5514	5527	5539	5551
36	5563	5575	5587	5599	5611	5623	5635	5647	5658	5670
37	5682	5694	5705	5717	5729	5740	5752	5763	5775	5786
38	5798	5809	5821	5832	5843	5855	5866	5877	5888	5899
39	5911	5922	5933	5944	5955	5966	5977	5988	5999	6010
40	6021	6031	6042	6053	6064	6075	6085	6096	6107	6117
41	6128	6138	6149	6160	6170	6180	6191	6201	6212	6222
42	6232	6243	6253	6263	6274	6284	6294	6304	6314	6325
43	6335	6345	6355	6365	6375	6385	6395	6405	6415	6425
44	6435	6444	6454	6464	6474	6484	6493	6503	6513	6522
45	6532	6542	6551	6561	6571	6580	6590	6599	6609	6618
46	6628	6637	6646	6656	6665	6675	6684	6693	6702	6712
47	6721	6730	6739	6749	6758	6767	6776	6785	6794	6803
48	6812	6821	6830	6839	6848	6857	6866	6875	6884	6893
49	6902	6911	6920	6928	6937	6946	6955	6964	6972	6981
50	6990	6998	7007	7016	7024	7033	7042	7050	7059	7067
51	7076	7084	7093	7101	7110	7118	7126	7135	7143	7152
52	7160	7168	7177	7185	7193	7202	7210	7218	7226	7235
53	7243	7251	7259	7267	7275	7284	7292	7300	7308	7316
54	7324	7332	7340	7348	7356	7364	7372	7380	7388	7396
N	0	1	2	3	4	5	6	7	8	9

N	0	1	2	3	4	5	6	7	8	9
55	7404	7412	7419	7427	7435	7443	7451	7459	7466	7474
56	7482	7490	7497	7505	7513	7520	7528	7536	7543	7551
57	7559	7566	7574	7582	7589	7597	7604	7612	7619	7627
58	7634	7642	7649	7657	7664	7672	7679	7686	7694	7701
59	7709	7716	7723	7731	7738	7745	7752	7760	7767	7774
60	7782	7789	7796	7803	7810	7818	7825	7832	7839	7846
61	7853	7860	7868	7875	7882	7889	7896	7903	7910	7917
62	7924	7931	7938	7945	7952	7959	7966	7973	7980	7987
63	7993	8000	8007	8014	8021	8028	8053	8041	8048	8055
64	8062	8069	8075	8082	8089	8096	8102	8109	8116	8122
65	8129	8136	8142	8149	8156	8162	8169	8176	8182	8189
66	8195	8202	8209	8215	8222	8228	8235	8241	8248	8254
67	8261	8267	8274	8280	8287	8293	8299	8306	8312	8319
68	8325	8331	8338	8344	8351	8357	8363	8370	8376	8382
69	8388	8395	8401	8407	8414	8420	8426	8432	8439	8445
70	8451	8457	8463	8470	8476	8482	8488	8494	8500	8506
71	8513	8519	8525	8531	8537	8543	8549	8555	8561	8567
72	8573	8579	8585	8591	8597	8603	8609	8615	8621	8627
73	8633	8639	8645	8651	8657	8663	8669	8675	8681	8686
74	8692	8698	8704	8710	8716	8722	8727	8733	8739	8745
75	8751	8756	8762	8768	8774	8779	8785	8791	8797	8802
76	8808	8814	8820	8825	8831	8837	8842	8848	8854	8859
77	8865	8871	8876	8882	8887	8893	8899	8904	8910	8915
78	8921	8927	8932	8938	8943	8949	8954	8960	8965	8971
79	8976	8982	8987	8993	8998	9004	9009	9015	9020	9025
80	9031	9036	9042	9047	9053	9058	9063	9069	9074	9079
81	9085	9090	9096	9101	9106	9112	9117	9122	9128	9133
82	9138	9143	9149	9154	9159	9165	9170	9175	9180	9186
83	9191	9196	9201	9206	9212	9217	9222	9227	9232	9238
84	9243	9248	9253	9258	9263	9269	9274	9279	9284	9289
85	9294	9299	9304	9309	9315	9320	9325	9330	9335	9340
86	9345	9350	9355	9360	9365	9370	9375	9380	9385	9390
87	9395	9400	9405	9410	9415	9420	9425	9430	9435	9440
88	9445	9450	9455	9460	9465	9469	9474	9479	9484	9489
89	9494	9499	9504	9509	9513	9518	9523	9528	9533	9538
90	9542	9547	9552	9557	9562	9566	9571	9576	9581	9586
91	9590	9595	9600	9605	9609	9614	9619	9624	9628	9633
92	9638	9643	9647	9652	9657	9661	9666	9671	9675	9680
93	9685	9689	9694	9699	9703	9708	9713	9717	9722	9727
94	9731	9736	9741	9745	9750	9754	9759	9763	9768	9773
95	9777	9782	9786	9791	9795	9800	9805	9809	9814	9818
96	9823	9827	9832	9836	9841	9845	9850	9854	9859	9863
97	9868	9872	9877	9881	9886	9890	9894	9899	9903	9908
98	9912	9917	9921	9926	9930	9934	9939	9943	9948	9952
99	9956	9961	9965	9969	9974	9978	9983	9987	9991	9996
N	0	1	2	3	4	5	6	7	8	9

Appendix B

Natural Trigonometric Functions

Angle	Sin	Tan	Cot	Cos	
0° 00′	.0000	.0000	—	1.0000	90° 00′
10	.0029	.0029	343.77	1.0000	50
20	.0058	.0058	171.89	1.0000	40
30	.0087	.0087	114.59	1.0000	30
40	.0016	.0116	85.940	.9999	20
50	.0145	.0145	68.750	.9999	10
1° 00′	.0175	.0175	57.290	.9998	89° 00′
10	.0204	.0204	49.104	.9998	50
20	.0233	.0233	42.964	.9997	40
30	.0262	.0262	38.188	.9997	30
40	.0291	.0291	34.368	.9996	20
50	.0320	.0320	31.242	.9995	10
2° 00′	.0349	.0349	28.636	.9994	88° 00′
10	.0378	.0378	26.432	.9993	50
20	.0407	.0407	24.542	.9992	40
30	.0436	.0437	22.904	.9990	30
40	.0465	.0466	21.470	.9989	20
50	.0494	.0495	20.206	.9988	10
3° 00′	.0523	.0524	19.081	.9986	87° 00′
10	.0552	.0553	18.075	.9985	50
20	.0581	.0582	17.169	.9983	40
30	.0610	.0612	16.350	.9981	30
40	.0640	.0641	15.605	.9980	20
50	.0669	.0670	14.924	.9978	10
4° 00′	.0698	.0699	14.301	.9976	86 °00′
10	.0727	.0729	13.727	.9974	50
20	.0756	.0758	13.197	.9971	40
30	.0785	.0787	12.706	.9969	30
40	.0814	.0816	12.251	.9967	20
50	.0843	.0846	11.826	.9964	10
5° 00′	.0872	.0875	11.430	.9962	85° 00′
10	.0901	.0904	11.059	.9959	50
20	.0929	.0934	10.712	.9957	40
30	.0958	.0963	10.385	.9954	30
40	.0987	.0992	10.078	.9951	20
50	.1016	.1022	9.7882	.9948	10
6° 00′	.1045	.1051	9.5144	.9945	84° 00′
10	.1074	.1080	9.2553	.9942	50
20	.1103	.1110	9.0098	.9939	40
30	.1132	.1139	8.7769	.9936	30
40	.1161	.1169	8.5555	.9932	20
50	.1190	.1198	8.3450	.9929	10
7° 00′	.1219	.1228	8.1443	.9925	83° 00′
10	.1248	.1257	7.9530	.9922	50
20	.1276	.1287	7.7704	.9918	40
30	.1305	.1317	7.5958	.9914	30
40	.1334	.1346	7.4287	.9911	20
50	.1363	.1376	7.2687	.9907	10
8° 00′	.1392	.1405	7.1154	.9903	82° 00′
	Cos	Cot	Tan	Sin	Angle

Angle	Sin	Tan	Cot	Cos	
8° 00′	.1392	.1405	7.1154	.9903	82° 00′
10	.1421	.1435	6.9682	.9899	50
20	.1449	.1465	6.8269	.9894	40
30	.1478	.1495	6.6912	.9890	30
40	.1507	.1524	6.5606	.9886	20
50	.1536	.1554	6.4348	.9881	10
9° 00′	.1564	.1584	6.3138	.9877	81° 00′
10	.1593	.1614	6.1970	.9872	50
20	.1622	.1644	6.0844	.9868	40
30	.1650	.1673	5.9758	.9863	30
40	.1679	.1703	5.8708	.9858	20
50	.1708	.1733	5.7694	.9853	10
10° 00′	.1736	.1763	5.6713	.9848	80° 00′
10	.1765	.1793	5.5764	.9843	50
20	.1794	.1823	5.4845	.9838	40
30	.1822	.1853	5.3955	.9833	30
40	.1851	.1883	5.3093	.9827	20
50	.1880	.1914	5.2257	.9822	10
11° 00′	.1908	.1944	5.1446	.9816	79° 00′
10	.1937	.1974	5.0658	.9811	50
20	.1965	.2004	4.9894	.9805	40
30	.1994	.2035	4.9152	.9799	30
40	.2022	.2065	4.8430	.9793	20
50	.2051	.2095	4.7729	.9787	10
12° 00′	.2079	.2126	4.7046	.9781	78° 00′
10	.2108	.2156	4.6382	.9775	50
20	.2136	.2186	4.5736	.9769	40
30	.2164	.2217	4.5107	.9763	30
40	.2193	.2247	4.4494	.9757	20
50	.2221	.2278	4.3897	.9750	10
13° 00′	.2250	.2309	4.3315	.9744	77° 00′
10	.2278	.2339	4.2747	.9737	50
20	.2306	.2370	4.2193	.9730	40
30	.2334	.2401	4.1653	.9724	30
40	.2363	.2432	4.1126	.9717	20
50	.2391	.2462	4.0611	.9710	10
14° 00′	.2419	.2493	4.0108	.9703	76° 00′
10	.2447	.2524	3.9617	.9696	50
20	.2476	.2555	3.9136	.9689	40
30	.2504	.2586	3.8667	.9681	30
40	.2532	.2617	3.8208	.9674	20
50	.2560	.2648	3.7760	.9667	10
15° 00′	.2588	.2679	3.7321	.9659	75° 00′
10	.2616	.2711	3.6891	.9652	50
20	.2644	.2742	3.6470	.9644	40
30	.2672	.2773	3.6059	.9636	30
40	.2700	.2805	3.5656	.9628	20
50	.2728	.2836	3.5261	.9621	10
16° 00′	.2756	.2867	3.4874	.9613	74° 00′
	Cos	Cot	Tan	Sin	Angle

Angle	Sin	Tan	Cot	Cos	
16° 00′	.2756	.2867	3.4874	.9613	74° 00′
10	.2784	.2899	3.4495	.9605	50
20	.2812	.2931	3.4124	.9596	40
30	.2840	.2962	3.3759	.9588	30
40	.2868	.2994	3.3402	.9580	20
50	.2896	.3026	3.3052	.9572	10
17° 00′	.2924	.3057	3.2709	.9563	73° 00′
10	.2952	.3089	3.2371	.9555	50
20	.2979	.3121	3.2041	.9546	40
30	.3007	.3153	3.1716	.9537	30
40	.3035	.3185	3.1397	.9528	20
50	.3062	.3217	3.1084	.9520	10
18° 00′	.3090	.3249	3.0777	.9511	72° 00′
10	.3118	.3281	3.0475	.9502	50
20	.3145	.3314	3.0178	.9492	40
30	.3173	.3346	2.9887	.9483	30
40	.3201	.3378	2.9600	.9474	20
50	.3228	.3411	2.9319	.9465	10
19° 00′	.3256	.3443	2.9042	.9455	71° 00′
10	.3283	.3476	2.8770	.9446	50
20	.3311	.3508	2.8502	.9436	40
30	.3338	.3541	2.8239	.9426	30
40	.3365	.3574	2.7980	.9417	20
50	.3393	.3607	2.7725	.9407	10
20 00′	.3420	.3640	2.7475	.9397	70° 00′
10	.3448	.3673	2.7228	.9387	50
20	.3475	.3706	2.6985	.9377	40
30	.3502	.3739	2.6746	.9367	30
40	.3529	.3772	2.6511	.9356	20
50	.3557	.3805	2.6279	.9346	10
21° 00′	.3584	.3839	2.6051	.9336	69° 00′
10	.3611	.3872	2.5826	.9325	50
20	.3638	.3906	2.5605	.9315	40
30	.3665	.3939	2.5386	.9304	30
40	.3692	.3973	2.5172	.9293	20
50	.3719	.4006	2.4960	.9283	10
22° 00′	.3746	.4040	2.4751	.9272	68° 00′
10	.3773	.4074	2.4545	.9261	50
20	.3800	.4108	2.4342	.9250	40
30	.3827	.4142	2.4142	.9239	30
40	.3854	.4176	2.3945	.9228	20
50	.3881	.4210	2.3750	.9216	10
23° 00′	.3907	.4245	2.3559	.9205	67° 00′
10	.3934	.4279	2.3369	.9194	50
20	.3961	.4314	2.3183	.9182	40
30	.3987	.4348	2.2998	.9171	30
40	.4014	.4383	2.2817	.9159	20
50	.4041	.4417	2.2637	.9147	10
24° 00′	.4067	.4452	2.2460	.9135	66° 00′
	Cos	Cot	Tan	Sin	Angle

Angle	Sin	Tan	Cot	Cos	
24° 00′	.4067	.4452	2.2460	.9135	66° 00′
10	.4094	.4487	2.2286	.9124	50
20	.4120	.4522	2.2113	.9112	40
30	.4147	.4557	2.1943	.9100	30
40	.4173	.4592	2.1775	.9088	20
50	.4200	.4628	2.1609	.9075	10
25° 00′	.4226	.4663	2.1445	.9063	65° 00′
10	.4253	.4699	2.1283	.9051	50
20	.4279	.4734	2.1123	.9038	40
30	.4305	.4770	2.0965	.9026	30
40	.4331	.4806	2.0809	.9013	20
50	.4358	.4841	2.0655	.9001	10
26° 00′	.4384	.4877	2.0503	.8988	64° 00′
10	.4410	.4913	2.0353	.8975	50
20	.4436	.4950	2.0204	.8962	40
30	.4462	.4986	2.0057	.8949	30
40	.4488	.5022	1.9912	.8936	20
50	.4514	.5059	1.9768	.8923	10
27° 00′	.4540	.5095	1.9626	.8910	63° 00′
10	.4566	.5132	1.9486	.8897	50
20	.4592	.5169	1.9347	.8884	40
30	.4617	.5206	1.9210	.8870	30
40	.4643	.5243	1.9074	.8857	20
50	.4669	.5280	1.8940	.8843	10
28° 00′	.4695	.5317	1.8807	.8829	62° 00′
10	.4720	.5354	1.8676	.8816	50
20	.4746	.5392	1.8546	.8802	40
30	.4772	.5430	1.8418	.8788	30
40	.4797	.5467	1.8291	.8774	20
50	.4823	.5505	1.8165	.8760	10
29° 00′	.4848	.5543	1.8040	.8746	61° 00′
10	.4874	.5581	1.7917	.8732	50
20	.4899	.5619	1.7796	.8718	40
30	.4924	.5658	1.7675	.8704	30
40	.4950	.5696	1.7556	.8689	20
50	.4975	.5735	1.7437	.8675	10
30° 00′	.5000	.5774	1.7321	.8660	60° 00′
10	.5025	.5812	1.7205	.8646	50
20	.5050	.5851	1.7090	.8631	40
30	.5075	.5890	1.6977	.8616	30
40	.5100	.5930	1.6864	.8601	20
50	.5125	.5969	1.6753	.8587	10
31° 00′	.5150	.6009	1.6643	.8572	59° 00′
10	.5175	.6048	1.6534	.8557	50
20	.5200	.6088	1.6426	.8542	40
30	.5225	.6128	1.6319	.8526	30
40	.5250	.6168	1.6212	.8511	20
50	.5275	.6208	1.6107	.8496	10
32° 00′	.5299	.6249	1.6003	.8480	58° 00′
	Cos	Cot	Tan	Sin	Angle

Angle	Sin	Tan	Cot	Cos	
32° 00′	.5299	.6249	1.6003	.8480	58° 00′
10	.5324	.6289	1.5900	.8465	50
20	.5348	.6330	1.5798	.8450	40
30	.5373	.6371	1.5697	.8434	30
40	.5398	.6412	1.5597	.8418	20
50	.5422	.6453	1.5497	.8403	10
33° 00′	.5446	.6494	1.5399	.8387	57° 00′
10	.5471	.6536	1.5301	.8371	50
20	.5495	.6577	1.5204	.8355	40
30	.5519	.6619	1.5108	.8339	30
40	.5544	.6661	1.5013	.8323	20
50	.5568	.6703	1.4919	.8307	10
34° 00′	.5592	.6745	1.4826	.8290	56° 00′
10	.5616	.6787	1.4733	.8274	50
20	.5640	.6830	1.4641	.8258	40
30	.5664	.6873	1.4550	.8241	30
40	.5688	.6916	1.4460	.8225	20
50	.5712	.6959	1.4370	.8208	10
35° 00′	.5736	.7002	1.4281	.8192	55° 00′
10	.5760	.7046	1.4193	.8175	50
20	.5783	.7089	1.4106	.8158	40
30	.5807	.7133	1.4019	.8141	30
40	.5831	.7177	1.3934	.8124	20
50	.5854	.7221	1.3848	.8107	10
36° 00′	.5878	.7265	1.3764	.8090	54° 00′
10	.5901	.7310	1.3680	.8073	50
20	.5925	.7355	1.3597	.8056	40
30	.5948	.7400	1.3514	.8039	30
40	.5972	.7445	1.3432	.8021	20
50	.5995	.7490	1.3351	.8004	10
37° 00′	.6018	.7536	1.3270	.7986	53° 00′
10	.6041	.7581	1.3190	.7969	50
20	.6065	.7627	1.3111	.7951	40
30	.6088	.7673	1.3032	.7934	30
40	.6111	.7720	1.2954	.7916	20
50	.6134	.7766	1.2876	.7898	10
38° 00′	.6157	.7813	1.2799	.7880	52° 00′
10	.6180	.7860	1.2723	.7862	50
20	.6202	.7907	1.2647	.7844	40
30	.6225	.7954	1.2572	.7826	30
40	.6248	.8002	1.2497	.7808	20
50	.6271	.8050	1.2423	.7790	10
39° 00′	.6293	.8098	1.2349	.7771	51° 00′
	Cos	Cot	Tan	Sin	Angle

Angle	Sin	Tan	Cot	Cos	
39° 00′	.6293	.8098	1.2349	.7771	51° 00′
10	.6316	.8146	1.2276	.7753	50
20	.6338	.8195	1.2203	.7735	40
30	.6361	.8243	1.2131	.7716	30
40	.6383	.8292	1.2059	.7698	20
50	.6406	.8342	1.1988	.7679	10
40° 00′	.6428	.8391	1.1918	.7660	50° 00′
10	.6450	.8441	1.1847	.7642	50
20	.6472	.8491	1.1778	.7623	40
30	.6494	.8541	1.1708	.7604	30
40	.6517	.8591	1.1640	.7585	20
50	.6539	.8642	1.1571	.7566	10
41° 00′	.6561	.8693	1.1504	.7547	49° 00′
10	.6583	.8744	1.1436	.7528	50
20	.6604	.8796	1.1369	.7509	40
30	.6626	.8847	1.1303	.7490	30
40	.6648	.8899	1.1237	.7470	20
50	.6670	.8952	1.1171	.7451	10
42° 00′	.6691	.9004	1.1106	.7431	48° 00′
10	.6713	.9057	1.1041	.7412	50
20	.6734	.9110	1.0977	.7392	40
30	.6756	.9163	1.0913	.7373	30
40	.6777	.9217	1.0850	.7353	20
50	.6799	.9271	1.0786	.7333	10
43° 00′	.6820	.9325	1.0724	.7314	47° 00′
10	.6841	.9380	1.0661	.7294	50
20	.6862	.9435	1.0599	.7274	40
30	.6884	.9490	1.0538	.7254	30
40	.6905	.9545	1.0477	.7234	20
50	.6926	.9601	1.0416	.7214	10
44° 00′	.6947	.9657	1.0355	.7193	46° 00′
10	.6967	.9713	1.0295	.7173	50
20	.6988	.9770	1.0235	.7153	40
30	.7009	.9827	1.0176	.7133	30
40	.7030	.9884	1.0117	.7112	20
50	.7050	.9942	1.0058	.7902	10
45° 00′	.7071	1.0000	1.0000	.7071	45° 00′
	Cos	Cot	Tan	Sin	Angle

Index

SCHAUM'S INTERACTIVE OUTLINE SERIES

Schaum's Outlines and Mathcad™ Combined. . .
The Ultimate Solution.

NOW AVAILABLE! Electronic, interactive versions of engineering titles from the Schaum's Outline Series:

- *Electric Circuits*
- *Electromagnetics*
- *Feedback and Control Systems*
- *Thermodynamics For Engineers*
- *Fluid Mechanics and Hydraulics*

McGraw-Hill has joined with MathSoft, Inc., makers of Mathcad, the world's leading technical calculation software, to offer you interactive versions of popular engineering titles from the Schaum's Outline Series. Designed for students, educators, and technical professionals, the *Interactive Outlines* provide comprehensive on-screen access to theory and approximately 100 representative solved problems. Hyperlinked cross-references and an electronic search feature make it easy to find related topics. In each electronic outline, you will find all related text, diagrams and equations for a particular solved problem together on your computer screen. Every number, formula and graph is interactive, allowing you to easily experiment with the problem parameters, or adapt a problem to solve related problems. The *Interactive Outline* does all the calculating, graphing and unit analysis for you.

These "live" *Interactive Outlines* are designed to help you learn the subject matter and gain a more complete, more intuitive understanding of the concepts underlying the problems. They make your problem solving easier, with power to quickly do a wide range of technical calculations. All the formulas needed to solve the problem appear in real math notation, and use Mathcad's wide range of built in functions, units, and graphing features. This interactive format should make learning the subject matter easier, more effective and even fun.

For more information about *Schaum's Interactive Outlines* listed above and other titles in the series, please contact:

> Schaum Division
> McGraw-Hill, Inc.
> 1221 Avenue of the Americas
> New York, New York 10020
> Phone: 1-800-338-3987

To place an order, please mail the coupon below to the above address or call the 800 number.

--✂--

Schaum's Interactive Outline Series
using Mathcad®

(Software requires 80386/80486 PC or compatibles, with Windows 3.1 or higher, 4 MB of RAM, 4 MB of hard disk space, and 3 1/2" disk drive.)

AUTHOR/TITLE	Interactive Software Only ($29.95 ea) ISBN	Quantity Ordered	Software and Printed Outline ($38.95 ea) ISBN	Quantity Ordered
MathSoft, Inc./DiStefano: Feedback & Control Systems	07-842708-8	_____	07-842709-6	_____
MathSoft, Inc./Edminister: Electric Circuits	07-842710-x	_____	07-842711-8	_____
MathSoft, Inc./Edminister: Electromagnetics	07-842712-6	_____	07-842713-4	_____
MathSoft, Inc./Giles: Fluid Mechanics & Hydraulics	07-842714-2	_____	07-842715-0	_____
MathSoft, Inc./Potter: Thermodynamics For Engineers	07-842716-9	_____	07-842717-7	_____

NAME_____ADDRESS_____

CITY _____ STATE_____ ZIP_____

ENCLOSED IS ❑ A CHECK ❑ MASTERCARD ❑ VISA ❑ AMEX (✓ONE)

ACCOUNT #_____ EXP. DATE _____

SIGNATURE_____

MAKE CHECKS PAYABLE TO McGRAW-HILL, INC. PLEASE INCLUDE LOCAL SALES TAX AND $1.25 SHIPPING/HANDLING

Schaum's Outlines and Solved Problems Books
in the
BIOLOGICAL SCIENCES

*SCHAUM OFFERS IN SOLVED-PROBLEM AND QUESTION-AND-ANSWER FORMAT
THESE UNBEATABLE TOOLS FOR SELF-IMPROVEMENT.*

✸ Fried **BIOLOGY** ORDER CODE 022401-3/$12.95
(including 888 solved problems)

✸ Jessop **ZOOLOGY** ORDER CODE 032551-0/$13.95
(including 1050 solved problems)

✸ Kuchel et al. **BIOCHEMISTRY** order code 035579-7/$13.95
(including 830 solved problems)

✸ Meislich et al. **ORGANIC CHEMISTRY, 2/ed** ORDER CODE 041458-0/$13.95
(including 1806 solved problems)

✸ Stansfield **GENETICS, 3/ed** ORDER CODE 060877-6/$12.95
(including 209 solved problems)

✸ Van de Graaff/Rhees **HUMAN ANATOMY AND PHYSIOLOGY** ORDER CODE 066884-1/$12.95
(including 1470 solved problems)

✸ Bernstein **3000 SOLVED PROBLEMS IN BIOLOGY** ORDER CODE 005022-8/$16.95

✸ Meislich et al. **3000 SOLVED PROBLEMS IN ORGANIC CHEMISTRY** ORDER CODE 056424-8/$22.95

Each book teaches the subject thoroughly through Schaum's pioneering solved-problem
format and can be used as a supplement to any textbook. If you want to excel in
any of these subjects, these books will help and they belong on your shelf.

Schaum's Outlines have been used by more than 25,000,000 student's worldwide!

PLEASE ASK FOR THEM AT YOUR LOCAL BOOKSTORE OR USE THE COUPON BELOW TO ORDER.

SCHAUM'S SOLVED PROBLEMS SERIES

- **Learn the best strategies for solving tough problems in step-by-step detail**
- **Prepare effectively for exams and save time in doing homework problems**
- **Use the indexes to quickly locate the types of problems you need the most help solving**
- **Save these books for reference in other courses and even for your professional library**

To order, please check the appropriate box(es) and complete the following coupon.

❑ **3000 SOLVED PROBLEMS IN BIOLOGY**
ORDER CODE 005022-8/**$16.95 406 pp.**

❑ **3000 SOLVED PROBLEMS IN CALCULUS**
ORDER CODE 041523-4/**$19.95 442 pp.**

❑ **3000 SOLVED PROBLEMS IN CHEMISTRY**
ORDER CODE 023684-4/**$20.95 624 pp.**

❑ **2500 SOLVED PROBLEMS IN COLLEGE ALGEBRA & TRIGONOMETRY**
ORDER CODE 055373-4/**$14.95 608 pp.**

❑ **2500 SOLVED PROBLEMS IN DIFFERENTIAL EQUATIONS**
ORDER CODE 007979-x/**$19.95 448 pp.**

❑ **2000 SOLVED PROBLEMS IN DISCRETE MATHEMATICS**
ORDER CODE 038031-7/**$16.95 412 pp.**

❑ **3000 SOLVED PROBLEMS IN ELECTRIC CIRCUITS**
ORDER CODE 045936-3/**$21.95 746 pp.**

❑ **2000 SOLVED PROBLEMS IN ELECTROMAGNETICS**
ORDER CODE 045902-9/**$18.95 480 pp.**

❑ **2000 SOLVED PROBLEMS IN ELECTRONICS**
ORDER CODE 010284-8/**$19.95 640 pp.**

❑ **2500 SOLVED PROBLEMS IN FLUID MECHANICS & HYDRAULICS**
ORDER CODE 019784-9/**$21.95 800 pp.**

❑ **1000 SOLVED PROBLEMS IN HEAT TRANSFER**
ORDER CODE 050204-8/**$19.95 750 pp.**

❑ **3000 SOLVED PROBLEMS IN LINEAR ALGEBRA**
ORDER CODE 038023-6/**$19.95 750 pp.**

❑ **2000 SOLVED PROBLEMS IN Mechanical Engineering THERMODYNAMICS**
ORDER CODE 037863-0/**$19.95 406 pp.**

❑ **2000 SOLVED PROBLEMS IN NUMERICAL ANALYSIS**
ORDER CODE 055233-9/**$20.95 704 pp.**

❑ **3000 SOLVED PROBLEMS IN ORGANIC CHEMISTRY**
ORDER CODE 056424-8/**$22.95 688 pp.**

❑ **2000 SOLVED PROBLEMS IN PHYSICAL CHEMISTRY**
ORDER CODE 041716-4/**$21.95 448 pp.**

❑ **3000 SOLVED PROBLEMS IN PHYSICS**
ORDER CODE 025734-5/**$20.95 752 pp.**

❑ **3000 SOLVED PROBLEMS IN PRECALCULUS**
ORDER CODE 055365-3/**$16.95 385 pp.**

❑ **800 SOLVED PROBLEMS IN VECTOR MECHANICS FOR ENGINEERS
Vol I: STATICS**
ORDER CODE 056582-1/**$20.95 800 pp.**

❑ **700 SOLVED PROBLEMS IN VECTOR MECHANICS FOR ENGINEERS
Vol II: DYNAMICS**
ORDER CODE 056687-9/**$20.95 672 pp.**

**ASK FOR THE *SCHAUM'S* SOLVED PROBLEMS SERIES AT YOUR LOCAL BOOKSTORE
OR CHECK THE APPROPRIATE BOX(ES) ON THE PRECEDING PAGE
AND MAIL WITH THIS COUPON TO:**

McGRAW-HILL, INC.
ORDER PROCESSING S-1
PRINCETON ROAD
HIGHTSTOWN, NJ 08520

OR CALL
1-800-338-3987

NAME (PLEASE PRINT LEGIBLY OR TYPE)

ADDRESS (NO P.O. BOXES)

CITY   **STATE**   **ZIP**

ENCLOSED IS   ❑ **A CHECK**   ❑ **MASTERCARD**   ❑ **VISA**   ❑ **AMEX**   (✓ ONE)

ACCOUNT # _____   **EXP. DATE** _____

SIGNATURE _____

MAKE CHECKS PAYABLE TO MCGRAW-HILL, INC. PLEASE INCLUDE LOCAL SALES TAX AND $1.25 SHIPPING/HANDLING
PRICES SUBJECT TO CHANGE WITHOUT NOTICE AND MAY VARY OUTSIDE THE U.S. FOR THIS INFORMATION,
WRITE TO THE ADDRESS ABOVE OR CALL THE 800 NUMBER.